Conic Sections

Circle:

$$(x - h)^2 + (y - k)^2 = r^2$$

Ellipse:

$$\frac{(x - h)^2}{a^2} + \frac{(y - k)^2}{b^2} = 1$$

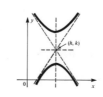

Hyperbola:

$$\frac{(x - h)^2}{a^2} - \frac{(y - k)^2}{b^2} = 1$$

Hyperbola:

$$\frac{(y - k)^2}{a^2} - \frac{(x - h)^2}{b^2} = 1$$

Parabola:

$$y - k = a(x - h)^2$$

Parabola:

$$x - h = a(y - k)^2$$

Sequences and Series

Arithmetic Sequence: $a_n = a_1 + (n - 1)d$

$$S_n = \frac{n}{2}(a_1 + a_n) = \frac{n}{2}[2a_1 + (n - 1)d]$$

Geometric Sequence: $a_n = a_1 r^{n-1}$

$$S_n = \frac{a_1(1 - r^n)}{1 - r} \quad \text{if } r \neq 1$$

$$= na_1 \qquad \text{if } r = 1$$

Infinite Geometric Series: $S = \dfrac{a_1}{1 - r}, \quad |r| < 1$

A Primer for Calculus

2nd Edition

A Primer for Calculus

2nd Edition

LEONARD I. HOLDER
Gettysburg College

Wadsworth Publishing Company
Belmont, California
A Division of Wadsworth, Inc.

Cover Credits and Comments: The color block graph on the cover is "Beast and Counter Beast," *Miles Color Art A1*, produced by E. P. Miles, Jr., on an Intecolor 8051 computer with programming by Billy Jasiniecki and photography by John Owen. The design was created using the function $(3x - y)^2/(300x + y^3 + 0.3)$, which was evaluated at each point in a 77×77 rectangular grid symmetric about the origin. From the eight colors available, one color was assigned to each of the integers $0, 1, \ldots, 7$. The function value obtained in each coordinate rectangle (x, y) was modified using truncation and/or modular arithmetic to fall in that range and determine the color assigned to that rectangle. In this graph the character (rectangle) is blue whenever the function f satisfies $0 \le f < 1$. Consider the spine of the book as the bottom of the graph with the origin at the center. The graph is largely blue because the fraction lies between 0 and 1 except at points near the zeros of the cubic denominator and away from the line $y = 3x$.

Mathematics Editor: Richard Jones
Editing and Production: Phyllis Niklas
Technical Illustrator: Carl Brown

Printed in the United States of America

2 3 4 5 6 7 8 9 10—85 84 83 82 81

Library of Congress Cataloging in Publication Data

Holder, Leonard Irvin,
  A primer for calculus.

  Includes index.
    1. Mathematics—1961–　I. Title.
QA39.2.H63　1981　　　512'.1　　　80-14029
ISBN 0-534-00855-0

Contents

Tables

Preface

The basic objectives and approach of the original book are retained in this edition. Those topics from algebra, trigonometry, and analytical geometry which I consider to be most important for the study of calculus are presented in an informal, readable style, yet in a way that is mathematically sound.

The Most Important Changes

1. A chapter (Chapter 13) on sequences and series has been added, including a section on mathematical induction. The binomial theorem, first introduced in Chapter 2, is proved in this chapter.
2. Sections have been added on linear programming, the algebra of matrices, irrational roots of polynomial equations, scientific notation, variation, and the trigonometric form of complex numbers.
3. Chapter 5 has been expanded to include quadratic as well as linear functions. Graphs of other basic second-degree equations also are introduced, although the in-depth treatment of conics continues to be in a separate chapter (Chapter 12). A brief treatment of rational functions has been added to Chapter 6, and this is expanded upon in Chapter 11.
4. The material on absolute value and inequalities, formerly a separate chapter, has been incorporated into Chapters 1, 3, and 10.
5. The exercise sets have been substantially expanded. There are now more than 4,600 problems.
6. Four cumulative review exercise sets have been added, following Chapters 3, 7, 9, and 13. Problems are in random order within these sets and are not separated into A and B type problems (although they tend to be of the B type). I am indebted to Romae Cormier of Northern Illinois University for suggesting this addition and for many other thoughtful suggestions.
7. The chapter on systems of equations and inequalities now occurs as Chapter 10, following the unit on trigonometry.

In addition, many sections have been rewritten, and the number of worked-out examples has been expanded. Where sections formerly were unusually long, these have been divided; and in other cases, sections have been combined in order to make for more uniformity of length.

Comments on the Use of Hand Calculators

There is a growing consensus in the mathematical community that for students of the '80's the use of hand calculators will be as common as the use of mathematical tables was for earlier generations of students. This point was forcefully made by James E. Hall of the University of Wisconsin at Madison, one of the reviewers of the manuscript for this text, who said, "*this is the age of the calculator* . . . I firmly believe that in five years they'll be a fixture in courses from remedial algebra through calculus." I have added problems throughout the text which can best be done with the aid of a calculator. The heaviest concentration continues to be in the chapters on exponential, logarithmic, and trigonometric functions (Chapters 7–9). I believe that where appropriate, students should be encouraged to use hand calculators, thus freeing them to spend more time on concepts and less in lengthy computations. I therefore have mentioned the use of tables only briefly in the text and as in the first edition, have omitted entirely discussion on computations with logarithms. Nevertheless, tables of squares, square roots, natural and common logarithms, exponentials, and trigonometric functions are included in the back of the book.

Organization

There is a considerable amount of flexibility possible in the use of this book. Chapters 1–3 are largely a review of basic algebraic concepts. It has been my experience in teaching calculus that it is precisely at this point that many students have their greatest difficulty. So a review of these concepts may be highly desirable. Chapters 4–9 are organized around the function concept, proceeding from a general discussion of functions to linear, quadratic, higher-degree polynomial, rational, exponential, logarithmic, trigonometric, and inverse trigonometric functions. Chapters 10–13 are largely independent of each other and may be taken in any order. The following scheme shows the dependence among chapters:

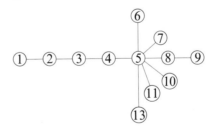

Finally, I wish to express my thanks to Wanda Bowers and Donna Cullison for the superior job they did of typing the manuscript, and to Richard Jones, Mathematics Editor of Wadsworth, for his support and encouragement throughout the preparation of this book.

Books that are effective teaching tools (for us) and learning tools (for our students) are not so much *written* as they are *developed* out of our classroom experience. The more experiences of colleagues an author is exposed to, the more the text is thoroughly developed into a reliable teaching tool. The first edition of these materials and the draft of the second edition received an invaluable contribution through the suggestions, criticisms, and encouragement of the following colleagues:

E. Ray Bobo
Georgetown University

Otha Britton
University of Tennessee, Martin

Anand M. Chak
University of West Virginia

James F. Chew
North Carolina A&T University

Romae Cormier
Northern Illinois University

Edwin Creasy
State University of New York
A&T, Canton

Robert Dahlin
University of Wisconsin, Superior

Robert Dean
Central Washington University

Marvin Eargle
Appalachian State University

Richard Fast
Mesa College

Karl Ray Gentry
University of North Carolina,
Greensboro

James E. Hall
University of Wisconsin, Madison

Ferdinand Haring
North Dakota State University

David Hoak
University of California,
Los Angeles

Louis Hoelzle
Bucks County Community College

Paul Hutchens
St. Louis Community College

Gail Jones
Delgado Junior College

William Jones
Xavier University

Charles Kinzek
Los Angeles Valley College

James McKim
University of Central Arkansas

Jerold C. Mathews
Iowa State University

Jane Morrison
Thornton Community College

Gary R. Penner
Richland College

John Petro
Western Michigan University

Edward Pettit
Augusta College

Glenn Prigge
University of North Dakota

Venice Scheurich
Del Mar College

Edwin Schulz
Elgin Community College

Donald Smith
Louisiana State University,
Shreveport

E. Genevieve Stanton
Portland State University

Bruno Wichnoski
University of North Carolina,
Charlotte

A Primer for Calculus

2nd Edition

1 The Real Number System

1 Introduction

Most of the so-called rules of algebra, so mysterious to many students on first exposure, have their justification in the fundamental properties of what is known as the **real number system.** We could start with a basic set of assumptions (or axioms) about real numbers, together with certain definitions, and from these we could derive other properties of real numbers. Then we could prove the rules of algebra from these properties. We will not do this, since to do so would almost be a course in itself. However, in the next two sections, we will list these assumptions and definitions, and we will indicate how certain further properties can then be derived.

The simplest real numbers are the **natural numbers,** which are just the ordinary counting numbers $1, 2, 3, \ldots$. Natural numbers are also called **positive integers,** and they, together with 0, and their negatives (the **negative integers** $-1, -2, -3, \ldots$), comprise the **integers.** The **rational numbers** are numbers that can be expressed as the ratio of two integers, that is, in the form m/n, where m and n are integers and n is not equal to 0. Rational numbers include ordinary fractions, such as $\frac{3}{4}$, but they also include the integers themselves, since any integer m can be written as $m/1$. Real numbers that are not rational are called **irrational.** These include such numbers as $\sqrt{2}$, π, and $\sqrt[3]{7}$.* Rational numbers and irrational numbers together form the set of real numbers.

One convenient way to visualize the real numbers is by way of a **number line,** as shown in Figure 1. After selecting a point as the **origin** to correspond to 0 and another point to correspond to 1, both a scale and a direction are determined. Points corresponding to both positive and negative integers can be determined. At least theoretically, we can see how to locate points corresponding to any rational number. For example, if we wanted to locate $\frac{13}{25}$, the unit interval from 0 to 1 could be divided into 25 equal intervals; then the right

Figure 1

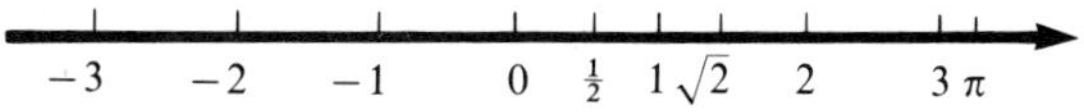

---

* Exponents and radicals will be studied in Sections 5 and 6, but some familiarity with these concepts is assumed.

1

end point of the thirteenth interval starting from 0 would be the desired point. It is more difficult to locate irrational numbers, but they are there (in abundance!). Irrationals are rather elusive, but they can always be approximated to any desired degree of accuracy by rationals. For example, the well-known number we designate by π is typically approximated by the rational number 3.1416 ($\frac{31,416}{10,000}$), and $\sqrt{2}$ is approximately 1.414. It should be kept in mind that these are approximations only; in fact, we can never write down in decimal form the exact value of any irrational number.

A significant feature of the real numbers is that they occupy *every* position on the number line. There are no gaps. This would not be true if we limited consideration to the rationals only, for while the rationals are infinite in number and are dense on the line (that is, between any two points, however close together, there are rationals), they do not occupy all of the line. In fact, the gaps (that is, the irrationals) are more numerous than the rational points.

The rational numbers have two forms when expressed as decimal expansions: (1) terminating, such as 1.25, or (2) repeating, such as 0.666 . . . . All other decimal expansions are irrational numbers. The real numbers, then, can be characterized as being the **set** of all decimal expansions.

We will frequently have occasion to refer to a **set** of things, by which we mean a *collection*. Sets are usually designated by using braces, { }, and inside the braces either listing the **members** of the set or giving a rule that describes the set. The members of a set are also called its **elements,** and we use a symbol such as $a \in A$ to indicate that a is an element of the set A. If every element of a set B is also an element of the set A, then B is said to be a **subset** of A, designated $B \subseteq A$. It is also convenient to be able to speak of a set that has no elements, and we call this the **empty set** and designate it by the symbol $\varnothing$. All other sets are said to be **nonempty.** The sets we will be primarily concerned with in this chapter are the set R of real numbers and its most important **subsets,** namely the set Q of rational numbers, the set I of irrational numbers, the set J of integers, and the set N of natural numbers. In Chapter 3 we will introduce the set C of complex numbers.

EXERCISE SET 1

A In Problems 1 and 2 express the numbers as repeating or terminating decimals.

1. **a.** $\frac{2}{3}$ **b.** $\frac{5}{8}$ **c.** $\frac{2}{11}$ **d.** $-\frac{8}{5}$ **e.** $\frac{10}{27}$

2. **a.** $\frac{15}{16}$ **b.** $-\frac{75}{8}$ **c.** $\frac{4}{9}$ **d.** $\frac{10}{7}$ **e.** $\frac{9}{37}$

3. The number π is sometimes approximated by the rational number $\frac{22}{7}$. How good is this approximation? (The value of π correct to ten decimal places is 3.1415926536.)

4. The decimal expansion of a certain number is 1.01001000100001 . . . , with the pattern continuing in the obvious way. Is this number rational or irrational? Give your reason.

5. State which of the following are rational and which are irrational:

    a. $\dfrac{\sqrt{81}}{4}$ **b.** -2 **c.** $\sqrt{5}$ **d.** $\sqrt[3]{-8}$ **e.** $\frac{1}{2}+\frac{2}{3}$

6. What can you conclude about $x + y$ if:
    a. x and y are rational?   **b.** x and y are irrational?
    c. x is rational and y is irrational?

7. Answer parts a, b, and c of Problem 6 for the product $x \cdot y$.

8. Determine the number $x = 1.242424 \ldots$ in the form m/n, where m and n are integers, by considering $100x - x$.

In Problems 9–11 show that each number is rational by expressing it in the form m/n, where m and n are integers. Use the idea of Problem 8.

9. a. $0.666\ldots$ **b.** $3.111\ldots$
10. a. $0.272727\ldots$ **b.** $0.0121212\ldots$
11. a. $0.243243243\ldots$ **b.** $5.132132132\ldots$

B   **12.** Without doing the division, state the maximum number of digits there can be in one cycle of the decimal expansion of $\frac{53}{137}$. Explain in general how you can be sure that m/n will always result in a repeating or terminating decimal.

    13. The point corresponding to $\sqrt{2}$ on the number line can be located geometrically by constructing a square on the unit interval from 0 to 1 and, with a compass, describing an arc of a circle with center at 0 and with radius equal to the diagonal of the square. The point at which the arc crosses the number line is the desired point.
    a. Explain the basis for this procedure.
    b. How can $\sqrt{5}$ be located in a similar way?
    Hint. For this problem you will need the Pythagorean theorem, which says that in a right triangle with legs a and b and hypotenuse c, $a^2 + b^2 = c^2$.

    14. Carry out the details outlined here to show that $\sqrt{2}$ is irrational. Begin by *assuming* $\sqrt{2}$ is rational. Then we can write $\sqrt{2} = m/n$ where m and n have no common factor other than 1. Square both sides of this equation and solve for m^2. Observe that m^2 is an even number, and explain why it follows that m itself must be even. From this conclude that n^2 is even, and also therefore that n is even. Why is this a contradiction? What do you conclude about the original assumption?

---

2 Basic Properties

In this section we list the basic properties (or axioms) of the real number system, R. From these, all other properties of real numbers can be derived. Since the concept of equality is fundamental to everything that follows, we will begin by stating the properties that provide the basis for working with all equalities.

In mathematics the symbol $A = B$ means A and B are two names for the same thing. Thus, we may replace B by A or A by B in any expression involving either symbol. So equality means "is identical to." Following are the fundamental properties, where the letters a, b, and c stand for real numbers:*

---

* They may also stand for complex numbers, and when we study complex numbers in Chapter 3, we will use these properties without further discussion.

Properties of Equality

1. $a = a$ Reflexive property
2. If $a = b$, then $b = a$.   Symmetric property
3. If $a = b$ and $b = c$, then $a = c$.   Transitive property
4. If $a = b$, then $a + c = b + c$ and $ac = bc$.
5. If $ac = bc$ and $c \neq 0$, then $a = b$.

Property 4 is sometimes read "equals may be added to, or multiplied by, equals."

The basic properties of the set of real numbers R can now be given:

1. The set R is **closed** with respect to addition and multiplication. That is, both $a + b$ and $a \cdot b$ are in R.

Here, the word *closed* means that any number that results from the addition or multiplication of members of R must also be a member of R.

2. Addition and multiplication in R are **commutative.** That is, $a + b = b + a$ and $a \cdot b = b \cdot a$.

Thus, the order of adding or multiplying two real numbers is immaterial.

3. Addition and multiplication in R are **associative.** That is, $a + (b + c) = (a + b) + c$ and $a \cdot (b \cdot c) = (a \cdot b) \cdot c$.

Without this property we would not know, for example, how to interpret $2 + 3 + 4$. The associative property asserts that we may add like this: $2 + (3 + 4) = 2 + 7 = 9$; or like this: $(2 + 3) + 4 = 5 + 4 = 9$. Similar comments apply to multiplication.

4. Multiplication is **distributive** over addition. That is, $a(b + c) = ab + ac$.

For example, we may evaluate $13(12 + 8)$ either this way: $13(12 + 8) = 13(20) = 260$; or by distributing the multiplication: $13(12 + 8) = 13(12) + 13(8) = 156 + 104 = 260$.

5. The number **0 is the additive identity** in R, and the number **1 is the multiplicative identity** in R. That is, $a + 0 = a$ and $a \cdot 1 = a$.

In simpler terms, when 0 is added to any number, this leaves the number unchanged, and when a number is multiplied by 1, the number is unchanged.

6. For each element a in R there is a unique **additive inverse** in R, designated by $-a$, and for each $a \neq 0$, there is a unique **multiplicative inverse** in R, designated by a^{-1}. Thus, $a + (-a) = 0$ and $a \cdot a^{-1} = 1$.

So the additive inverse of a number is that number which must be added to it to give 0, and the multiplicative inverse of a number is that number by which it must be multiplied to give 1. For example, the additive inverse of 2 is -2, and the additive inverse of $-\frac{2}{3}$ is $\frac{2}{3}$; that is, $2 + (-2) = 0$ and $-\frac{2}{3} + \frac{2}{3} = 0$.

The multiplicative inverse of 2 is $\frac{1}{2}$, and the multiplicative inverse of $\frac{2}{3}$ is $\frac{3}{2}$; that is, $2 \cdot 2^{-1} = 2 \cdot \frac{1}{2} = 1$ and $\frac{2}{3} \cdot \left(\frac{2}{3}\right)^{-1} = \frac{2}{3} \cdot \frac{3}{2} = 1$. It is important to note that 0 has no multiplicative inverse. (Why?)

7. The **positive real numbers,** designated by R^+, constitute a subset of R with the following properties:
   a. **If a and b are in R^+, so are $a + b$ and $a \cdot b$.** That is, R^+ is closed with respect to addition and multiplication.
   b. Every real number falls into exactly one of three distinct categories: **It is in R^+, it is zero, or its additive inverse is in R^+.**

Property 7a says that if you add or multiply two positive real numbers, the result is positive. Property 7b divides R into three disjoint classes: (1) the positive real numbers, (2) zero, and (3) the negative real numbers (that is, numbers with additive inverses that are positive).

Before stating the final basic property, it is necessary to give some definitions. You may have wondered why the properties stated so far do not mention subtraction or division. This is because these operations are defined in terms of addition and multiplication.

DEFINITION 1   The **difference** between a and b is defined by

$$a - b = a + (-b)$$

This explains the origin of the rule: "To subtract, change the sign of the subtrahend and add."

DEFINITION 2   If $b \neq 0$, the **quotient** of a by b is defined by

$$\frac{a}{b} = a \cdot b^{-1}$$

The notation $a \div b$ is also used for the quotient of a by b. Note that $a/0$ is not defined, since 0 has no multiplicative inverse. This should be stressed. **The denominator of a fraction cannot be 0; that is, you cannot divide by 0.**

DEFINITION 3   The number a is said to be **less than b,** written $a < b$, provided $b - a$ is positive. Alternately, we may say b is **greater than a,** and write $b > a$. If a is **either less than or equal to b,** we write $a \leq b$. Alternately, we say b is **greater than or equal to a,** and write $b \geq a$.

DEFINITION 4   A subset S of R is said to be **bounded above** if there is a real number k such that every number in S is less than or equal to k. Such a number k is called an **upper bound** for S. If no upper bound of S is less than k, then k is said to be the **least upper bound** of S.

EXAMPLE 1   The set $\{1, 1\frac{1}{2}, 1\frac{3}{4}, 1\frac{7}{8}, 1\frac{15}{16}, \ldots\}$ is bounded above by 4, 5, 100, and many other numbers. But 2 is the least upper bound.

We can now state the final basic property of the real numbers:

8.   The set R is **complete** in the sense that every nonempty subset of R that is bounded above has a least upper bound in R.

This property is admittedly rather difficult to understand, and in this course it will not be necessary to explore its meaning in depth. But you should be aware that this property guarantees that every point on the number line corresponds to some real number and also that every infinite decimal represents a real number.

A host of properties can be derived from the basic ones. We list some of these below for reference.

Further Properties of Real Numbers

1.   The right-hand distributive property: $(a + b)c = ac + bc$.

2.   The extended associative properties: The result is the same regardless of the order of adding or of multiplying any finite collection of real numbers. For example, consider

$$2 + 3 + 4 + 5$$

We may perform this addition in any of the following ways:

$$(2 + 3) + (4 + 5) = 5 + 9 = 14$$
$$[2 + (3 + 4)] + 5 = [2 + 7] + 5 = 9 + 5 = 14$$
$$2 + [3 + (4 + 5)] = 2 + [3 + 9] = 2 + 12 = 14$$

and so on.

3.   When 0 is multiplied by any number, the result is 0, that is: $a \cdot 0 = 0$.

4. $-a = (-1)a$

5. $(-a)(b) = -(ab)$

6. $(-a)(-b) = ab$

7.   If $ab = 0$,   then either $a = 0$ or $b = 0$.

More properties will be given after introducing more definitions.

EXAMPLE 2   Show that $(-a)(b) = -(ab)$.

Note.   The instruction "show that" is equivalent to "prove that."

Solution   In the proof we will indicate the justification for each step to the right of the step.

$$a + (-a) = 0 \qquad \text{Definition of additive inverse}$$
$$[a + (-a)] \cdot b = 0 \cdot b \qquad \text{Equality property 4}$$
$$ab + (-a)(b) = 0 \cdot b \qquad \text{Right-hand distributive property}$$
$$ab + (-a)(b) = 0 \qquad \text{Property of 0 (see Problem 10, Exercise Set 2)}$$

The last equation says that $(-a)(b)$ is a number which when added to ab gives 0, and since additive inverses are unique (see Problem 11, Exercise Set 2), it follows that $(-a)(b)$ is the additive inverse of ab, that is,

$$(-a)(b) = -(ab)$$

EXERCISE SET 2

A   1.   State the basic property of R used in each of the following:

   a. $2(3 + 4) = 2 \cdot 3 + 2 \cdot 4$ **b.** $4 + (-4) = 0$
   c. $(3 \cdot 3^{-1}) \cdot 6 = 1 \cdot 6$ **d.** $1 + 0 = 1$
   e. $1 \cdot 0 = 0$ **f.** $(-2 + 3) + 5 = -2 + (3 + 5)$
   g. $2 \cdot (3 \cdot 4) = (2 \cdot 3) \cdot 4$ **h.** $(3 + 2) \cdot 5 = (2 + 3) \cdot 5$
   i. $(5 + 7) \cdot 6 = 6 \cdot (5 + 7)$ **j.** $(2 + 0) \cdot 7 = 2 \cdot 7$

   2.   **a.**   Give the additive inverse of each of the following: 2,   8, -2, $\frac{3}{4}$, $-\frac{5}{2}$.
       b.   Give the multiplicative inverse of each of the following: 5, -3,   2, $\frac{2}{5}$, $-\frac{7}{3}$.

   3.   Evaluate each of the following:
       a. $2(-3)$ **b.** $(-3)(-5)$ **c.** $-(-2)$ **d.** 3^{-1} **e.** $3 + [4 + (5 + 7)]$

   4.   Evaluate each of the following:
       a. $a - 0$ **b.** $0 - a$ **c.** $\frac{0}{1}$ **d.** $(-3)^{-1}$ **e.** $\left(\frac{3}{4}\right)^{-1}$

   5.   **a.**   What is the additive inverse of 0?
       b.   What is the multiplicative inverse of 1?
       c.   What is the multiplicative inverse of -1?
       d.   Are there any other "self-inverses"?

   6.   Does the commutative property hold for:
       a.   Subtraction?   **b.**   Division?
       Justify your answers.

   7.   Does the associative property hold for:
       a.   Subtraction?   **b.**   Division?
       Justify your answers.

   8.   Is addition distributive over multiplication? Justify your answer.

   9.   Determine the least upper bound of each of the following sets:
       a. $\{1, 2, 5, 3, -2\}$ **b.** $\{1, \frac{3}{2}, \frac{7}{4}, \frac{15}{8}, \frac{31}{16}, \frac{63}{32}, \ldots\}$
       c. $\{-1, -\frac{1}{2}, -\frac{1}{3}, -\frac{1}{4}, -\frac{1}{5}, \ldots\}$
       d. $\{x \in R: \ x < 5\}$
           Note.   This is read "the set of all x in R such that x is less than 5."
       e. $\left\{3 - \dfrac{1}{n}: \ n \in N\right\}$

10. State the basic property of R employed in each step of the following proof that $a \cdot 0 = 0$:

$$a \cdot (0 + 0) = a \cdot 0$$
$$a \cdot 0 + a \cdot 0 = a \cdot 0$$
$$-(a \cdot 0) + (a \cdot 0 + a \cdot 0) = -(a \cdot 0) + a \cdot 0 \qquad \text{Property of equality}$$
$$-(a \cdot 0) + (a \cdot 0 + a \cdot 0) = 0$$
$$[-(a \cdot 0) + a \cdot 0] + a \cdot 0 = 0$$
$$0 + a \cdot 0 = 0$$
$$a \cdot 0 = 0$$

11. The following proof shows that there can be only one additive inverse of a given real number, that is, additive inverses are unique. State the basic property of R used in each step of the proof.*

    Let $a \in R$. By basic property 6 we know there is an additive inverse of a, which we designate by $-a$. Suppose b is also an additive inverse of a, so that

    $$a + b = 0$$

    We add $-a$ to both sides and proceed as shown.

    $$-a + (a + b) = -a + 0$$
    $$-a + (a + b) = -a$$
    $$(-a + a) + b = -a$$
    $$0 + b = -a$$
    $$b = -a$$

    So b is the same as $-a$, which shows there is only one additive inverse of a.

12. Use basic properties of R to prove the right-hand distributive law:

    $$(a + b)c = ac + bc$$

B 13. Follow a procedure similar to that used in Problem 11 to prove that the multiplicative inverse of a nonzero real number a is unique.

14. Show that R is closed with respect to subtraction. Is R closed with respect to division? Show why or why not.

15. Show that $a(b - c) = ab - ac$.

16. Show that $-(a + b) = -a - b$.

17. Show that if $b \neq 0$, then $1/b = b^{-1}$.

18. Show that if $b \neq 0$, then $0/b = 0$.

19. Show that if $a \neq 0$ and $b \neq 0$, then $(ab)^{-1} = a^{-1}b^{-1}$.

20. A **lower bound** of a set S of real numbers is a number that is less than or equal to every element of S, and the **greatest lower bound** of S is a lower bound of S that is greater than or equal to all other lower bounds of S. Show that the number m is the greatest lower bound of a set S of real numbers provided the following two conditions are satisfied: (1) $x \geq m$ for all $x \in S$ and (2) if $x \geq k$ for all $x \in S$, then $k \leq m$.

21. Find the greatest lower bound of each of the following sets (see Problem 20):

    a. $\{3, -1, 2, 4, 0\}$ **b.** $\{1, \frac{1}{2}, \frac{1}{3}, \frac{1}{4}, \frac{1}{5}, \ldots\}$ **c.** $\left\{ \dfrac{n+1}{n} : \quad n \in N \right\}$

    d. $\{x \in R: \quad x > 5\}$ **e.** $\{y \in R: \quad y > -2\}$

---

* We included uniqueness in the statement of the basic property on additive inverses. This problem shows that uniqueness need not be assumed but is a consequence of the other properties.

3 Order Properties and Absolute Value

The following properties of **inequality** can be deduced as consequences of the properties of R given in Section 2.

1. $a > 0$ if and only if a is positive

So, henceforth, the statements "a is positive" and "a is greater than 0" may be used interchangeably.

2. If $a \neq 0$, then $a^2 > 0$.

In particular, $1 > 0$, since $1^2 = 1$.

3. If $a < b$, then $a + c < b + c$.

This says that the same number may be added to both sides of an inequality without changing the sense of the inequality (that is, the symbol $<$ is unchanged in direction).

4. If $a < b$, then
$$\begin{cases} ac < bc & \text{if } c > 0 \\ ac > bc & \text{if } c < 0 \end{cases}$$

The situation in multiplication is more complicated. Multiplying by a positive number leaves the sense of an inequality unchanged, but multiplying by a negative number reverses the sense.

5. If a is any real number, then precisely one of the following is true: $a > 0$, $a = 0$, $a < 0$.

This is the property of **trichotomy**.

6. If $a < b$ and $b < c$, then $a < c$.

This is the property of **transitivity**.

7. If $a < b$ and $c < d$, then $a + c < b + d$.

Note the distinction between this and property 3. It says that as long as both inequalities are of the same type (both "less than" as shown), we may add the respective sides and retain the same sense of inequality.

8. If $a < b$ and $a \cdot b > 0$, then $1/a > 1/b$.

Thus, taking reciprocals reverses the sense of the inequality *provided* both numbers are positive or both numbers are negative.

These properties provide the basis for working with more complicated inequalities in algebra, as we will see in Chapter 3. There are times when it is useful to write a combined inequality of the form

$$a < b < c$$

This means that both $a < b$ *and* $b < c$. It is convenient sometimes to read from the middle and say "b is less than c and greater than a."

We next introduce the concept of the **absolute value** of a number, which is, roughly speaking, the magnitude of the number without regard to sign. The precise definition follows.

DEFINITION 5   The **absolute value** of a real number a is equal to a itself if $a \geq 0$ and is equal to $-a$ if $a < 0$. The absolute value of a is designated by $|a|$.

From this we see that $|a|$ is always nonnegative. For example, $|2| = 2$ and $|-3| = -(-3) = 3$. Note that when a is negative, $|a| = -a$, and since a itself is negative, $-a$ is positive (despite appearances).

Geometrically, the absolute value of a number a can be interpreted as the distance between a and 0 on a number line, as illustrated in Figure 2.

Figure 2

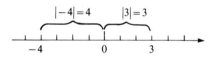

Absolute values and inequalities are often used in combination. For example, consider the set*

$$S = \{x: \ |x| < 1\}$$

The set S consists of all real numbers whose distance from 0 is less than 1, namely the numbers between -1 and 1. So another way of writing S is

$$S = \{x: \ -1 < x < 1\}$$

This can be shown on a number line, as in Figure 3. The parentheses at -1 and 1 indicate that these points are not included. If the end points are to be included, we use square brackets, as in Figure 4, which depicts the set $\{x: \ |x| \leq 1\}$.

Figure 3

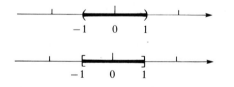

Figure 4

More generally, we can say that the numbers x that satisfy the inequality

$$|x| < a$$

are the same as those that satisfy

$$-a < x < a$$

Similarly, the inequality

$$|x| > a$$

---

* In descriptions of sets such as this we will understand x to be real unless otherwise specified.

is satisfied by those numbers x whose distance from 0 is greater than a. When x is positive, this means that x is to the right of a, that is, $x > a$. When x is negative, x is to the left of $-a$, that is, $x < -a$. So

$$\text{either} \quad x > a \quad \text{or} \quad x < -a$$

For example, consider the set

$$T = \{x: \quad |x| > 2\}$$

We can write this as

$$T = \{x: \quad x > 2 \quad \text{or} \quad x < -2\}$$

It is convenient here to use the idea of **the union of two sets, which is defined as the set of all elements in either, or both, of the two given sets.** The union of two sets A and B is designated by $A \cup B$. We can therefore write the set T above as

$$T = \{x: \quad x > 2\} \cup \{x: \quad x < -2\}$$

Figure 5 shows how this can be indicated on a number line.

Figure 5

$$-2 \qquad 0 \qquad 2$$

EXERCISE SET 3

A In Problems 1 and 2 replace each question mark by the appropriate inequality sign.

1. **a.** $-10 ? 2$ **b.** $-7 ? -9$ **c.** $0 ? 4$ **d.** $0 ? -1$ **e.** $-100 ? 1$

2. **a.** $\frac{2}{3} ? \frac{5}{9}$ **b.** $-\frac{1}{2} ? -\frac{1}{3}$ **c.** $\sqrt{5} ? 2.4$
    d. $3.6 ? \frac{15}{4}$ **e.** $-5 ? -\sqrt{17}$

3. Use Definition 3 to show that each of the following is true:
    a. $3 < 7$ **b.** $-2 < 4$ **c.** $-4 < -1$ **d.** $5 > 2$ **e.** $-3 > -5$

In Problems 4 and 5 write each statement using inequality symbols.

4. **a.** x is greater than 5 **b.** x is less than or equal to 2
    c. x is nonnegative **d.** x is negative
    e. x is not positive

5. **a.** x is less than 5 and greater than 2
    b. x is less than or equal to 0 and greater than -1
    c. x is greater than or equal to 0 and less than 2
    d. x is nonnegative and less than or equal to 6
    e. x is greater than 3 but does not exceed 4

In Problems 6 and 7 show each set on a number line.

6. **a.** $\{x: \quad x < 2\}$ **b.** $\{x: \quad x \geq 3\}$
    c. $\{x: \quad 2 < x < 4\}$ **d.** $\{x: \quad x < 0\} \cup \{x: \quad x > 3\}$
    e. $\{x \in R^+: \quad x < 5\}$

7. **a.** $\{x: \quad x < 3\} \cup \{x: \quad x \geq 5\}$ **b.** $\{x: \quad x \leq 0\} \cup \{x: \quad x > 2\}$
    c. $\{x: \quad -1 \leq x < 2\}$ **d.** $\{x: \quad 2 < x \leq 6\}$
    e. $\{x: \quad x < 0\} \cup \{x: \quad x > 0\}$

8. Give the value of each of the following:
   a. $|-6|$ b. $|\pi|$ c. $|-1.5|$ d. $|0|$ e. $|a|$, where $a \in R$

9. Write the following inequalities in equivalent ways without absolute value signs:
   a. $|x| < 2$ b. $|y| > 1$ c. $|t| \leq 3$ d. $|w| \geq 2$

10. Give an equivalent representation of each of the following sets that does not use absolute value signs, and show each set on a number line:
    a. $\{x: \ |x| < 3\}$ b. $\{x: \ |x| > 2\}$
    c. $\{x: \ |x| \leq 1\}$ d. $\{x: \ |x| \geq 3\}$

11. Give an equivalent representation of each of the following sets that makes use of absolute value signs:
    a. $\{x: \ -5 < x < 5\}$ b. $\{x: \ x > 6\} \cup \{x: \ x < -6\}$
    c. $\{x: \ -4 \leq x \leq 4\}$ d. $\{x: \ x \leq -8\} \cup \{x: \ x \geq 8\}$

B   Problems 12–17 refer to the properties of inequalities given at the beginning of this section.

12. Prove property 1.   13. Prove property 2.
14. Prove property 3.   15. Prove property 4.
16. Prove property 6.   17. Prove property 7.
                                                      Hint.   Use properties 3 and 6

18. Prove that if $a > 0$, then $a^{-1} > 0$.
   Hint.   Consider the product of a and a^{-1} and suppose a^{-1} is not positive.

19. Use the result of Problem 18 to give a rule for dividing both sides of an inequality by the same nonzero number, and justify your result.

20. Prove inequality property 8.
   Hint.   Multiply both sides of $a < b$ by $a^{-1}b^{-1}$ and explain why this is valid based on Problem 18.

21. Prove or disprove: If $a < b$ and $c < d$, then $a - c < b - d$.

22. If A and B are sets, **the intersection of A and B, designated $A \cap B$, is defined to be the set of all elements in both A and B.** Write each of the following in an equivalent form as the intersection of two sets:
    a. $\{x: \ -3 < x < 2\}$ b. $\{x: \ 2 < x \leq 6\}$
    c. $\{x: \ a < x < b\}$ d. $\{x: \ |x| < a\}$

4   The Arithmetic of Rational Numbers

Our primary goal in this section is to derive the rules for adding and multiplying rational numbers. Subtraction and division can then be accomplished by making use of Definitions 1 and 2. Since rational numbers are ratios of integers, we begin by stating rules for adding and multiplying integers.

The arithmetic of the natural numbers (or positive integers) will be taken as given. Let m and n designate arbitrary positive integers. Then $-m$ and $-n$ are negative integers, and the rules listed below apply. These can be proved on the basis of the properties of R given in Section 3.

1. $m + (-n) = \begin{cases} +(m - n) & \text{if } m \geq n \\ -(n - m) & \text{if } n > m \end{cases}$

2. $(-m) + (-n) = -(m + n)$

3. $m(-n) = -(mn)$

4. $(-m)(-n) = mn$

By rule 1, when we add integers of unlike sign, we actually subtract the smaller absolute value from the larger and give the result the sign of the larger. (When writing positive integers the plus sign is usually omitted.) Adding numbers of unlike sign can be thought of as finding the net value.

EXAMPLE 3   Find the indicated sums and products.

 a. $5 + (-2)$ **b.** $2 + (-5)$ **c.** $(-2) + (-5)$
 d. $3(-4)$ **e.** $(-3)(-4)$

 Solution **a.** By rule 1, $5 + (-2) = 5 - 2 = 3$.
 b. By rule 1, $2 + (-5) = -(5 - 2) = -3$.
 c. By rule 2, $(-2) + (-5) = -(2 + 5) = -7$.
 d. By rule 3, $3(-4) = -(3 \cdot 4) = -12$.
 e. By rule 4, $(-3)(-4) = 3 \cdot 4 = 12$.

Subtraction is accomplished by using Definition 1. That is, $a - b = a + (-b)$. Here it should be emphasized that the additive inverse of a negative integer is the corresponding positive integer. This says that $-(-m) = m$. Thus, for example,

$$3 - (-5) = \mathbf{3 + [-(-5)]} = 3 + 5 = 8$$

In the future, we will omit the second step in discussions of this type of problem.

EXAMPLE 4   Evaluate:

 a. $4 - 7$ **b.** $3 - (-4)$ **c.** $-2 - 7$ **d.** $-2 - (-7)$

 Solution **a.** $4 - 7 = 4 + (-7) = -3$ **b.** $3 - (-4) = 3 + 4 = 7$
 c. $-2 - 7 = -2 + (-7) = -9$ **d.** $-2 - (-7) = -2 + 7 = 5$

Now let a, b, c, and d denote arbitrary integers, with c and d not equal to 0. Then a/b and c/d are rational numbers. Their product is found as follows:

$$\frac{a}{b} \cdot \frac{c}{d} = (a \cdot b^{-1}) \cdot (c \cdot d^{-1}) = (a \cdot c)(b^{-1} \cdot d^{-1})$$

$$= (ac)(bd)^{-1} = \frac{ac}{bd}$$

You will be asked to supply the reasons for these steps in Problem 26, Exercise Set 4. The result,

$$\frac{a}{b} \cdot \frac{c}{d} = \frac{ac}{bd}$$

says that "the product of two fractions is the product of their numerators over the product of their denominators."

Before giving the rule for addition, we observe that for any number $c \neq 0$,

$$\frac{a}{b} = \frac{a}{b} \cdot 1 = \frac{a}{b} \cdot (cc^{-1}) = \frac{a}{b} \cdot \frac{c}{c} = \frac{ac}{bc}$$

or, stated in reverse,

$$\frac{ac}{bc} = \frac{a}{b}$$

This is sometimes referred to as the **fundamental property of fractions** and is the basis for such simplifications as $\frac{6}{8} = \frac{3}{4}$.

A Word of Caution.   The fundamental property of fractions applies only when *products* are involved in the numerator and denominator. It is not true for sums. For example,

$$\frac{a+c}{b+c} \neq \frac{a}{b} \qquad \text{unless } c = 0$$

For addition, we have

$$\frac{a}{b} + \frac{c}{d} = \frac{ad}{bd} + \frac{bc}{bd} = (ad)(bd)^{-1} + (bc)(bd)^{-1}$$

$$= (ad + bc)(bd)^{-1}$$

$$= \frac{ad + bc}{bd}$$

Thus,

$$\frac{a}{b} + \frac{c}{d} = \frac{ad + bc}{bd}$$

While this formula always leads to the correct result, it is not always the most efficient way of arriving at the simplest form of the answer. Suppose the two fractions have the same denominator, say, $a/c + b/c$. Then an application of the above formula, together with the fundamental property, gives

$$\frac{a}{c} + \frac{b}{c} = \frac{ac + bc}{cc} = \frac{(a+b)c}{c \cdot c} = \frac{a+b}{c}$$

Thus,

$$\frac{a}{c} + \frac{b}{c} = \frac{a+b}{c}$$

This says: "To add two fractions with the same denominator, add the numerators and write the sum over the common denominator."

When the fractions have different denominators, the trick is to express each fraction with the smallest possible **common denominator,** called the **lowest common denominator (LCD).** With relatively small denominators this can usually be done mentally. For example, consider $\frac{3}{4} + \frac{5}{6}$. It is easy to see that

12 is the smallest natural number that is a multiple of both 4 and 6. So we write

$$\frac{3}{4} + \frac{5}{6} = \frac{3 \cdot 3}{4 \cdot 3} + \frac{5 \cdot 2}{6 \cdot 2} = \frac{9}{12} + \frac{10}{12} = \frac{19}{12}$$

For larger denominators it may be useful to write each denominator as a product of **primes.** You may recall that **a prime number is a natural number other than 1 that cannot be written as a product of natural numbers other than itself and 1.** Examples of primes are 2, 3, 5, 7, 11, 13, ... All other natural numbers except 1 are called **composite,** and they can be expressed as products of primes.

EXAMPLE 5 Find the following sums:

  a. $\frac{3}{4} + \frac{-7}{6}$ **b.** $\frac{-5}{16} + \frac{3}{28}$ **c.** $\frac{7}{780} + \frac{11}{504}$

Solution **a.** $\dfrac{3}{4} + \dfrac{-7}{6} = \dfrac{9}{12} + \dfrac{-14}{12} = \dfrac{9 + (-14)}{12} = \dfrac{-5}{12}$

  b. To find the LCD of 16 and 28 we note that $16 = 4 \cdot 4$ and $28 = 4 \cdot 7$. So the LCD $= 4 \cdot 4 \cdot 7$, since this is the smallest number that both $4 \cdot 4$ and $4 \cdot 7$ will divide evenly.

$$\frac{-5}{16} + \frac{3}{28} = \frac{-5}{4 \cdot 4} \cdot \frac{7}{7} + \frac{3}{4 \cdot 7} \cdot \frac{4}{4} = \frac{-35 + 12}{4 \cdot 4 \cdot 7} = \frac{-23}{112}$$

  c. This time we write each denominator as a product of primes.

2	780			2	504	
2	390	So	$780 = 2^2 \cdot 3 \cdot 5 \cdot 13$	2	252	So
3	195			2	126	$504 = 2^3 \cdot 3^2 \cdot 7$
5	65			3	63	
	13			3	21	
					7	

Thus, the LCD $= 2^3 \cdot 3^2 \cdot 5 \cdot 7 \cdot 13$. Notice that the LCD contains each prime number occurring in either number to the *highest* power to which it appears. So we multiply the numerator and denominator of each fraction by the primes that do not appear in its denominator to make both denominators equal to the LCD:

$$\frac{7}{780} + \frac{11}{504} = \frac{7}{780} \cdot \frac{2 \cdot 3 \cdot 7}{2 \cdot 3 \cdot 7} + \frac{11}{504} \cdot \frac{5 \cdot 13}{5 \cdot 13} = \frac{294 + 715}{32,760}$$

$$= \frac{1,009}{32,760}$$

The handling of signs is made easier by the following considerations. Since

$$\frac{a}{b} + \frac{(-a)}{b} = \frac{a + (-a)}{b} = \frac{0}{b} = 0$$

it follows that $-a/b$ is the additive inverse of a/b; that is,

$$\frac{-a}{b} = -\frac{a}{b}$$

Also,

$$\frac{a}{-b} = \frac{a}{-b} \cdot \frac{-1}{-1} = \frac{-(a \cdot 1)}{b \cdot 1} = \frac{-a}{b}$$

We also have that

$$\frac{-a}{-b} = \frac{a(-1)}{b(-1)} = \frac{a}{b}$$

Combining these results, we obtain

$$\frac{-a}{b} = \frac{a}{-b} = -\frac{a}{b} = -\frac{-a}{-b}$$

We can often save some steps in handling products of fractions by anticipating an application of the fundamental property. Since

$$\frac{a}{b} \cdot \frac{c}{a} = \frac{ac}{ba} = \frac{ca}{ba} = \frac{c}{b}$$

we can skip the middle step by "dividing out" the a before multiplication, as follows:

$$\frac{\not a}{b} \cdot \frac{c}{\not a} = \frac{c}{b}$$

More generally, whenever two or more fractions are to be multiplied, any number that occurs as a **factor** in one of the numerators and also as a factor in one of the denominators can (and should) be divided out before multiplication. For example,

$$\frac{\overset{2}{\not 4} \cdot 7}{9 \ \underset{5}{\not{10}}} = \frac{14}{45}$$

A Word of Caution.   The key word in the above discussion is *factor*. **A factor is one member of a product.** What we have said about dividing out a **common factor,** then, does not apply to sums and differences. For example, it would be *incorrect* to "divide out" the 2 as shown in

$$\frac{\not 2}{3} + \frac{5}{\not 2}$$

A fraction is said to be in **lowest terms** if the numerator and denominator have no integer factor in common other than 1 or -1. A fraction may be reduced to lowest terms by using the fundamental property.

EXAMPLE 6   Perform the indicated operations and reduce answers to **lowest terms.**

a. $\frac{-3}{5} - \frac{1}{2} - \frac{-7}{-4}$ b. $\frac{2}{3}(\frac{3}{5} - \frac{1}{2})$

Solution

a. $\dfrac{-3}{5} - \dfrac{1}{2} - \dfrac{-7}{-4} = \dfrac{-3}{5} + \dfrac{-1}{2} + \dfrac{-7}{4} = \dfrac{-12}{20} + \dfrac{-10}{20} + \dfrac{-35}{20} = \dfrac{-57}{20}$

b. $\dfrac{2}{3}\left(\dfrac{3}{5} - \dfrac{1}{2}\right) = \dfrac{2}{3}\left(\dfrac{6}{10} + \dfrac{-5}{10}\right) = \dfrac{2}{3}\left(\dfrac{1}{10}\right) = \dfrac{2}{30} = \dfrac{1}{15}$

Following is a summary of the rules for the arithmetic of rational numbers:

$$\frac{a}{b} + \frac{c}{d} = \frac{ad + bc}{bd} \tag{1}$$

$$\frac{a}{c} + \frac{b}{c} = \frac{a + b}{c} \tag{2}$$

$$\frac{a}{b} \cdot \frac{c}{d} = \frac{ac}{bd} \tag{3}$$

$$\frac{ac}{bc} = \frac{a}{b} \tag{4}$$

$$-\frac{a}{b} = \frac{-a}{b} = \frac{a}{-b} = -\frac{-a}{-b} \tag{5}$$

EXERCISE SET 4

A   Perform the indicated operations in Problems 1–18.

1. a. $-5 + 7$ b. $3 + (-6)$ c. $-2 + (-4)$
    d. $6 - 8$ e. $5 - (-3)$

2. a. $3 + (-4)$ b. $-4 + (-7)$ c. $-3 - (-5)$
    d. $10 - (-4)$ e. $-8 - (-2)$

3. a. $2(-3)$ b. $(-4)(5)$ c. $(-5)(-6)$
    d. $3 - 2(4)$ e. $5 + 2(-3)$

4. a. $(-3)(-4)$ b. $(-5)(9)$ c. $-2 + [(-7) + (-9)]$
    d. $(-3)[(-2) + (-5)]$ e. $(-5)(-2 + 8)$

5. a. $|-3 + 7|$ b. $|8 - 15|$ c. $|-3(-5)|$
    d. $|-6 - 9|$ e. $|2(-6)|$

6. a. $|2 - (-3)(4)|$ b. $|5(-7) + (-2)(-6)|$
    c. $|8(-2) - (-3)(4)|$ d. $|3(-2) - (-4)(-5) + (-7)(-6)|$

7. a. $\frac{5}{6} + \frac{3}{4}$ b. $\frac{7}{8} - \frac{5}{12}$

8. a. $\frac{2}{3} - \frac{1}{2}$ b. $\frac{3}{8} + \frac{7}{10}$

9. a. $\frac{1}{2} - \frac{2}{3} + \frac{5}{4}$ b. $\frac{3}{4} + \frac{5}{6} - \frac{7}{12}$

10. a. $\frac{7}{10} + \frac{-2}{15}$ b. $\frac{-7}{24} + \frac{11}{18}$

11. a. $2 - \frac{-3}{5}$ b. $\frac{5}{-6} - 1$

12. a. $-\frac{3}{4} - \frac{-1}{2} + 2$ b. $\frac{2}{3} - 1 - \frac{7}{-4}$

13. a. $\frac{2}{3}(\frac{-6}{7})$ b. $(\frac{-5}{2})(\frac{-9}{10})$

14. **a.** $3(\frac{5}{12})$ **b.** $(-\frac{9}{16})(-2)$

15. **a.** $\frac{11}{36} - \frac{19}{54}$ **b.** $\frac{7}{126} + \frac{5}{84} - \frac{17}{210}$

16. **a.** $\frac{3}{140} + \frac{5}{126}$ **b.** $\frac{7}{540} - \frac{17}{324}$

17. **a.** $-\frac{3}{4}(\frac{2}{3} - \frac{4}{5})$ **b.** $\frac{5}{6}\left(\frac{3}{4} - \frac{-4}{15}\right)$

18. **a.** $\frac{2}{5}(\frac{5}{8} - \frac{3}{4})$ **b.** $\frac{3}{11}\left(-\frac{5}{6} - \frac{-3}{8}\right)$

19. If a, b, and c are natural numbers such that $a = bc$, then both b and c are said to be **divisors** of a. The **greatest common divisor (GCD)** of two natural numbers is the largest natural number that is a divisor of both numbers. Find the GCD of each of the following pairs:

    a. 32, 48   **b.** 26, 78   **c.** 84, 56   **d.** 128, 320   **e.** 540, 315

B **20.** A definition of subtraction that is valid for the natural numbers N is as follows: "If a and b are natural numbers, then $a - b$ equals that natural number c, if it exists, which when added to b gives a." Show that when a, b, and c are elements of R, this definition is equivalent to Definition 1.

    21. A definition of division applicable to the integers J is as follows: "If a and b are integers with $b \neq 0$, then $a \div b$ equals that integer c, if it exists, which when multiplied by b gives a." Show that when a, b, and c are elements of R, this definition is equivalent to Definition 2.

    22. Prove each of the following:

      a. The set N is not closed with respect to subtraction, but J is.

      b. The set J is not closed with respect to division by numbers other than 0, but Q is.

In Problems 23–25 perform the indicated operations.

23. **a.** $3(-\frac{2}{5})(-\frac{1}{4}) + \frac{5}{8}(-\frac{2}{3} + 6)$ **b.** $-(-\frac{2}{3})(-\frac{7}{8}) - 2(\frac{5}{-6} + \frac{-3}{8}) - 1$

24. **a.** $\frac{5}{32} - \frac{3}{5}(\frac{7}{-8}) + \frac{2}{3}(3 - \frac{8}{5})$ **b.** $1 - \frac{9}{14} + \frac{2}{3}(\frac{-5}{-7}) - (-\frac{2}{7})(\frac{5}{12})$

25. **a.** $-\frac{5}{16} - \frac{3}{4}(\frac{2}{3} - \frac{5}{6}) + \frac{5}{2}(-\frac{3}{8})$ **b.** $(-\frac{3}{5})(\frac{-7}{8} - \frac{5}{6}) - 2 + \frac{7}{12}(-\frac{4}{5} + \frac{3}{2})$

26. Justify each step of the derivation of the rule for the product of two rational numbers found on page 13.

---

5 Complex Fractions

Rules (1)–(5) in Section 4 are not limited to rational numbers of the form a/b, in which a and b are integers ($b \neq 0$). In fact, an examination of their derivation will show that, as long as the denominators involved are not 0, these rules hold for all real numbers. In particular, the numerators and denominators themselves may contain fractions, giving rise to **complex fractions.**

The first such situation to consider is the quotient of two fractions:*

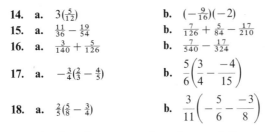

---

*Note that for this to be defined we must not only have $b \neq 0$ and $d \neq 0$, but also $c \neq 0$, so that $c/d \neq 0$.

The definition of division requires that we have

$$\frac{\frac{a}{b}}{\frac{c}{d}} = \frac{a}{b} \cdot \left(\frac{c}{d}\right)^{-1}$$

But what is $(c/d)^{-1}$? Since $(c/d) \cdot (d/c) = (cd/dc) = 1$, and multiplicative inverses are unique, we see that

$$\left(\frac{c}{d}\right)^{-1} = \frac{d}{c}$$

$$\frac{\frac{a}{b}}{\frac{c}{d}} = \frac{a}{b} \cdot \frac{d}{c} = \frac{ad}{bc} \tag{6}$$

This is the familiar rule: "To divide fractions, invert the denominator and multiply."

The following examples illustrate how to handle more complicated complex fractions.

EXAMPLE 7 Simplify: $\dfrac{\frac{2}{3} + \frac{3}{4}}{\frac{5}{6} + \frac{1}{2}}$

Solution In general, the best approach to this sort of problem is first to multiply the numerator and denominator of the main fraction by the LCD of all **minor denominators.** This is justified by rule (4). In this case, the LCD is 12. So we have

$$\frac{(\frac{2}{3} + \frac{3}{4})}{(\frac{5}{6} + \frac{1}{2})} \cdot \frac{12}{12} = \frac{8 + 9}{10 + 6} = \frac{17}{16}$$

The omitted steps can be done mentally. It is important to note that when multiplying the numerator and denominator by 12, *every* term must be multiplied. Notice, too, that in multiplying we have divided out common factors *before* multiplication. For example, the operations done mentally on the numerator are shown in more detail as

$$\left(\frac{2}{3} + \frac{3}{4}\right)12 = \frac{2}{3} \cdot 12 + \frac{3}{4} \cdot 12 = \frac{2}{\cancel{3}} \cdot \frac{\cancel{12}^{4}}{1} + \frac{3}{\cancel{4}} \cdot \frac{\cancel{12}^{3}}{1} = 8 + 9$$

As an alternative way of doing this problem, we could perform the additions on the numerator and denominator separately and then use rule (6). Doing it this way we get

$$\frac{\frac{2}{3} + \frac{3}{4}}{\frac{5}{6} + \frac{1}{2}} = \frac{\frac{8}{12} + \frac{9}{12}}{\frac{5}{6} + \frac{3}{6}} = \frac{\frac{17}{12}}{\frac{8}{6}} = \frac{17}{\cancel{12}} \cdot \frac{\cancel{6}^{1}}{8} = \frac{17}{16}$$

The first method is usually easier, but there are exceptions.

EXAMPLE 8   Simplify: $\dfrac{\frac{2}{3} - \frac{1}{4}(\frac{5}{6})}{\frac{5}{12} - \frac{-3}{2}}$

Solution   Here we must be careful to note that $(\frac{1}{4})(\frac{5}{6}) = \frac{5}{24}$, so that the desired LCD is 24.

$$\frac{\frac{2}{3} - \frac{1}{4}(\frac{5}{6})}{\frac{5}{12} - \frac{-3}{2}} \cdot \frac{24}{24} = \frac{16 - 5}{10 + 36} = \frac{11}{46}$$

EXAMPLE 9   Simplify: $\dfrac{2}{1 - \dfrac{1}{2 - \frac{3}{4}}}$

Solution   We do this in stages, beginning with the complex fraction in the denominator.

$$\frac{2}{1 - \dfrac{1}{2 - \frac{3}{4}}} = \frac{2}{1 - \dfrac{1}{2 - \frac{3}{4}} \cdot \dfrac{4}{4}} = \frac{2}{1 - \dfrac{4}{8 - 3}}$$

$$= \frac{2}{1 - \frac{4}{5}} \cdot \frac{5}{5} = \frac{10}{5 - 4} = \frac{10}{1} = 10$$

EXERCISE SET 5

Express each complex fraction as a simple fraction in lowest terms.

A   1. $\dfrac{\frac{1}{2} - \frac{1}{3}}{\frac{3}{4} + \frac{1}{6}}$

2. $\dfrac{\frac{2}{3} - \frac{3}{5}}{1 - \frac{1}{5}}$

3. $\dfrac{\frac{3}{8} + \frac{5}{6}}{\frac{5}{12} - \frac{3}{4}}$

4. $\dfrac{2 + \frac{3}{8}}{\frac{1}{4} - \frac{1}{6}}$

5. $\dfrac{3 - (-\frac{1}{2})}{-2 - \frac{3}{4}}$

6. $\dfrac{\frac{-3}{2} - \frac{-4}{3}}{\frac{3}{4} - \frac{5}{6}}$

7. $\dfrac{\frac{5}{8} - 2}{\frac{2}{3} + 1}$

8. $\dfrac{\frac{4}{15} - \frac{3}{5}}{1 - \frac{7}{3}}$

9. $\dfrac{\frac{3}{2} - \frac{4}{3}}{-\frac{5}{6} - 1}$

10. $\dfrac{\frac{3}{5} - \frac{9}{4}}{\frac{-7}{2} + \frac{6}{5}}$

11. $\dfrac{\frac{3}{4} + \frac{1}{2}}{1 - (\frac{3}{4})(\frac{1}{2})}$

12. $\dfrac{\frac{1}{4} + \frac{1}{2}(\frac{2}{3})}{1 - (\frac{3}{4})(\frac{2}{3})}$

13. $\dfrac{\frac{5}{12} - \frac{3}{4}}{1 + (\frac{3}{4})(\frac{5}{12})}$

14. $\dfrac{\frac{1}{3} - \frac{5}{6}}{\frac{2}{7} - \frac{3}{14}}$

15. $1 - \dfrac{1}{2 - \frac{1}{2}}$

16. $\dfrac{1}{\dfrac{1}{1 - \frac{2}{3}} - \dfrac{1}{2}}$

B **17.** $1 - \dfrac{1}{1 - \dfrac{3}{1 - \frac{2}{3}}}$

18. $\dfrac{\dfrac{2}{3} - \dfrac{1 + \frac{4}{9}}{2 \cdot \frac{2}{3}}}{\frac{4}{9}}$

19. $\dfrac{\dfrac{10}{9}\left(\dfrac{1}{3}\right) - \dfrac{\frac{25}{81}}{4(\frac{1}{3})}}{2 \cdot \frac{5}{9} - 1}$

20. $\dfrac{-\dfrac{3}{2} - \left(\dfrac{-27}{8} - 1\right) \cdot \dfrac{1}{3 \cdot \frac{9}{4}}}{\frac{9}{4}}$

---

6 Integral Exponents

Exponents originated as a shorthand for repeated multiplication of a number by itself. For example, $2^3 = 2 \cdot 2 \cdot 2$. In general, if a is any real number and n is a positive integer, then

$$a^n = \underbrace{a \cdot a \cdot a \cdot \cdots \cdot a}_{n \text{ factors}}$$

In particular, $a^1 = a$. The familiar rules for multiplication and division are immediate:

$$a^m \cdot a^n = (\overbrace{a \cdot a \cdot \cdots \cdot a}^{m \text{ factors}})(\overbrace{a \cdot a \cdot \cdots \cdot a}^{n \text{ factors}})$$

$$= \overbrace{a \cdot a \cdot a \cdot a \cdot \cdots \cdot a}^{m + n \text{ factors}} = a^{m+n}$$

If $m > n$, then

$$\frac{a^m}{a^n} = \frac{\cancel{a} \cdot \cancel{a} \cdot \cdots \cdot \cancel{a} \cdot \overbrace{a \cdot a \cdot \cdots \cdot a}^{m - n \text{ factors}}}{\cancel{a} \cdot \cancel{a} \cdot \cdots \cdot \cancel{a}} = a^{m-n} \qquad (a \neq 0)$$

If $m = n$, then

$$\frac{a^m}{a^n} = 1 \qquad (a \neq 0)$$

Whereas if $n > m$, then

$$\frac{a^m}{a^n} = \frac{\cancel{a} \cdot \cancel{a} \cdot \cdots \cdot \cancel{a}}{\cancel{a} \cdot \cancel{a} \cdot \cdots \cdot \cancel{a} \cdot \underbrace{a \cdot a \cdot \cdots \cdot a}_{n - m \text{ factors}}} = \frac{1}{a^{n-m}} \qquad (a \neq 0)$$

These relationships are summarized as follows:

$$a^m \cdot a^n = a^{m+n} \tag{7}$$

$$\frac{a^m}{a^n} = \begin{cases} a^{m-n} & \text{if } m > n \\ 1 & \text{if } m = n \qquad (a \neq 0) \\ \dfrac{1}{a^{n-m}} & \text{if } n > m \end{cases} \tag{8}$$

For example,

$$2^3 \cdot 2^5 = 2^8 \qquad \frac{2^5}{2^3} = 2^2 \qquad \frac{2^3}{2^5} = \frac{1}{2^2}$$

Rule (7) can be used to obtain the formula $(a^m)^n = a^{mn}$. To illustrate, consider $(2^3)^4$. We have, by definition of the outer exponent,

$$(2^3)^4 = 2^3 \cdot 2^3 \cdot 2^3 \cdot 2^3$$

and by extending rule (7),

$$2^3 \cdot 2^3 \cdot 2^3 \cdot 2^3 = 2^{12}$$

So, $(2^3)^4 = 2^{12}$. More generally,

$$(a^m)^n = \overbrace{a^m \cdot a^m \cdot \ \cdots \ \cdot a^m}^{n \text{ factors}} = a^{mn} \tag{9}$$

since the sum of n m's is $nm = mn$.

One additional property of exponents should be noted, namely, that $(a \cdot b)^n = a^n b^n$. This follows from the definition and from commutativity and associativity:

$$(ab)^n = (ab)(ab)(ab) \cdots (ab)$$
$$= (a \cdot a \cdot a \ \cdots \ \cdot a)(b \cdot b \cdot b \ \cdots \ \cdot b) = a^n b^n \tag{10}$$

This would be about all we could say concerning exponents if consideration were limited to exponents that are positive integers. Their usefulness is greatly extended, however, by giving meaning to negative and zero exponents, and even more by permitting rational and finally irrational exponents.

To arrive at a definition of negative integral exponents, we define a^{-1} for $a \neq 0$ as the unique multiplicative inverse of a. (This is not new, but the symbol a^{-1} as used previously was not thought of as "a raised to the power negative one.") Since

$$a \cdot \frac{1}{a} = \frac{a}{1} \cdot \frac{1}{a} = \frac{a}{a} = 1$$

it follows that $1/a$ is the multiplicative inverse of a. So $a^{-1} = 1/a$. For consistency with rule (9), we then have

$$a^{-n} = a^{(-1)n} = (a^{-1})^n = \left(\frac{1}{a}\right)^n = \underbrace{\frac{1}{a} \cdot \frac{1}{a} \cdot \ \cdots \ \cdot \frac{1}{a}}_{n \text{ factors}} = \frac{1}{a^n}$$

Here, again, it is essential that $a \neq 0$.

To see how we might reasonably define a^0, consider a^m/a^m with $a \neq 0$. This clearly equals 1 (by the definition of division). However, if we extend the first part of rule (8) to $m = n$, we obtain $a^m/a^m = a^{m-m} = a^0$. So we are led to the definition $a^0 = 1$.

It can now be shown that rules (7)–(10) remain true with m and n positive, negative, or zero integral exponents, but we will omit the details. The important definitions and results are summarized in the box.

Laws of Exponents

1. $a^0 = 1$ 2. $a^{-n} = \dfrac{1}{a^n}$

3. $a^m \cdot a^n = a^{m+n}$ 4. $\dfrac{a^m}{a^n} = a^{m-n} = \dfrac{1}{a^{n-m}}$

5. $(a^m)^n = a^{mn}$ 6. $(ab)^n = a^n b^n$

The proviso that $a \neq 0$ must be inserted whenever zero or negative exponents are involved. Note that because zero and negative exponents are permitted, it is not necessary to give any conditions on m and n in the three forms of the quotient rule (8).

A special case involving a negative power should be mentioned. Consider $(a/b)^{-n}$. Since $(a/b)(b/a) = ab/ab = 1$, it follows that $(a/b)^{-1} = b/a$. Thus,

$$\left(\frac{a}{a}\right)^{-n} = \left[\left(\frac{a}{b}\right)^{-1}\right]^n = \left(\frac{b}{a}\right)^n$$

This result is easy to remember and should be used whenever a fraction is raised to a negative power. So rather than writing, for example,

$$\left(\frac{2}{3}\right)^{-2} = \frac{1}{(\frac{2}{3})^2} = \frac{1}{\frac{4}{9}} = \frac{9}{4}$$

we write immediately

$$\left(\frac{2}{3}\right)^{-2} = \left(\frac{3}{2}\right)^2 = \frac{9}{4}$$

The following examples illustrate how the laws of exponents can be used to simplify certain complex expressions. In all cases answers are written with positive exponents only.

EXAMPLE 10   Simplify:

    a. $(3)^{-2}$ **b.** $\left(\dfrac{2}{3}\right)^0$ **c.** $\left(\dfrac{1}{2}\right)^{-3} \cdot 4^{-2}$ **d.** $\dfrac{2^4 \cdot 3^{-2}}{2^2 \cdot 3^{-5}}$

Solution   **a.** $\dfrac{1}{3^2} = \dfrac{1}{9}$ **b.** 1 **c.** $2^3 \cdot \dfrac{1}{4^2} = \dfrac{8}{16} = \dfrac{1}{2}$ **d.** $2^2 \cdot 3^3 = 4 \cdot 27 = 108$

EXAMPLE 11   Simplify: $\dfrac{2^0 \cdot a^2 \cdot b^{-3} \cdot c^7}{3^2 \cdot a^5 \cdot b \cdot c^{-2}}$

Solution

$$\frac{2^0 \cdot a^2 \cdot b^{-3} \cdot c^7}{3^2 \cdot a^5 \cdot b \cdot c^{-2}} = \frac{c^9}{9a^3b^4}$$

Note that the form of the quotient rule to be used is determined by which produces a final answer with positive exponents, since this is preferred to an answer with negative exponents. Thus, in solving the problem we wrote

$$\frac{a^2}{a^5} = \frac{1}{a^3} \qquad \frac{b^{-3}}{b} = \frac{1}{b^{1-(-3)}} = \frac{1}{b^4} \qquad \frac{c^7}{c^{-2}} = c^{7-(-2)} = c^9$$

EXAMPLE 12   Simplify: $\dfrac{x^{-2} \cdot y^{-3} \cdot z^4 \cdot x^5}{y^{-2} \cdot z^{-3} \cdot x^8}$

Solution

$$\frac{x^{-2} \cdot y^{-3} \cdot z^4 \cdot x^5}{y^{-2} \cdot z^{-3} \cdot x^8} = \frac{x^3 y^{-3} z^4}{y^{-2} z^{-3} x^8} = \frac{z^7}{x^5 y}$$

In dealing with very large or very small numbers it is often convenient to use what is called **scientific notation.** A number is expressed in scientific notation when it is written as a product of a number between 1 and 10, multiplied by a power of 10. For example,

$$2{,}500{,}000{,}000 = 2.5 \times 10^9$$

(It is customary to use "×" as the multiplication symbol in scientific notation.) The exponent of 10, when it is positive, indicates how many places to the *right* the decimal should go. A negative exponent indicates how many places to the *left* the decimal should go:

$$0.00000000003 = 3 \times 10^{-11}$$

Scientific notation, combined with the laws of exponents, can often greatly simplify complicated arithmetic problems. The next example illustrates this.

EXAMPLE 13   Evaluate: $\dfrac{(72{,}000{,}000)(0.0000000036)}{(0.12)(27{,}000)}$

Solution

$$\frac{(72{,}000{,}000)(0.0000000036)}{(0.12)(27{,}000)} = \frac{(7.2 \times 10^7)(3.6 \times 10^{-9})}{(1.2 \times 10^{-1})(2.7 \times 10^4)}$$

$$= \frac{(7.2)(3.6)}{(1.2)(2.7)} \times \frac{10^{-2}}{10^3} = 8.0 \times 10^{-5}$$

$$= 0.000080$$

When numbers are expressed in scientific notation, they can be added or subtracted provided they involve the same power of 10. For example,

$$2.3 \times 10^{-5} + 5.8 \times 10^{-5} = 8.1 \times 10^{-5}$$

If they do not involve the same power of 10, then we can alter the form of one of them to make them agree. For example, consider

$$3 \times 10^2 + 5 \times 10^3$$

We can multiply and divide the first number by 10 to get

$$3 \times 10^2 = 3 \times 10^2 \times \tfrac{10}{10} = \tfrac{3}{10} \times 10^3 = 0.3 \times 10^3$$

Note that we can go immediately from the first step to the last by observing that dividing by 10 is equivalent to moving the decimal one place to the left. So we can now do the problem as follows:

$$3 \times 10^2 + 5 \times 10^3 = 0.3 \times 10^3 + 5 \times 10^3 = 5.3 \times 10^3$$

Many hand calculators have scientific notation capability which is automatically employed when the size of the number would otherwise exceed the capacity of the calculator. For example, if you enter the number 23,200 and then multiply by 4,500,000, the answer shown would be

$$1.044 \qquad 11$$

which is to be interpreted as 1.044×10^{11}. Numbers can also be entered in scientific notation on such calculators.

Remarks on Accuracy in Computations. When calculations are made with numbers that are approximations to true values, it is important to realize that **the answer can be no more accurate than the least accurate of the numbers involved.** It is necessary, therefore, in many instances to **round off the answer** to make it conform to this restriction. A number is rounded up or down according to whether the amount dropped is greater than or less than one-half unit in the last place retained. When it is exactly one-half, we will follow the convention of rounding so that the last digit retained is even. Here are some examples, then, of numbers rounded to two decimal places:

Number	Rounded number	
32.576	32.58	
5.2349	5.23	
163.045	163.04	**We rounded to the**
77.875	77.88	*even* **number**
2.045001	2.05	

In addition and subtraction it is the number of decimal places that is important, whereas in multiplication and division it is the number of **significant digits** that controls the accuracy. All digits other than 0 are always significant. The digit 0 is significant except when it is used solely to place the decimal, as in 0.0021. To avoid ambiguity about the significance of the zeros in whole numbers ending in 0, we can use scientific notation. For example, if by 1,200 we mean only two significant digits (accuracy to the nearest hundred), we can write 1.2×10^3. On the other hand, if we mean three, or four, digits to be significant, we can write 1.20×10^3, or 1.200×10^3, respectively. Here are some more examples:

Number	Number of significant digits
237	3
1.02	3
0.045	2
0.0450	3
2,001	4
1.234×10^{12}	4
3.00×10^{-5}	3

Before stating the rules for computation, we introduce the concept of the comparative **precision** of two numbers. A number A is more precise than a number B if the last significant digit of A occurs in a position to the right of the last significant digit of B. For example, 0.0032 is more precise than 2.753 (even though the latter has more significant digits).

Rule for Addition and Subtraction of Approximate Data. Carry out the computations with the data given and round the answer back so that its precision is that of the least precise of the data. The more precise data may first be rounded to *one place greater* precision than will be retained in the final answer, but they should not be rounded all the way back to that of the least precise data until the calculations have been performed. (An exception is when only two numbers are being added or subtracted.)

Rule for Multiplication and Division of Approximate Data. Carry out the computations with the given data and round off the answer so that it has the same number of significant digits as the number in the data with the least number of significant digits. Again, it is permissible to round the more accurate data initially to *one more* significant digit than will be retained ultimately.

In computations involving powers and roots, the rule for multiplication and division should be used.

The subject of errors in computed results is a deep one, requiring far more analysis than we can go into here, but the rules stated provide reasonable results. It should be noted, however, that even when these rules are followed, there can be some doubt about the reliability of the answer, especially in the last digit. It is especially important to be aware of the limitations on accuracy when using a hand calculator, since it is easy to have delusions of accuracy. For example, to find the area of a circle with radius that has been measured to be 4.53 cm, on a calculator (using $A = \pi r^2$) we could get $A = 64.46830869 \text{ cm}^2$, which is an absurd answer. Our best approximation to the answer would be 64.5 cm^2.

EXERCISE SET 6

A   In Problems 1–9 simplify each of the expressions and give answers with positive exponents only. (All letters represent real numbers that are understood to be nonzero whenever they occur in a denominator or to a negative or zero power.)

1. **a.** 3^4
   d. $(15)^0$
   b. 4^{-1}
   e. $(\frac{1}{2})^{-2}$
   c. $(-2)^3$

2. **a.** $2^7 \cdot 2^3$
   d. $\dfrac{3^2}{3^{-4}}$
   b. $\dfrac{3^8}{3^5}$
   e. $\dfrac{4^{-2}}{4^{-3}}$
   c. $\dfrac{5^4}{5^6}$

3. **a.** 2^{-4}
   d. $\dfrac{3^3 \cdot 5^{-2}}{3^{-2} \cdot 5^{-1}}$
   b. $(\frac{3}{2})^{-1}$
   e. $(\frac{3}{4})^{-2} \cdot 2^{-3}$
   c. $5^0 \cdot (-3)^3$

4. **a.** $2^0 \cdot 3^{-1}$
   d. $\dfrac{2^{-3} \cdot 3^2 \cdot 5^{-1}}{2^2 \cdot 3^{-4} \cdot 5^{-2}}$
   b. $3^{-2}(-2)^3$
   e. $\dfrac{4^3 \cdot 3^{-4} \cdot 7^2}{2^4 \cdot 3^{-2} \cdot 7^3}$
   c. $(\frac{3}{2})^{-1}(\frac{2}{3})^2$

5. **a.** $x^5 \cdot x^2$
   d. $(x^2y^3)^2$
   b. $\dfrac{x^7}{x^3}$
   e. $(2x^3)^0$
   c. $(x^3)^4$

6. **a.** $(2x^2)^3$
   d. $\left(\dfrac{2}{3x}\right)^{-3}$
   b. $(2x^4)(3x^5)$
   e. $\dfrac{x^{-2}}{x^5}$
   c. $(5^0x^{-2})(x^4)$

7. **a.** $\dfrac{a^2b^{-4}}{a^{-3}b}$
   b. $\dfrac{a^{-2}b^4c^0}{a^{-5}b^{-2}c^3}$
   c. $\dfrac{6s^{-2}t^{-3}}{9s^4t^{-7}}$

8. **a.** $\dfrac{2^3a^2b^{-3}c^0}{2^{-5}a^4b^{-7}}$
   b. $\dfrac{(-2)^3a^{-1}b^2c^{-3}}{4^2a^{-2}b^{-3}c^4}$

9. **a.** $\left(\dfrac{a}{b^2}\right)^{-3} \cdot \left(\dfrac{c^3}{d^{-2}}\right)^{-2}$
   b. $\dfrac{\left(\dfrac{x^2}{y^3}\right)^4}{\left(\dfrac{x}{y^2}\right)^{-2}}$

10. Substitute $x = 3$, $y = 2$ in the following expression and simplify the result:
$$\frac{2x^{-1} - 3y^{-1}}{x^{-2} - y^{-2}}$$

11. Substitute $a = 2$, $b = 3$ in the following expression and simplify the result:
$$\frac{a^2b^{-1} - a^{-2}b}{2a^{-1} - 3b^{-2}}$$

12. Substitute $x = \frac{1}{2}$, $y = \frac{3}{4}$ in the following expression and simplify the result:
$$\frac{x^{-1}y - 2x^{-2}y^{-1}}{3xy^{-2} - (2y)^{-1}}$$

13. Express in scientific notation:
 a. 2,000,000,000,000 b. 0.000007 c. 3,100,000
 d. 0.13   e. 254,000,000

In Problems 14–16 perform the indicated calculations using scientific notation.

14. a. $(3.2 \times 10^4)(2 \times 10^{-2})$ b. $\dfrac{5.6 \times 10^7}{7 \times 10^2}$

15. a. $6.2 \times 10^4 + 3.8 \times 10^3$ b. $5 \times 10^2 - 3 \times 10^3$

16. a. $8 \times 10^6 + 4 \times 10^4$ b. $2.7 \times 10^{-3} - 5.2 \times 10^{-2}$

17. Use scientific notation to calculate

$$\frac{(0.0015)(25,000,000)}{(300,000,000)(0.00005)}$$

18. The nearest star is approximately 4 light-years away. Light travels at approximately 186,000 miles per second. Taking 365 as the number of days in a year, find the distance in miles to the nearest star. Use scientific notation.

19. a. The mass of a hydrogen atom is approximately 0.00000000000000000000001673 gram. Express this number in scientific notation.
  b. There are approximately 6.023×10^{23} atoms in a mole (a unit of measure used in chemistry). Express this number as an integer.

B In Problems 20 and 21 m and n are integers.

20. Prove that the formula $a^m \cdot a^n = a^{m+n}$ holds true:
 a. When m is positive and n is negative
 b. When both m and n are negative
 c. When $n = 0$
 Hint. For part a let $n = -p$, where p is positive.

21. Prove that $(a^m)^n = a^{mn}$ under the following conditions:
 a. m is positive and n is negative
 b. m is negative and n is positive
 c. m and n are both negative
 d. $n = 0$

22. Use scientific notation to calculate

$$\frac{(120,000)^2 \sqrt{0.0000000009}}{50,000,000}$$

23. Calculate the following using scientific notation and a hand calculator:

$$\frac{(129,000,000)^2 \sqrt{0.0005873}}{(0.0002597)^2 \sqrt[3]{32,460}}$$

---

7 Radicals and Fractional Exponents

For $a \geq 0$, the symbol $\sqrt{a}$ means the nonnegative real number whose square is a. This is called the **principal square root of** a. It is also true that $-\sqrt{a}$ is a square root of a. It is a common mistake to suppose that $\sqrt{a}$ means both the principal square root and its negative, but this would lead to ambiguity and confusion. Thus, $\sqrt{4} = 2$, whereas $\sqrt{4} \neq -2$. The symbol $\sqrt{}$ is called a **radical sign.**

The fact that square roots of all nonnegative numbers exist can be proved using the basic properties of R, but the proof is beyond the level of this course. If $a < 0$, then $\sqrt{a}$ does not exist in R (however, it does exist in the complex number system, which we will discuss briefly in Chapter 3).

In a similar way, $\sqrt[n]{a}$, where n is any natural number greater than 1 and $a \geq 0$, is defined to mean the (unique) nonnegative real number whose nth power is a. If $a < 0$ and n is an odd natural number greater than 1, then $\sqrt[n]{a}$ is the negative real number whose nth power is a. For example, $\sqrt[3]{-8}, = -2$, since $(-2)^3 = (-2)(-2)(-2) = -8$. If n is even and $a < 0$, then $\sqrt[n]{a}$ does not exist in R. The number designated by $\sqrt[n]{a}$ is called the **principal nth root of a.**

From the laws of exponents in Section 6 we see that if $a \geq 0$ and $b \geq 0$, then

$$(\sqrt[n]{a}\ \sqrt[n]{b})^n = (\sqrt[n]{a})^n(\sqrt[n]{b})^n = ab$$

It follows that

$$\sqrt[n]{ab} = \sqrt[n]{a}\ \sqrt[n]{b} \tag{11}$$

This can be extended to arbitrary finite products.

By using equation (11) we can often simplify expressions involving radicals. The object is to write the quantity under the radical (called the **radicand**) as a product of the largest possible perfect nth power, times whatever is left. The nth root of the perfect nth power can be written without the radical sign. This is what we will mean by the instruction "simplify" in connection with problems involving radicals.

EXAMPLE 14   Simplify:
   a. $\sqrt{50}$ **b.** $\sqrt{45}$ **c.** $\sqrt{32} - \sqrt{18}$ **d.** $\sqrt[3]{81} + \sqrt[3]{24} + \sqrt[3]{192}$

Solution   **a.** $\sqrt{50} = \sqrt{25 \cdot 2} = \sqrt{25}\sqrt{2} = 5\sqrt{2}$
   b. $\sqrt{45} = \sqrt{9 \cdot 5} = \sqrt{9}\sqrt{5} = 3\sqrt{5}$
   c. $\sqrt{32} - \sqrt{18} = \sqrt{16 \cdot 2} - \sqrt{9 \cdot 2} = 4\sqrt{2} - 3\sqrt{2} = \sqrt{2}$
   d. $\sqrt[3]{81} + \sqrt[3]{24} + \sqrt[3]{192} = \sqrt[3]{27 \cdot 3} + \sqrt[3]{8 \cdot 3} + \sqrt[3]{64 \cdot 3}$
   $$= 3\sqrt[3]{3} + 2\sqrt[3]{3} + 4\sqrt[3]{3}$$
   $$= 9\sqrt[3]{3}$$

EXAMPLE 15   Simplify the following, where $x \geq 0$ and $y \geq 0$:
   a. $\sqrt{112x^3y^6}$ **b.** $\sqrt{6xy^3}\ \sqrt{2xy}$

Solution   **a.** $\sqrt{112x^3y^6} = \sqrt{16 \cdot 7x^2 \cdot x(y^3)^2} = 4xy^3\sqrt{7x}$
   b. $\sqrt{6xy^3}\ \sqrt{2xy} = \sqrt{12x^2y^4} = \sqrt{3 \cdot 4x^2y^4} = 2xy^2\sqrt{3}$

If $b \neq 0$ and $\sqrt[n]{a}$ and $\sqrt[n]{b}$ exist, then the following formula for division, analogous to rule (11), holds true:

$$\sqrt[n]{\frac{a}{b}} = \frac{\sqrt[n]{a}}{\sqrt[n]{b}} \tag{12}$$

You will be asked to prove this in Problem 28, Exercise Set 7. We can use this to remove the radical sign from the denominator, as we show in the next example. This process is called **rationalizing the denominator.**

EXAMPLE 16 Rationalize the denominators in each of the following:

a. $\sqrt{\dfrac{1}{2}}$ **b.** $\sqrt{\dfrac{4}{3}}$ **c.** $\sqrt{\dfrac{5}{18}}$ **d.** $\dfrac{3}{\sqrt{5}}$ **e.** $\sqrt[3]{\dfrac{2}{9}}$

Solution We can make each denominator a perfect power by multiplying the numerator and denominator by an appropriate number. Then rule (12) can be applied.

a. $\sqrt{\dfrac{1}{2}} = \sqrt{\dfrac{1}{2} \cdot \dfrac{2}{2}} = \sqrt{\dfrac{2}{4}} = \dfrac{\sqrt{2}}{\sqrt{4}} = \dfrac{\sqrt{2}}{2}$

b. $\sqrt{\dfrac{4}{3}} = \sqrt{\dfrac{4}{3} \cdot \dfrac{3}{3}} = \sqrt{\dfrac{4 \cdot 3}{9}} = \dfrac{\sqrt{4} \cdot \sqrt{3}}{\sqrt{9}} = \dfrac{2\sqrt{3}}{3}$

c. $\sqrt{\dfrac{5}{18}} = \sqrt{\dfrac{5}{18} \cdot \dfrac{2}{2}} = \sqrt{\dfrac{10}{36}} = \dfrac{\sqrt{10}}{\sqrt{36}} = \dfrac{\sqrt{10}}{6}$

d. $\dfrac{3}{\sqrt{5}} = \dfrac{3}{\sqrt{5}} \cdot \dfrac{\sqrt{5}}{\sqrt{5}} = \dfrac{3\sqrt{5}}{5}$

e. $\sqrt[3]{\dfrac{2}{9}} = \sqrt[3]{\dfrac{2}{9} \cdot \dfrac{3}{3}} = \sqrt[3]{\dfrac{6}{27}} = \dfrac{\sqrt[3]{6}}{\sqrt[3]{27}} = \dfrac{\sqrt[3]{6}}{3}$

We are now in a position to give a reasonable definition of fractional exponents. First consider $a^{1/n}$, where n is a natural number. In order to be consistent with the laws of exponents in Section 5, we must have

$$(a^{1/n})^n = a^{(1/n)n} = a^1 = a$$

So, we are led to the definition

$$a^{1/n} = \sqrt[n]{a}$$

The restriction that $a \geq 0$ when n is even must be observed. Also, for any integer m,

$$a^{m/n} = a^{(1/n)m} = (a^{1/n})^m = (\sqrt[n]{a})^m \tag{13}$$

Alternately, we can write

$$a^{m/n} = a^{m(1/n)} = (a^m)^{1/n} = \sqrt[n]{a^m} \tag{14}$$

Again, if n is even, we must have $a \geq 0$. In words, the fractional (rational) exponent m/n means: "Take the nth root and raise to the mth power, in either order." Taking the root first, as in (13), is more desirable than raising to the power first, as in (14), whenever a is a perfect nth power. For example, using (13), we have

$$(27)^{4/3} = (\sqrt[3]{27})^4 = 3^4 = 81$$

whereas using (14),

$$(27)^{4/3} = \sqrt[3]{(27)^4} = \sqrt[3]{531,441}$$

This again yields the answer 81, but it is obviously much more difficult to calculate.

All the laws of exponents in Section 6 continue to hold true for rational exponents, but the details of showing this are tedious and will be omitted here.

EXERCISE SET 7

In the problems that follow we will understand that all letters represent nonnegative numbers when even roots are involved and are not equal to zero when in the denominator or raised to a nonpositive power.

A In Problems 1–14 simplify each expression.

1. **a.** $\sqrt{125}$ **b.** $\sqrt[3]{40}$

2. **a.** $\sqrt{147}$ **b.** $\sqrt[3]{-250}$

3. **a.** $\sqrt{45} + \sqrt{80}$ **b.** $2\sqrt{50} - 3\sqrt{162}$

4. **a.** $\sqrt[3]{-108} + 3\sqrt[3]{32}$ **b.** $4\sqrt{75} - 3\sqrt{48} - \sqrt{108}$

5. $\sqrt{50x^4y^7}$ 6. $\sqrt{75a^3b^{-4}}$

7. $\sqrt{18x^3y^0z^5}$ 8. $\sqrt{72a^5b^4c^{-2}}$

9. $\sqrt{\dfrac{300a^3b^{-1}c^6}{ab^5c^2}}$ 10. $\sqrt[3]{-54x^6y^7z^5}$

11. $\sqrt{3xy^3}\sqrt{6x^5y}$ 12. $\sqrt[3]{4a^2b^4}\sqrt[3]{144ab}$

13. $\sqrt[3]{-81a^5b^6c^0}$ 14. $\sqrt{15s^{-3}t}\sqrt{30s^6t^5}$

In Problems 15–18 rationalize the denominator and simplify.

15. **a.** $\sqrt{\frac{2}{3}}$ **b.** $\sqrt{\frac{8}{5}}$ **c.** $\dfrac{2}{\sqrt{7}}$ **d.** $\sqrt[3]{\frac{1}{4}}$ **e.** $\sqrt{\frac{27}{50}}$

16. **a.** $\sqrt{\frac{2}{7}}$ **b.** $\sqrt{\frac{3}{8}}$ **c.** $\sqrt{\frac{5}{28}}$ **d.** $\dfrac{1}{\sqrt{50}}$ **e.** $\sqrt[3]{\frac{7}{18}}$

17. $\sqrt{\dfrac{32x^{-5}y^3}{3xy^{-7}}}$ 18. $\sqrt{\dfrac{5x^{-3}yz^3}{18x^{-2}y^{-3}z^{-1}}}$

In Problems 19–26 simplify the expressions. Write answers without using negative exponents.

19. **a.** $(2)^{1/3} \cdot (2)^{3/4}$ **b.** $(3)^{-1/2} \cdot (3)^{2/3}$ **c.** $(8)^{-2/3} \cdot (\frac{9}{16})^{-1/2}$
     d. $(32)^{-2/5} \cdot (8)^{2/3}$

20. **a.** $(-8)^{5/3}(4)^{-3/2}$ **b.** $(16)^{-1/2}(4^0)$ **c.** $(-3)^{-2}(\frac{4}{9})^{-1/2}$
     d. $[3^2 \cdot (2^0) + (\frac{1}{32})^{-4/5}]^{-1/2}$

21. **a.** $(2x^{1/3})(3x^{1/2})$ **b.** $\dfrac{x^{3/4}y^{1/2}}{x^{2/3}y^{1/3}}$ **c.** $(8x)^{1/3}(4x)^{1/2}$

22. **a.** $(x^{2/3})^{3/4}$ **b.** $(x^{-3/2})^{-4/3}$ **c.** $(4x^{4/3}y^{8/9})^{3/2}$

23. **a.** $(8x^6)^{1/3}$ **b.** $(16a^4b^{-2})^{1/2}$ **c.** $(-125x^{-9}y^{12})^{1/3}$

24. **a.** $(27a^3b^6)^{2/3}$ **b.** $(16x^4y^6)^{3/2}$ **c.** $(-8a^{-3}b^6)^{5/3}$

25. **a.** $(64a^6b^9)^{4/3}$ **b.** $(-8a^3b^{-6})^{7/3}$ **c.** $(36a^2b^4c^0)^{3/2}$

26. **a.** $\left(\dfrac{81x^{-2}y^6}{16x^6y^{-10}}\right)^{3/4}$ **b.** $\left(\dfrac{49x^2y^{-6}}{25x^{-4}y^{-2}}\right)^{3/2}$

27. Give an equivalent expression that does not involve radicals.
    a. $\sqrt[3]{8a^4b^2}$ **b.** $(\sqrt{2a^3b^5})^3$ **c.** $(\sqrt[3]{5x^2y^4})^4$

B 28. Prove rule (12).

29. Prove that for any natural number k, $\sqrt[kn]{a^{km}} = \sqrt[n]{a^m}$, where $a \geq 0$, and m and n are natural numbers greater than or equal to 2.

30. Use the result of Problem 29 to simplify each of the following:
    a. $\sqrt[4]{4}$ **b.** $\sqrt[6]{8x^9}$ **c.** $\sqrt[8]{9x^6y^{12}}$ **d.** $\sqrt[4]{64a^{10}b^8c^2}$

31. Prove that $\sqrt[n]{\sqrt[m]{a}} = \sqrt[mn]{a}$, where $a \geq 0$ and m and n are natural numbers greater than or equal to 2.
    Hint. Use fractional exponents.

32. Using the results of Problems 29 and 31, write each of the following with only one radical sign, and simplify the result:
    a. $\sqrt{\sqrt{32}}$ **b.** $\sqrt[3]{\sqrt{64x^9}}$ **c.** $\sqrt{8\sqrt[3]{16a^4b^6}}$ **d.** $\sqrt[3]{27\sqrt{8x^6y^{10}}}$

33. Prove that
$$\left(\frac{a}{b}\right)^{-m/n} = \left(\frac{b}{a}\right)^{m/n}$$

34. Using the result of Problem 33, simplify the following:
    a. $\left(\dfrac{27a^9b^6}{343c^{12}}\right)^{-2/3}$ **b.** $\left(\dfrac{256x^{12}}{81y^8z^4}\right)^{-3/4}$

Simplify each of the expressions in Problems 35–41. Express answers without negative exponents and without radical signs in the denominator or fractions under radical signs.

35. **a.** $(8x^{3/4}y^{-9/2})^{-2/3}$ **b.** $(16a^{4/3}b^{-2/3}c^{4/9})^{-3/2}$

36. **a.** $(32x^{-5/4}y^{10/3})^{-2/5}$ **b.** $(27a^{3/4}b^{-3/2}c^{-9/8})^{-4/3}$

37. $\dfrac{2\sqrt{72}}{3} - \dfrac{3\sqrt{128}}{4} + 5\sqrt{\dfrac{1}{2}}$

38. $\dfrac{1}{2}\sqrt{\dfrac{4}{3}} - \dfrac{3}{4}\sqrt{\dfrac{1}{12}} + \dfrac{\sqrt{75}}{2}$

39. $\sqrt[3]{32} - \sqrt[3]{\dfrac{27}{2}} + \dfrac{1}{3}\sqrt[3]{256}$

40. $\sqrt{\dfrac{125x^4y^{-2}z^5}{72x^{-2}y^{-7}z}}$

41. $\sqrt[3]{\dfrac{729a^0b^7c^{-4}}{128a^{-6}b^{-2}c}}$

Review Exercise Set

A 1. Express as a repeating or terminating decimal.
    a. $\frac{11}{16}$ **b.** $\frac{25}{8}$ **c.** $\frac{3}{7}$ **d.** $\frac{5}{27}$ **e.** $-\frac{13}{6}$

2. Express in the form m/n, where m and n are integers.
    a. $0.111\ldots$ **b.** $0.242424\ldots$ **c.** $1.0363636\ldots$ **d.** $0.216216216\ldots$

3. Identify which numbers are rational and which are irrational.

   a. $-\sqrt[3]{27}$ **b.** $\dfrac{5}{\sqrt{121}}$ **c.** $\dfrac{1}{\sqrt{2}}$ **d.** $\sqrt[3]{-125}$ **e.** $(\sqrt{3})^6$

   f. $\sqrt{3}+2$ **g.** $\dfrac{0}{\sqrt{7}}$ **h.** $\left(\dfrac{2}{3}\right)^{-1}$ **i.** π^2 **j.** $\sqrt{32}$

4. Evaluate each of the following:
   a. $(-3)(-4)(-2)$ **b.** $(-2)-(-6)$
   c. $-3+(-2)[(-4)-(-7)]$ **d.** $[5+(-2)](-6)$

5. Evaluate:
   a. $|-3-(-6)|$ **b.** $|4+(-9)|$ **c.** $|-10+(-2)|$
   d. $|(-2)(-7)+3(-4)|$ **e.** $|-(-6)(-8)-(5)(-7)|$

6. Write an equivalent expression without using absolute value signs.
   a. $|x|<7$ **b.** $|x|\le 4$ **c.** $|x|>8$ **d.** $|x|\ge 3$

7. Show each of the following sets on a number line:
   a. $\{x:\ x>-1\}$ **b.** $\{x:\ 0<x\le 2\}$ **c.** $\{x:\ |x|<3\}$
   d. $\{x:\ |x|\ge 2\}$ **e.** $\{x:\ x<-2\}\cup\{x:\ x\ge 1\}$

In Problems 8–13 perform the indicated operations. Reduce final answers to lowest terms.

8. a. $\frac{3}{7}+\frac{5}{9}$ **b.** $\frac{2}{3}-\frac{3}{4}+\frac{5}{6}$

9. a. $\frac{8}{9}+\frac{-5}{6}$ **b.** $\frac{5}{24}-\frac{11}{-18}$

10. a. $\frac{11}{180}-\frac{5}{168}$ **b.** $\frac{1}{792}+\frac{1}{1134}$

11. a. $\frac{5}{96}-\frac{7}{160}+\frac{3}{112}$ **b.** $(-\frac{15}{16})(-\frac{7}{10}+\frac{3}{2})$

12. a. $\dfrac{\frac{2}{3}-\frac{1}{2}}{\frac{5}{6}+\frac{3}{2}}$ **b.** $\dfrac{1-\frac{3}{4}}{\frac{1}{3}-\frac{1}{2}}$

13. a. $\dfrac{-2+\frac{3}{8}}{\frac{3}{4}-\frac{5}{6}}$ **b.** $\dfrac{\frac{2}{3}-\frac{1}{4}}{1+(\frac{2}{3})(\frac{1}{4})}$

14. Write in scientific notation:
   a. 132,000,000   **b.** 0.00023   **c.** 207,000   **d.** 0.00000035

15. Evaluate using scientific notation:
   a. $3.7\times 10^5+2.4\times 10^4$ **b.** $1.5\times 10^{-7}-3.6\times 10^{-5}$

   c. $\dfrac{(260,000)(0.0003)}{(0.013)(12,000,000)}$

In Problems 16–20 simplify the expressions, writing answers without using negative exponents.

16. a. $(3x^4y^2)(4x^3y^5)$ **b.** $(5x^3y^{-2}z^0)(2x^{-1}y^5)$

17. a. $\dfrac{3^2x^4y^{-8}z^0}{3^{-4}x^6y^{-5}z^{-2}}$ **b.** $\dfrac{2^{-3}a^{-2}b^5}{2^{-5}a^{-4}b^3}$

18. a. $\dfrac{24x^5y^{-2}z^{-4}}{64x^{-3}y^{-8}z^2}$ **b.** $\dfrac{42a^{-3}b^4c^0}{24a^{-5}b^{-2}}$

19. a. $(3x^{-2}y^3z)^3$ **b.** $\left(\dfrac{x^2y^{-1}}{2x^{-3}y^2}\right)^2$

20. a. $(81x^{-12}y^4z^{-8})^{3/4}$ **b.** $\left(\dfrac{-8a^2b^{-5}}{27a^{-1}b}\right)^{-2/3}$

Simplify the expressions in Problems 21–23.

21. a. $\sqrt{50x^3y^4z^0}$ **b.** $\sqrt[3]{-32a^6b^{-3}}$

22. a. $\sqrt{6xy^3}\sqrt{12x^3y^5}$ **b.** $\sqrt{\dfrac{98x^{-3}y^5}{125x^{-1}y^{-3}}}$

23. a. $\sqrt{\dfrac{2}{3}}$ **b.** $\dfrac{2}{\sqrt{5}}$ **c.** $\sqrt[3]{\dfrac{16}{9}}$ **d.** $\sqrt{\dfrac{3x}{8y}}$

B **24.** By using the meaning of the additive inverse of a number, together with its uniqueness, show that $-(-a) = a$.

25. Prove that if a, b, and c are real numbers, with $a > b$, then $a - c > b - c$.

26. Sometimes the definition for $a < b$ is given as follows: "$a < b$ means there is a positive real number c such that $b = a + c$." Show that this is equivalent to Definition 3.

27. Show that if $a \cdot b = 0$, then either $a = 0$ or $b = 0$.
Hint. Suppose $a \neq 0$, so that a^{-1} exists. Multiply both sides by a^{-1}.

28. Simplify:

   a. $\sqrt{18} - \dfrac{2\sqrt{200}}{3} + \sqrt{\dfrac{81}{2}}$ **b.** $\sqrt{\dfrac{2}{3}} - \dfrac{\sqrt{24}}{5} + \sqrt{\dfrac{27}{2}}$

29. Evaluate, using scientific notation:

$$\frac{(320,000)(0.000025)^{1/2}}{(0.00008)(5,000,000)}$$

30. Use a hand calculator to evaluate:

$$\frac{(2,637,000)^3 \sqrt{0.0000000003514}}{(734,000,000)^{3/2}}$$

Simplify the expressions in Problems 31 and 32.

31.

$$\frac{\dfrac{5}{6}\left(2 + \dfrac{\frac{4}{3} + \frac{3}{4}}{\frac{5}{6}}\right) - \dfrac{25}{12}\left(2 \cdot \dfrac{\frac{25}{12}}{\frac{3}{2} - \frac{2}{3}} - 1\right)}{\left(\dfrac{3}{2} - \dfrac{2}{3}\right)^2}$$

32. a. $(9x^{-2/3}y^{4/9}z^{-1/3})^{-3/2}$ **b.** $\sqrt{\dfrac{576x^{-5}y^{-2}z^3}{343x^3y^{-7}z^0}}$

2

Basic Algebraic Concepts

A mastery of the techniques of elementary algebra is essential to success in calculus. Open any calculus book to any page and you are almost sure to see algebra employed in one form or another. To illustrate, consider the following problem taken directly from a calculus textbook.*

$$y = \frac{x + 2}{x^{1/2}}$$

$$y' = \frac{x^{1/2} \cdot 1 - (x + 2) \cdot \frac{1}{2}x^{-1/2}}{x}$$

$$= \frac{\sqrt{x} - \frac{x + 2}{2\sqrt{x}}}{x}$$

$$= \frac{2x - (x + 2)}{2x^{3/2}}$$

$$= \frac{x - 2}{2x^{3/2}}$$

At this stage you are not expected to understand this but simply to observe the use of algebra. Only in the second step is an operation of calculus (called *differentiation*) employed. The rest is algebraic simplification, involving radicals, fractional exponents, and complex fractions. Other examples could easily be given that involve factoring, multiplication, solution of equations, and every conceivable algebraic technique.

---

* Douglas F. Riddle, *Calculus and Analytic Geometry*, 3d ed. (Belmont, Cal.: Wadsworth, 1979), p. 84. Reprinted by permission.

1 Introduction

One of the distinctive features of algebra is the use of letters to represent numbers. Among other things, this permits us to state general rules about numbers in a clear and concise way, as we saw in Chapter 1. Also, letters are used to represent unknown quantities in equations and inequalities, the object being to find the correct value or values that make the relation true. The point to be emphasized is that whenever letters are used in elementary algebra, they are used to represent numbers—usually real numbers. This fact enables us to use all the properties of real numbers given in Chapter 1 in algebra problems. In the discussion that follows, unless specifically stated otherwise, all letters used will be understood to represent real numbers. When a letter is used in this way, it is called a **variable. A variable is a letter that may be assigned any value from a given set of numbers.** Letters may also be used to designate fixed, but unspecified, numbers, and in this case they are called **constants.** Specific numbers also are called constants. Thus, in the expressions $2x + 3$ and $ax + b$, the specific numbers 2 and 3 are constants, and the letters a and b represent constants, even though they are unspecified. The letter x is a variable. Usually, the context will make clear whether a letter represents a constant or a variable. Frequently, letters from the latter part of the alphabet, such as x, y, or z, are used for variables, and letters such as a, b, c, from the first part are used for constants.

We use the term **algebraic expression** to mean any combination of variables and constants that is formed using a finite number of the operations of addition, subtraction, multiplication, division, raising to a power, or taking roots. Thus,

$$ax^3 \qquad \frac{x^{-1} - y^{-1}}{x^{-2} - y^{-2}} \qquad \frac{-10\sqrt{x - y}}{(u + v)^8}$$

are examples of algebraic expressions.

Of particular interest in this chapter will be algebraic expressions called **polynomials.** Examples of polynomials are

$$2x^3 - 3x^2 + 5x - 8 \qquad x^{10} - y^{10} \qquad 4x^2y^3z - 3xyz^3 + 5x^3yz^5$$

A **term** of a polynomial consists of a constant, called the **coefficient,** multiplied by one or more variables each raised to a nonnegative integral power. So a term of a polynomial in one variable has the form ax^m, and in two variables, ax^my^n, and so on. A polynomial consists of the sum of finitely many such terms. This includes differences since we may write, for example,

$$2x^3 - 3x^2 + 5x - 8 = 2x^3 + (-3)x^2 + 5x + (-8)$$

The **degree of a term** of a polynomial is the sum of the exponents on the variables in that term. For example, the degree of $2x^3$ is 3, whereas the degree of $4x^2y^3z$ is $2 + 3 + 1 = 6$ (remember that $z = z^1$). The **degree of a polynomial** is the largest of the degrees of its terms. A polynomial of degree n in one variable x can always be written in the form

$$a_nx^n + a_{n-1}x^{n-1} + \cdots + a_1x + a_0 \qquad (a_n \neq 0)$$

If $n = 0$ and $a_0 \neq 0$, the polynomial is a **constant polynomial** and has degree 0. The constant 0 has no degree. A polynomial having only one term is called a **monomial,** one with two terms is a **binomial,** and one with three terms is a **trinomial.**

2 Sums and Products of Polynomials

Consider the sum

$$(2x^2 + 3x + 5) + (3x^2 - x + 7)$$

The addition is carried out term-by-term, where **like terms** are added, that is, terms involving the same variable (or variables) raised to the same power are added. Therefore, the sum is $5x^2 + 2x + 12$. This is justified by use of the commutative, associative, and distributive properties, as can be seen by the following calculations:

$$
\begin{aligned}
(2x^2 + 3x + 5) + (3x^2 - x + 7) &= (2x^2 + 3x^2) + (3x - x) + (5 + 7) \\
&= (2 + 3)x^2 + (3 - 1)x + (5 + 7) \\
&= 5x^2 + 2x + 12
\end{aligned}
$$

Subtraction is also carried out term-by-term. For example,

$$(2x^2 + 3x + 5) - (3x^2 - x + 7) = -x^2 + 4x - 2$$

Multiplication of polynomials is carried out by repeated use of the distributive property of real numbers. Consider a product of the form

$$(a + b)(c + d)$$

We first treat $(a + b)$ as a single term and distribute the multiplication of it over the sum $(c + d)$:

$$(a + b)(c + d) = (a + b)c + (a + b)d$$

Now we can apply the right-hand distributive law to each of the products on the right to obtain

$$(a + b)(c + d) = ac + bc + ad + bd$$

Observe that **the result is the sum of all possible products involving one term from the first factor and one term from the second.** This result holds true in general. For example,

$$(a + b)(c + d + e) = ac + ad + ae + bc + bd + be$$

and similarly for any number of terms in either of the factors.

If we apply this result to two binomials in x, say, $(2x + 3)(4x + 5)$, we have

$$(2x + 3)(4x + 5) = 8x^2 + 12x + 10x + 15 = 8x^2 + 22x + 15$$

In this case the combining of like terms is possible, and this situation is typical. This combining of terms can be anticipated when multiplying two *like* binomials, for example, both binomials in x or in x and y. The result is generalized by the following:

$$(ax + b)(cx + d) = acx^2 + (ad + bc)x + bd \qquad (1)$$

This should not be memorized as a formula, but the process of arriving at the answer is important. The diagram below illustrates how each term on the right is obtained.

First term: $(ax + b)(cx + d)$

First times First $= acx^2$

First times Last $= adx$

Middle term: $(ax + b)(cx + d)$ ——Combine: $adx + bcx = (ad + bc)x$

Last times First $= bcx$

Last term: $(ax + b)(cx + d)$

Last times Last $= bd$

All this can be done mentally in some problems. For example,

$$(2x + 3)(4x + 5) = 8x^2 + 22x + 15$$

Here are some other examples:

$$(3x + 2)(5x - 4) = 15x^2 - 2x - 8$$

(Notice that the middle term is the sum of $-12x$ and $10x$.)

$$(2 + 3t)(4 + t) = 8 + 14t + 3t^2$$
$$(x - 2y)(x + 3y) = x^2 + xy - 6y^2$$

Two special cases of products that should be memorized are:

$$(a + b)(a - b) = a^2 - b^2 \qquad (2)$$

and

$$(a + b)^2 = a^2 + 2ab + b^2 \qquad (3)$$

The first of these is immediately seen to be true, since the middle term drops out ($-ab + ba = 0$). The second, called a **perfect square,** is verified by writing

$$(a + b)^2 = (a + b)(a + b) = a^2 + (ab + ba) + b^2 = a^2 + 2ab + b^2$$

Similarly, we have

$$(a - b)^2 = a^2 - 2ab + b^2 \qquad (4)$$

It is useful to express these formulas in words. We can express equation (2) by saying: "The sum of two numbers times their difference is the difference of their squares." For the perfect square we can say: "The square of the sum (difference) of two numbers is the square of the first, plus (minus) twice the product of the two, plus the square of the last."

These results are illustrated by the following examples:

**Product of sum and
difference of two terms:**

$$(x + 2)(x - 2) = x^2 - 4$$
$$(2x - 3)(2x + 3) = 4x^2 - 9$$
$$(4s + 5t)(4s - 5t) = 16s^2 - 25t^2$$

Perfect square:

$$(x + 5)^2 = x^2 + 10x + 25$$
$$(2x - 3)^2 = 4x^2 - 12x + 9$$
$$(3x^2 - 8y)^2 = 9x^4 - 48x^2y + 64y^2$$

EXERCISE SET 2

A   Perform the indicated operations.

1. $(2x^2 + 3x + 5) + (3x^2 + 7x + 8)$ 2. $(3x^2 - 2x + 10) + (5x^2 + x - 6)$
3. $(5y^3 - 2y^2 + 3y - 4) + (3y^3 + 4y^2 - y - 7)$
4. $(x^3 - 3x^2 + 5) + (2x^3 + 9x - 8)$ 5. $(3t^2 - 6t + 10) - (2t^2 - 3t - 4)$
6. $(x^2 + 2x - 1) - (4x^2 - 7x + 2)$
7. $(2x^2 - 3xy + 4y^2) + (5x^2 + 7xy - 2y^2)$
8. $(8u^2 + 2uv - 3v^2) - (6u^2 - 3uv + 4v^2)$
9. $(x^2 - 2x + 4) + (3x^2 + 2x - 7) - (2x^2 - x - 6)$
10. $(5x^2 - 3x + 4) - (2x^2 - 7) + (6x^2 - x + 3)$
11. $(x + 3)(x + 4)$ 12. $(x - 1)(x - 2)$
13. $(x - 3)(x + 4)$ 14. $(x + 3)(x - 5)$
15. $(x + y)(x + 2y)$ 16. $(x + 3y)(x - 2y)$
17. $(1 - x)(2 + x)$ 18. $(x + 2)(3x + 4)$
19. $(2x + 1)(3x + 7)$ 20. $(4x - 2)(x + 3)$
21. $(2t - 3)(4t + 2)$ 22. $(3x - 2y)(4x + 3y)$
23. $(5y - 4)(2y - 3)$ 24. $(x^2 - 2y)(3x^2 + 5y)$
25. $(2 - 3x)(4 - x)$ 26. $(x - 1)(x + 1)$
27. $(x + 4)(x - 4)$ 28. $(t - 3)(t + 3)$
29. $(2x - 1)(2x + 1)$ 30. $(x - 3y)(x + 3y)$
31. $(a - 2b)(a + 2b)$ 32. $(5x + 4)(5x - 4)$
33. $(3a - 7bc)(3a + 7bc)$ 34. $(x + 7)^2$
35. $(x - 2)^2$ 36. $(3x + 1)^2$
37. $(x - 2y)^2$ 38. $(1 - 2t)^2$
39. $(2x - 3)^2$ 40. $(5x + 2y)^2$
41. $(a^2 - 3b^2)^2$

B   42. $(x + 1)(x^2 + x + 2)$ 43. $(x - 2)(x^2 + 2x - 1)$
44. $(2x - 1)(3x^2 - 2x + 4)$ 45. $(x^2 - 3x + 4)(x^2 - x + 2)$
46. $(2a^2 - a - 1)(3a^2 + 2a - 4)$ 47. $[(x + y) + 2][(x + y) - 2]$
48. $(3x - 2y + 4)(3x - 2y - 4)$ 49. $[x + (y - 1)][x - (y - 1)]$
50. $(x + 2y + 3)(x - 2y - 3)$ 51. $(4 - x + y)(4 + x - y)$
52. $(a + b + c)^2$ 53. $(2x - 3y + 4)^2$
54. $(a - 2b + 4c)^2$

3   The Binomial Theorem

The formula for $(a + b)^2$ is often used in applications. Other powers, such as $(a + b)^3$ and $(a + b)^4$, are used less often but nevertheless are important to know. One by one these powers can be obtained through multiplication, but the process rapidly becomes tedious, as the following calculations show:

$$(a + b)^3 = (a + b)(a + b)^2 = (a + b)(a^2 + 2ab + b^2)$$
$$= a^3 + 2a^2b + ab^2 + ba^2 + 2ab^2 + b^3$$
$$= a^3 + 3a^2b + 3ab^2 + b^3$$

$$(a + b)^4 = (a + b)(a + b)^3 = (a + b)(a^3 + 3a^2b + 3ab^2 + b^3)$$
$$= a^4 + 3a^3b + 3a^2b^2 + ab^3 + ba^3 + 3a^2b^2 + 3ab^3 + b^4$$
$$= a^4 + 4a^3b + 6a^2b^2 + 4ab^3 + b^4$$

We might hope for a pattern to emerge that permits the result to be written down at once for any power. And it turns out that the powers of a and b in the successive terms on the right *can* easily be anticipated. For example, we would expect that in the expansion of $(a + b)^5$ the respective terms on the right would be of the form

$$a^5, \quad a^4b, \quad a^3b^2, \quad a^2b^3, \quad ab^4, \quad b^5$$

But what are the coefficients? If you do not already know the answer, you may wish to carry out a few more multiplications and try to conjecture the general pattern. The answer is given in the **binomial theorem.**

BINOMIAL THEOREM

For any positive integer n,

$$(a + b)^n = a^n + na^{n-1}b + \frac{n(n - 1)}{1 \cdot 2} a^{n-2}b^2 + \frac{n(n - 1)(n - 2)}{1 \cdot 2 \cdot 3} a^{n-3}b^3$$

$$+ \cdots + \frac{n(n - 1)(n - 2) \cdots (n - k + 2)}{1 \cdot 2 \cdot 3 \cdots (k - 1)} a^{n-(k-1)}b^{k-1}$$

$$+ \cdots + nab^{n-1} + b^n \tag{5}$$

We defer a proof of this until later in this book. Meanwhile, the following observations can be made:

1. In the expansion of $(a + b)^n$ there are $n + 1$ terms.
2. Beginning with a^n in the first term, the powers of a diminish by consecutive integers in each succeeding term, reaching the first power in the next-to-last term. In the second term, b is to the first power, and the powers of b increase by consecutive integers in each succeeding term, reaching b^n in the last term.
3. In each term, the sum of the exponents on a and b is n.

4. For $1 < k \leq n + 1$, the kth term is given by

$$\frac{n(n-1)(n-2)\cdots(n-k+2)}{1\cdot 2\cdot 3\cdots(k-1)} a^{n-(k-1)}b^{k-1} \tag{6}$$

The coefficient has $k-1$ factors in the numerator starting with n and decreasing by 1 in each succeeding factor, and $k-1$ factors in the denominator starting with 1 and increasing by 1 in each succeeding factor.

5. An alternate way of calculating the coefficient of any term after the first is to multiply the coefficient of the *preceding* term by the exponent of a in that term and divide the result by 1 more than the exponent of b in that term.

EXAMPLE 1   Use equation (5) to expand $(a + b)^4$.

Solution

$$(a + b)^4 = a^4 + 4a^3b + \frac{4\cdot 3}{1\cdot 2}a^2b^2 + \frac{4\cdot 3\cdot 2}{1\cdot 2\cdot 3}ab^3 + \frac{4\cdot 3\cdot 2\cdot 1}{1\cdot 2\cdot 3\cdot 4}b^4$$

$$= a^4 + 4a^3b + 6a^2b^2 + 4ab^3 + b^4$$

EXAMPLE 2   Expand $(a + b)^5$, using item 5 above to calculate the coefficients.

Solution   We know that the first term is a^5, which can be written $1 \cdot a^5b^0$. So to get the next coefficient we multiply 1 by 5 and divide by $0 + 1$. The result is 5. So the second term is $5a^4b^1$. To get the third coefficient we multiply 5 by 4 and divide by $1 + 1 = 2$. The result is $20 \div 2 = 10$. Up to this point we have

$$(a + b)^5 = a^5 + 5a^4b + 10a^3b^2 + \cdots$$

The next coefficient is $10 \cdot 3 \div 3 = 10$. Continuing in this way we get the complete expansion:

$$(a + b)^5 = a^5 + 5a^4b + 10a^3b^2 + 10a^2b^3 + 5ab^4 + b^5$$

EXAMPLE 3   Expand $(x + 2y)^3$.

Solution   Using item 5 above, we get

$$(x + 2y)^3 = x^3 + 3x^2(2y) + 3x(2y)^2 + (2y)^3$$

$$= x^3 + 6x^2y + 12xy^2 + 8y^3$$

EXAMPLE 4   Expand $(3x - y)^5$.

Solution

$$(3x - y)^5 = [3x + (-y)]^5 = (3x)^5 + 5(3x)^4(-y) + 10(3x)^3(-y)^2$$
$$+ 10(3x)^2(-y)^3 + 5(3x)(-y)^4 + (-y)^5$$

$$= 243x^5 - 405x^4y + 270x^3y^2 - 90x^2y^3 + 15xy^4 - y^5$$

EXAMPLE 5   Find the eighth term of the expansion of $(a + b)^{12}$.

Solution   We let $k = 8$ in formula (6). So $k - 1 = 7$. Thus, the eighth term is

$$\frac{12 \cdot 11 \cdot 10 \cdot 9 \cdot 8 \cdot 7 \cdot 6}{1 \cdot 2 \cdot 3 \cdot 4 \cdot 5 \cdot 6 \cdot 7} a^{12-7}b^7 = 792a^5b^7$$

EXAMPLE 6   Find the sixth term in the expansion of $(x - 2y)^{10}$.

Solution   We write this first as $[x + (-2y)]^{10}$. We let $k = 6$, so that $k - 1 = 5$, and by formula (6) we obtain

$$\frac{10 \cdot 9 \cdot 8 \cdot 7 \cdot 6}{1 \cdot 2 \cdot 3 \cdot 4 \cdot 5} x^{10-5}(-2y)^5 = 252x^5(-32y^5)$$

$$= -8,064x^5y^5$$

If we write the coefficients of $(a + b)^n$ for consecutive values of n in rows, an interesting observation can be made. For completeness, we will include $(a + b)^0 = 1$ and $(a + b)^1 = a + b$.

Coefficients

$n = 0$	1
$n = 1$	1 1
$n = 2$	1 2 1
$n = 3$	1 3 3 1
$n = 4$	1 4 6 4 1
$n = 5$	1 5 10 10 5 1
$n = 6$	1 6 15 20 15 6 1

Do you see how to get the next row? Think about it before you read on. Each time the leading 1 is written one space farther to the left. Then, each succeeding coefficient through the next-to-last one is obtained by adding the two above. For example, we get the coefficients for $n = 7$ from those for $n = 6$ as shown:

$n = 6$ 1   6   15   20   15   6   1

$n = 7$ 1   7   21   35   35   21   7   1

This triangular pattern is known as **Pascal's triangle,** after the brilliant French mathematician Blaise Pascal (1623–1662), who played a major role in the development of probability theory. Pascal's triangle is interesting, but if you wanted to expand $(a + b)^{20}$ (which admittedly is not very likely), it would be necessary to construct a rather large triangle.

EXERCISE SET 3

A   In Problems 1–5 use equation (5) to expand, and simplify the results.

1. $(a + b)^7$ 2. $(x + y)^8$ 3. $(a - b)^6$
4. $(x - y)^5$ 5. $(1 + x)^6$

In Problems 6–15 expand using rule 5 to calculate the coefficients, and simplify the results.

6. $(x + y)^5$ 7. $(2x + y)^4$ 8. $(x - 3y)^3$
9. $(2x + 3y)^4$ 10. $(3x - 2y)^4$ 11. $(5x - y^2)^3$
12. $(3 - t)^6$ 13. $\left(\dfrac{2}{s} - 3\right)^5$ 14. $(1 + \sqrt{x})^8$
15. $(x^2 - 2y)^5$

In Problems 16–20 find the indicated term of the expansion, using formula (6).

16. Fifth term of $(x + y)^{12}$ 17. Eighth term of $(a - b)^{10}$
18. Sixth term of $(2x + y)^9$ 19. Fourth term of $(2x - 3y)^8$
20. Seventh term of $(x^2 - 2)^{11}$

B   21. Show that the signs in the expansion of $(a - b)^n$ alternate between plus and minus.

22. The symbol $n!$ is read **n factorial** and means $1 \cdot 2 \cdot 3 \cdots n$. For example, $4! = 1 \cdot 2 \cdot 3 \cdot 4 = 24$. We define $0! = 1$. Show that formula (6) can be put in the form

$$\frac{n!a^{n-k+1}b^{k-1}}{(k-1)!(n-k+1)!}$$

Hint.   In formula (6) multiply the numerator and denominator by $(n - k + 1)!$

23. Use the formula given in Problem 22 to find:
    a.   The eighth term in the expansion of $(2x^2 - y)^{12}$
    b.   The seventh term in the expansion of $(\sqrt{x} - 3)^9$

24. Find the term in the expansion of

$$\left(2x - \frac{1}{3x}\right)^{12}$$

that does not involve x.

25. Find the term in the expansion of $(4a^2 + b^3)^8$ that involves b^{12}.
26. Find the term involving y^4 in the expansion of $(xy - 2y^{-1})^{10}$
27. Find the first four terms in the expansion of $(2a^2 - 3b^{-2})^{15}$.
28. Find the first four terms in the expansion of $(3a^{-1} + 4\sqrt{b})^{20}$.

---

4   Factoring

To factor an algebraic expression means to write it as a product. So, factoring is just the reverse of multiplication. One might reasonably ask why it is important to factor anything. Some of the reasons are: (1) to simplify a fraction by dividing out common factors, (2) to help recognize the lowest common denominator in adding fractions, (3) to facilitate the solution of equations and inequalities, and (4) to help analyze the sign of an algebraic expression.

Most of the factoring problems we will consider will involve polynomials. We say a polynomial is **irreducible** if it cannot be written as the product of two polynomials of positive degree. Whether a polynomial is irreducible will depend on the restriction placed on the coefficients. For example, $x^2 - 2$ is irreducible if the coefficients are restricted to the integers, but it can be factored as $(x + \sqrt{2})(x - \sqrt{2})$ if we allow irrational coefficients. In the discussion that follows, factors are limited to those with integer coefficients unless specifically stated otherwise.

The polynomial $4x^2 + 8$ is irreducible over the integers. However, by the distributive property, this can be written as the product $4(x^2 + 2)$, and we say we have factored out the common factor 4. We will understand in all factorization problems that common integer factors such as this are to be factored out. With this understanding, we will consider an expression completely factored if it is written as a product of irreducible factors.

The most common factoring problems fall into one of the categories described below.

Common Factors.   The distributive law provides justification for an important factoring technique. When read from right to left, this law says,

$$ab + ac = a(b + c)$$

Since a appears as a factor of each term on the left, it is referred to as a **common factor.** Applying the distributive law to obtain the right-hand side is referred to as "factoring out the common factor." This technique is illustrated in the following examples.

EXAMPLE 7 $$4xy^2 + 6x^3y = 2xy(2y + 3x^2)$$

EXAMPLE 8 $$3a^2b^3 - 6a^4b^2c + 9a^3b^4c^2 = 3a^2b^2(b - 2a^2c + 3ab^2c^2)$$

EXAMPLE 9 $$2x^{3/2} - 4x^{1/2} = 2x^{1/2}(x - 2)$$

It is almost always desirable to factor out *all* common factors, that is, to take out the greatest common factor.

In Examples 7 and 8 the common factors were monomials. The expression in Example 9 is not a polynomial, since the exponents are not integers, but we still factored out a single term. It often happens that common factors consist of more than one term. Common binomial factors occur in the next three examples.

EXAMPLE 10

$$(a + 2)b + (a + 2)c = (a + 2)(b + c)$$

The common factor is the binomial $(a + 2)$. It is treated as a unit and factored out just as if it were a single term.

EXAMPLE 11

$$2x^2 + 4xy + xy^2 + 2y^3 = (2x^2 + 4xy) + (xy^2 + 2y^3)$$
$$= 2x(x + 2y) + y^2(x + 2y) = (2x + y^2)(x + 2y)$$

The initial step here is called **grouping.** We grouped the first two terms together and the last two together. After factoring out the common factor $2x$ from the first group and y^2 from the second, we recognize the binomial factor $(x + 2y)$ as being common to the two groups.

EXAMPLE 12

$$ax + 9by - 3ay - 3bx = ax - 3ay - 3bx + 9by$$
$$= a(x - 3y) - 3b(x - 3y)$$
$$= (a - 3b)(x - 3y)$$

Here it was necessary to group the first and third terms together and the second and fourth together. A common error occurs in problems such as this because of the negative signs. The troublesome step is

$$-3bx + 9by = -3b(x - 3y)$$

The point to remember is that when a negative term is factored out, the signs inside the parentheses must be changed.

EXAMPLE 13

$$(1 - x^2)^{3/2} - 3x^2\sqrt{1 - x^2} = (1 - x^2)^{3/2} - 3x^2(1 - x^2)^{1/2}$$
$$= (1 - x^2)^{1/2}(1 - x^2)^{2/2} - 3x^2(1 - x^2)^{1/2}$$
$$= (1 - x^2)^{1/2}(1 - x^2 - 3x^2)$$
$$= \sqrt{1 - x^2}(1 - 4x^2)$$

The common factor in this case is the irrational factor $(1 - x^2)^{1/2}$.

The Perfect Square and the General Trinomial. Equations (3) and (4) of Section 2 provide useful formulas for factoring. Written in reverse order, these become

$$a^2 + 2ab + b^2 = (a + b)^2$$
$$a^2 - 2ab + b^2 = (a - b)^2$$

Consider, for example, the problem of factoring $4x^2 + 20x + 25$. This might be a perfect square, with $a = 2x$ and $b = 5$. We check the middle term and find that the trinomial is a perfect square. So

$$4x^2 + 20x + 25 = (2x + 5)^2$$

Here are some other examples.

EXAMPLE 14
$$4x^2 + 4x + 1 = (2x)^2 + 2(2x) \cdot 1 + 1^2 = (2x + 1)^2$$

EXAMPLE 15
$$9x^2 - 12xy + 4y^2 = (3x)^2 - 2(3x)(2y) + (2y)^2 = (3x - 2y)^2$$

EXAMPLE 16
$$a^4 + 8a^2 + 16 = (a^2)^2 + 2(a^2) \cdot 4 + 4^2 = (a^2 + 4)^2$$

EXAMPLE 17
$$18x^3y - 60x^2y^2 + 50xy^3 = 2xy(9x^2 - 30xy + 25y^2)$$
$$= 2xy[(3x)^2 - 2(3x)(5y) + (5y)^2]$$
$$= 2xy(3x - 5y)^2$$

Here we first factored out the common factor $2xy$ and then recognized the other factor as a perfect square. Any existing common factor should always be factored out first.

A trinomial that is a perfect square of a binomial is, of course, a special situation. Certain trinomials that are not perfect squares can also be factored. We rely here on the multiplication formula (1):

$$(ax + b)(cx + d) = acx^2 + (ad + bc)x + bd$$

To illustrate, consider $6x^2 + 17x + 12$. To put this into the form $(ax + b)(cx + d)$, there are various possibilities to consider. The product of a and c must be 6, and b times d must be 12. The key is the middle term. In this case, we find (after several attempts) that the correct combination is $(2x + 3)(3x + 4)$. This is a trial-and-error procedure, but the number of trials is finite. Often, trials can be made mentally.

EXAMPLE 18
$$x^2 - 2x - 8 = (x - 4)(x + 2)$$

EXAMPLE 19
$$4x^2 + 5x - 6 = (4x - 3)(x + 2)$$

EXAMPLE 20
$$2x^2 + xy - 15y^2 = (2x - 5y)(x + 3y)$$

EXAMPLE 21
$$10x^4 - 11x^2 - 6 = (5x^2 + 2)(2x^2 - 3)$$

Difference of Squares and Sum and Difference of Cubes. Equation (2) in reverse order reads

$$a^2 - b^2 = (a + b)(a - b)$$

and this is an important factoring formula called the **difference of squares.** The examples below illustrate its use.

EXAMPLE 22 Factor each of the following:
a. $x^2 - 4$ b. $4x^2 - 9y^2$ c. $m^2 - 4k^2$
d. $9a^2 - 100b^2$ e. $3y^2 - 5$ (Factor over the reals.)

Solution a. $x^2 - 4 = (x - 2)(x + 2)$ b. $4x^2 - 9y^2 = (2x - 3y)(2x + 3y)$
c. $m^2 - 4k^2 = (m - 2k)(m + 2k)$ d. $9a^2 - 100b^2 = (3a - 10b)(3a + 10b)$
e. $3y^2 - 5 = (\sqrt{3}y + \sqrt{5})(\sqrt{3}y - \sqrt{5})$

EXAMPLE 23 Factor:
a. $x^2 - 4x + 4 - y^2$ b. $a^2 - b^2 + 2b - 1$

Solution In both of these it is necessary to group the terms first.

a. $x^2 - 4x + 4 - y^2 = (x^2 - 4x + 4) - y^2$
$= (x - 2)^2 - y^2$
$= [(x - 2) - y][(x - 2) + y]$
$= (x - y - 2)(x + y - 2)$

b. $a^2 - b^2 + 2b - 1 = a^2 - (b^2 - 2b + 1)$
$= a^2 - (b - 1)^2$
$= [a - (b - 1)][a + (b - 1)]$
$= (a - b + 1)(a + b - 1)$

Since the difference of two squares factors in such a nice way, we might be led to consider the difference of two cubes, $a^3 - b^3$. A reasonable guess is that one of the factors is $a - b$. This is, in fact, correct, and the complete factorization is given by

$$a^3 - b^3 = (a - b)(a^2 + ab + b^2)$$

You can verify this by multiplying the factors on the right.
There is a similar result for the sum of cubes:

$$a^3 + b^3 = (a + b)(a^2 - ab + b^2)$$

So while the sum of squares, $a^2 + b^2$, is irreducible, the sum of cubes can be factored.

EXAMPLE 24 Factor:
  a. $x^3 - 8$ **b.** $a^3 + 27$ **c.** $8x^3 - 125y^3$ **d.** $x^6 - 1$ **e.** $x^6 + 1$

Solution **a.** $x^3 - 8 = (x - 2)(x^2 + 2x + 4)$
  b. $a^3 + 27 = (a + 3)(a^2 - 3a + 9)$
  c. $8x^3 - 125y^3 = (2x)^3 - (5y)^3 = (2x - 5y)(4x^2 + 10xy + 25y^2)$
  d. This problem is best handled by first factoring $x^6 - 1$ as the difference of squares, resulting in the difference and sum of cubes:

$$x^6 - 1 = (x^3 - 1)(x^3 + 1) = (x - 1)(x^2 + x + 1)(x + 1)(x^2 - x + 1)$$

  e. We treat this as $(x^2)^3 + 1$ to get

$$x^6 + 1 = (x^2 + 1)(x^4 - x^2 + 1)$$

The second factor can be further factored over the reals by adding and subtracting $3x^2$ as follows:

$$x^4 - x^2 + 1 = (x^4 + 2x^2 + 1) - 3x^2 = (x^2 + 1)^2 - 3x^2$$
$$= (x^2 + 1 - \sqrt{3}x)(x^2 + 1 + \sqrt{3}x)$$

The Forms $a^n - b^n$ and $a^n + b^n$. The formulas for the difference and sum of cubes are special cases of the following:

1. **For any integer $n > 2$,**

$$a^n - b^n = (a - b)(a^{n-1} + a^{n-2}b + a^{n-3}b^2 + a^{n-4}b^3 + \cdots + ab^{n-2} + b^{n-1}) \quad (7)$$

2. **For any *odd* integer $n > 2$,**

$$a^n + b^n = (a + b)(a^{n-1} - a^{n-2}b + a^{n-3}b^2 - a^{n-4}b^3 + \cdots - ab^{n-2} + b^{n-1}) \quad (8)$$

Note. When n is even, $a^n + b^n$ is irreducible over the rationals.

EXAMPLE 25 Factor completely:
  a. $x^5 - y^5$ **b.** $x^5 + y^5$ **c.** $a^4 - b^4$
  d. $x^9 + y^9$ **e.** $x^7 - 128y^7$

Solution **a.** $x^5 - y^5 = (x - y)(x^4 + x^3y + x^2y^2 + xy^3 + y^4)$
  b. $x^5 + y^5 = (x + y)(x^4 - x^3y + x^2y^2 - xy^3 + y^4)$
  c. While formula (6) could be applied to obtain

$$a^4 - b^4 = (a - b)(a^3 + a^2b + ab^2 + b^3)$$

it is better to factor this as the difference of squares.*

$$a^4 - b^4 = (a^2 - b^2)(a^2 + b^2) = (a - b)(a + b)(a^2 + b^2)$$

---

* This is a general procedure; for example, $x^{2n} - y^{2n} = (x^n - y^n)(x^n + y^n)$.

This suggests that the second factor obtained in the first method must be factorable itself. This is the case, and the two answers can be shown to be the same. (Try this.)

d. We could apply equation (7) directly, but it is better to consider this first as the sum of cubes:

$$x^9 + y^9 = (x^3 + y^3)(x^6 - x^3y^3 + y^6)$$
$$= (x + y)(x^2 - xy + y^2)(x^6 - x^3y^3 + y^6)$$

e. $x^7 - 128y^7 = (x - 2y)(x^6 + 2x^5y + 4x^4y^2 + 8x^3y^3 + 16x^2y^4 + 32xy^5 + 64y^6)$

Use of the Binomial Theorem. There are occasions when a polynomial can be recognized as being a perfect nth power of a binomial. For example, we saw in Section 3 that

$$a^3 + 3a^2b + 3ab^2 + b^3 = (a + b)^3$$

and that

$$a^4 + 4a^3b + 6a^2b^2 + 4ab^3 + b^4 = (a + b)^4$$

The best approach here is to be alert to the possibility that a polynomial might be a perfect nth power by observation of the exponents; then check by means of the binomial theorem to see if this is indeed the case.

Caution. Do not confuse $(a + b)^3$ with $a^3 + b^3$, or $(a - b)^3$ with $a^3 - b^3$.

EXAMPLE 26   Factor: $x^3 + 6x^2y + 12xy^2 + 8y^3$

Solution   By observing the pattern of the exponents, we see that this has a chance of being the expansion of $(x + 2y)^3$, and upon checking we find that this is correct. So the answer is $(x + 2y)^3$.

EXAMPLE 27   Factor: $x^5 - 5x^4 + 10x^3 - 10x^2 + 5x - 1$

Solution   A possibility is $(x - 1)^5$, and this does work.

EXERCISE SET 4

Factor all problems completely with integer coefficients unless otherwise specified.

A   Problems 1–13 involve common factors.

1. **a.** $2a^2b - 4ab^2$ **b.** $6a^3b^4c + 9a^2b^3c^2$

2. **a.** $4a^2b^3 - 10a^4b^4$ **b.** $36x^5y^4z - 42x^4y^3$

3. **a.** $24x^3y^2 - 40x^5y^3z + 48x^2y^4z^2$ **b.** $156a^4b - 260a^3b^2$

4. **a.** $3\sqrt{x} - 6x$ **b.** $10x^{3/2} + 15x^{5/2}$

5. **a.** $2x^{4/3} - 4x^{1/3}$ **b.** $12\sqrt[3]{x} + 18x^{4/3}$

6. **a.** $x(y-2)+3(y-2)$ **b.** $(2a+3)b-(2a+3)c$
7. **a.** $(x+2)y+(x+2)$ **b.** $(b-1)-a(b-1)$
8. **a.** $xy-x+2y-2$ **b.** $ab-ac+bd-cd$
9. **a.** $x^2-x+xy-y$ **b.** $x^2y-xy+x-1$
10. **a.** $a^2b+2ab^2-2ac-4bc$ **b.** $x^3-x^2y-y^3+xy^2$
11. **a.** $xy-x-y+1$ **b.** $a^2-ab-ac+bc$
12. **a.** $2m^2+3n-mn-6m$ **b.** $2x^2+6y^3-4xy-3xy^2$
13. **a.** $12+xy-4x-3y$ **b.** $a^3-b^2+ab^2-a^2$

Problems 14–26 involve perfect squares and general trinomials.

14. **a.** x^2-4x+4 **b.** $a^2+8a+16$
15. **a.** $t^2-12t+36$ **b.** $9m^2+6m+1$
16. **a.** $9x^2-30x+25$ **b.** $49a^2+28ab+4b^2$
17. **a.** $4x^2-12xy+9y^2$ **b.** $16t^2-72st+81s^2$
18. **a.** x^2-4x+3 **b.** $a^2+7a+12$
19. **a.** x^2+3x+2 **b.** x^2-x-2
20. **a.** t^2-t-12 **b.** $x^2-3x-10$
21. **a.** $y^2+5y-14$ **b.** $a^2+5a-24$
22. **a.** $2x^2+7x+3$ **b.** $3x^2+11x+6$
23. **a.** $6x^2-5x-6$ **b.** $8x^2+6xy-9y^2$
24. **a.** $16x^2-58x+25$ **b.** $18a^2+37ab-20b^2$
25. **a.** $2x^4+9x^2+4$ **b.** $24t^2-5t-36$
26. **a.** $10+x-2x^2$ **b.** $45m^2+3mn-18n^2$

Problems 27–38 involve difference of squares and sum and difference of cubes.

27. **a.** $4x^2-9$ **b.** $16m^2-25n^2$
28. **a.** x^4-16 **b.** b^3-a^2b
29. **a.** a^8-1 **b.** $x^2y^4-y^2$
30. **a.** $81x^4-256y^4$ **b.** $121-49t^2$
31. **a.** $(x-2)^2-y^2$ **b.** $x^2-(y-2)^2$
32. **a.** $x^2-2x+1-4y^2$ **b.** $4x^2+4x+1-y^2$
33. **a.** x^2-y^2+4x+4 **b.** $16a^2-25b^2-40a+25$
34. **a.** x^2-16-y^2-8y **b.** $a^2-4b-1-4b^2$
35. **a.** x^3-64 **b.** $27a^3+8$
36. **a.** $1-8x^3$ **b.** $27x^3+125y^3$
37. **a.** x^3-125 **b.** a^3+216
38. **a.** $8u^3-27v^3$ **b.** $64t^3+8$

Problems 39–45 involve a^n-b^n and a^n+b^n.

39. x^5-1 40. a^5+32b^5
41. x^7+y^7 42. t^5+1
43. $1-32x^5$ 44. x^7+128
45. $x^{10}-243y^5$

Problems 46–53 involve use of the binomial theorem.

46. $x^3+3x^2y+3xy^2+y^3$ 47. $8x^3-12x^2y+6xy^2-y^3$
48. $a^4-4a^3b+6a^2b^2-4ab^3+b^4$ 49. $x^4-8x^3+24x^2-32x+16$

50. $x^3 + 9x^2 + 27x + 27$

51. $a^5 - 5a^4b + 10a^3b^2 - 10a^2b^3 + 5ab^4 - b^5$

52. $t^3 - 12t^2 + 48t - 64$ **53.** $8x^3 + 36x^2y^2 + 54xy^4 + 27y^6$

In Problems 54–69 factor in whatever way is appropriate.

54. $12x^2 + 14x - 40$ **55.** $9x^3 - 25x$

56. $x^4 - 5x^2 + 4$ **57.** $6x^3 - 20x^2 + 6x$

58. $a^3 + a^2b + ab^2 + b^3$ **59.** $x^4 - 8x^2 + 16$

60. $x^2 - 2xy + y^2 - 4$ **61.** $32x^5 + 243$

62. $48a^2 + 37ab - 36b^2$ **63.** $x^2 - y^2 - 2x + 1$

64. $x^3 + x^2 - 4x - 4$ **65.** $8t^3 - 28t^2 + 24t$

66. $x^6 - 64$ **67.** $27a^3 + 64b^3$

68. $a^2 - b^2 - 2ab^2 + 2b^3$ **69.** $2a^3b + 6a^2b^2 - 8a^2b$

B In Problems 70–73 factor out a negative power. Write each answer with positive exponents.

70. $x(1 - x)^{-1/2} + (1 - x)^{1/2}$

71. $(x + 1)^{2/3}(x - 1)^{-2/3} + 2(x + 1)^{-1/3}(x - 1)^{1/3}$

72. $(x^2 + 1)^{1/3}(x - 1)^{-1/3} - (x^2 + 1)^{-2/3}(x - 1)^{2/3}$

73. $(x - 2)^{-1/3}(2x + 1)^{1/3} - 2(x - 2)^{2/3}(2x + 1)^{-2/3}$

In Problems 74–92 factor completely with integer coefficients unless otherwise specified.

74. $75x^3y^2 + 60x^2y^3 - 36xy^4$ **75.** $32a^3b + 144a^2b^2 + 162ab^3$

76. $48x^2yz - 30y^3z - 36xy^2z$ **77.** $40x^2 + 18x - 63$

78. $a^2 + b^2 + c^2 + 2ab + 2ac + 2bc$ **79.** $x^2 + 4y^2 + 1 - 4xy + 2x - 4y$

80. $4x^4 - 9$ (Factor over the reals.)

81. a. $x^6 - 64$ **b.** $x^6 + 64$ (Factor over the reals.)

82. $x^2 + 2xy + y^2 - a^2 - 2ab - b^2$
 Hint. Use groups of three.

83. $x^2 - z^2 + y^2 - 2xy - 4 + 4z$
 Hint. Rearrange and use groups of three.

84. $x^8 - y^8$ **85.** $a^{12} - b^{12}$

86. $8x^3 + 36x^2y + 54xy^2 + 27y^3$

87. $x^5 - 15x^4 + 90x^3 - 270x^2 + 405x - 243$

88. $27a^3 - 54a^2 + 36a - 8$ **89.** $a^3 - b^3 + a - b$

90. $x^8 - 256$ **91.** $x^2 - 2xy + y^2 - 4x + 4y$

92. $a^6 + 7a^3b^3 - 8b^6$

---

5 Rational Fractions

Fractions that are quotients of polynomials are called **rational**. Examples are

$$\frac{x + 3}{x^2 - 2x + 5} \qquad \frac{1}{x + 2} \qquad \frac{x^6 - y^6}{x^2 + 2xy + y^2}$$

The arithmetic associated with these is governed by rules already established in Sections 4 and 5 of Chapter 1. Factoring plays a key role in carrying out the steps involved. When a common factor appears in the numerator and denominator, we make use of the fundamental property of fractions to reduce the fraction to lowest terms. To "simplify" an expression involving rational fractions means to perform the indicated operations and express the result as a simple fraction in which no common factor (other than 1) exists in the numerator and denominator.

The examples below illustrate the most important techniques.

EXAMPLE 28   Perform the indicated operations and simplify.

a. $\dfrac{x^2 - x - 6}{x^2 - 6x + 9}$ b. $\dfrac{x^2 + 4x + 4}{x^2 - 3x - 4} \cdot \dfrac{2x^2 + x - 1}{x^2 - 3x - 10}$

c. $\dfrac{2x^2 - 6x}{2x^2 - x - 10} \div \dfrac{x^2 - 5x + 6}{x^2 - 4}$

Solution   a. $\dfrac{x^2 - x - 6}{x^2 - 6x + 9} = \dfrac{\overset{1}{\cancel{(x-3)}}(x+2)}{(x-3)^{\cancel{2}^{1}}} = \dfrac{x+2}{x-3}$

b. $\dfrac{x^2 + 4x + 4}{x^2 - 3x - 4} \cdot \dfrac{2x^2 + x - 1}{x^2 - 3x - 10} = \dfrac{(x+2)^{\cancel{2}}}{(x-4)\underset{1}{\cancel{(x+1)}}} \cdot \dfrac{(2x-1)\overset{1}{\cancel{(x+1)}}}{(x-5)\underset{1}{\cancel{(x+2)}}}$

$= \dfrac{(x+2)(2x-1)}{(x-4)(x-5)} = \dfrac{2x^2 + 3x - 2}{x^2 - 9x + 20}$

c.   From Section 4 of Chapter 1 we know that to divide, we invert the denominator and multiply:

$\dfrac{2x^2 - 6x}{2x^2 - x - 10} \div \dfrac{x^2 - 5x + 6}{x^2 - 4} = \dfrac{2x^2 - 6x}{2x^2 - x - 10} \cdot \dfrac{x^2 - 4}{x^2 - 5x + 6}$

$= \dfrac{2x\overset{1}{\cancel{(x-3)}}}{(2x-5)\cancel{(x+2)}} \cdot \dfrac{\overset{1}{\cancel{(x+2)}}\overset{1}{\cancel{(x-2)}}}{\underset{1}{\cancel{(x-3)}}\underset{1}{\cancel{(x-2)}}} = \dfrac{2x}{2x-5}$

EXAMPLE 29   Perform the indicated operations.

a. $\dfrac{1}{x-1} + \dfrac{3}{x+2}$ b. $\dfrac{3}{x^2 - 1} + \dfrac{2}{x^2 - x - 2}$

Solution   a.   This is done by a direct application of the addition rule, $a/b + c/d = (ad + bc)/bd$:

$\dfrac{1}{x-1} + \dfrac{3}{x+2} = \dfrac{(x+2) + 3(x-1)}{(x-1)(x+2)} = \dfrac{x+2+3x-3}{(x-1)(x+2)} = \dfrac{4x-1}{x^2 + x - 2}$

b. Here it is best to bring each fraction to an equivalent form having the same (lowest) denominator:

$$\frac{3}{x^2 - 1} + \frac{2}{x^2 - x - 2} = \frac{3}{(x + 1)(x - 1)} + \frac{2}{(x + 1)(x - 2)}$$

$$= \frac{3}{(x + 1)(x - 1)} \cdot \frac{x - 2}{x - 2} + \frac{2}{(x + 1)(x - 2)} \cdot \frac{x - 1}{x - 1}$$

$$= \frac{3x - 6 + 2x - 2}{(x + 1)(x - 1)(x - 2)}$$

$$= \frac{5x - 8}{(x + 1)(x - 1)(x - 2)}$$

With a little practice, the second and third steps can be combined. It is often useful to leave the denominator in factored form, as we have done here.

EXAMPLE 30   Combine and simplify: $\dfrac{2}{x + y} - \dfrac{x - y}{x^2 + 3xy + 2y^2} + \dfrac{3}{x + 2y}$

Solution   We have, on factoring the second denominator:

$$\frac{2}{x + y} - \frac{x - y}{(x + 2y)(x + y)} + \frac{3}{x + 2y} = \frac{2(x + 2y) - (x - y) + 3(x + y)}{(x + y)(x + 2y)}$$

$$= \frac{2x + 4y - x + y + 3x + 3y}{(x + y)(x + 2y)}$$

$$= \frac{4x + 8y}{(x + y)(x + 2y)}$$

$$= \frac{\overset{1}{4(x + 2y)}}{(x + y)\underset{1}{(x + 2y)}}$$

$$= \frac{4}{x + y}$$

This illustrates one reason for retaining the denominator in factored form; we recognized a common factor between the numerator and denominator and so were able to apply the fundamental property to simplify the result.

EXAMPLE 31   Combine and simplify: $\dfrac{1}{2x + 4} - \dfrac{2}{2 - x} + \dfrac{4}{3x - 6}$

Solution   The first thing to observe is that the denominator of the second fraction would be easier to deal with if it were $x - 2$ instead of $2 - x$. Since these factors are negatives of each other, this change can be accomplished using the fact that $a/(-b) = -a/b$. Thus,

$$\frac{2}{2 - x} = -\frac{2}{x - 2}$$

so that the problem becomes

$$\frac{1}{2(x+2)} + \frac{2}{x-2} + \frac{4}{3(x-2)} = \frac{3(x-2) + 12(x+2) + 8(x+2)}{6(x+2)(x-2)}$$

$$= \frac{3x - 6 + 12x + 24 + 8x + 16}{6(x+2)(x-2)}$$

$$= \frac{23x + 34}{6(x^2 - 4)}$$

EXAMPLE 32   Simplify: $\dfrac{\dfrac{1}{a^2} - \dfrac{2}{ab}}{\dfrac{3}{b^2} + \dfrac{1}{2ab}}$

Solution   Just as in Section 5 of Chapter 1, we multiply the numerator and denominator of the main fraction by the LCD of all minor denominators—in this case $2a^2b^2$.

$$\frac{\dfrac{1}{a^2} - \dfrac{2}{ab}}{\dfrac{3}{b^2} + \dfrac{1}{2ab}} \cdot \frac{\dfrac{2a^2b^2}{1}}{\dfrac{2a^2b^2}{1}} = \frac{2b^2 - 4ab}{6a^2 + ab}$$

EXAMPLE 33   Simplify: $\dfrac{\dfrac{1}{x-1} + \dfrac{1}{x^2 - 1}}{x - \dfrac{2}{x+1}}$

Solution   Here the LCD of the minor denominators is $(x+1)(x-1) = x^2 - 1$. Proceeding as above, we obtain

$$\frac{\dfrac{1}{x-1} + \dfrac{1}{x^2 - 1}}{x - \dfrac{2}{x+1}} \cdot \frac{\dfrac{(x+1)(x-1)}{1}}{\dfrac{(x+1)(x-1)}{1}} = \frac{x + 1 + 1}{x(x+1)(x-1) - 2(x-1)}$$

$$= \frac{x + 2}{(x-1)(x^2 + x - 2)}$$

$$= \frac{\overset{1}{\cancel{x+2}}}{(x-1)\underset{1}{\cancel{(x+2)}}(x-1)}$$

$$= \frac{1}{(x-1)^2}$$

Note that in the second step the factors appearing in the terms of the denominator were not multiplied out. This enabled us to recognize the common factor

$(x - 1)$. Had the multiplication been done, the result would have been $x^3 - 3x + 2$, and factoring this would have been far more difficult. (Try it!)

EXAMPLE 34   Simplify: $\dfrac{\dfrac{x}{x-1} - 2}{x - 2}$

Solution

$$\frac{\dfrac{x}{x-1} - 2}{x - 2} \cdot \frac{\dfrac{x-1}{1}}{\dfrac{x-1}{1}} = \frac{x - 2(x-1)}{(x-2)(x-1)}$$

$$= \frac{x - 2x + 2}{(x-2)(x-1)}$$

$$= \frac{-x + 2}{(x-2)(x-1)}$$

$$= -\frac{\overset{1}{\cancel{x-2}}}{\underset{1}{\cancel{(x-2)}}(x-1)}$$

$$= -\frac{1}{x - 1}$$

Note carefully the handling of signs in this problem.

EXERCISE SET 5

Perform the indicated operations. Reduce final answers to simplest form.

A   **1. a.** $\dfrac{x^2 - 1}{x^2 + 3x + 2}$ **b.** $\dfrac{2x^2 + x - 6}{x^2 + 4x + 4}$

2. a. $\dfrac{t^2 - 4}{t^3 + 4t^2 + 4t}$ **b.** $\dfrac{6x^2 + 5x - 4}{2x^2 - 9x + 4}$

3. $\dfrac{x^2 - 2x - 8}{x^2 - 1} \cdot \dfrac{2x^2 - x - 3}{x^2 - 8x + 16}$ **4.** $\dfrac{2x^2 + 4x}{3x^2 + 2x - 1} \cdot \dfrac{3x^2 - 10x + 3}{x^3 - x^2 - 6x}$

5. $\dfrac{x^3 - 1}{x^2 - 5x - 6} \cdot \dfrac{x^2 + 3x + 2}{x^2 + x - 2}$ **6.** $\dfrac{6x^2 + xy - 12y^2}{2x^2 + xy - 3y^2} \div \dfrac{3x^2 - xy - 4y^2}{x^2 - 2xy + y^2}$

7. $\dfrac{2x^2 + x - 1}{x^2 - 1} \div \dfrac{x^2 + 6x + 9}{x^2 + 2x - 3}$ **8.** $\dfrac{3x^2 + 2x - 8}{2x^2 - x - 6} \div \dfrac{x^2 - 4}{2x^2 + 3x}$

9. $\dfrac{3}{x + 4} + \dfrac{1}{x - 2}$ **10.** $\dfrac{x}{x - 3} - \dfrac{1}{x + 1}$

11. $\dfrac{2}{x - 1} + \dfrac{3}{x + 1} - \dfrac{x - 5}{x^2 - 1}$ **12.** $\dfrac{3}{a + 2} - \dfrac{2a - 1}{a^2 - a - 6} + \dfrac{1}{a - 3}$

13. $1 - \dfrac{2}{x} - \dfrac{3}{x-2}$

14. $\dfrac{x-3}{x^2+3x+2} + \dfrac{2x-5}{x^2+x-2} - \dfrac{x-3}{x^2-1}$

15. $\dfrac{x^2-10}{x+4} - x + 7$

16. $\dfrac{x^2+1}{x-2} - x + 3$

17. $\dfrac{1}{x-2} - \dfrac{3}{2-x} + 1$

18. $\dfrac{a}{x+a} - \dfrac{x}{x-a} - \dfrac{2x^2}{a^2-x^2}$

19. $\dfrac{3}{x^2-x-2} + \dfrac{5}{2+3x-2x^2}$

20. $\dfrac{\dfrac{a}{b} - \dfrac{b}{a}}{\dfrac{1}{b^2} - \dfrac{1}{a^2}}$

21. $\dfrac{\dfrac{1}{x^2} - \dfrac{1}{4}}{x-2}$

22. $\dfrac{\dfrac{1}{x-1} - \dfrac{8}{x^2-1}}{\dfrac{2}{x+1} - \dfrac{1}{x-1}}$

23. $\dfrac{\dfrac{x}{x-2} - 2}{x-4}$

24. $\dfrac{\dfrac{y}{x-1} - \dfrac{y}{x+1}}{1 + \dfrac{y^2}{x^2-1}}$

25. $\dfrac{\dfrac{x^2}{y} - \dfrac{y^2}{x}}{\dfrac{x}{y} + \dfrac{y}{x} + 1}$

26. $\dfrac{y+3 - \dfrac{16}{y+3}}{y-6 + \dfrac{20}{y+3}}$

27. $\dfrac{\dfrac{2}{a+h} - \dfrac{2}{a}}{h}$

28. $\dfrac{\dfrac{2+h}{1+h} - 2}{h}$

29. $\dfrac{\dfrac{1}{z} + 2}{5 + 2\left(z + \dfrac{1}{z}\right)}$

30. $\dfrac{\dfrac{x-1}{x+1} - \dfrac{1}{3}}{x-2}$

31. $\dfrac{\dfrac{x^2}{x+2} - \dfrac{a^2}{a+2}}{x-a}$

32. $\dfrac{\dfrac{2x+3}{x^2} - 1}{x+1}$

33. $\dfrac{\dfrac{2}{(x+h)^2} - \dfrac{2}{x^2}}{h}$

B **34.** $\dfrac{x-3}{2x^2-x-3} + \dfrac{x+2}{2x^2-7x+6} - \dfrac{x-1}{x^2-x-2}$

35. $\dfrac{1}{x^2-1} - \dfrac{x^2+3}{x^4+x}$

36. $\dfrac{2(x-2)^3(x+1) - 3(x+1)^2(x-2)^2}{(x-2)^6}$

37. $\dfrac{3(2x-3)^2(x+4)^2 - 4(x+4)^3(2x-3)}{(2x-3)^4}$

38. $\dfrac{\dfrac{x+h-2}{x+h+3} - \dfrac{x-2}{x+3}}{h}$

39. $\dfrac{\dfrac{x^2}{1-2x}-\dfrac{a^2}{1-2a}}{x-a}$

40. $\dfrac{1-x^3}{1+\dfrac{x}{1-\dfrac{x}{1+x}}}$

41. $\dfrac{3x^2+10x-8}{2-\dfrac{3}{1-\dfrac{x}{x-2}}}$

Hint. First work with complex fractions in the denominator.

6 Irrational Fractions

All the rules for working with rational fractions continue to hold for fractions involving negative and fractional exponents. To simplify such expressions means to perform indicated operations, to divide out common factors, and to write the answer with positive exponents.

EXAMPLE 35 Simplify: $\dfrac{x^{-1}-y^{-1}}{x^{-2}-y^{-2}}$

Solution We replace each term by its equivalent expressed with a positive exponent and then proceed as in Section 5.

$$\frac{x^{-1}-y^{-1}}{x^{-2}-y^{-2}} = \frac{\dfrac{1}{x}-\dfrac{1}{y}}{\dfrac{1}{x^2}-\dfrac{1}{y^2}} \cdot \frac{\dfrac{x^2y^2}{1}}{\dfrac{x^2y^2}{1}} = \frac{xy^2-x^2y}{y^2-x^2} = \frac{xy(\overset{1}{\cancel{y-x}})}{(y+x)(\underset{1}{\cancel{y-x}})} = \frac{xy}{y+x}$$

EXAMPLE 36 Simplify: $x(1-x)^{-3}+(1-x)^{-2}$

Solution There are two ways to proceed.
a. Write with positive exponents first.

$$x(1-x)^{-3}+(1-x)^{-2} = \frac{x}{(1-x)^3}+\frac{1}{(1-x)^2}$$

$$= \frac{x+(1-x)}{(1-x)^3} = \frac{1}{(1-x)^3}$$

b. Factor out the common factor $(1-x)^{-3}$. It is important to take out the factor with the negative exponent which is largest in absolute value, so that the remaining factor will have positive exponents only.

$$x(1-x)^{-3}+(1-x)^{-2} = (1-x)^{-3}(x+1-x) = \frac{1}{(1-x)^3}$$

EXAMPLE 37   Simplify: $\dfrac{(x^2 - 1)^{1/2} - x^2(x^2 - 1)^{-1/2}}{x^2 - 1}$

Solution   Just as with a complex fraction, we multiply the numerator and denominator by the same quantity. This time we choose the factor that will eliminate all negative powers, namely, $(x^2 - 1)^{1/2}$:

$$\frac{(x^2 - 1)^{1/2} - x^2(x^2 - 1)^{-1/2}}{x^2 - 1} \cdot \frac{(x^2 - 1)^{1/2}}{(x^2 - 1)^{1/2}} = \frac{x^2 - 1 - x^2}{(x^2 - 1)^{3/2}} = \frac{-1}{(x^2 - 1)^{3/2}}$$

EXAMPLE 38   Simplify: $\dfrac{\dfrac{x^2}{\sqrt{x^2 - 1}} - \sqrt{x^2 - 1}}{x^2}$

Solution   We multiply the numerator and denominator by $\sqrt{x^2 - 1}$ to obtain

$$\frac{\dfrac{x^2}{\sqrt{x^2 - 1}} - \sqrt{x^2 - 1}}{x^2} \cdot \frac{\dfrac{\sqrt{x^2 - 1}}{1}}{\dfrac{\sqrt{x^2 - 1}}{1}} = \frac{x^2 - (x^2 - 1)}{x^2 \sqrt{x^2 - 1}} = \frac{1}{x^2 \sqrt{x^2 - 1}}$$

EXAMPLE 39   Rationalize the numerator: $\dfrac{\sqrt{x + 2} - 2}{x - 2}$

Solution   This type of problem frequently occurs early in the study of calculus. To "rationalize the numerator" means to rewrite the fraction so that the numerator contains no radicals. This is accomplished by multiplying the numerator and denominator by $\sqrt{x + 2} + 2$:*

$$\frac{\sqrt{x + 2} - 2}{x - 2} \cdot \frac{\sqrt{x + 2} + 2}{\sqrt{x + 2} + 2} = \frac{(x + 2) - 4}{(x - 2)(\sqrt{x + 2} + 2)}$$

$$= \frac{\overset{1}{\cancel{x - 2}}}{\underset{1}{\cancel{(x - 2)}}(\sqrt{x + 2} + 2)} = \frac{1}{\sqrt{x + 2} + 2}$$

EXAMPLE 40   Rationalize the numerator: $\dfrac{\sqrt{x + 1 + h} - \sqrt{x + 1}}{h}$

---

* The terms $(\sqrt{x + 2} - 2)$ and $(\sqrt{x + 2} + 2)$ are called **conjugate surds**. More generally, $(\sqrt{a} + \sqrt{b})$ and $(\sqrt{a} - \sqrt{b})$ are conjugate surds. The key fact is that their product is the rational expression $a - b$.

Solution The idea is the same as in the preceding example.

$$\frac{\sqrt{x+1+h}-\sqrt{x+1}}{h} \cdot \frac{\sqrt{x+1+h}+\sqrt{x+1}}{\sqrt{x+1+h}+\sqrt{x+1}} = \frac{(x+1+h)-(x+1)}{h(\sqrt{x+1+h}+\sqrt{x+1})}$$

$$= \frac{\overset{1}{\cancel{h}}}{\underset{1}{\cancel{h}}(\sqrt{x+1+h}+\sqrt{x+1})}$$

$$= \frac{1}{\sqrt{x+1+h}+\sqrt{x+1}}$$

EXERCISE SET 6

A Simplify the expressions in Problems 1–20. Express answers with positive exponents only.

1. $x^{-1}-y^{-1}$

2. $\dfrac{x^{-1}+y^{-1}}{x^{-2}-y^{-2}}$

3. $\dfrac{a^{-2}-b^{-2}}{a^{-1}+b^{-1}}$

4. $x^{-1}-x^{-2}(1+x)$

5. $x^{-2}-2(x+2)x^{-3}$

6. $2x(x-2)^{-1}-x^2(x-2)^{-2}$

7. $2(x+1)(x-1)^{-1}-(x+1)^2(x-1)^{-2}$

8. $\dfrac{2x^{-1}-y^{-1}}{4x^{-2}-4x^{-1}y^{-1}+y^{-2}}$

9. $\dfrac{2x^{-2}-3x^{-1}+1}{1-x^{-2}}$

10. $\dfrac{x}{\sqrt{1+x}}+2\sqrt{1+x}$

11. $\dfrac{1+x}{2\sqrt{x}}-\sqrt{x}$

12. $\dfrac{2x^2}{\sqrt{x^2+1}}+\sqrt{x^2+1}$

13. $\sqrt{1-2x}+\dfrac{x}{\sqrt{1-2x}}$

14. $\dfrac{\sqrt{1-x}+\dfrac{x}{2\sqrt{1-x}}}{1-x}$

15. $\dfrac{\sqrt{x^2+4}-\dfrac{5}{\sqrt{x^2+4}}}{x+1}$

16. $\dfrac{(x-1)^{1/2}-2x(x-1)^{-1/2}}{x+1}$

17. $\dfrac{\sqrt{4-x^2}+x^2(4-x^2)^{-1/2}}{4-x^2}$

18. $(x-1)^{-1/2}(x+1)^{1/2}-(x+1)^{-1/2}(x-1)^{1/2}$

19. $\dfrac{2y(1-y^2)^{1/2}+\dfrac{y^3}{\sqrt{1-y^2}}}{1-y^2}$

20. $\dfrac{\dfrac{1+x^2}{2\sqrt{x}}-2x\sqrt{x}}{(1+x^2)^2}$

In Problems 21–23 rationalize the numerator.

21. $\dfrac{\sqrt{x+1}-1}{x}$

22. $\dfrac{\sqrt{x-3}-1}{x-4}$

23. $\dfrac{\sqrt{2(x+h)+1}-\sqrt{2x+1}}{h}$

In Problems 24–26 rationalize the denominator.

24. $\dfrac{1}{\sqrt{x}-\sqrt{a}}$

25. $\dfrac{1}{\sqrt{x+1}-\sqrt{x+2}}$

26. $\dfrac{1}{x+\sqrt{x-1}}$

B Simplify the expressions completely. Express answers with positive exponents only.

27. $-\dfrac{8x^2}{9}(x^2 - 1)^{-4/3} + \frac{4}{3}(x^2 - 1)^{-1/3}$

28. $\dfrac{(3 - x)^{1/2}(2 + x)^{-1/2} + (3 - x)^{-1/2}(2 + x)^{1/2}}{2(3 - x)}$

29. $\dfrac{(1 - 2x)^{3/2} \cdot 2x - x^2 \cdot \frac{3}{2}(-2)(1 - 2x)^{1/2}}{(1 - 2x)^3}$

30. $\dfrac{x^{-3} + y^{-3}}{xy^{-1} + x^{-1}y - 1}$

31. $\dfrac{\dfrac{1}{\sqrt{x + h}} - \dfrac{1}{\sqrt{x}}}{h}$

32. $\dfrac{\sqrt[3]{x} - \sqrt[3]{a}}{x - a}$

Hint. Multiply the numerator and denominator by a factor that will rationalize the numerator.

33. $\dfrac{\sqrt[3]{x^2 - 1} - x^2 \cdot \frac{2}{3}(x^2 - 1)^{-2/3}}{(x^2 - 1)^{2/3}}$

34. $\dfrac{\frac{1}{3}(2x + 3)^{1/4}(x + 1)^{-2/3} - (x + 1)^{1/3} \cdot \frac{1}{2}(2x + 3)^{-3/4}}{\sqrt{2x + 3}}$

35. $\dfrac{(x^2 - 4)^{1/2} \cdot \frac{1}{3}(x^2 - 1)^{-2/3}(2x) - (x^2 - 1)^{1/3} \cdot \frac{1}{2}(x^2 - 4)^{-1/2}(2x)}{(x^2 - 4)}$

Review Exercise Set

A In Problems 1 and 2 perform the indicated operations.

1. **a.** $(2x^2 - 3x - 4) + (3x^2 + 7x - 5)$
 b. $(7x^2 - 2xy - 3y^2) - (5x^2 - 2xy - y^2)$

2. **a.** $(2x^3 - 3x^2 + 4x - 8) + (x^3 - 5x + 7)$
 b. $(3t^2 - 4t + 6) - (t^2 - 3t + 2)$

3. Multiply:
 a. $(2x - 3y)(5x + 4y)$ **b.** $(x + 2y)(x - 2y)$
 c. $(3x - 5)^2$ **d.** $(x - 1)(x^3 + 2x^2 - 3x + 4)$

4. Expand:
 a. $(x + 2y)^5$ **b.** $(3a - 2b)^4$

5. Expand:
 a. $(3a + 4b)^3$ **b.** $(x^2 - 3)^6$

6. Expand:
 a. $(1 - 2\sqrt{x})^4$ **b.** $\left(2t + \dfrac{3}{t}\right)^5$

Find the indicated term of the expansion in Problems 7 and 8.

7. **a.** Sixth term of $(3a - b^2)^{11}$ **b.** Fifth term of $(2\sqrt{x} + 3)^{20}$

8. **a.** Eighth term of $(t^{-1} - 3t)^{12}$ **b.** Seventh term of $\left(1 - \dfrac{2}{x}\right)^{15}$

In Problems 9–21 factor completely with integer coefficients.

9. **a.** $3x^2 - 5x - 2$ **b.** $x^3y - 12x^2y^2 + 36xy^3$

10. **a.** $18x^2 + 33x - 40$ **b.** $a^3 - a^2b - ab^2 + b^3$

11. **a.** $2x^4 - 5x^2 - 12$ **b.** $x^2 - y^2 + 4x + 4$

12. **a.** $x^2 - 4y^2 - 2x + 4y$ **b.** $a^3 + 27$

13. **a.** $24x^3 - 23x^2y - 12xy^2$ **b.** $a^3 + 3a^2 - 9a - 27$

14. **a.** $2x^5y^3 - 16x^2$ **b.** $a^5 + 243$

15. **a.** $4 - x^2 - y^2 + 2xy$ **b.** $6x^4 - 5x^2y - 6y^2$

16. **a.** $a^2b - 4b - a^2 + 4$ **b.** $x^2 - x - y^2 + y$

17. **a.** $x^6 - \dfrac{y^3}{8}$ **b.** $x^8 - 1$

18. **a.** $15x^2 - xy - 28y^2$ **b.** $16a^4 + 32a^3b^2 + 24a^2b^4 + 8ab^6 + b^8$

19. **a.** $12a^2 - 36ab + 27b^2$ **b.** $s^{10} + 32t^{10}$

20. **a.** $2x^{7/2} + 16x^{1/2}$ **b.** $3x^2(x^2 - 4)^{3/2} + (x^2 - 4)^{5/2}$

21. **a.** $3x^2(x^2 - 1)^{1/2} + 2x(x^2 - 1)^{3/2}$ **b.** $6x(1 - 2x)^{4/3} - 8x^2\sqrt[3]{1 - 2x}$

In Problems 22–31 perform the indicated operations and simplify the result. Express all answers with positive exponents only.

22. **a.** $\dfrac{2x^2 + 5x - 12}{x^2 - 25} \cdot \dfrac{3x^2 + 13x - 10}{6x^2 - 13x + 6}$ **b.** $\dfrac{3x^2 - 7x - 20}{x^2 - 8x + 16} \div \dfrac{3x^2 + 14x + 15}{2x^2 - 9x + 4}$

23. $\dfrac{1}{x + 2} + \dfrac{3}{2 - x} + \dfrac{4}{x^2 - 4}$ **24.** $\dfrac{4x + 5}{2x - 3} - \dfrac{x + 7}{x + 4} - \dfrac{25x + 23}{2x^2 + 5x - 12}$

25. $\dfrac{\dfrac{2x}{x - 3} + 4}{x - 2}$ **26.** $\dfrac{\dfrac{1}{x - 2} - \dfrac{3}{x + 2}}{\dfrac{3x}{x^2 - 4} - \dfrac{2}{x - 2}}$

27. $\dfrac{2\sqrt{x^2 - 4} - \dfrac{x(2x + 3)}{\sqrt{x^2 - 4}}}{x^2 - 4}$ **28.** $3x^2(1 - 2x)^{-2} + 4x^3(1 - 2x)^{-3}$

29. $\dfrac{\sqrt{1 - x^2} + \dfrac{x^2}{\sqrt{1 - x^2}}}{1 - x^2}$ **30.** $-(x^{-2} + x^{-1})^{-2}(-2x^{-3} - x^{-2})$

31. $\dfrac{2}{3}\left(\dfrac{x^2 + 1}{x^2 - 1}\right)^{-1/3} \cdot \dfrac{(x^2 - 1)2x - (x^2 + 1)2x}{(x^2 - 1)^2}$

32. Rationalize the numerator:

a. $\dfrac{\sqrt{2x - 3} - 1}{x - 2}$ **b.** $\dfrac{\sqrt{x + h - 2} - \sqrt{x - 2}}{h}$

33. Rationalize the denominator:

a. $\dfrac{1}{\sqrt{x + 3} - 2}$ **b.** $\dfrac{\sqrt{a + 4} - \sqrt{a - 4}}{\sqrt{a + 4} + \sqrt{a - 4}}$

B **34.** Multiply:
a. $(x^2 - 2x + 3)(2x^2 - 5x - 4)$ **b.** $(x - 2y + 5)^2$

In Problems 35–37 factor completely with integer coefficients.

35.   **a.** $x^2 - 4 + 4y^2 - 4xy$ **b.** $x^4 - 2x^3 - x + 2$

36.   **a.** $4x^2 - 3xy - 10y^2 - 3x + 6y$ **b.** $8x^3 - 60x^2y + 150xy^2 - 125y^3$

37.   **a.** $4x^2 + 9y^2 - z^2 - 12xy - 6z - 9$

     b. $(2x - 3)^{-1/3}(x + 1)^{-2/3} - 2(2x - 3)^{-4/3}(x + 1)^{1/3}$

In Problems 38–42 perform the indicated operations and simplify the result. Express answers with positive exponents only.

38. $\dfrac{2x - 1}{x^2 + x - 2} - \dfrac{x + 5}{x^2 - 1} - \dfrac{x + 3}{x^2 + 3x + 2}$

39. $\dfrac{9(2x - 5)^2(3x + 2)^2 - 4(3x + 2)^3(2x - 5)}{(2x - 5)^3}$

40. $\dfrac{2x\sqrt[3]{1 - x^2} + \frac{2}{3}x^3(1 - x^2)^{-2/3}}{(1 - x^2)^{2/3}}$

41. $\dfrac{\sqrt[3]{3x - 1} \cdot \dfrac{1}{2\sqrt{x + 1}} - \sqrt{x + 1} \cdot (3x - 1)^{-2/3}}{(3x - 1)^{2/3}}$

42. $\dfrac{(x + 1)\left[\dfrac{-x^2}{\sqrt{1 - x^2}} + \sqrt{1 - x^2}\right] - x\sqrt{1 - x^2}}{(x + 1)^2}$

3

Equations and Inequalities of the First and Second Degree

The following excerpt illustrates how a quadratic equation arises in an application of calculus:*

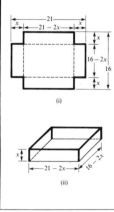

(i)

(ii)

An open box with a rectangular base is to be constructed from a rectangular piece of cardboard 16 inches wide and 21 inches long by cutting out a square from each corner and then bending up the sides. Find the size of the corner square which will produce a box having the largest possible volume.

Solution. We begin by drawing a picture of the cardboard box as shown in (i). . . , where we have introduced a variable x to denote the length of the side of the square to be cut out of each corner of the cardboard. Our goal is to maximize the volume V of the box to be constructed by folding along the dashed lines (see (ii). . .).

The volume of the box shown in (ii) of the figure is given by

$$V = x(16 - 2x)(21 - 2x) = 2(168x - 37x^2 + 2x^3).$$

This equation expresses V as a function of x and we proceed to find its critical numbers. Differentiating with respect to x we obtain

$$D_x V = 2(168 - 74x + 6x^2) = 4(3x^2 - 37x + 84) = 4(3x - 28)(x - 3).$$

The possible critical numbers are $\frac{28}{3}$ and 3. Since $\frac{28}{3}$ is outside the domain of x, the only critical number is 3. . . . Consequently, a three-inch square should be cut from each corner of the cardboard in order to maximize the volume of the resulting box.

Observe that the two possible *critical numbers*, $\frac{28}{3}$ and 3, were found by solving the equation obtained by setting $D_x V = 0$. This was accomplished by factoring the trinomial $3x^2 - 37x + 84 = (3x - 28)(x - 3)$, and we will show in this chapter that this is one of the main techniques for solving quadratic equations. Also observe how the volume of the box was found, because this is similar to a problem you will be asked to do later in this chapter.

* Earl W. Swokowski, *Calculus with Analytic Geometry*, 2d ed. (Boston: Prindle, Weber, & Schmidt, 1979), pp. 177, 178. Reprinted by permission.

1 Introduction

The solution of equations lies at the very heart of algebra and its applications. Much of what we have done up to this point has provided tools for solving equations. The importance of equations stems from the fact that they provide a means by which many complicated relationships in real-world problems can be written down in a concise and precise form. In the terminology used by mathematicians (and others), equations often provide **mathematical models** of real-world problems.

Equations that arise as mathematical models often involve more than algebraic concepts. Some involve trigonometry, and many involve calculus. Even in these more complicated equations, however, a knowledge of the techniques of algebra is essential. In this chapter we consider algebraic equations involving only one unknown. Equations with more than one unknown will be discussed later, when we consider systems of several equations.

Equations of special importance are **polynomial equations in one unknown.** Some examples are

$$2x - 3 = 7$$
$$x^2 - 3x + 2 = 0$$
$$x^3 - 2x^2 - 3 = 0$$

In general, if n is a positive integer, a polynomial equation of the nth degree in the variable x is an equation that can be written in the form

$$a_n x^n + a_{n-1} x^{n-1} + \cdots + a_1 x + a_0 = 0$$

where $a_0, a_1, \ldots, a_n$ are constants, with $a_n \neq 0$. If $n = 1$, the equation is called **linear or first degree,** and for $n = 2$, it is called **quadratic or second degree.** Higher-degree equations are sometimes given names (for example cubic, quartic, and quintic for $n = 3$, 4, and 5, respectively), but we usually treat these as a group. In this chapter we consider the two simplest cases—linear and quadratic equations. A distinction should be made between a polynomial in x and a polynomial equation in x. For example, consider the polynomial

$$2x^2 - 3x - 5$$

and the polynomial equation

$$2x^2 - 3x - 5 = 0$$

In the first case, no restriction is placed on the variable x, and in fact, x is simply left unspecified unless further instructions are given. In the polynomial equation, however, we are seeking the specific values of x that make the equation true. For this reason, we often refer to the variable in such an equation as an **unknown;** it is something with a value that is yet to be determined.

By a **solution** to an equation we mean a number which when substituted for the unknown makes the equation a true statement. A solution is also referred to as a **root** of the equation. For example, the number 2 is a solution, or root, of $x^2 - 4 = 0$, since when 2 is substituted for x we get $4 - 4 = 0$, which is true. The set of all solutions to an equation is called its **solution set.** Since $(-2)^2 = 4$, we see that -2 is also a solution to the equation $x^2 - 4 = 0$, and we will soon see that 2 and -2 are the only solutions. So we can say that the solution set of this equation is $\{2, -2\}$. Two equations are said to be **equivalent** if they have exactly the same solution set. To **solve** an equation means to find its solution set.

2 Linear and Quadratic Equations

By definition, **a linear equation is of the form**

$$ax + b = 0 \qquad (a \neq 0)$$

and by adding $-b$ to both sides and dividing by a we obtain

$$x = -\frac{b}{a}$$

Thus, the only possible solution is $x = -b/a$. We can verify that this is the solution by substitution:

$$a\left(-\frac{b}{a}\right) + b = -b + b = 0$$

We could say that the solution set is $\{-b/a\}$.

There is no point, of course, in memorizing the result just obtained. It is simple enough in each situation to carry out the steps leading to the answer.

EXAMPLE 1 Solve the equation: $3x - 4 = x + 7$

Solution
$$
\begin{aligned}
3x - 4 &= x + 7 \qquad &&\text{Add } -x + 4 \text{ to both sides} \\
2x &= 11 \qquad &&\text{Divide both sides by 2} \\
x &= \tfrac{11}{2}
\end{aligned}
$$

EXAMPLE 2 Solve the equation: $2(x - 3) = 5x - 8$

Solution
$$
\begin{aligned}
2(x - 3) &= 5x - 8 \qquad &&\text{Perform multiplication on left} \\
2x - 6 &= 5x - 8 \qquad &&\text{Add } -2x + 8 \text{ to both sides} \\
2 &= 3x \qquad &&\text{Divide by 3 and interchange right and left sides} \\
x &= \tfrac{2}{3}
\end{aligned}
$$

In solving equations we make repeated use of the properties of equality given in Section 2 of Chapter 1. In particular, we may interchange the two sides of an equation as in Example 2, using the reflexive property. We may add the same quantity to both sides (this includes subtraction, since this is the same as adding a negative quantity), and we may multiply by the same *nonzero* quantity.

EXAMPLE 3   Solve the equation: $\dfrac{2x}{3} - \dfrac{5}{4} = 2$

Solution   We first multiply both sides of the equation by the LCD of the fractions involved. It is essential that *every* term on both sides be multiplied by the LCD. In this case, the LCD is 12. So we have

$$12\left(\frac{2x}{3} - \frac{5}{4}\right) = 12 \cdot 2$$

$$\frac{\overset{4}{\cancel{12}}}{1} \cdot \frac{2x}{\underset{1}{\cancel{3}}} - \frac{\overset{3}{\cancel{12}}}{1} \cdot \frac{5}{\underset{1}{\cancel{4}}} = 24$$

$$8x - 15 = 24$$

$$8x = 39$$

$$x = \tfrac{39}{8}$$

EXAMPLE 4   Solve for t: $at + d = b - ct$

Solution
$$at + d = b - ct$$

$$at + ct = b - d$$

$$(a + c)t = b - d$$

$$t = \frac{b - d}{a + c} \qquad (c \neq -a)$$

Quadratic equations require different approaches. There are three principal techniques for solving them: (1) by factoring, (2) by completing the square, and (3) by the quadratic formula. In this section we consider solution by factoring, and in Section 3 we will study the other two methods.

Of fundamental importance in the solution by factoring is the fact that if a and b are real numbers, then $a \cdot b = 0$ if and only if $a = 0$ or $b = 0$. We will prove this important result. Suppose that $a \cdot b = 0$. If $a \neq 0$, we may multiply both sides by a^{-1} to obtain

$$a^{-1}(ab) = a^{-1} \cdot 0$$

$$(a^{-1}a)b = 0$$

$$1 \cdot b = 0$$

$$b = 0$$

So if $a \neq 0$, then b must be 0. On the other hand, if either $a = 0$ or $b = 0$, then the equation $ab = 0$ is satisfied. So the proof is complete.

A Word of Caution.   The reasoning used in the above argument is not valid if the right-hand side is any number other than 0. For example, consider the following *fallacious argument*:

$$x^2 - 2x = 4$$
$$x(x - 2) = 4$$
$$x = 4 \quad \text{or} \quad x - 2 = 4, \quad \text{so that} \quad x = 6$$

A check of these results reveals that they are wrong. The trouble lies in the fact that if $ab = 4$, we cannot infer anything about the values of a and b.

The next three examples illustrate how to solve a quadratic equation by factoring.

EXAMPLE 5 Solve the equation: $x^2 - x - 2 = 0$

Solution
$$x^2 - x - 2 = 0$$
$$(x - 2)(x + 1) = 0$$

Now, from what we have just shown we know that the product $(x - 2)(x + 1)$ will be 0 if and only if either $x - 2 = 0$ or $x + 1 = 0$, that is, if $x = 2$ or $x = -1$. So these two numbers are solutions, and they are the only solutions. Therefore, the solution set is $\{2, -1\}$.

EXAMPLE 6 Solve: $2x^2 - 3x - 5 = 0$

Solution We have
$$2x^2 - 3x - 5 = 0$$
$$(2x - 5)(x + 1) = 0$$
$$
\begin{array}{c|c}
2x - 5 = 0 & x + 1 = 0 \\
x = \tfrac{5}{2} & x = -1
\end{array}
$$

So the solution set is $\{\tfrac{5}{2}, -1\}$. Again, either of these numbers satisfies the original equation, but no other numbers do so.

EXAMPLE 7 Solve: $x(x - 3) = 4$

Solution As we have seen, it is essential for one side of the equation to be 0. So we add -4 to both sides. This gives
$$x(x - 3) - 4 = 0$$
$$x^2 - 3x - 4 = 0$$
$$(x - 4)(x + 1) = 0$$
$$
\begin{array}{c|c}
x - 4 = 0 & x + 1 = 0 \\
x = 4 & x = -1
\end{array}
$$

So the solution set is $\{4, -1\}$.

EXERCISE SET 2

A   Find the complete solution for each of the following equations:

1. **a.** $2x - 3 = 5$ **b.** $3x + 4 = 10$
2. **a.** $2x - 3 = 8 + 3x$ **b.** $3x + 7 = 8 - 2x$
3. **a.** $2 - x = 6 + x$ **b.** $4 - 2x = x + 7$
4. **a.** $2(x - 1) = 3x + 4$ **b.** $3 - 4x = 5(3x + 2)$
5. **a.** $4(x - 2) = 6$ **b.** $3(2 - x) = 2x + 1$

6. **a.** $\dfrac{x}{2} + \dfrac{1}{3} = 1$ **b.** $\dfrac{2x}{3} - \dfrac{1}{4} = \dfrac{x}{2}$

7. **a.** $\dfrac{x}{3} - \dfrac{1}{2} = \dfrac{3}{4}$ **b.** $\dfrac{3x + 4}{2} = 1 - \dfrac{x}{3}$

8. **a.** $\dfrac{3}{5} - \dfrac{7x}{10} = \dfrac{1}{2}$ **b.** $2 - \dfrac{3x}{8} = \dfrac{5x}{2} + \dfrac{3}{4}$

9. **a.** $\dfrac{2x - 1}{3} = \dfrac{4x + 2}{5}$ **b.** $\dfrac{1 - 3t}{4} - \dfrac{1}{2} = \dfrac{t + 6}{3}$

10. **a.** $\dfrac{5s + 1}{6} - \dfrac{2 - 3s}{8} = 1$ **b.** $\dfrac{3}{4}\left(\dfrac{m}{2} - \dfrac{1}{3}\right) = \dfrac{2 - m}{6}$

11. **a.** $\dfrac{2}{3}\left(\dfrac{x}{4} - 1\right) = \dfrac{3x}{4} - \dfrac{1}{2}$ **b.** $\dfrac{3}{5}\left(\dfrac{1}{2} - \dfrac{s}{3}\right) = \dfrac{5s}{6} + \dfrac{7}{15}$

12. **a.** $\dfrac{1}{2}\left(t - \dfrac{1}{3}\right) = \dfrac{3}{4} - \dfrac{t}{6}$ **b.** $\dfrac{2y - 3}{15} - 1 = \dfrac{1}{3} - y$

13. **a.** $\dfrac{4}{3}\left(\dfrac{2x - 5}{3}\right) - \dfrac{5}{6} = \dfrac{5x}{2} - \dfrac{4}{3}$ **b.** $\dfrac{3}{2}\left(\dfrac{1}{4} - \dfrac{3x}{8}\right) = \dfrac{x - 5}{12} + \dfrac{3}{16}$

14. **a.** $x^2 - 3x + 2 = 0$ **b.** $y^2 + 5y - 6 = 0$
15. **a.** $t^2 + 5t + 6 = 0$ **b.** $2t^2 - 3t - 2 = 0$
16. **a.** $3x^2 - 13x + 4 = 0$ **b.** $6t^2 + t - 2 = 0$
17. **a.** $3x^2 + 10x - 8 = 0$ **b.** $4x^2 - 3x = 0$
18. **a.** $4m^2 - 5m - 6 = 0$ **b.** $12v^2 + 11v - 15 = 0$
19. **a.** $4r^2 - 9 = 0$ **b.** $4r^2 - 9r = 0$
20. **a.** $15y^2 - 26y + 8 = 0$ **b.** $16t^2 - 8t + 1 = 0$
21. **a.** $2x^2 = x + 3$ **b.** $3t^2 - t = 10$
22. **a.** $12x^2 + 8x - 15 = 0$ **b.** $25 - 16s^2 = 0$
23. **a.** $t(2t + 1) = 1$ **b.** $6 + 7x - 3x^2 = 0$
24. **a.** $x^2 - 3x = 10$ **b.** $t(t + 1) = 6$
25. **a.** $2x^2 = 1 + x$ **b.** $x(6x + 1) = 35$
26. **a.** $11x - 4 = 6x^2$ **b.** $x(2 - x) = 3(x - 4)$
27. **a.** $(x - 1)(2x + 3) = 3$ **b.** $t(5t + 2) = 2t(1 - t) + 7$
28. **a.** $3x(x + 1) = 2(x + 2)$ **b.** $2a(a + 3) = a(1 - a) + 2$

B   Solve for the unknown indicated.

29. x: $ax - b = cx + d$ 30. s: $rs - at = bt - s$

31. x: $\dfrac{3xy - 4}{2y + 3} = 1 - x$ 32. n: $\dfrac{2mn - 3}{1 + m} = 4 - n$

33. $x:$ $8a^2x^2 - 2abx - 15b^2 = 0$ **34.** $r:$ $2s^2t^2 - 5rst - 3r^2 = 0$

35. $n:$ $n(6n - m) = m^2$ **36.** $t:$ $3(k^2t^2 - 5s^2) = 4kst$

37. Find k so that $x = 2$ is a solution of $2(k - 3) + 3kx - 2k^2 = 0$.

38. Find k so that $x = \frac{3}{2}$ is a solution of $2k^2x^2 + (3k - 4)x + 12k = 0$.

3   Completing the Square and the Quadratic Formula

The technique of factoring a quadratic polynomial to solve an equation is applicable only when the factors can be readily determined. There is a more general technique by which *all* quadratic equations can be solved, which leads to the **quadratic formula.** This is the technique known as **completing the square.** In order to illustrate it, we need to recall the form of the square of a binomial $(x + a)$. This is

$$(x + a)^2 = x^2 + 2ax + a^2$$

Now, suppose we are confronted with an expression such as

$$x^2 + 6x$$

and we wish to determine what should be added to this to make it a perfect square. We see that $2a = 6$, so that $a = 3$. Therefore, $a^2 = 9$, which is the correct term to add, yielding

$$x^2 + 6x + 9 = (x + 3)^2$$

More generally, since the coefficient of x in the expansion of $(x + a)^2$ is $2a$, it follows that **the correct constant term to add is the square of one-half of the coefficient of x.** Here are some examples.

EXAMPLE 8   Determine what number should be added to each of the following to make it a perfect square:

a. $x^2 + 4x$ **b.** $x^2 + 12x$ **c.** $x^2 + 3x$ **d.** $x^2 - 8x$

e. $x^2 - 5x$ **f.** $x^2 + \frac{3}{4}x$ **g.** $x^2 - \frac{5}{3}x$

Solution   **a.** $\left(\frac{1}{2} \cdot 4\right)^2 = 2^2 = 4$ **b.** $\left(\frac{1}{2} \cdot 12\right)^2 = 6^2 = 36$

c. $\left(\frac{1}{2} \cdot 3\right)^2 = \frac{9}{4}$ **d.** $\left[\frac{1}{2} \cdot (-8)\right]^2 = (-4)^2 = 16$

e. $\left[\frac{1}{2} \cdot (-5)\right]^2 = \frac{25}{4}$ **f.** $\left(\frac{1}{2} \cdot \frac{3}{4}\right)^2 = \left(\frac{3}{8}\right)^2 = \frac{9}{64}$

g. $\left[\frac{1}{2} \cdot \left(-\frac{5}{3}\right)\right]^2 = \left(\frac{-5}{6}\right)^2 = \frac{25}{36}$

One additional fact is needed before proceeding to the solution of equations by this method. This is that $a^2 = b^2$ if and only if $a = b$ or $a = -b$ (or, more briefly, $a = \pm b$). If either $a = b$ or $a = -b$, then squaring both sides shows that $a^2 = b^2$. On the other hand, if $a^2 = b^2$, we have

$$a^2 - b^2 = 0$$

$$(a - b)(a + b) = 0$$

$a - b = 0$	$a + b = 0$
$a = b$	$a = -b$

The next two examples illustrate the technique of solution by completing the square.

EXAMPLE 9   Solve: $x^2 + 3x - 2 = 0$

Solution   We determine that there are no rational factors and proceed as follows: Add 2 to both sides:

$$x^2 + 3x = 2$$

Complete the square on the left and add the same number on the right:

$$x^2 + 3x + \tfrac{9}{4} = 2 + \tfrac{9}{4}$$

Factor the left side and simplify the right:

$$\left(x + \frac{3}{2}\right)^2 = \frac{17}{4}$$

Use fact that $a^2 = b^2$ if and only if $a = \pm b$:

$$x + \frac{3}{2} = \pm \frac{\sqrt{17}}{2}$$

Solve for x:

$$x = -\frac{3}{2} \pm \frac{\sqrt{17}}{2} = \frac{-3 \pm \sqrt{17}}{2}$$

EXAMPLE 10   Solve: $2x^2 - 8x + 3 = 0$

Solution   We will carry out steps similar to those in Example 9 but will omit the explanation. The fact that the coefficient of x^2 is not 1 is a slight complication, but this is handled by dividing both sides of the equation by 2. In solving a quadratic equation by completing the square, this step of dividing both sides of the equation by the coefficient of x^2, if it is not 1 to begin with, should always be done first.

$$2x^2 - 8x + 3 = 0$$
$$x^2 - 4x + \tfrac{3}{2} = 0$$
$$x^2 - 4x = -\tfrac{3}{2}$$
$$x^2 - 4x + 4 = -\tfrac{3}{2} + 4$$
$$(x - 2)^2 = \tfrac{5}{2}$$
$$x - 2 = \pm \sqrt{\tfrac{5}{2}}$$
$$x = 2 \pm \sqrt{\frac{5}{2} \cdot \frac{2}{2}} = 2 \pm \frac{\sqrt{10}}{2} = \frac{4 \pm \sqrt{10}}{2}$$

We now apply this technique to a general quadratic equation, that is, one in which we do not specify the coefficients. Such an equation may be written in the form

$$ax^2 + bx + c = 0 \qquad (a \neq 0) \tag{1}$$

Remark. In defining a polynomial equation of degree n we used a_0, a_1, $a_2, \ldots, a_n$ for the coefficients because to use different letters of the alphabet would be complicated by not knowing how many letters are needed. But in the present case we know we need three coefficients, so we choose to use a, b, and c rather than a_0, a_1, and a_2.

We proceed as in Example 9:

$$ax^2 + bx + c = 0$$

Divide both sides by a:

$$x^2 + \frac{b}{a}x + \frac{c}{a} = 0$$

Add $(-c)/a$ to both sides:

$$x^2 + \frac{b}{a}x = -\frac{c}{a}$$

Complete the square by adding $[\frac{1}{2} \cdot (b/a)]^2 = b^2/(4a^2)$ to both sides:

$$x^2 + \frac{b}{a}x + \frac{b^2}{4a^2} = -\frac{c}{a} + \frac{b^2}{4a^2}$$

Factor the left side and collect terms on the right:

$$\left(x + \frac{b}{2a}\right)^2 = \frac{b^2 - 4ac}{4a^2}$$

Take square roots:

$$x + \frac{b}{2a} = \frac{\pm\sqrt{b^2 - 4ac}}{2a}$$

Finally, solve for x, thus obtaining the **quadratic formula:**

$$x = \frac{-b \pm \sqrt{b^2 - 4ac}}{2a}$$

By committing this formula to memory, you can write down at once the solutions to any quadratic equation. It should be emphasized, however, that factoring is usually easier when the factors can be readily found.

It is essential before attempting to apply the quadratic formula that the equation be written precisely in the form given in equation (1) so that the correct values of a, b, and c can be determined. We call this the **standard form.**

EXAMPLE 11   Solve the equation $2x^2 - 3x - 7 = 0$ by the quadratic formula.

Solution   We see that $a = 2$, $b = -3$, and $c = -7$.* So

$$x = \frac{-(-3) \pm \sqrt{(-3)^2 - 4(2)(-7)}}{2 \cdot 2} = \frac{3 \pm \sqrt{9 + 56}}{4}$$

$$= \frac{3 \pm \sqrt{65}}{4}$$

There is a difficulty that an alert reader might have noticed. The quantity $(b^2 - 4ac)$ appearing under the radical in the quadratic formula could be negative, in which case there is no solution in the set of real numbers. As we have seen, if a is any nonzero real number, then $a^2 > 0$, so that there is no square root of a negative number in R. To overcome this difficulty, it is necessary to consider a wider class of numbers called the **complex numbers.** These will be discussed in the next section.

EXERCISE SET 3

A   Solve Problems 1–16 by completing the square.

1. $x^2 + 4x - 12 = 0$ 2. $x^2 - 6x + 8 = 0$
3. $x^2 - 2x - 3 = 0$ 4. $x^2 + 6x - 7 = 0$
5. $y^2 + 3y - 10 = 0$ 6. $x^2 - 7x + 6 = 0$
7. $t^2 - 5t = 24$ 8. $2x^2 - 7x + 6 = 0$
9. $t^2 - 4t + 2 = 0$ 10. $x^2 - 8x - 4 = 0$
11. $3x^2 - 4x - 32 = 0$ 12. $t^2 + 2t - 4 = 0$
13. $m^2 - 5m = 3$ 14. $n^2 + 2 = 3n$
15. $2x^2 = 3(x + 3)$ 16. $2x(x - 2) = 3$

---

* We think of the equation in the equivalent form $2x^2 + (-3)x + (-7) = 0$; that is, the negative signs must be considered as part of the coefficients.

Solve Problems 17–32 by the quadratic formula.

17. $x^2 - 2x - 2 = 0$ 18. $x^2 - 3x - 6 = 0$
19. $t^2 + 5t = 9$ 20. $x^2 + x = 1$
21. $2x^2 - x = 5$ 22. $2x^2 - 3x - 5 = 0$
23. $3x^2 = 2x + 2$ 24. $3x^2 + 10x + 2 = 0$
25. $5y^2 - y = 3$ 26. $5t^2 + 10t + 4 = 0$
27. $2x + 3 = 4x^2$ 28. $6x = 4x^2 - 1$
29. $x(2x - 3) = 4$ 30. $3x(x - 2) = 4(x + 1)$
31. $(t - 2)(2t + 1) = 1$ 32. $3t^2 = 2(2 - t)$

B   Solve Problems 33–39 for the unknown indicated.

33. t: $\frac{1}{2}gt^2 - v_0 t - s_0 = 0$ 34. x: $x^2 + 2t(x + 1) = 1$

35. R: $w = \dfrac{mgR^2}{(x + R)^2}$ $(R > 0, x > 0)$ 36. r: $S = \pi r^2 + 2\pi r l$

37. s: $2s^2 - \sqrt{3}\,ks - 9k^2 = 0$ $(k > 0)$ 38. r: $Lr^2 + Rr + \dfrac{1}{C} = 0$

39. x: $2x^2 + 3kx - (3k + 2) = 0$ $(k > 0)$

In Problems 40 and 41 solve for x with the aid of a hand calculator.

40. $2.38x^2 - 5.47x + 1.32 = 0$ 41. $32.43 + 2.054x - 0.03172x^2 = 0$

---

4   Complex Numbers

We have already noted that certain quadratic equations cannot be solved in the domain of real numbers. Even the simple equation $x^2 + 1 = 0$ falls in this category. If x is any real number, its square is nonnegative; so adding it to the positive integer 1 results in a positive number, which is therefore greater than zero. This difficulty persists in higher-degree equations. The resolution of the problem lies in expanding our number system.

Number systems have evolved over thousands of years, and each step has met with resistance from segments of the intellectual community. The introduction of negative and irrational numbers were certainly two of the most controversial stages. As a matter of fact, all numbers are creations of the human mind (even the integers, Kronecker notwithstanding*). Numbers do not exist in nature, waiting to be discovered. They exist only in the realm of ideas. We are free, then, to invent new numbers if a need exists. To make them useful, additions should be consistent with our known number systems and should, in some sense, include these as subsystems. The real numbers, for example, include the rational numbers as a subsystem, which include the integers, which in turn include the natural numbers.

---

* Leopold Kronecker (1823–1891) said that "God made the integers, and all the rest is the work of man."

The **complex numbers** provide just such an extension of the reals. **They consist of elements of the form $a + bi$, where a and b are arbitrary real numbers, and i is the "imaginary unit" defined by $i^2 = -1$.** These complex numbers exist just as surely as any numbers do; there is nothing imaginary about them. It is an unfortunate fact of history that the terms *real* and *imaginary* are used, because of their connotations. One might better refer to real numbers as one-dimensional numbers and complex numbers as two-dimensional numbers. This terminology reflects the fact that real numbers can be associated with points on a line, whereas complex numbers can be made to correspond to points in a plane.

The real numbers are included among the complex numbers, since a real number a may be thought of as being of the form $a + 0i$. If $b \neq 0$, the complex number $a + bi$ is called an **imaginary number.** Complex numbers of the form $0 + bi = bi$ with $b \neq 0$ are called **pure imaginary numbers.**

Two complex numbers $a + bi$ and $c + di$ are equal if and only if $a = c$ and $b = d$. Addition is defined by

$$(a + bi) + (c + di) = (a + c) + (b + d)i$$

and multiplication by

$$(a + bi)(c + di) = (ac - bd) + (ad + bc)i$$

The latter formula is the result of multiplying as with two real binomials and then substituting $i^2 = -1$:

$$(a + bi)(c + di) = ac + bic + adi + bdi^2 = ac + i(bc + ad) + bd(-1)$$
$$= (ac - bd) + (ad + bc)i$$

Multiplication is usually carried out in this way rather than by simply memorizing the definition. For example,

$$(2 + 3i)(4 - 7i) = 8 - 2i - 21i^2 = 8 - 2i + 21 = 29 - 2i$$

The identity elements are $0 = 0 + 0i$ and $1 = 1 + 0i$. The additive inverse of $a + bi$ is seen to be $(-a) + (-b)i$. When a and b are not both equal to 0 (that is, when $a + bi \neq 0$), we can show that the number $a + bi$ has a multiplicative inverse within the complex number system. To do this we must show that $(a + bi)^{-1}$ is a number of the form $A + Bi$, where A and B are real. The following calculations show that this is the case, with $A = a/(a^2 + b^2)$ and $B = -b/(a^2 + b^2)$:

$$(a + bi)^{-1} = \frac{1}{a + bi} = \frac{1}{a + bi} \cdot \frac{a - bi}{a - bi} = \frac{a - bi}{a^2 - b^2 i^2} = \frac{a - bi}{a^2 + b^2}$$

$$= \frac{a}{a^2 + b^2} + \frac{-b}{a^2 + b^2} i$$

Rather than learning this result, it is better to learn the procedure. For example,

$$\frac{1}{2 + 3i} = \frac{1}{2 + 3i} \cdot \frac{2 - 3i}{2 - 3i} = \frac{2 - 3i}{4 - 9i^2} = \frac{2 - 3i}{4 + 9} = \frac{2}{13} + \frac{-3}{13} i$$

We define the conjugate of a complex number $a + bi$ to be the number $a - bi$. So to find the inverse of a complex number, we write its reciprocal and then multiply the numerator and denominator by its conjugate.

Division is accomplished in a similar way. For example:

$$\frac{3 + 2i}{5 - 4i} = \frac{3 + 2i}{5 - 4i} \cdot \frac{5 + 4i}{5 + 4i} = \frac{15 + 22i + 8i^2}{25 - 16i^2} = \frac{15 + 22i - 8}{25 + 16} = \frac{7}{41} + \frac{22}{41} i$$

More generally,

$$\frac{a + bi}{c + di} = \frac{a + bi}{c + di} \cdot \frac{c - di}{c - di} = \frac{(ac + bd) + (bc - ad)i}{c^2 + d^2} = \frac{ac + bd}{c^2 + d^2} + \frac{bc - ad}{c^2 + d^2} i$$

To indicate the operation of taking the conjugate, a bar is used above the number:

$$\overline{a + bi} = a - bi$$

For example:

$$\overline{2 + 3i} = 2 - 3i$$
$$\overline{5 - 4i} = 5 + 4i$$
$$\overline{-7i} = \overline{0 - 7i} = 7i$$
$$\overline{-2} = \overline{-2 + 0i} = -2 - 0i = -2$$

Often the letter z is used to designate a complex number. Then $\bar{z}$ indicates its conjugate. The following facts about conjugates can be shown:

1. $\overline{z_1 + z_2} = \overline{z_1} + \overline{z_2}$ Conjugate of a sum = Sum of conjugates

2. $\overline{z_1 \cdot z_2} = \overline{z_1} \cdot \overline{z_2}$ Conjugate of a product = Product of conjugates

3. $\overline{\left(\dfrac{z_1}{z_2}\right)} = \dfrac{\overline{z_1}}{\overline{z_2}}$ Conjugate of a quotient = Quotient of conjugates

4. $z\bar{z}$ is real

5. If z is real, then $\bar{z} = z$.

Facts 4 and 5 result from the following reasoning: Suppose $z = a + bi$. Then $z\bar{z} = (a + bi)(a - bi) = a^2 - b^2i^2 = a^2 + b^2$, which is real. If $z = a$, where a is real, we have $\bar{z} = \bar{a} = \overline{a + 0i} = a - 0i = a = z$.

It is useful to know the various powers of i:

$$i = \sqrt{-1}$$
$$i^2 = -1$$
$$i^3 = i^2 \cdot i = (-1)i = -i$$
$$i^4 = i^2 \cdot i^2 = (-1)(-1) = 1$$

Now, higher powers can be readily calculated. For example, $i^{22} = i^{20} \cdot i^2 = (i^4)^5 \cdot i^2 = (1)^5 \cdot i^2 = i^2 = -1$. Also, $i^{35} = i^{32} \cdot i^3 = (i^4)^8 \cdot i^3 = (1)^8 \cdot i^3 = i^3 = -i$. The object is to factor out the highest power of i^4, which is 1, and the remaining factor can be evaluated using the results just obtained, that is, $i^1 = i$, $i^2 = -1$, $i^3 = i$, $i^4 = 1$. More generally, if $n > 4$, and we write $n = 4k + r$, where $0 \leq r \leq 3$, then $i^n = i^{4k+r} = (i^{4k}) \cdot i^r = (i^4)^k \cdot i^r = 1^k \cdot i^r = i^r$.

In applications of the quadratic formula, and elsewhere, we sometimes encounter expressions of the form $\sqrt{-p}$, where p is a positive real number. We define this as follows:

$$\sqrt{-p} = \sqrt{(-1)p} = \sqrt{-1}\sqrt{p} = i\sqrt{p}$$

Whenever such an expression occurs, it is important to change it immediately to the form $i\sqrt{p}$. To see why, consider $\sqrt{-2} \cdot \sqrt{-5}$. If we carelessly applied formula (11) from Chapter 1, we would get

$$\sqrt{-2} \cdot \sqrt{-5} = \sqrt{(-2)(-5)} = \sqrt{10}$$

This is incorrect. Instead, we should do as follows:

$$\sqrt{-2} \cdot \sqrt{-5} = (i\sqrt{2})(i\sqrt{5}) = i^2(\sqrt{2} \cdot \sqrt{5}) = -\sqrt{10}$$

Let us return now to the solution of quadratic equations. Recall that by the quadratic formula, the solutions of

$$ax^2 + bx + c = 0 \qquad (a \neq 0)$$

are given by

$$\frac{-b + \sqrt{b^2 - 4ac}}{2a} \qquad \text{and} \qquad \frac{-b - \sqrt{b^2 - 4ac}}{2a}$$

Although the quadratic formula holds true even when a, b, and c are complex, for the present we assume them to be real. The nature of these two solutions depends on the number under the radical. This number, $b^2 - 4ac$, **is called the discriminant of the quadratic equation.** When the discriminant is positive, the square root is a positive real number, so the solutions are real and distinct. When the discriminant is 0, the square root is 0, so the solutions are real and equal. (There is only one solution in this case.) Finally, when the discriminant is negative, the square root is imaginary, so the solutions are imaginary; they are, in fact, conjugates of each other. We summarize these results in the box.

$$\text{If} \quad b^2 - 4ac \quad \begin{cases} > 0 \\ = 0 \\ < 0 \end{cases} \quad \text{the solutions are} \quad \begin{cases} \text{real and unequal} \\ \text{real and equal} \\ \text{imaginary} \end{cases}$$

EXAMPLE 12   Without solving, determine the nature of the solutions of each of the following.
a. $2x^2 - 3x - 4 = 0$ **b.** $9x^2 - 24x + 16 = 0$ **c.** $3x^2 - 2x + 4 = 0$

Solution   **a.** $b^2 - 4ac = (-3)^2 - 4(2)(-4) = 9 + 32 = 41 > 0$. So the solutions are real and unequal.
b. $b^2 - 4ac = (-24)^2 - 4(9)(16) = 576 - 576 = 0$. So the solutions are real and equal.
c. $b^2 - 4ac = (-2)^2 - 4(3)(4) = 4 - 48 = -44 < 0$. So the solutions are imaginary.

EXAMPLE 13   Solve: $x^2 - 6x + 10 = 0$

Solution   By the quadratic formula we obtain

$$x = \frac{6 \pm \sqrt{36 - 40}}{2} = \frac{6 \pm \sqrt{-4}}{2} = \frac{6 \pm i\sqrt{4}}{2}$$

$$= \frac{6 \pm 2i}{2} = 3 \pm i$$

EXERCISE SET 4

A   In Problems 1–8 perform the indicated operations. Express answers in the form $a + bi$.

1. **a.** $(2 - 5i) + (7 + 3i)$ **b.** $(3 + 2i) + (-5 - 7i)$
    c. $(5 + 8i) - (6 - i)$ **d.** $(3 - i) - (2i - 3)$

2. **a.** $(3 + 7i) + (2 + 3i)$ **b.** $(5 - 3i) + (2 + i)$
    c. $(6 - 2i) - (3 + 4i)$ **d.** $(4i + 2) - (3 - 8i)$

3. **a.** $(4 + 3i)(3 - 2i)$ **b.** $(2 - i)(2 + i)$
    c. $(5i - 9)(4i + 7)$ **d.** $(i + 3)(4 - 3i)$

4. **a.** $(5 + 4i)(6 + 7i)$ **b.** $(2 - 5i)(3 + i)$
    c. $(3 + 4i)(3 - 4i)$ **d.** $(2i - 3)(4 + 5i)$

5. **a.** $\dfrac{1}{1 + i}$ **b.** $(3 - 2i)^{-1}$ **c.** $\dfrac{1}{i}$ **d.** $\dfrac{1}{7 + 8i}$

6. **a.** $\dfrac{1}{2 + 3i}$ **b.** $(4 - i)^{-1}$ **c.** $\dfrac{1}{i - 1}$ **d.** $(5 - 4i)^{-1}$

7. **a.** $\dfrac{2 + 3i}{3 - 2i}$ **b.** $\dfrac{1 - i}{1 + i}$ **c.** $\dfrac{3 + 4i}{5 - 3i}$ **d.** $\dfrac{3i - 1}{6 + 5i}$

8. **a.** $\dfrac{3 - i}{2 + 3i}$ **b.** $\dfrac{4 + 3i}{4 - 3i}$ **c.** $\dfrac{i}{i + 1}$ **d.** $\dfrac{2i - 3}{5 - 2i}$

In Problems 9–12 simplify the expressions by using properties of i.

9. **a.** $\sqrt{-4}$ **b.** $\sqrt{-162}$ **c.** $\sqrt{-40}$
    d. $\sqrt{-4}\sqrt{-9}$ **e.** $\sqrt{-2}\sqrt{-8}$

10. **a.** $\sqrt{-75}$ **b.** $\dfrac{1}{\sqrt{-12}}$ **c.** $\sqrt{-3}\sqrt{-6}$

    d. $\dfrac{\sqrt{20}}{\sqrt{-5}}$ **e.** $\sqrt{-8}\sqrt{2}$

11. **a.** i^7 **b.** i^9 **c.** i^{12} **d.** i^{25} **e.** i^{18}

12. **a.** $\dfrac{1}{i^2}$ **b.** $\dfrac{1}{i^3}$ **c.** $\dfrac{1}{i^9}$ **d.** i^{-5} **e.** i^{-10}

In Problems 13–20 solve the quadratic equations.

13. **a.** $x^2 + 4 = 0$ **b.** $x^2 - x + 2 = 0$
14. **a.** $x^2 = -9$ **b.** $4x^2 + 25 = 0$
15. **a.** $x^2 + 3x + 3 = 0$ **b.** $2x^2 + 5 = 6x$

16. **a.** $x^2 - 2x + 6 = 0$ **b.** $t^2 + 3t = -4$
17. **a.** $3m^2 - 4m + 6 = 0$ **b.** $5t^2 + 2 = 4t$
18. **a.** $2s^2 - 5s + 6 = 0$ **b.** $3s^2 + 2s + 1 = 0$
19. **a.** $2t^2 = 6t - 7$ **b.** $t(5t + 6) + 4 = 0$
20. **a.** $5 + 7x^2 = 8x$ **b.** $x(8 - 5x) = 4$

In Problems 21–23 determine whether the solutions are real and unequal, real and equal, or imaginary by using the discriminant. Do not solve.

21. **a.** $3x^2 - 9x + 5 = 0$ **b.** $7x^2 + 6x + 3 = 0$
22. **a.** $2t^2 - 8 = 3t$ **b.** $9m^2 = 5(6m - 5)$
23. **a.** $y(4y - 7) + 3 = 0$ **b.** $8 + 6s^2 = 13s$

B   In Problems 24 and 25 expand and simplify.

24. **a.** $(2 + 3i)^3$ **b.** $(1 + i)^5$
25. **a.** $(2 - i)^4$ **b.** $(3 - 4i)^3$

26. Evaluate the polynomial $x^3 - 5x^2 + 17x - 13$ when $x = 2 - 3i$.
27. Evaluate the polynomial $2x^3 - 7x^2 - 10x - 6$ when $x = 1 - i$.
28. Solve the following quadratic equations:
    a. $3x^2 - 8ix - 4 = 0$ **b.** $x^2 - x + 1 + i = 0$
29. Prove the following, where $z_1 = a + bi$ and $z_2 = c + di$.

    a. $\overline{z_1 z_2} = \overline{z_1} \cdot \overline{z_2}$ **b.** $\overline{\left(\dfrac{z_1}{z_2}\right)} = \dfrac{\overline{z_1}}{\overline{z_2}}$

30. For what values of k will the equation $k^2 x^2 + (k - 1)x + 4 =$ have equal solutions?

---

5   Equations that Are Convertible to Linear or Quadratic Forms

Some equations, although they are not linear or quadratic, can nevertheless be solved using techniques developed in this chapter. We consider first equations that are **quadratic in form.** These are equations that can be written in the form

$$a(\quad)^2 + b(\quad) + c = 0$$

where the same expression appears in each set of parentheses. For example,

$$x^4 - 5x^2 + 4 = 0$$

is quadratic in form, since it can be written

$$(x^2)^2 - 5(x^2) + 4 = 0$$

We could say it is quadratic in the unknown x^2. Another example is

$$3x^{-2} - 4x^{-1} + 1 = 0$$

which can be written

$$3(x^{-1})^2 - 4(x^{-1}) + 1 = 0$$

In the next example we solve these two equations.

EXAMPLE 14   Solve the equations:
a. $x^4 - 5x^2 + 4 = 0$ **b.** $3x^{-2} - 4x^{-1} + 1 = 0$

Solution   **a.**   We solve by factoring, first treating x^2 as the unknown.

$$x^4 - 5x^2 + 4 = 0$$
$$(x^2 - 1)(x^2 - 4) = 0$$

$x^2 - 1 = 0$	$x^2 - 4 = 0$
$x^2 = 1$	$x^2 = 4$
$x = \pm 1$	$x = \pm 2$

So the solution set is $\{\pm 1, \pm 2\}$.

b.

$$3x^{-2} - 4x^{-1} + 1 = 0$$
$$(3x^{-1} - 1)(x^{-1} - 1) = 0$$

$3x^{-1} - 1 = 0$	$x^{-1} - 1 = 0$
$3x^{-1} = 1$	$x^{-1} = 1$
$\dfrac{3}{x} = 1$	$\dfrac{1}{x} = 1$
$x = 3$	$x = 1$

So the solution set is $\{3, 1\}$.

As an alternate procedure we may make an appropriate substitution so that the equation becomes quadratic. In part a, for example, we could let $t = x^2$. Then the equation becomes

$$t^2 - 5t + 4 = 0$$

Solving for t yields $t = 1$ or $t = 4$. Finally, we replace t by x^2 in each case and solve for x:

$t = x^2 = 1$	$t = x^2 = 4$
$x = \pm 1$	$x = \pm 2$

Similarly in part b, we could let $t = x^{-1}$. The equation would become

$$3t^2 - 4t + 1 = 0$$

yielding $t = \frac{1}{3}$ or $t = 1$, from which we would get

$x^{-1} = \frac{1}{3}$	$x^{-1} = 1$
$\dfrac{1}{x} = \dfrac{1}{3}$	$\dfrac{1}{x} = 1$
$x = 3$	$x = 1$

The next type of equation we consider is one involving fractions, where the unknown appears in a denominator. Our procedure will be to clear of fractions by multiplying both sides by the LCD. Now we know that multiplying an equation by a nonzero number results in an equivalent equation. But if the LCD involves the unknown, then we do not know in advance if the number we are multiplying by is nonzero. The way out of this dilemma is to proceed with the clearing of fractions and solve the resulting equation but then to check the answer. A complete check is to substitute each value found for x into the *original* equation to see if the equation is satisfied. This complete check is recommended, but sometimes it can be quite tedious. A *minimum requirement* is to check each value of x obtained by substituting it in the LCD to see that the LCD $\neq 0$. If the LCD $= 0$ for a value of x, this value must be discarded. It should be noted that simply showing that the LCD $\neq 0$ for a value of x which has been found as a possible solution shows that our procedure was justified for that value of x, but it does not guarantee that the solution is correct (an arithmetic error may have been made).

EXAMPLE 15   Solve for x: $\dfrac{1}{x-2} - \dfrac{2}{x+1} = \dfrac{7}{x^2-x-2}$

Solution   The procedure is to multiply both sides of the equation by the LCD, which in this case is $(x-2)(x+1)$. It is essential to multiply every term on both sides of the equation. In so doing, all denominators cancel.

$$\left(\frac{1}{x-2} - \frac{2}{x+1} \right) \frac{(x-2)(x+1)}{1} = \frac{7}{x^2-x-2} \cdot \frac{(x-2)(x+1)}{1}$$

$$x + 1 - 2(x-2) = 7$$

$$x + 1 - 2x + 4 = 7$$

$$-x = 2$$

$$x = -2$$

We check $x = -2$ in the LCD and see that it is not 0. So (assuming no arithmetic error) the solution is $x = -2$.

EXAMPLE 16   Solve for x: $\dfrac{3}{x-3} - \dfrac{2}{x-2} = \dfrac{3}{x^2-5x+6}$

Solution   The LCD is $(x-3)(x-2)$. Multiplying both sides by this yields

$$3(x-2) - 2(x-3) = 3$$

$$3x - 6 - 2x + 6 = 3$$

$$x = 3$$

Checking the LCD, we find that it is 0 for this value of x. Since this is the only possible solution and it is not admissible, we conclude that the equation has **no solution.** The solution set is therefore the empty set.

EXAMPLE 17   Solve for x: $\dfrac{1}{x-1} + \dfrac{3}{x+3} = 1$

Solution   We multiply both sides by the LCD, $(x-1)(x+3)$, to obtain

$$x + 3 + 3(x - 1) = (x - 1)(x + 3)$$
$$x + 3 + 3x - 3 = x^2 + 2x - 3$$
$$0 = x^2 - 2x - 3$$
$$(x - 3)(x + 1) = 0$$
$$x = 3 \quad | \quad x = -1$$

Since the LCD is nonzero for each of these, the solution set is $\{-1, 3\}$.

EXAMPLE 18   Solve for y and check the results: $\dfrac{2}{3y} + \dfrac{1}{y-1} = \dfrac{1}{2}$

Solution   The LCD is $6y(y - 1)$. Multiplying by this yields

$$4(y - 1) + 6y = 3y(y - 1)$$
$$4y - 4 + 6y = 3y^2 - 3y$$
$$3y^2 - 13y + 4 = 0$$
$$(3y - 1)(y - 4) = 0$$
$$3y - 1 = 0 \quad | \quad y - 4 = 0$$
$$3y = 1 \quad \quad \quad y = 4$$
$$y = \tfrac{1}{3} \quad |$$

The LCD is nonzero, so we know the equation obtained after multiplication is equivalent to the original. To verify that these actually are solutions, we substitute in the original equation.

Check. $y = \tfrac{1}{3}$

$$\frac{2}{3 \cdot \frac{1}{3}} + \frac{1}{\frac{1}{3} - 1} = \frac{2}{1} + \frac{3}{1 - 3} = 2 + \frac{3}{-2} = 2 - \frac{3}{2} = \frac{4}{2} - \frac{3}{2} = \frac{1}{2}$$

So $y = \tfrac{1}{3}$ is a solution.

Check. $y = 4$

$$\frac{2}{3 \cdot 4} + \frac{1}{4 - 1} = \frac{2}{12} + \frac{1}{3} = \frac{1}{6} + \frac{1}{3} = \frac{1}{6} + \frac{2}{6} = \frac{3}{6} = \frac{1}{2}$$

So $y = 4$ is also a solution. The solution set is therefore $\{\tfrac{1}{3}, 4\}$.

Remark. Multiplying both sides of an equation by an expression involving the unknown may introduce extraneous roots (as in Example 16). On the other hand, dividing both sides by such an expression may result in the loss of solutions. To illustrate the latter, if in the equation $x^2 = 4x$, we divide both sides by x, getting $x = 4$, we lose the solution $x = 0$.

The next three examples involve squaring both sides of an equation. This may or may not lead to an equivalent equation, as the following considerations show: If $a = b$, then clearly $a^2 = b^2$. But conversely, if $a^2 = b^2$, then we have either $a = b$ or $a = -b$. So squaring both sides of an equation results in the possibility of additional solutions not contained in the original (namely, those contained in $a = -b$). So, when squaring is desirable, the appropriate way to proceed is to square both sides, solve the resulting problem, and check all answers in the original equation. The squared equation contains all solutions of the original, but it may contain additional solutions that are not valid.

EXAMPLE 19   Solve for x: $\sqrt{x - 1} = 2$

Solution   On squaring, we obtain
$$x - 1 = 4$$
$$x = 5$$

Check. $\sqrt{5 - 1} = \sqrt{4} = 2$
So the answer is 5.

EXAMPLE 20   Solve for x: $\sqrt{2x - 3} + x = 3$

Solution   Before squaring we wish to isolate the radical in order to eliminate it after squaring. So, we first add $-x$ to both sides:
$$\sqrt{2x - 3} = 3 - x$$
$$2x - 3 = 9 - 6x + x^2$$
$$x^2 - 8x + 12 = 0$$
$$(x - 2)(x - 6) = 0$$
$$x = 2 \ \mid \ x = 6$$

Check. $x = 2$
$$\sqrt{2 \cdot 2 - 3} + 2 = \sqrt{1} + 2 = 1 + 2 = 3$$

So $x = 2$ is a solution.

Check. $x = 6$

$$\sqrt{2 \cdot 6 - 3} + 6 = \sqrt{9} + 6 = 3 + 6 = 9 \neq 3$$

So $x = 6$ is not a solution.

EXAMPLE 21   Solve for x: $\sqrt{2x + 3} - \sqrt{x + 1} = 1$

Solution   In a situation such as this it is usually best to isolate the more complicated radical, square once, isolate the remaining radical, and square again.

$$\sqrt{2x + 3} = 1 + \sqrt{x + 1}$$
$$(\sqrt{2x + 3})^2 = (1 + \sqrt{x + 1})^2$$
$$2x + 3 = 1 + 2\sqrt{x + 1} + x + 1$$
$$x + 1 = 2\sqrt{x + 1}$$
$$x^2 + 2x + 1 = 4(x + 1)$$
$$x^2 - 2x - 3 = 0$$
$$(x - 3)(x + 1) = 0$$
$$x = 3 \quad | \quad x = -1$$

Check. $x = 3$

$$\sqrt{2 \cdot 3 + 3} - \sqrt{3 + 1} = \sqrt{9} - \sqrt{4} = 3 - 2 = 1$$

So $x = 3$ checks.

Check. $x = -1$

$$\sqrt{2(-1) + 3} - \sqrt{-1 + 1} = \sqrt{-2 + 3} - 0 = \sqrt{1} = 1$$

So $x = -1$ also checks. The complete solution set is therefore $\{-1, 3\}$.

EXERCISE SET 5

A   Solve the equations.

1. $4x^4 - 13x^2 + 9 = 0$
2. $x^4 - 13x^2 + 36 = 0$
3. $x^{-2} - x^{-1} - 6 = 0$
4. $2x^{-2} - 3x^{-1} - 5 = 0$
5. $x^4 + 3x^2 - 4 = 0$
6. $4x^{-4} - 13x^{-2} + 9 = 0$
7. $4x^4 - 25x^2 + 36 = 0$
8. $3x^{-2} - 10x^{-1} + 8 = 0$
9. $x^{2/3} - 2x^{1/3} - 3 = 0$
10. $6x^{-2/3} + 7x^{-1/3} - 3 = 0$
11. $\dfrac{1}{x - 1} = \dfrac{2}{x + 3}$
12. $\dfrac{3}{x} - \dfrac{4}{x - 2} = 0$
13. $\dfrac{3}{x} + \dfrac{1}{2} = \dfrac{x}{x + 4}$
14. $\dfrac{4}{2x - 3} = \dfrac{5}{3x - 4}$

(Check your answer to see that it works.)

15. $\dfrac{1}{x+2} - \dfrac{2}{x-1} = \dfrac{3}{x^2+x-2}$

16. $\dfrac{2}{x+2} - \dfrac{3}{x} = \dfrac{1}{x^2+2x}$

17. $\dfrac{1}{x} - \dfrac{1}{x-3} = \dfrac{2}{x^2}$

18. $\dfrac{1}{x+1} - \dfrac{2}{2x-3} = \dfrac{4x+5}{2x^2-x-3}$

(Check your answer to see that it works.)

19. $\dfrac{3}{2x} - \dfrac{1}{x-5} = 1$

20. $\dfrac{2}{x-1} + \dfrac{3}{x+1} = 3$

21. $\dfrac{1}{x-3} + \dfrac{6}{x+2} = 2$

22. $\dfrac{5}{2x+1} - \dfrac{x-4}{x+2} = 0$

23. $\dfrac{4}{2x+3} + \dfrac{2}{x-6} = 1$

24. $\dfrac{3x+2}{x-1} + \dfrac{7}{2x} = 0$

25. $\dfrac{x}{x-2} + \dfrac{3}{x} = \dfrac{4}{x^2-2x}$

26. $\sqrt{2x-5} = 1$

27. $\sqrt{x+3} = -1$

28. $\sqrt{2x-1} = 1$

29. $\sqrt{x+2} = x$

30. $\sqrt{2x+5} - 1 = x$

31. $2\sqrt{x+4} = \sqrt{1-x}$

32. $\sqrt{x+5} = 1 + \sqrt{x}$

33. $\sqrt{2x-1} = x-2$

B **34.** $\dfrac{2x+1}{x-1} - \dfrac{x}{x+3} = 3$

35. $\dfrac{x}{x+2} - \dfrac{x-1}{2x+5} + 1 = 0$

36. $\dfrac{2x+3}{x+2} - \dfrac{x-2}{x-1} = \dfrac{2x+7}{x^2+x-2}$

37. $\dfrac{x+3}{2x-3} - \dfrac{2x-5}{x-2} = 1$

38. $\dfrac{2x}{x-3} + \dfrac{3}{1-x} = \dfrac{2x+4}{x^2-4x+3}$

39. $\dfrac{x}{2x+5} - \dfrac{4x-1}{1-x} = \dfrac{2x+1}{2x^2+3x-5}$

40. $\sqrt{x+4} = \sqrt{1-x} - 1$

41. $\sqrt{x+1} = 2 - \sqrt{3-x}$

42. $\sqrt{2x-1} - \sqrt{x-1} = 1$

43. $\dfrac{1}{\sqrt{x-2}} + \dfrac{3}{2} = \sqrt{x-2}$

44. $\dfrac{3}{\sqrt{3x-2}} - \dfrac{1}{\sqrt{x-1}} = \dfrac{1}{\sqrt{3x^2-5x+2}}$

45. $\sqrt{1-x} - \sqrt{4+x} = \sqrt{2x+7}$

46. $\sqrt{2x+3} - 2\sqrt{x-2} = \sqrt{4-x}$

47. $x^4 + 2x^2 - 2 = 0$

48. $x^6 - 7x^3 - 8 = 0$

49. $36x^{4/3} + 7x^{2/3} - 4 = 0$

50. $8x^{-6} + 19x^{-3} - 27 = 0$

6 Applications

In the application of mathematics to real-life situations problems are typically stated in words, not mathematical symbols. Learning to translate such verbal statements into mathematical terms is as important an aspect of the study of mathematics as is learning how to solve problems after they are formulated. The examples and exercises in this section provide practice in setting up these so-called "word problems" which lead to linear or quadratic equations.

Unfortunately, there is no magical road to success in setting up word problems. Guidelines can be given, but each problem has its own unique features. There are, however, general classes into which many problems can be categorized, and learning how to do one problem of a given class will usually help in solving other problems of that class. For example, in physical applications many problems deal with motion in a straight line in which the relation "distance equals rate times time" is applicable (where the rate, or velocity, is constant). Thus, the formula Distance = (Rate)(Time) or $d = rt$ is used in all such situations. Another class of problems has to do with the mixing of substances, as in chemistry experiments. These and other classes will be illustrated in the examples.

The following guidelines are suggested:

1. Read the problem through and identify what unknown quantity is to be determined.
2. Introduce some letter to designate the unknown quantity. The letter x is often used for historical reasons, but sometimes it is helpful to use a suggestive letter, such as d for distance or v for velocity. Be very specific about this; for example, a statement like "Let $t =$ the time in hours required to complete the job" is typical.
3. Begin a careful rereading of the problem, phrase-by-phrase, and write down the relevant information that is given, expressing relationships with the unknown when appropriate. Often, a sketch can be useful in looking for relationships.
4. Try to identify in the problem some equation that relates the unknown and the known information and write this down in terms of the symbols you have introduced. Here it is useful to observe that the words *is* and *are* often translate as *equals* in an equation.
5. Solve the equation obtained in Step 4, and check your answers to see if they are reasonable in the context of the original problem. Sometimes, one or more solutions must be eliminated because of physical limitations imposed by the problem.

When more than one unknown is involved, obvious modifications in these steps should be made. Also, there are occasions when it is best to introduce as the unknown a quantity related to, but not the same as, the quantity to be determined.

The importance of Steps 2 and 3 should be emphasized. A thorough and careful listing of the relevant given information and a precise identification of the unknown are often the keys to success in solving problems.

EXAMPLE 22   **Motion Problem.**   A man leaves town A by car and travels at a constant speed of 40 miles per hour toward town B, which is 100 miles away. One hour later a woman leaves town B and travels on the same highway toward town A at a constant speed of 50 miles per hour. Find how much time elapses before they meet.

Solution   Let

$$t = \text{Number of hours elapsed from time man leaves until they meet}$$

Then

$$t - 1 = \text{Number of hours woman travels}$$

We use the fundamental relationship

$$\text{Distance} = (\text{Average rate})(\text{Time}) \qquad \text{or} \qquad d = rt$$

to determine the distances for the man and woman.

$$40t = \text{Distance man travels before they meet}$$

$$50(t - 1) = \text{Distance woman travels before they meet}$$

Since the total distance is 100, we see from Figure 1 that

$$40t + 50(t - 1) = 100$$
$$40t + 50t - 50 = 100$$
$$90t = 150$$
$$t = \tfrac{150}{90} = \tfrac{5}{3}$$

So the man travels $1\frac{2}{3}$ hours, or 1 hour and 40 minutes.

Figure 1

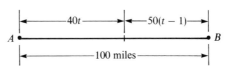

EXAMPLE 23   **Geometry Problem.**   A flower bed is in the shape of a rectangle. Its length is twice its width. The bed is surrounded by a walkway 4 feet wide. If the area of the walk is exactly twice the area of the bed, find the dimensions of the bed.

Solution   In problems such as this a sketch such as Figure 2 is definitely helpful.

Figure 2

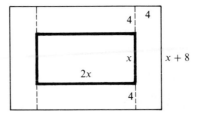

Let

$$x = \text{Width of bed in feet}$$

Then
$$2x = \text{Length of bed in feet}$$
$$\text{Area of bed} = (\text{Length})(\text{Width}) = (2x)(x) = 2x^2$$
$$\text{Area of walk} = 2 \cdot 4(x + 8) + 2 \cdot 4(2x)$$
$$= 8x + 64 + 16x$$
$$= 24x + 64$$

(This is obtained by dividing the walk into rectangles as shown in Figure 2.) Therefore,
$$24x + 64 = 2(2x^2)$$

or, on dividing both sides by 4 and rearranging,
$$x^2 - 6x - 16 = 0$$
$$(x - 8)(x + 2) = 0$$
$$x - 8 = 0 \quad \bigg| \quad x + 2 = 0$$
$$x = 8 \quad \bigg| \quad x = -2$$

Clearly, x cannot be negative, so the width of the bed is 8 feet and the length is $2x = 16$ feet.

EXAMPLE 24

Mixing Problem.   A large tank contains 100 gallons of pure water. How many gallons of a saline solution containing 25% salt must be added to obtain a solution that is 10% salt?

Solution Let
$$x = \text{Number of gallons of 25\% solution that must be added}$$
Then
$$0.25x = \text{Number of gallons of salt added}$$
$$100 + x = \text{Total number of gallons of final solution}$$
$$0.10(100 + x) = \text{Number of gallons of salt in final solution}$$

Since the salt added is the only source of salt, it follows that the amount of salt added equals the amount of salt in the final solution. So we have the equation
$$0.25x = 0.10(100 + x)$$

Decimals can be eliminated by multiplying both sides by 100.
$$25x = 10(100 + x)$$
$$25x = 1{,}000 + 10x$$
$$15x = 1{,}000$$
$$x = \frac{1{,}000}{15} = \frac{200}{3}$$

So $66\frac{2}{3}$ gallons of the mixture must be added.

EXAMPLE 25   **Mixing Problem.** A solution containing 80% sulfuric acid is to be mixed with a 65% solution to obtain a 75% solution. If 10 gallons of the final solution are desired, how many gallons of each of the original solutions should be used?

Solution   Let

$$x = \text{Number of gallons of 80\% solution}$$

Then

$$10 - x = \text{Number of gallons of 65\% solution}$$

$$0.80x = \text{Number of gallons of pure sulfuric acid in first solution}$$

$$0.65(10 - x) = \text{Number of gallons of pure sulfuric acid in second solution}$$

$$0.75(10) = \text{Number of gallons of pure sulfuric acid in final solution}$$

So, since the amount of pure sulfuric acid in the first solution plus the amount in the second solution must equal the amount in the final solution, we have

$$0.80x + 0.65(10 - x) = 0.75(10)$$
$$80x + 65(10 - x) = 750$$
$$80x + 650 - 65x = 750$$
$$15x = 100$$
$$x = \frac{100}{15} = \frac{20}{3}$$

And

$$10 - x = 10 - \frac{20}{3} = \frac{10}{3}$$

Thus, $6\frac{2}{3}$ gallons of the 80% solution and $3\frac{1}{3}$ gallons of the 65% solution should be used.

The basic idea in mixture problems is:

$$\begin{pmatrix} \textbf{Amount of pure} \\ \textbf{substance in the} \\ \textbf{first solution} \end{pmatrix} + \begin{pmatrix} \textbf{Amount of pure} \\ \textbf{substance in the} \\ \textbf{second solution} \end{pmatrix} = \begin{pmatrix} \textbf{Amount of pure} \\ \textbf{substance in the} \\ \textbf{final mixture} \end{pmatrix}$$

EXAMPLE 26   **Falling Body Problem.** In calculus it is proved that if air resistance is not considered, the distance s above the earth of a falling body after t seconds is given by the law

$$s = -\tfrac{1}{2}gt^2 + v_0 t + s_0$$

where g is the acceleration due to gravity (approximately 32 feet per second per second), v_0 is the initial velocity of the object, and s_0 is the initial distance above the earth. Here, s is measured positively upward from ground level. (This accounts for the negative sign on the first term, since acceleration is directed downward.) If a ball is thrown upward from the top edge of a 128 foot building with a velocity of 32 feet per second, find how long it will be before the ball strikes the ground.

Solution   In this problem the unknown t has already been introduced, and in fact the equation relating t to known information is given. All that remains is to interpret the data given in the problem in terms of the constants in the equation of motion.

It is evident that $v_0 = 32$ and $s_0 = 128$. Furthermore, at the time the ball strikes the ground $s = 0$; so we have

$$0 = -\tfrac{1}{2}(32)t^2 + 32t + 128$$
$$16t^2 - 32t - 128 = 0$$
$$t^2 - 2t - 8 = 0$$
$$(t - 4)(t + 2) = 0$$
$$t = 4 \quad | \quad t = -2$$

Clearly, $t = 4$ is the only admissible solution. So the ball will strike the ground after 4 seconds.

EXAMPLE 27

Economics Problem. A theater has an average daily attendance of 400 with the current ticket price of $2.00. It is estimated that for each 10¢ decrease in the ticket price the average attendance will increase by 40 persons. What price should be charged to increase gross income by $100? What will be the new average attendance?

Solution

This is a situation in which it is best to use as the unknown something related to the ticket price rather than the ticket price itself.

Let

$$x = \text{Number of 10¢ reductions in price needed to achieve desired increase in income}$$

Then

$$2.00 - 0.10x = \text{Price of each ticket in dollars}$$
$$40x = \text{Increase in average attendance}$$
$$400 + 40x = \text{New average attendance}$$
$$\text{Income in dollars} = \text{Price of each ticket in dollars times number attending}$$
$$\text{Present income in dollars} = (2.00)(400) = 800$$
$$\text{Income desired in dollars} = 800 + 100 = 900$$

So,

$$900 = (2.00 - 0.10x)(400 + 40x)$$
$$900 = 800 + 40x - 4x^2$$
$$4x^2 - 40x + 100 = 0$$
$$x^2 - 10x + 25 = 0$$
$$(x - 5)^2 = 0$$
$$x = 5$$

So the ticket price should be reduced by 5 increments of 10¢ for a new price of $1.50 per ticket. The new average attendance will be $400 + (40)(5) = 600$. (It can be shown by calculus that this is the optimum price, that is, any other price, either higher or lower, would produce less income under the stated assumptions.)

EXAMPLE 28   **Motion Problem.**   An airplane flies from city A to city B in 1 hour and 40 minutes against a headwind of 80 miles per hour. The return trip, with no shift in the wind, requires 1 hour. The pilot flew at the same indicated airspeed going and coming. Find his indicated airspeed.

Note.   The indicated airspeed is the speed the plane would move relative to the ground if there were no wind.

Solution   Let

$$v = \text{Indicated airspeed in miles per hour}$$

Then

$$v - 80 = \text{Actual speed going}$$
$$v + 80 = \text{Actual speed returning}$$
$$\text{Time going} = 1 \text{ hour } 40 \text{ minutes} = \tfrac{5}{3} \text{ hours}$$
$$\text{Time returning} = 1 \text{ hour}$$
$$\text{Distance} = (\text{Rate})(\text{Time})$$
$$\text{Distance from } A \text{ to } B = (v - 80)(\tfrac{5}{3})$$
$$\text{Distance from } B \text{ to } A = (v + 80)(1)$$

These distances are equal, so

$$(v - 80)\tfrac{5}{3} = (v + 80)(1)$$
$$5v - 400 = 3v + 240$$
$$2v = 640$$
$$v = 320$$

The indicated airspeed is 320 miles per hour.

EXAMPLE 29   **Work Problem.**   A pump can fill a reservoir in 12 days. A second pump, operating independently, can fill the same reservoir in 8 days. How long will it take to fill the reservoir if both pumps operate simultaneously?

Solution   Let

$x = $ Number of days required to fill the reservoir with both pumps in operation

$\frac{1}{12} = $ Fractional part of the job first pump does per day

$\frac{1}{8} = $ Fractional part of the job second pump does per day

So the fractional part of the job done by first pump in x days $= x/12$, and the fractional part of the job done by second pump in x days $= x/8$. Working together they do the entire job.

$$\frac{x}{12} + \frac{x}{8} = 1$$
$$8x + 12x = 96$$
$$20x = 96$$
$$x = 4.8$$

Thus, 4.8 days would be required to fill the reservoir with both pumps in operation.

EXAMPLE 30   **Economics Problem.**   A vendor purchased a shipment of lamps for a total cost of $1,000. Company clerks damaged 4 of the lamps so that they could not be sold. The remaining lamps were sold at a profit of $25.00 each, and a total profit of $200 was realized when all the lamps were sold. How many lamps were purchased?

Solution   Let

$$x = \text{Total number of lamps purchased}$$

Then

$$x - 4 = \text{Number of lamps sold}$$

$$\frac{1,000}{x} = \text{Cost of each lamp in dollars}$$

$$\frac{1,200}{x - 4} = \text{Selling price of each lamp in dollars}$$

(Since total profit was $200, total selling price was $1,200.)

$$\text{Profit per lamp} = (\text{Selling price per lamp}) - (\text{Cost per lamp})$$
$$25 = \frac{1,200}{x - 4} - \frac{1,000}{x}$$
$$25(x - 4)x = 1,200x - 1,000(x - 4)$$
$$x^2 - 4x = 48x - 40x + 160 \qquad \text{Dividing by 25}$$
$$x^2 - 12x - 160 = 0$$
$$(x - 20)(x + 8) = 0$$
$$x = 20 \quad | \quad x = -8$$

So 20 lamps were purchased.

In Example 30 we used the following fundamental property of business:

Profit = (Selling price) − (Cost)

This applies to all problems involving the sale of manufactured or purchased items.

It should be evident to the reader that the degree of practicality of these problems varies considerably. Some are designed solely for the purpose of giving practice in translating from a verbal statement to a mathematical one; others accomplish this purpose and at the same time show useful applications of algebra. The latter would seem to be more desirable, but sometimes efforts to achieve relevance to real-world situations result in such complication as to obscure the basic intention. Furthermore, without the use of calculus we are limited in the degree of practicality that can be achieved. Nevertheless, these problems and those that follow, however contrived some may seem, develop facility in setting up and solving mathematical models, which will prove useful in more complicated situations encountered later.

EXERCISE SET 6

A

1. Train A leaves Washington for New York and goes at an average speed of 60 miles per hour. On a parallel track, train B (the Metroliner) leaves Washington a half hour later and travels at an average speed of 90 miles per hour. At what point will train B pass train A?

2. Two cars leave from the same point and go in opposite directions on a straight road. The average speed of one car is 8 miles per hour greater than the other. After 2 hours the cars are 196 miles apart. Find the average speed of each car.

3. A cyclist and a jogger leave at the same time and follow the same route for a distance of 4 kilometers. The cyclist arrives 15 minutes before the jogger. If the cyclist goes twice as fast as the jogger, find the speed of the jogger.

4. A girl walked from her home into town at the rate of 4 miles per hour. She decided to return by bus, over the same route. The bus averaged 20 miles per hour, and the entire trip took 1 hour and 40 minutes, including a 10 minute wait for the bus. How far did the girl walk?

5. A field is in the shape of a rectangle twice as long as it is wide. A fence costing $2.50 per foot is to be placed around the field, and another fence costing $1.00 per foot is to be placed across the middle so as to divide the area into two squares. If the total cost of the fencing is $584, find the dimensions of the field.

6. The length of a rectangular flower bed is 10 feet greater than its width. The bed is surrounded by a walk that is 2 feet wide. The overall area of the bed and walk is 231 square feet. Find the dimensions of the bed.

7. The length of a room is 3 feet less than twice its width, and the perimeter is 72 feet. Find its dimensions.

8. A pasture is bounded by a river on one side. The farmer has 400 feet of fencing for enclosing a rectangular area of 15,000 square feet. Only three sides will be fenced, using the river as the boundary for the fourth side. There are two ways to do this. Find them.

9. Find three consecutive odd numbers whose sum is 375.

10. A student has grades of 63, 82, and 90 on the first three exams in a math course. What grade does the student need on the fourth exam to obtain an average of 80?

11. A student's average grade on three exams is 78. What would the student need to make on the fourth exam to bring the average grade up to 80?

12. A boy in a town 32 miles from his home got a ride for all but the last 2 miles, which he had to walk. The average speed of the car was 40 miles per hour, and the whole trip took 1 hour and 15 minutes. Find his rate of walking.

13. At a baseball game the gross receipts were $63,943 from 22,796 paid admissions. If grandstand seats cost $3.50 and bleacher seats cost $2.00, how many of each kind of ticket were sold?

14. Admission to an amusement park is $5.00 for adults and $2.50 for children. The total receipts on a certain day were $17,500 for 5,000 paid admissions. How many tickets were sold to adults and how many to children?

15. By radar a highway patrolman observes a speeder going 80 miles per hour. The patrolman has difficulty starting his car so that 5 minutes elapse before he is able to begin pursuit. It is 20 miles to the state line. What average speed would the patrolman have to go to overtake the speeder before he crosses the state line?

16. Coffee costing $3.50 per pound is to be mixed with coffee costing $4.25 per pound to obtain 30 pounds of a blend worth $4.00 per pound. How much of each type should be used?

17. How many cubic centimeters of a 50% alcohol solution must be added to 200 cubic centimeters of a mixture that is 25% alcohol to obtain a 30% alcohol solution?

18. A dairy wishes to obtain 200 gallons of milk that is 3.8% butterfat by mixing lowfat milk, which is 2% butterfat, with half-and-half, which is 11% butterfat. How much of each should be used?

19. From a helicopter hovering 6,400 feet above the ground a projectile is fired straight down with an initial velocity of 1,200 feet per second. When will the projectile strike the ground? (See Example 26.)

20. A room has an area of 180 square feet. If the length is 3 feet more than the width, find its dimensions.

21. A lady paid $87.50 for a dress that was reduced by 30%. What was the original price of the dress?

22. A 10 gallon container is filled with a salt solution containing 20% salt. How many gallons should be drained and replaced by pure water to reduce the salinity to 15%?

23. A woman invests $10,000 in two accounts, one an ordinary savings account yielding $5\frac{1}{2}$% annual interest and the other a certificate of deposit yielding 7% annual interest. At the end of 1 year the combined interest is $652. How much did she invest in each account?

24. An open-top box is to be made by cutting squares from the corners of a flat 10 × 12 inch piece of cardboard and folding up the sides (see sketch). What size square should be removed to produce a box having a total surface area of 95 square inches?

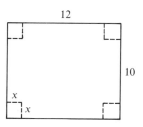

25. A farmer can buy feeder calves weighing 530 pounds on the average for $150 each. If it costs an average of 30¢ per pound added to fatten a calf, and they can be sold at an average of 42¢ per pound, how much weight would have to be added to each calf on the average to realize a profit of $150 per animal?

26. A group of 40 people charter a bus for an excursion. They are told that if they can get 15 additional people to go, the price of each ticket will be reduced by $7.50 (while the total cost for chartering the bus remains the same). Find what the cost of each ticket will be if they get the additional 15 passengers. What will it be if only the original 40 people go?

27. A plumber charges $9.00 per hour, and his assistant, who is an apprentice, receives $5.00 per hour. The bill for a certain job was $81.00. This included $25.00 for parts and $12.00 for making the house call. If the assistant worked 1 hour less than the plumber, find how long each worked.

28. Find two consecutive positive odd integers whose product is 195.

29. The sum of the reciprocals of two consecutive even integers is $\frac{5}{12}$. Find the numbers.

30. The sum of the reciprocals of two consecutive even integers is $\frac{13}{84}$. Find the numbers.

31. Forty people charter a boat at a cost of $15 each. They are told that for every additional person they can get to go, the price of each ticket will be reduced by 25¢, provided the total does not exceed 60, which is the capacity of the boat. The net income to the boat operator is $625. How many people are going to take the boat ride?

32. A bookstore purchased a number of copies of a new book for a total cost of $750. The selling price per book was $2 greater than the cost. At the end of 1 month all but 5 copies had been sold, and the profit was $90. How many books did the bookstore purchase?

33. A picture is 10 inches longer than it is wide. It is framed with a matting 2 inches wide at each end and 3 inches wide at the top and bottom. The area of the picture plus the matting is 468 square inches. Find the dimensions of the picture.

34. A man rows 6 miles down a river and back in 2 hours and 15 minutes. If the rate of the current is 2 miles per hour, find how fast he can row in still water.

B   **35.** A railroad company agrees to run an excursion train for a group under the following conditions. If 200 people or less go, the rate is $10 per ticket. The rate for all tickets will be reduced by 2¢ per ticket for each person in excess of 200. If the total intake by the railroad was $2,450, find how many people took the trip.

36. Machine A can produce parts at the rate of 3,000 per hour, and machine B can produce the same parts at the rate of 5,000 per hour. Machine A is started at the beginning of the day, and 2 hours later machine B is started. How long will it take for 50,000 parts to be produced?

37. A certain city bus line charges 50¢ per ticket and has an average daily ridership of 200,000. The capacity is double this amount, and in order to provide an incentive for people to use public transportation, the company experiments with reducing the fare by 1¢ increments. They find that for each 1¢ fare reduction the number of riders each day increases by an average of 7,500. They also find that up to a certain point their income rises, but further reductions cause the income to decline. Their maximum income exceeds what it was before any fare reduction by $10,200. What fare did they charge to produce this income, and what was the average daily ridership with this fare?

38. It takes a small pump 4 hours longer to fill a tank than a large pump. If both pumps work together, the tank can be filled in 2 hours and 40 minutes. How long will it take each pump working alone to fill the tank?

39. The octane rating of gasoline is a number that measures its antiknock value. The octane rating of a fuel consisting of a mixture of normal heptane (which has a decided knocking tendency) and isooctane (which has a decided antiknocking tendency) is the percentage of isooctane in the mixture. Other fuels are rated in comparison with this mixture. For example, gasoline rated at 90 octane has the same antiknock qualities as a mixture of heptane and isooctane that is 90% isooctane. In working with octane ratings, then, we may treat the fuel as if it were a mixture of isooctane and heptane. If two batches of gasoline, one rated at 96 octane and the other at 87 octane, are to be mixed to obtain 200 gallons of 93 octane gasoline, find how many gallons of each should be used.

40. Fifty gallons of 88 octane gasoline are mixed with 75 gallons of 94 octane gasoline. What is the octane rating of the mixture? (See Problem 39.)

41. One pump can fill a reservoir in 8 days and another can do it in 6 days. If both pumps begin working together but the faster pump breaks down at the end of the third day, how long will it take to fill the reservoir?

42. A girl is in a boat on a lake 4 miles from the point nearest to her on the shore (point A). She wants to get to point B, 9 miles along the straight shore from A. She can row at an average of 6 miles per hour and walk at 4 miles per hour. She decides to row directly to a point C between A and B and walk from there. The total trip takes 2 hours and 20 minutes. How far is it from A to C?

43. Point A is on one bank of a river and B is on the other, directly opposite A. The river is 120 feet wide and is essentially straight at the area in question. An underground telephone cable is to go from A to a point C, which is at the river's edge on the same side as B and 150 feet from B. It costs \$4 per foot to run the cable under the river and \$2 per foot to place it underground. The cable is run under the river from A to a point D between B and C, and from there to C underground. If the total cost is \$720, how far is it from B to D?

44. Bill can mow a large lawn with his rider mower in $3\frac{1}{2}$ hours, and Joanne can do the same job with her hand-operated power mower in 6 hours. How long will it take them working together?

45. A room containing 1,990 cubic feet is originally free of carbon monoxide. Cigarette smoke, containing 2.4% carbon monoxide, is then introduced into the room at the rate of 0.1 cubic feet per minute.
a. Find a general expression for the concentration of carbon monoxide in the room in terms of the elapsed time t in minutes that the smoke has been entering the room.
b. Extended exposure to concentrations of carbon monoxide as low as 0.00012 is harmful to the human body. Find the critical value of t at which this concentration is reached.

7 Linear Inequalities

It may come as a surprise to you that inequalities are used in mathematics almost as much as equations. The only difference in appearance of an inequality and an equation is that the equals sign is replaced by an inequality sign ($<$, $>$, $\leq$, or $\geq$). For example, $2x - 3 > 4$ is a linear inequality in one unknown, and $x^2 - 2x - 3 \leq 0$ is a quadratic inequality in one unknown. By the solution set of an inequality we mean the set of all values of the unknown (or unknowns) for which the inequality holds true. In this section and the next we consider the solution of linear inequalities. Quadratic inequalities will be discussed in Section 9, and inequalities involving more than one unknown will be taken up later, when we discuss systems of inequalities.

The order properties given in Section 3 of Chapter 1 provide the basis for solving inequalities of the first degree in one unknown. Briefly, these state that we can add the same number to, or subtract the same number from, both sides of an inequality. Also, we can multiply or divide by any *positive* number and retain the sense (that is, the direction of the inequality symbol) of the inequality. But multiplying or dividing by a *negative* number reverses the sense of the inequality. This last point requires special care, and this feature distinguishes the solution of linear inequalities from that of linear equations. We illustrate the technique with several examples.

EXAMPLE 31   Solve for x: $2x - 3 < 5x + 6$

Solution   We solve just as if it were an equation, collecting the x's together on one side of the inequality sign and the constants on the other. One way to do this is to add 3 to both sides and also to subtract $5x$, obtaining

$$-3x < 9$$

Now to get x we want to divide by -3 (or multiply by $-\frac{1}{3}$), but remember that dividing by a negative number reverses the sense of the inequality. So we get

$$x > -3$$

We write the solution set as

$$\{x: \quad x > -3\}$$

It consists not of just one value of x but *all* numbers greater than -3. This can be depicted on a number line, as shown in Figure 3.

Figure 3

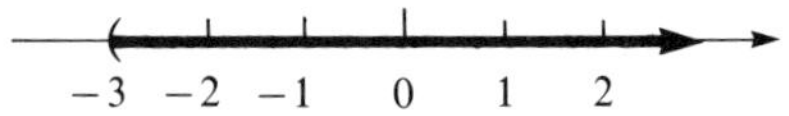

Some people prefer to avoid dividing by negative numbers and so collect terms in such a way as to result in a positive coefficient for x. Thus, we could have added $-2x$ and -6 to each member in this example to get:

$$-9 < 3x$$
$$-3 < x$$

which can be turned around to read $x > -3$.

EXAMPLE 32   Solve the inequality: $\dfrac{2x - 3}{5} \geq \dfrac{x}{2} - 1$

Solution   First, we clear fractions by multiplying by the LCD, which is 10.

$$4x - 6 \geq 5x - 10$$
$$4 \geq x$$
$$x \leq 4$$

So the solution set is $\{x: \quad x \leq 4\}$. This can be depicted graphically as shown in Figure 4.

Figure 4

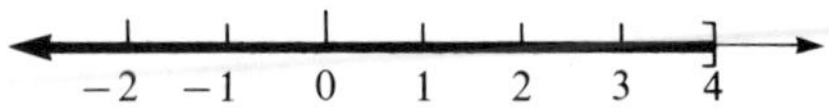

EXAMPLE 33   Find all x such that $-4 < 2x - 3 \leq 5$.

Solution   This is really the combination of two inequalities, $-4 < 2x - 3$ and $2x - 3 \leq 5$, and we are seeking the values of x that satisfy both inequalities at once. We could solve each inequality individually and find those values of x common to the two solution sets (that is, the intersection of the two solution sets). An easier way, however, is to work with the combined inequality as follows. Add 3 to every member and then divide by 2:

$$-1 < 2x \leq 8$$
$$-\tfrac{1}{2} < x \leq 4$$

The solution set is $\{x: \quad -\tfrac{1}{2} < x \leq 4\}$. We can read this from the middle: x is greater than $-\tfrac{1}{2}$ and less than or equal to 4. Graphically, we show this in Figure 5.

Figure 5

EXERCISE SET 7

A   In Problems 1–25 find the solution sets and depict these on a number line.

1. $2x + 3 < 5$

2. $x - 3 < 4$

3. $3x + 4 < x + 8$

4. $3x + 5 < 11$

5. $2x - 3 > 5x + 4$

6. $2x + 1 < 3x - 2$

7. $4 - 2x < 6 - 3x$

8. $2(x + 3) \leq 5$

9. $3(x - 2) > 1$

10. $\dfrac{x + 2}{3} \quad \dfrac{5}{6}$

11. $\dfrac{x - 4}{3} < \dfrac{3x}{4}$

12. $\dfrac{x}{3} - \dfrac{1}{2} \leq \dfrac{3x}{4} + \dfrac{5}{6}$

13. $\dfrac{2x + 5}{4} > \dfrac{4 - 3x}{6}$

14. $\dfrac{x}{2} - \dfrac{3}{4} > \dfrac{5x}{6} - \dfrac{1}{3}$

15. $\dfrac{3x}{2} + \dfrac{1}{4} > \dfrac{5x}{8} - \dfrac{1}{2}$

16. $1 < x - 4 < 3$

17. $3 < 2x - 1 < 5$

18. $-1 < 3x - 4 \leq 7$

19. $-3 < 2x + 1 < 5$

20. $-2 \leq \dfrac{3x - 4}{5} < 3$

21. $0 < \dfrac{3x - 2}{4} \leq 2$

22. $-\tfrac{1}{2} < x - 2 < \tfrac{1}{2}$

23. $0 \leq \dfrac{3 - 2x}{4} \leq 1$

24. $-3 \leq \dfrac{5 - 3x}{2} < 6$

25. $-0.01 < 2x - 3 < 0.01$

26. A student has grades of 72, 83, and 78 on three tests. How high must he score on the fourth test to have an average of 80 or greater?

27. The temperature in degrees Fahrenheit (°F) ranges between 77 and 86 on a certain South Sea island. Celsius temperature (°C) is related to Fahrenheit temperature by the formula $°C = (\tfrac{5}{9})(°F - 32)$. What is the Celsius temperature range on this island?

28.   Apples keep best in cold storage if they are held above freezing but at no greater than 5°C, that is, $0 < °C \le 5$. Find the corresponding range for the Fahrenheit temperature (°F). (See Problem 27.)

29.   The perimeter of a rectangle with length 15 inches must be at least 42 inches but no greater than 60 inches. What is the allowable range of values of the width?

B   In Problems 30 and 31 solve the inequalities.

30. $\dfrac{2x-5}{3} + \dfrac{3-2x}{12} \ge \dfrac{3x}{4}$

31. $3 > \dfrac{2(3-4x)}{7} \ge -2$

32.   Explain why each of the following has no solution:
       a. $5 < x < 3$ **b.** $6 < 2x + 5 < -1$ **c.** $3 > x - 2 > 4$

33.   Find all values of x that satisfy *either* $3x - 2 < 4$ *or* $4 + 3x > x + 10$, and depict this solution graphically. Write the solution set as the union of two sets.

34.   Find the set of all numbers that satisfy either

$$\frac{3-2x}{4} \ge 5$$

or

$$\frac{4-x}{2} < \frac{2x-8}{3}$$

Depict the solution graphically. Write the solution set as the union of two sets.

35.   The cost of manufacturing x microcomputers of a certain kind each week is given by $C = 2,000 + 50x$, and the revenue from selling these is given by $R = 75x$. How many microcomputers must be produced and sold in order to realize a profit?

8   Inequalities and Absolute Value

Inequalities often occur in combination with absolute value, such as $|x - 3| < 1$ or $|x - 2| > 3$. In order to see how to solve these, we first consider some further properties of absolute value.

Recall that when viewed geometrically $|a|$ can be interpreted as the distance between a and 0 on the number line. For example, $|-2| = 2$, and the point corresponding to -2 on the number line is 2 units from the origin. Similarly, $|a - b|$ can be interpreted as the distance between a and b on the number line. You can convince yourself of this by considering several possibilities of a and b on a number line.

The following further properties of absolute value are often useful:

Properties of Absolute Values

1. $|-x| = |x|$ 2. $|xy| = |x||y|$
3. $|x| = \sqrt{x^2}$ 4. $|x + y| \le |x| + |y|$

We will not prove these in detail. You can convince yourself of the truth of properties 1 and 2 by considering the definitions of the quantities involved. For property 3, remember that $\sqrt{N}$, where $N \geq 0$, means the unique *nonnegative* number whose square is N. Since both $x \cdot x = x^2$ and $(-x) \cdot (-x) = x^2$, then $\sqrt{x^2} = x$ when x is nonnegative, and $\sqrt{x^2} = -x$ when x is negative, that is, $\sqrt{x^2} = |x|$. Property 4 is known as the **triangle inequality,** and is probably the most used inequality in all mathematics. Because of its importance, we will prove it. First, though, you should note that it is intuitively evident, because when x and y are both positive or both negative, the two sides are the same, but when x and y are opposite in sign, the left-hand side really involves a subtraction [for example, $5 + (-2)$], whereas the right-hand side always involves addition (for example, $|5| + |-2| = 5 + 2$) and hence is greater than the left.

To prove property 4, observe that $-|x| \leq x \leq |x|$. In fact, this is an extreme inequality in the sense that the quantity in the middle always equals one or the other of the end points. For example, suppose $x = 2$. Then

$$-|2| \leq 2 \leq |2|$$

since this is the same as

$$-2 \leq 2 \leq 2$$

Or suppose $x = -3$. Then

$$-|-3| \leq -3 \leq |-3|$$

since this is the same as

$$-3 \leq -3 \leq 3$$

Now we write a similar inequality for y and add:

$$
\begin{array}{r}
-|x| \leq x \leq |x| \\
-|y| \leq y \leq |y| \\
\hline
-|x| - |y| \leq x + y \leq |x| + |y|
\end{array}
$$

or

$$-(|x| + |y|) \leq x + y \leq |x| + |y| \tag{2}$$

The left-hand inequality can be rewritten after multiplying by -1:

$$-(x + y) \leq |x| + |y|$$

Since, by the right-hand inequality in (2), $(x + y) \leq |x| + |y|$, it follows that both $x + y$ and its negative are always less than or equal to $|x| + |y|$. But $|x + y|$ is either $x + y$ or its negative. So

$$|x + y| \leq |x| + |y|$$

Let us return now to the problems posed at the beginning of this section.

EXAMPLE 34   Find the solution set for the inequality $|x - 3| < 1$.

Solution   Since $|x - 3|$ can be interpreted as the distance between x and 3, we are seeking all numbers x whose distance from 3 is less than 1 unit. These are the numbers between 2 and 4, as shown on the number line in Figure 6. We can also obtain this without reference to geometry. Suppose first that $x > 3$. Then $x - 3$ is positive, and $|x - 3| = x - 3$. So $x - 3 < 1$ when $x < 4$, and all numbers

Figure 6

between 3 and 4 are in the solution set. Next suppose $x < 3$, so that $x - 3$ is negative. Then $|x - 3| = -(x - 3) = 3 - x$, and $3 - x < 1$ if $x > 2$. So all numbers between 2 and 3 are in the solution set. When $x = 3$, then $|x - 3| = 0$, and this is certainly less than 1. We therefore conclude that all numbers between 2 and 4 constitute the solution set, that is, $\{x: \ 2 < x < 4\}$.

EXAMPLE 35   Find the solution set for the inequality $|x - 2| > 3$.

Solution   Again we could reason in terms of distance. We are seeking all numbers whose distance from 2 is greater than 3. These are numbers that are either greater than 5 or else less than -1, as shown in Figure 7. So we can write the solution set as

$$\{x: \ x < -1\} \cup \{x: \ x > 5\}$$

Figure 7

Without reference to geometry, we can solve this inequality by considering two cases. First, if $x \geq 2$, then $|x - 2| = x - 2$. So the inequality becomes $x - 2 > 3$, or $x > 5$. On the other hand, if $x < 2$, then $x - 2$ is negative, so that $|x - 2| = -(x - 2) = 2 - x$, and the inequality becomes

$$2 - x > 3$$
$$x < -1$$

So, again, we see that the inequality is satisfied if either $x > 5$ or $x < -1$.

We wish to be able to solve inequalities like those in Examples 34 and 35 more rapidly, and toward this end we first consider the simple cases $|x| < b$ and $|x| > b$. As we saw in Chapter 1, the solution set to $|x| < b$ is the set of all numbers x whose distance from 0 on the number line is less than b. These are the numbers between $-b$ and b. That is,

$$|x| < b \qquad \text{is equivalent to} \qquad -b < x < b \qquad (3)$$

Similarly, the solution set of the inequality $|x| > b$ is the set of all numbers x whose distance from 0 on the number line is greater than b. These are the numbers that are greater than b, together with the numbers that are less than $-b$. That is,

$$|x| > b \qquad \text{is equivalent to} \qquad x > b \quad \text{or} \quad x < -b \qquad (4)$$

So in the future, if we have to solve an inequality of the form

$$|x - a| < b$$

we replace x by $x - a$ in inequality (3) and solve the equivalent inequality

$$-b < x - a < b$$

by adding a to all three members:

$$a - b < \quad x \quad < a + b$$

For example, let us again consider $|x - 3| < 1$, which is equivalent to

$$-1 < x - 3 < 1$$

We add 3 to all three members to get

$$2 < \quad x \quad < 4$$

So the solution set is $\{x: \quad 2 < x < 4\}$.

Similarly, an inequality of the form $|x - a| > b$ can be solved using inequality (4) and replacing x by $x - a$. The inequality $|x - a| > b$ is therefore equivalent to

$$x - a > b \qquad \text{or} \qquad x - a < -b$$

which, on adding a to both sides of each inequality, gives $x > a + b$ or $x < a - b$. For example, $|x - 2| > 3$ is equivalent to

$$x - 2 > 3 \qquad \text{or} \qquad x - 2 < -3$$

from which we get

$$x > 5 \qquad \text{or} \qquad x < -1$$

So the solution set is $\{x: \quad x > 5\} \cup \{x: \quad x < -1\}$.

There are times when it is helpful to consider various cases in solving a problem involving absolute values. The next example illustrates this.

EXAMPLE 36 Solve the inequality: $|x| + |x - 1| \leq 2$

Solution We divide the number line into three regions, as shown in Figure 8. The division points are those for which either of the terms on the left is 0.

Figure 8

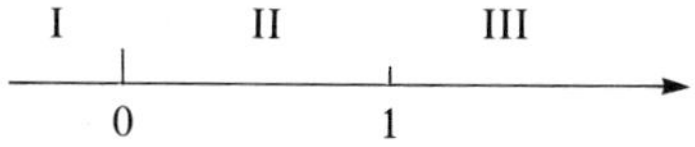

Region I. $x < 0$ and $x - 1 < 0$

$$-x - (x - 1) \leq 2$$
$$-2x \leq 1$$
$$x \geq -\tfrac{1}{2}$$

So all x such that $-\tfrac{1}{2} \leq x < 0$ are in the solution set.

Region II. $x > 0$ and $x - 1 < 0$

$$x - (x - 1) \leq 2$$
$$1 \leq 2$$

So all of region II is in the solution set.

Region III. $x > 0$ and $x - 1 > 0$

$$x + (x - 1) \leq 2$$
$$2x \leq 3$$
$$x \leq \tfrac{3}{2}$$

So all x such that $1 < x \leq \tfrac{3}{2}$ are in the solution set.

Finally, we check the end points $x = 0$ and $x = 1$ and see that both lie in the solution set. So the complete solution set is $\{x: \quad -\tfrac{1}{2} \leq x \leq \tfrac{3}{2}\}$, as shown in Figure 9.

Figure 9

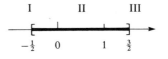

A set of the form $\{x: \quad a < x < b\}$ is called an **open interval** and is sometimes designated as (a, b). A set of the form $\{x: \quad a \leq x \leq b\}$ is called a **closed interval** and may be designated as $[a, b]$. In a closed interval the end points are included, but in an open interval they are not.

EXAMPLE 37   Illustrate on a number line the intervals $(-2, 1)$ and $[3, 4]$.

Solution   The interval $(-2, 1)$ is the set $\{x: \quad -2 < x < 1\}$, and $[3, 4]$ is the set $\{x: 3 \leq x \leq 4\}$. These are shown in Figure 10.

Figure 10

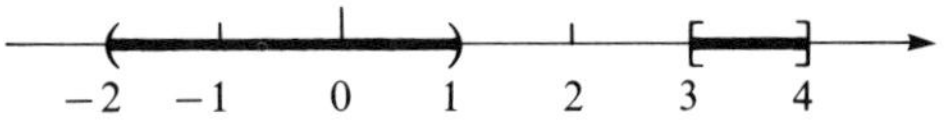

An interval that includes only one of the end points is called **half-open** (or **half-closed**). There are two types:

$$[a, b) = \{x: \quad a \leq x < b\} \qquad \text{and} \qquad (a, b] = \{x: \quad a < x \leq b\}$$

By a **neighborhood** of a real number a is meant an open interval centered on a.* So if r designates any positive number, then $(a - r, a + r)$ is a neighborhood of a; r is called the **radius** of the neighborhood. Since the set of all x's satisfying $|x - a| < r$ is the same as those satisfying $-r < x - a < r$, and hence also of $a - r < x < a + r$, we see that this is another description of the neighborhood of a having radius r. To have a better intuitive feel for this, recall that $|x - a|$ can be interpreted as the distance between x and a, and so to require that x satisfy $|x - a| < r$ is to require that x be less than r units from a to either side.

---

* This is sometimes called a **symmetric neighborhood**, and any open interval with a in its interior is called a neighborhood. However, it is sufficient to work only with symmetric neighborhoods.

If we consider the set of all such numbers x, this describes the entire open interval $(a - r, a + r)$, that is, the neighborhood of a of radius r.

Sometimes we wish to speak of a **deleted neighborhood** (or **punctured neighborhood**) of a. This means a neighborhood of a with a itself removed. The inequality

$$0 < |x - a| < r$$

describes such a deleted neighborhood. The fact that $|x - a| > 0$ guarantees that $x \neq a$, and $|x - a| < r$ guarantees that x is less than r units from a.

Consider, for example, the set of all x's satisfying

$$0 < |x - 2| < 1$$

We show this on a number line in Figure 11. It is a deleted neighborhood of 2 having radius 1, and so it extends 1 unit to the left and right of 2, deleting 2 itself.

Figure 11

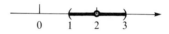

When we wish to emphasize the radius of a neighborhood, we use the designations **r-neighborhood** or **deleted r-neighborhood,** where r is the radius. In calculus, because of tradition, the two Greek letters ε and δ are frequently used to designate radii of neighborhoods. So we say, for example, "ε-neighborhood" or "deleted δ-neighborhood." The concept of neighborhood is very useful in the study of limits and continuity, which are fundamental ideas in calculus.

EXERCISE SET 8

A   **1.** Verify the four absolute value properties for each of the following:
  a. $x = 7, \quad y = 3$ **b.** $x = -12, \quad y = 5$
  c. $x = -4, \quad y = 10$ **d.** $x = -5, \quad y = -7$

2. Verify on a number line that the distance between a and b is $|a - b|$ for each of the following:
  a. $a = 3, \quad b = 8$ **b.** $a = 6, \quad b = -2$
  c. $a = -8, \quad b = 2$ **d.** $a = -3, \quad b = -10$

3. Show that $|a|^2 = a^2$. Is it also true that $|a|^3 = a^3$? Explain why or why not.

Solve the inequalities in Problems 4–25.

4. $|x - 2| < 3$ **5.** $|x - 4| < 7$
6. $|x + 5| < 8$ **7.** $|x + 1| < \frac{1}{2}$
8. $|x - 1| > 3$ **9.** $|x - 2| > 1$
10. $|x + 3| \geq 2$ **11.** $|3x - 2| \leq 4$
12. $|4 - x| > 3$ **13.** $|2x - 3| > 4$
14. $|3 - 2x| \geq 2$ **15.** $|4x - 5| > 3$
16. $|x - a| < 5$ **17.** $|x - b| \geq 3$
   (a is a constant)   (b is a constant)

18. $\left|\dfrac{x + 2}{3}\right| < 4$ **19.** $\left|\dfrac{2x - 5}{3}\right| \leq 1$

20. $\left|\dfrac{3-x}{4}\right| > 2$

21. $\left|\dfrac{x-5}{3}\right| < 6$

22. $\left|\dfrac{2-3x}{4}\right| \geq 3$

23. $|2 - x| < 0.01$

24. $|-3(2x - 4)| < 7$

25. $|-4(3 - 2x)| > 3$

26. Write out the meaning in set notation of each of the following intervals:
    a. $[2, 5]$ **b.** $(-1, 3]$ **c.** $(2, 4)$ **d.** $[-4, -1)$

27. Write in terms of an inequality involving absolute values:
    a. x lies in a neighborhood of 2 of radius 1.
    b. The distance between x and 2 is less than 1.
    c. x lies in a deleted neighborhood of 2 of radius 1.
    d. x is closer than 1 unit to 2 but is not equal to 2.
    e. x lies in a δ-neighborhood of 2.

28. Describe in terms of neighborhood and distance what the following inequalities say:
    a. $|x - 3| < 0.05$ **b.** $0 < |x - 3| < 0.05$
    c. $|x - 3| < \delta$ **d.** $0 < |x - 3| < \delta$

B 29. Show that $|a - b| \leq |a| + |b|$.
       Hint. Write $a - b = a + (-b)$.

30. Show that $|a^{-1}| = |a|^{-1}$.
    Hint. Show that $|a^{-1}|$ fulfills the requirement of being the multiplicative inverse of $|a|$.

31. Use the result of Problem 30 to show that $|a/b| = |a|/|b|$.

32. Show that $|a + b + c| \leq |a| + |b| + |c|$.
    Hint. Write $a + b + c = (a + b) + c$.

33. **a.** In the triangle inequality, let $x = a$ and $y = b - a$ to obtain

$$|b - a| \geq |b| - |a|$$

b. Similarly, let $x = b$ and $y = a - b$ to obtain

$$|a - b| \geq |a| - |b|$$

c. Use the results of parts a and b to obtain the inequality

$$|a - b| \geq ||a| - |b||$$

Solve the inequalities in Problems 34–41.

34. $|x - a| < \delta$
    (a and δ are constants)

35. $0 < |x - 2| < 1$

36. $0 < |2x + 3| \leq 5$

37. $0 < |x - 3| < \delta$
    (δ is a constant)

38. $|x - 1| + |x + 2| \leq 5$

39. $|x| + |2x - 1| \geq 2$

40. $|3x - 2| - |4 - x| > 4$

41. $|x - 2| - |x + 1| < 3$

---

9 Quadratic and Other Nonlinear Inequalities

Quadratic inequalities can be solved by several methods. We illustrate two here, and in Chapter 5 we will present a third method, employing graphs.

Consider the problem $x^2 - 3x - 4 < 0$. Just as with quadratic equations, it is essential that one side of the inequality be 0. In both methods we factor the expression if possible. In this case we get $(x - 4)(x + 1) < 0$.

Method 1. Since $(x - 4)(x + 1) < 0$, it follows that the factors $(x - 4)$ and $(x + 1)$ must be *unlike* in sign. So either $x - 4 > 0$ *and* $x + 1 < 0$, or else $x - 4 < 0$ *and* $x + 1 > 0$. We see that the first condition is impossible, since $x - 4 > 0$ implies $x > 4$, whereas $x + 1 < 0$ implies $x < -1$. There is no value of x that satisfies both conditions at once (the intersection of the two solution sets is the empty set). The second condition requires that $x < 4$ and $x > -1$, which is possible. Thus, the solution is $\{x: \quad -1 < x < 4\}$.

Method 2. We show on the number line in Figure 12 the points for which the left-hand side of the inequality equals 0. These determine three regions to be tested: region I, $x < -1$; region II, $-1 < x < 4$; region III, $x > 4$. We check each region to see if the inequality is satisfied there. This is best done by observing the sign of each factor and then the sign of the product. In region I, the factor $(x - 4)$ is negative and $(x + 1)$ is negative, so the product is positive. Thus, region I is not a part of the solution set. In region II, $(x - 4)$ is negative and $(x + 1)$ is positive, so the product is negative, and region II is in the solution set. In region III, both factors are positive and hence also their product is positive, so region III is not in the solution set.

Figure 12

The solution set is, therefore, all x for which $-1 < x < 4$. Note the correspondence between the solution set as depicted on the number line in Figure 13 and the way it is written in terms of inequalities. The 4 and -1 appear on the same sides of x as shown on the number line, and the two inequality signs are "less than" when read from left to right. (The answer could have been written $4 > x > -1$, but it would not have this convenient relationship with the picture.)

Figure 13

The second method of solution has the advantage that it can be extended to any inequality in which one side is 0 and the other can be written as a product and/or quotient of linear factors. Points are marked off on a number line corresponding to values of the unknown for which **either the numerator or the denominator is 0.** These are called **critical numbers.** They divide the number line into regions that are then tested for sign to see if the inequality is satisfied there or if it is not satisfied. The critical numbers themselves also have to be tested in order to see if they do or do not belong to the solution set.

Remark. In testing the sign of the given expression in a region, it is sufficient to substitute any one value of x that falls in that region. The sign of each factor can then be determined and shown with a $+$ or a $-$ sign. If the number of minus signs is *even*, the resulting sign is $+$ for the entire region, and the expression is greater than 0 there. If the number of minus signs is *odd*, the resulting sign is $-$, and the expression is less than 0 for the entire region.

EXAMPLE 38   Solve the inequality: $x^2 - 2x > 8$

Solution   First add -8 to both sides to get 0 on the right, and factor the quadratic expression on the left:

$$x^2 - 2x - 8 > 0$$

$$(x - 4)(x + 2) > 0$$

The critical numbers are $x = 4$ and $x = -2$. Since for these values of x the left side is 0, the inequality is not satisfied for either value. The regions into which these divide the number line, and the signs of the factors in each region are shown in Figure 14. In regions I and III, the product is positive (>0), so the inequality is satisfied for all x values in either region. In region II, the product is negative, so the given inequality is not satisfied there. Hence, the solution set is $\{x: \ x < -2\} \cup \{x: \ x > 4\}$.

Figure 14

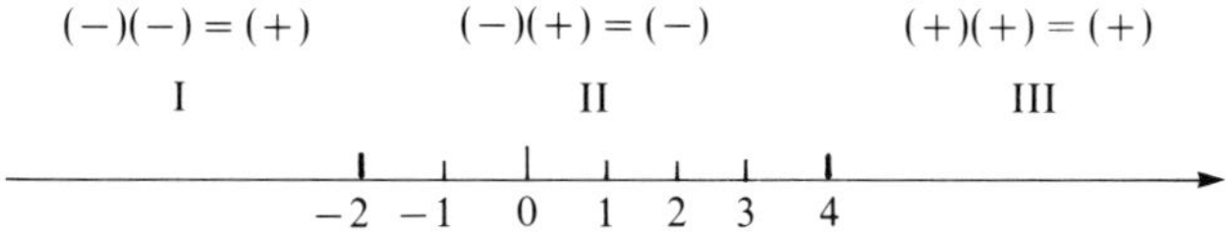

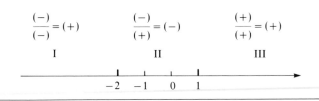

$$(-)(-) = (+) \qquad (-)(+) = (-) \qquad (+)(+) = (+)$$

$$\text{I} \qquad\qquad \text{II} \qquad\qquad \text{III}$$

$$-2 \ -1 \quad 0 \quad 1 \quad 2 \quad 3 \quad 4$$

EXAMPLE 39   Solve the inequality: $\dfrac{x - 1}{x + 2} < 0$

Solution   The critical numbers are $x = 1$ and $x = -2$. These divide the number line into three regions, as shown in Figure 15. For any value of x in region I, both the numerator and denominator are negative, so the quotient is positive (>0). Therefore, region I is not a part of the solution set. In region II, the numerator is negative and the denominator is positive, so the quotient is less than 0. Therefore, region II is a part of the solution set. In region III, the numerator and denominator are both positive, so the quotient is positive, and region III is not in the solution set. Finally, we test the critical numbers themselves. For $x = -2$, the expression on the left is not defined, so $x = -2$ is not in the solution set. When $x = 1$, the expression is 0, so it is not less than 0, and $x = 1$ is not a part of the solution set. It follows, then, that the complete solution set is region II, that is, the interval $(-2, 1)$, or $\{x: \ -2 < x < 1\}$.

Figure 15

$$\frac{(-)}{(-)} = (+) \qquad \frac{(-)}{(+)} = (-) \qquad \frac{(+)}{(+)} = (+)$$

$$\text{I} \qquad\qquad \text{II} \qquad\qquad \text{III}$$

$$-2 \quad -1 \quad 0 \quad 1$$

Note.   Since the solution set to an inequality typically involves one or more intervals, we may use the notation for intervals introduced in Section 8 to simplify the writing of the answer. A set of the form $\{x: \ x > a\}$ is called an **unbounded interval** and is designated (a, ∞). The symbol ∞ is read "infinity," and can be roughly translated "beyond all bound." Similarly, $(-\infty, a)$ is the set $\{x: \ x < a\}$, $[a, \infty)$ is the set $\{x: \ x \geq a\}$, and $(-\infty, a]$ is the set $\{x: \ x \leq a\}$. Since ∞ is not a number, the interval is never closed at the unbounded end. Using this notation, the solution to Example 38 can be written $(-\infty, -2) \cup (4, \infty)$.

EXAMPLE 40   Solve the inequality: $\dfrac{3x - 4}{x + 2} \leq 1$

Solution   We cannot simply clear of fractions by multiplying by the LCD as in an equation, because since the LCD involves the unknown x, we do not know if this LCD is positive or negative—and in multiplying both sides of an inequality, it is essential to know the sign of the multiplier. We can, however, combine the terms into one fraction. The first step is to add -1 to both sides, since for our procedure to work, we must have 0 on one side of the inequality:

$$\frac{3x - 4}{x + 2} - 1 \leq 0$$

Then we combine the left-hand side into one fraction:

$$\frac{3x - 4 - (x + 2)}{x + 2} \leq 0$$

$$\frac{2x - 6}{x + 2} \leq 0$$

$$\frac{2(x - 3)}{x + 2} \leq 0$$

The critical numbers are $x = 3$ and $x = -2$, as indicated in Figure 16. We observe at the outset that $x = 3$ is a part of the solution set, and $x = -2$ is not. Since region II is the only one for which the inequality is satisfied, it follows that the complete solution set is the interval $(-2, 3]$.

Figure 16

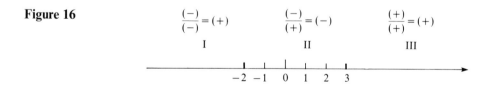

EXAMPLE 41   Solve the inequality: $\dfrac{3}{x} - \dfrac{2x}{x + 2} + 1 < 0$

Solution   As in Example 40, we combine into one fraction:

$$\frac{3(x + 2) - 2x^2 + x(x + 2)}{x(x + 2)} < 0$$

$$\frac{-x^2 + 5x + 6}{x(x + 2)} < 0$$

We multiply both sides by -1, which reverses the sense of the inequality, and then factor the numerator:

$$\frac{x^2 - 5x - 6}{x(x + 2)} > 0$$

$$\frac{(x - 6)(x + 1)}{x(x + 2)} > 0$$

The critical numbers are $x = 6$, $x = -1$, $x = 0$, and $x = -2$, none of which lies in the solution set. The regions that form the solution set are I, III, and V, as shown in Figure 17. So we have the solution $(-\infty, -2) \cup (-1, 0) \cup (6, \infty)$.

Figure 17

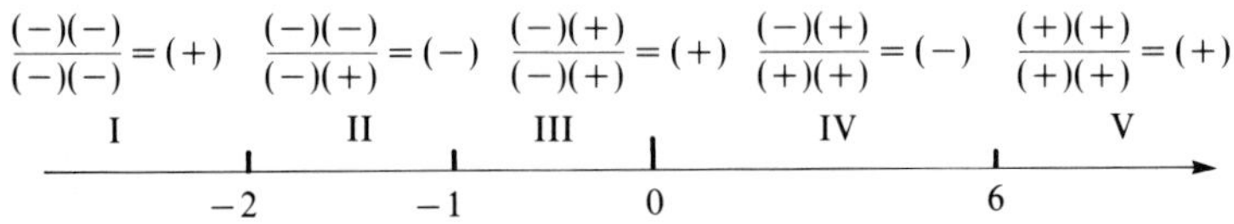

$$\frac{(-)(-)}{(-)(-)} = (+) \quad \frac{(-)(-)}{(-)(+)} = (-) \quad \frac{(-)(+)}{(-)(+)} = (+) \quad \frac{(-)(+)}{(+)(+)} = (-) \quad \frac{(+)(+)}{(+)(+)} = (+)$$

I   II   III   IV   V

-2 -1 0 6

EXAMPLE 42 Find the set of values of x for which the following expression has real values:

$$\sqrt{\frac{x(x - 2)}{x + 3}}$$

Solution The expression will be real provided

$$\frac{x(x - 2)}{x + 3} \geq 0$$

so we wish to solve this inequality. The critical numbers are $x = 0$, $x = 2$, and $x = -3$, as indicated in Figure 18. We note that $x = 0$ and $x = 2$ lie in the solution set, but $x = -3$ does not. Regions II and IV are parts of the solution set, so the complete solution set is $(-3, 0] \cup [2, \infty)$.

Figure 18

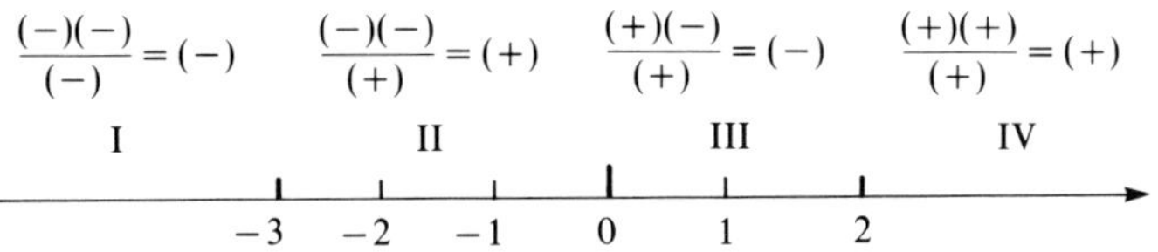

$$\frac{(-)(-)}{(-)} = (-) \quad \frac{(-)(-)}{(+)} = (+) \quad \frac{(+)(-)}{(+)} = (-) \quad \frac{(+)(+)}{(+)} = (+)$$

I   II   III   IV

-3 -2 -1 0 1 2

EXERCISE SET 9

A Solve Problems 1–6 by method 1 of this section.

1. $x^2 - 2x - 3 < 0$
2. $x^2 - x - 6 > 0$
3. $x^2 - 3x - 4 \leq 0$
4. $2x^2 - 9x + 4$
5. $x^2 - 4 < 0$
6. $x^2 - 2x < 8$

Solve Problem 7–34 by method 2 of this section.

7. $x^2 + x - 12 < 0$
8. $2x^2 - 3x - 5 > 0$
9. $3x^2 - x - 2 \leq 0$
10. $12x^2 - 7x - 10 \geq 0$

11. $x(x-1) > 2$

12. $x(2x-3) < 5$

13. $3x^2 \geq 4(1-x)$

14. $6(x^2+3) \leq 31x$

15. $\dfrac{x+2}{x-3} < 0$

16. $\dfrac{x}{x+4} > 0$

17. $\dfrac{3-x}{x+2} < 0$

18. $(x-1)(x+2)(x-4) \leq 0$

19. $(2x-1)(x+4)(x-3) \geq 0$

20. $\dfrac{x^2-16}{x+1} < 0$

21. $\dfrac{x^2-9}{(x+1)^2} < 0$

22. $\dfrac{x^2-1}{x^2-9} > 0$

23. $\dfrac{x^2-2x-3}{x+5} < 0$

24. $\dfrac{x-2}{x^2-2x-8} \geq 0$

25. $\dfrac{x^2-x-6}{2x^2-7x-15} < 0$

26. $\dfrac{2x^2-3x-2}{x^2+x-12} \leq 0$

27. $\dfrac{3x+1}{x-2} > 2$

28. $\dfrac{2x-3}{x-1} \leq 4$

29. $\dfrac{x+2}{2x-3} < 1$

30. $\dfrac{x+1}{x-2} < \dfrac{x+3}{x-4}$

31. $\dfrac{3}{x} + \dfrac{2x}{x-2} < 1$

32. $\dfrac{x}{x-1} + \dfrac{2}{x+2} < 1$

33. $1 + \dfrac{x}{x+1} < \dfrac{9}{x^2-2x-3}$

34. $\dfrac{(x-2)(x+3)^2}{(x-6)(x+1)} \geq 0$

B 35. Find all values of k for which the equation below will have real solutions.

$$kx^2 - 3kx + k - 2 = 0$$

36. Find all values of a for which the solutions of the equation below will be real.

$$ax^2 - (a-3)x + 1 = 0$$

37. For what values of c will the solutions of the equation below be imaginary?

$$x^2 - 2cx - 2c + 3 = 0$$

38. Solve the inequality:

$$\frac{x-3}{x^2+3x+2} + \frac{2x-5}{x^2+x-2} < \frac{x-3}{x^2-1}$$

39. Solve the inequality:

$$\frac{x-3}{2x^2-x-3} + \frac{x+2}{2x^2-7x+6} > \frac{x-1}{x^2-x-2}$$

40. For what values of x will the expression below be real?

$$\sqrt{\frac{(x-2)(x+1)}{x^3(x-4)}}$$

41. An object is thrown downward from the top of a building 256 feet high with an initial velocity of 16 feet per second. The formula for the distance s above the ground

of the object at any time t is given by $s = -16t^2 - 16t + 256$ (see Example 26). Find the time interval during which the object will be at least 64 feet above the ground.

42. The intensity I, in candlepower, of a certain light is given by $I = 6,400/d^2$, where d is the distance in feet from the source. For what range of distances will I satisfy the inequality $900 \le I \le 3,600$?

43. The load L that can be safely supported by a wooden beam of length l with rectangular cross-section of width w and depth d is given by the formula

$$L = \frac{kwd^2}{l}$$

where k is a constant depending on the type of wood. A beam is to be 15 feet long and 4 inches wide, and it is known that for the material being used, $k = 120$. If the maximum load the beam will have to support is 3,200 pounds, find the minimum depth d that should be used.

Note.   The given value of k is for both w and d in inches and l in feet.

Review Exercise Set

A   In Problems 1–22 find the solution set.

1. **a.** $2(3x - 4) = 6 - 7x$
    b. $\dfrac{x}{2} - \dfrac{3}{4} = \dfrac{5(2 - 3x)}{6}$

2. **a.** $4(x - 3) = 2(3 - 4x)$
    b. $\dfrac{2x}{5} - \dfrac{1}{3} = \dfrac{2(4 - x)}{15}$

3. **a.** $\dfrac{2x - 7}{3} = \dfrac{3}{8} - \dfrac{5x}{6}$
    b. $\dfrac{3}{5}\left(\dfrac{1}{4} - \dfrac{x}{2}\right) = \dfrac{3x - 7}{4}$

4. **a.** $x^2 - x - 6 = 0$
    b. $3x^2 + 10x - 8 = 0$

5. **a.** $6x^2 + x - 15 = 0$
    b. $12x^2 - 25x + 12 = 0$

6. **a.** $x(23 - 6x) = 20$
    b. $4x^2 - 25 = 0$

7. **a.** $x^2 - 4x + 3 = 0$
    b. $3y^2 - 2y - 4 = 0$

8. **a.** $9x^2 + 16 = 0$
    b. $10x^2 - 3x = 18$

9. **a.** $2r^2 - 5r + 4 = 0$
    b. $3y^2 + 2 = 4y$

10. **a.** $x^2 = 2(x - 2)$
    b. $t^2 = \dfrac{4(t - 1)}{5}$

11. **a.** $2(4 + m) = 15m^2$
    b. $x^2 = 1 + \dfrac{4x}{3}$

12. **a.** $\dfrac{2}{x - 2} - \dfrac{3}{x + 1} = 0$
    b. $\dfrac{2}{x} - \dfrac{5}{x + 1} = \dfrac{4}{x^2 + x}$

13. **a.** $\dfrac{4}{2x - 3} - \dfrac{5}{4 - x} = 0$
    b. $\dfrac{x}{x + 1} - \dfrac{2}{x - 1} = \dfrac{x^2 + 3}{x^2 - 1}$

14. **a.** $\dfrac{x - 1}{x + 2} = \dfrac{x - 3}{x + 1} + \dfrac{2x}{x^2 + 3x + 2}$
    b. $\dfrac{x}{3x + 4} - \dfrac{5}{2x - 3} = \dfrac{7 - 3x}{6x^2 - x - 12}$

15. **a.** $\dfrac{x + 1}{x - 2} - \dfrac{x}{2x - 3} = \dfrac{3x + 5}{2x^2 - 7x + 6}$
    b. $\dfrac{x - 1}{x - 3} - \dfrac{2x}{x + 2} = \dfrac{4x}{x^2 - x - 6}$

16. **a.** $\dfrac{x+1}{x-2} = 3 - \dfrac{x-1}{x+2}$ **b.** $\dfrac{x-1}{2x-3} - \dfrac{x}{x+1} = \dfrac{1}{3}$

17. **a.** $\sqrt{3-2x} = 4$ **b.** $\sqrt{x+2} - 2 = x$

18. **a.** $\sqrt{5x-2} = 3$ **b.** $1 - \sqrt{x-1} = x$

19. **a.** $\sqrt{x-2} = \sqrt{x+3} - 1$ **b.** $\dfrac{1}{\sqrt{2x-1}} + \sqrt{2x-1} = 2$

20. **a.** $\sqrt{2+x} - \sqrt{5-2x} + 3 = 0$ **b.** $\sqrt{x-1} = \sqrt{2x-1} - 1$

21. **a.** $9x^4 + 32x^2 - 16 = 0$ **b.** $12x^{-2} + x^{-1} - 6 = 0$

22. **a.** $2x^{2/3} + x^{1/3} - 15 = 0$ **b.** $4x^{-4} - 37x^{-2} + 9 = 0$

In Problems 23 and 24 perform the indicated operations and write the answer in the form $a + bi$.

23. **a.** $(2+3i)(4-5i)$ **b.** $\dfrac{3+2i}{4-3i}$ **c.** $(4-5i)^3$

24. **a.** $(3+4i)^{-1}$ **b.** $(3-\sqrt{-4})^4$ **c.** $\dfrac{2i-4}{3+2i}$

Solve the equations in Problems 25 and 26 by completing the square.

25. **a.** $2x^2 - 5x + 4 = 0$ **b.** $3t^2 + 2t - 6 = 0$

26. **a.** $2x^2 + 4x + 3 = 0$ **b.** $4x^2 - 3 = 6x$

27. Without solving the equations below, determine whether the solutions are real and unequal, real and equal, or imaginary.
    a. $3x^2 - 7x + 12 = 0$ **b.** $9x^2 - 15x - 13 = 0$
    c. $12x(5-3x) = 25$ **d.** $8t^2 + 9t + 3 = 0$

28. Evaluate each of the following:
    a. i^{43} **b.** i^{-10} **c.** $\sqrt{-5}\sqrt{-10}$ **d.** $\dfrac{1}{i^{17}}$ **e.** $\dfrac{\sqrt{24}}{\sqrt{-6}}$

29. The length of a room is 6 feet more than its width, and its perimeter is 64 feet. Find the length and width.

30. After selling four-fifths of her stock of a certain type of winter jacket at $55 each, a store owner placed the others on sale at $40 each. If all the jackets were sold and the gross amount realized from them was $3,900, how many jackets did she sell at the regular price and at the sale price?

31. Two cars leave from the same point and travel the same route. The first car travels at an average speed of 65 kilometers per hour, and the second car, which leaves 30 minutes after the first, travels at an average speed of 90 kilometers per hour. How many kilometers will each car have gone when the second car overtakes the first?

32. By combining two types of candy, one costing $3.00 per pound and the other $4.00 per pound, a 2 pound box is prepared that costs $6.50. How much of each kind of candy is used?

33. A man has money invested in two types of savings accounts, one paying 5% and the other 6% annual interest. He has $5,000 more in the account paying 6% than in the account paying 5%. The interest at the end of 1 year is $1,180. How much does he have in each account?

34. A jetliner takes off from Washington for San Francisco and flies at an average speed of 440 miles per hour. Forty-five minutes later an Air Force jet leaves from the same airport and follows the same flight path. If the Air Force jet flies at an average speed

of 800 miles per hour, how long will it take to overtake the jetliner? (Assume no change in the wind.)

35. A quantity of brine containing 25% salt is on hand, and it is desired to reduce the salinity to 10% by adding pure water. If 7 gallons of pure water are required, how many gallons of brine were there originally?

36. Two pumps working simultaneously can fill a water tank in 20 hours. The first one alone can fill it in 50 hours. How long would it take the second pump alone?

37. Driver M makes a trip of 220 miles. Driver N makes the same trip in 20 minutes less time, averaging 5 miles per hour faster than driver M. What is the average speed of each driver?

38. Pumps A and B working together can pump all the water out of a reservoir in 20 hours. Pump A could do the job alone 9 hours faster than pump B alone. How long would it take each pump alone?

39. A rain gutter is to be formed from a long, flat piece of tin, 10 inches wide, by bending up at a right angle a certain amount on each side (see the sketch). By calculus it is determined that the greatest amount of water can be accommodated when the cross-sectional area is 12.5 square inches. How many inches on each side should be turned up to produce the gutter?

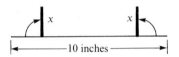

40. The length of a rectangle is 14 inches greater than the width, and the diagonal is 26 inches. Find the dimensions of the rectangle.

41. How many kilograms of silver alloy containing 35% silver should be mixed with 12 kilograms of an alloy having 43% silver in order to obtain an alloy with 38% silver?

Find the solution sets for the inequalities in Problems 42–50.

42. a. $\dfrac{2x-5}{3} \geq \dfrac{x}{4} - 1$
    b. $\dfrac{4-3x}{2} < \dfrac{2x+5}{3}$

43. a. $-1 \leq \dfrac{x-2}{3} < 1$
    b. $0 < \dfrac{2x}{3} - \dfrac{1}{4} \leq 2$

44. a. $9x^2 - 6x - 8 < 0$
    b. $x(2x+5) \geq 3$

45. a. $\dfrac{x+2}{x^2-9} \geq 0$
    b. $\dfrac{x-4}{3} < \dfrac{x-6}{x+4}$

46. a. $\dfrac{3x^2+2x-8}{x^2-3x} \geq 0$
    b. $\dfrac{2}{x+3} - \dfrac{1}{x-2} < 1$

47. a. $(x-2)(x+4)(3x-10) \leq 0$
    b. $\dfrac{(x-1)(x-2)}{5-x} < 1$

48. a. $|x-2| < \frac{1}{2}$
    b. $|2x+1| > 3$

49. a. $|3x-4| \leq 1$
    b. $|2-x| > 3$

50. a. $\left|\dfrac{x+1}{3}\right| \geq 1$
    b. $\left|\dfrac{1}{2} - \dfrac{x}{3}\right| \leq \dfrac{2}{3}$

51. Write in terms of an inequality involving absolute values:
    a. x lies in a neighborhood of 3 of radius $\frac{1}{2}$.
    b. The distance between x and 5 is less than 2.
    c. x lies in a deleted neighborhood of -2 of radius 1.
    d. x lies in a δ-neighborhood of a.
    e. y lies in an ε-neighborhood of y_0.

52. A student has grades of 72, 63, 68, and 75 on the four hour exams in a course. The final exam is to count the equivalent of two hour exams. In what range should the final exam grade fall so that the student's average grade for the course will be greater than or equal to 70? Is it possible for the average to be 80 or better?

53. The temperature on a certain day in Los Angeles ranged between 59 and 86°F. What was the range in degrees Celsius?
    Note. $°C = (\frac{5}{9})(°F - 32)$

B   **54.** Solve for x:
    a. $2a^2x - 3b = b^2x + 4a$ **b.** $10a^2x^2 + 9abxy - 36b^2y^2 = 0$

  55. **a.** Solve for y: $y^2 - 2ky + 2k^2 = 1, \quad k > 1$

    b. Solve for x: $\dfrac{2}{x} - \dfrac{1}{x-a} = \dfrac{1}{x-b}$

56. Find all values of k for which the equation below will have equal solutions.

$$kx^2 - 2(k-3)x + 3k + 1 = 0$$

For each value of k so determined, find the solutions of the equation.

57. For what values of k will the solutions of the equation below be real?

$$kx^2 + (k-2)x + 2 = k$$

58. Use a hand calculator to solve for x:
    a. $34.27x^2 + 28.39x - 52.76 = 0$ **b.** $3.258x^2 - 1.237x + 0.965 = 0$

In Problems 59–61 find the solution set.

59. $\sqrt{x+3} + \sqrt{2x+5} = 2$ **60.** $\sqrt{x-1} + \sqrt{x-4} = \sqrt{2x-1}$

61. $\dfrac{x+3}{2x^2+x-3} - \dfrac{3x}{2x^2+7x+6} = \dfrac{5x+1}{x^2+x-2}$ (Check all answers.)

62. Find the solution set for each and depict it graphically.

    a. $-\dfrac{3}{2} < \dfrac{3(2x-4)}{5} \le 2$ **b.** $\dfrac{x-1}{x+2} - \dfrac{x-3}{x+1} < 1$

63. Solve the inequality:

$$\frac{2x}{x^2 - 2x - 3} + \frac{x+2}{1-x^2} \ge \frac{6}{x^2 - 4x + 3}$$

64. Find the values of x for which the expression below will be real.

$$\sqrt{\frac{(x-1)^3(x+2)}{x^3 + x^2}}$$

65. Find the solution set for each.
    a. $|x-1| - |x+2| > 1$ **b.** $|2x| + |4-x| < 7$

66. A group of 25 students rents a bus for a field trip at a cost of $8.00 per person. They learn that the price of each ticket will be reduced by 10¢ for each additional student they can recruit to go on the trip, provided that the total number does not exceed 45 (the capacity of the bus). If the final total cost of the bus was $251.60, find out how many students actually went and the price of each ticket.

67. A container to hold a piece of machinery is to be constructed with a square base and with height 2 feet more than the side of the base. The material for the top and bottom costs 50¢ per square foot and the material for the sides costs 25¢ per square foot. If the total cost is $24, find the dimensions of the container.

68. A motorist drives 300 kilometers, and after a 2 hour visit returns over the same route. His average speed on the return trip is 20 kilometers per hour greater than his average speed going. If the total elapsed time from the time he left until he returned was 8 hours and 45 minutes, find his average speed going and coming.

69. An 8 inch by 10 inch picture is to be surrounded by a matting of the same width on all sides. The total area of the picture plus matting is not to exceed 195 square inches. What is the maximum width of matting that may be used?

70. A group of people in New Orleans arrange for an all-day boat excursion on the Gulf for a total price of $1,080. They learn that the boat will accommodate 15 additional people, and they succeed in getting 15 more to go, which results in the cost per person being $1.00 less than it would have been without the extra people. How many people actually went on the excursion, and what was the cost per person?

Cumulative Review Exercise Set I (Chapters 1–3)

1. Perform the indicated operations and simplify:
$$\frac{(2x - 1)(x + 2) - 2(4x + 5)}{x + 4 - (x - 1)(4x + 5)}$$

2. Simplify the expressions. Answers should not involve negative exponents, and there should be no fractions under the radical. All letters represent positive numbers.

   a. $\sqrt{\dfrac{2x^{-3}y^5z^0}{3x^{-2}y^3}}$ b. $\sqrt{a^{-1}b - 2 + b^{-1}a}$

3. Prove that if $a < b$ and $c < d$, then $ac < bd$ provided b and c are positive. Show by examples that the conclusion is not necessarily true if b and c are not both positive.

4. A 50% alcohol solution is to be combined with a 25% alcohol solution to obtain 50 cubic centimeters of a solution having 35% alcohol. How much of each solution should be used?

5. a. Express $0.2135135135\ldots$ as the ratio of two integers.
   b. If a, b, and c are integers, with $a \neq 0$, show that regardless of the nature of the roots of the equation $ax^2 + bx + c = 0$, their sum and their product are always rational numbers.

6. Solve for x:
   a. $3x^2 - 8x + 6 = 0$ b. $(3 + x)(14 - 3x) = 30$

7. Perform the indicated operations and simplify:
$$\frac{x^2 - 2x + 1}{2x^2 + x - 3} \cdot \frac{8x^2 + 6x - 9}{3x^2 + x - 4} - \frac{6x^2 + 5x - 4}{4x^2 + 5x - 6} \cdot \frac{2x^2 + 5x + 2}{1 - 4x^2}$$

8. Solve for t: $\sqrt{2t + 5} - \sqrt{2 - t} = 1$

9. A flower garden in the shape of a rectangle 8 feet wide by 12 feet long is surrounded by a walkway of uniform width. If the combined area of the bed and walk is 221 square feet, find the width of the walk.

10. Perform the indicated operations and simplify:

   a. $\dfrac{\dfrac{3}{2} - \left(-\dfrac{5}{4}\right)\left(1 - \dfrac{2}{3}\right)}{\dfrac{5}{6}\left(\dfrac{3}{4} - 2\right)}$ b. $\dfrac{\dfrac{x}{2} - 1}{x - 1 - \dfrac{2}{x}}$

11. a. Rationalize the numerator: $\dfrac{\sqrt{2x + 2h - 3} - \sqrt{2x - 3}}{h}$

    b. Simplify and combine terms. Leave no fraction under a radical and no radical in a denominator.
$$\sqrt{\frac{3}{2}} - \frac{1}{3}\sqrt{24} + \frac{3}{2}\sqrt{\frac{32}{3}} - \frac{5}{2\sqrt{6}}$$

12. Solve for x and give a careful justification for each step based on fundamental properties:
$$\frac{2(1 - x)}{3} + \frac{1}{2} = \frac{3x}{4}$$

13. Factor completely with integer coefficients:
    a. $4xy + 3x^2 - x^3y - 12$ b. $72x^3 + 10x^2y - 48xy^2$

14. a. Expand and simplify: $\left(x^2 - \dfrac{2}{x}\right)^6$

    b. Find the first four terms of the expansion of $(2x + 3y)^{10}$.

15. a. Solve for s: $F = \dfrac{rs - 1}{r + 2s}$

b. Solve for t: $s = \frac{1}{2}gt^2 + v_0 t \quad (s \geq 0, t \geq 0)$

16. Solve the inequalities and show the solution sets on a number line.

a. $x(3x + 14) \geq 24$ **b.** $\left| \dfrac{3x - 4}{5} - \dfrac{3x}{4} \right| < \dfrac{1}{2}$

17. a. Simplify, and express the answer with positive exponents only:

$$\left(\frac{25a^{-2}b^{4/3}}{\sqrt[3]{16}\, a^{8/3}b^2 c^0} \right)^{-3/2}$$

b. Evaluate using scientific notation:

$$\frac{(0.000003)\sqrt{250{,}000{,}000{,}000}}{2 \times 10^{-3} + 5 \times 10^{-4}}$$

18. Solve for x: $\dfrac{2x}{x - 2} - \dfrac{3x - 1}{x + 1} = 4$

19. Simplify, and write the answer with positive exponents only:

$$\frac{\sqrt[3]{(x^2 - 1)^2} - \dfrac{4x^2}{3}(x^2 - 1)^{-1/3}}{(x^2 - 1)^{4/3}}$$

20. A woman drove a distance of 120 kilometers to do some research in the Library of Congress. She stayed 4 hours and 24 minutes before returning home. On the return trip she encountered rush hour traffic, and as a result her average speed on the return trip was 15 kilometers per hour less than on the trip going. If the total elapsed time from leaving home until returning was 8 hours, find her average speed going and returning.

21. Solve the inequalities:

a. $\dfrac{x}{x - 1} - \dfrac{1}{x + 3} \leq 1$ **b.** $\dfrac{2}{x - 1} \geq \dfrac{x + 3}{x + 11}$

22. For $a = \pm 1$ and $b = \pm 2$ show that $a + bi$ satisfies the equation $x^4 + 6x^2 + 25 = 0$. Also, solve the equation by the quadratic formula. Verify that the solutions are the same.

23. A tank can be filled by two pumps working together in 6 hours. The larger pump alone can fill the tank in 5 hours less time than it would take the smaller pump alone. How long would it take each pump working alone to fill the tank?

24. Perform the indicated operations and simplify:

a. $\dfrac{\dfrac{1}{x - 1} - \dfrac{2}{x + 2}}{\dfrac{18}{x^2 + x - 2} - 1}$ **b.** $\dfrac{x + 2}{x} - \dfrac{8}{4x - x^3} - \dfrac{x}{x + 2}$

25. A men's clothing store owner had a month long 20% off sale on suits of a certain type. His gross income from the sale of these suits during the month of the sale was $15,120. During the preceding month, when the suits were full price, his gross income from selling 30 fewer of these suits was $12,600. Find how many of this type of suit he sold during the month of the sale. What was the original price of the suit?

4

Functions and Graphs

An understanding of the notion of a function and of the use of functional notation are basic to an understanding of calculus. The language and symbolism associated with functions are found on almost every page of any calculus textbook. The following example illustrates one usage:*

Find $D_x \dfrac{1}{(x^3 - 4x^2 + 1)^2}$.

Solution   In our mind, we see

$$\frac{1}{(x^3 - 4x^2 + 1)^2} = (x^3 - 4x^2 + 1)^{-2} = f(g(x))$$

where

$$f(x) = x^{-2}, \qquad g(x) = x^3 - 4x^2 + 1$$

Since $f'(x) = -2x^{-3}$, we have

$$f'(g(x)) = -2(x^3 - 4x^2 + 1)^{-3}, \qquad g'(x) = 3x^2 - 8x$$

Therefore, by the chain rule,

$$D_x \frac{1}{(x^3 - 4x^2 + 1)^2} = -2(x^3 - 4x^2 + 1)^{-3}(3x^2 - 8x) = \frac{-6x^2 + 16x}{(x^3 - 4x^2 + 1)^3}$$

We will learn in this chapter about such expressions as $f(x)$, $g(x)$, and $f(g(x))$. The symbols D_x, $f'(x)$, and $g'(x)$ signify the operation of *differentiation*, which will be studied in calculus.

---

* R. E. Johnson, F. L. Kiokemeister, and E. S. Wolk, *Calculus with Analytic Geometry*, 6th ed. (Boston: Allyn and Bacon, 1978), p. 111. Reprinted by permission.

1 The Cartesian Coordinate System

The French mathematician and philosopher René Descartes (1596–1650) is usually given credit for the idea of representing points in a plane by ordered pairs of numbers. By an **ordered pair** is meant a pair of numbers (a, b) in which the position of each number is important.* Thus, $(2, 3)$ and $(3, 2)$ are two different ordered pairs. As is so often the case in mathematical discoveries, the same concept was obtained independently at about the same time by Pierre de Fermat (1601–1665), also a Frenchman, and a jurist as well as a mathematician. Descartes's book, *Geometry*, appeared in 1635, and was his only book on mathematics, but it had a profound influence on the subsequent development of mathematics because it provided an essential prerequisite for the invention of calculus. The approach to geometry introduced by Descartes forms the basis of what is now referred to as **analytic geometry.**

The method employed by Descartes makes use of the so-called **rectangular** or **cartesian** (after Descartes) **coordinate system.** Although this system is familiar to many students from high school mathematics, we will review its essential elements. We consider two number lines perpendicular to each other, one vertical and one horizontal, so that the point of intersection corresponds to 0 on each line; this point is called the **origin.** It is customary to have positive numbers to the right on the horizontal line and along the top part of the vertical line. The two lines are called the **horizontal and vertical axes.** It is also customary (but not essential) to name the horizontal axis the x **axis** and the vertical axis the y **axis.** These axes divide the plane into four parts, called **quadrants,** numbered I, II, III, and IV counterclockwise, as shown in Figure 1. Corresponding to

Figure 1

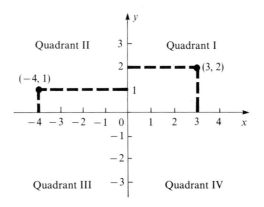

each ordered pair (x, y) of real numbers is a point in the plane determined as follows: Starting from the origin, we move x units along the x axis (right or left, depending on whether x is positive or negative) and from there move y units parallel to the y axis (up or down, depending on whether y is positive or

---

* The context will usually make clear whether (a, b) designates an open interval or a point in a plane.

negative). The resulting point is said to have **coordinates** (x, y); for brevity we often say "the point (x, y)." The points having coordinates $(3, 2)$ and $(-4, 1)$ are pictured in Figure 1. The first number of the ordered pair is called the **x coordinate,** or **abscissa,** of the point, and the second number is called the **y coordinate,** or **ordinate.**

If a point in the plane is given, then that point has unique coordinates (x, y) determined by erecting lines through the point and perpendicular to the x and y axes. So the coordinate system provides a one-to-one correspondence between all points in the plane (geometric objects) and all ordered pairs of real numbers (algebraic objects).

2 The Distance Formula

In this section we develop an important formula that gives the distance between two points in the plane in terms of their coordinates. The basis for this is the Pythagorean theorem, which you will recall, states that in a right triangle the square of the hypotenuse equals the sum of the squares of the legs. If we designate the lengths of the sides of a right triangle as in Figure 2, then the Pythagorean theorem can be written as $c^2 = a^2 + b^2$.

Figure 2

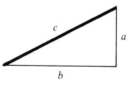

Now let (x_1, y_1) and (x_2, y_2) designate any two points in the plane. Construct a right triangle as shown in Figure 3. For reference we have labeled the vertices. We wish to determine the distance d between P_1 and P_2, designated $\overline{P_1 P_2}$. The line segment $P_1 Q$ is horizontal, so the distance $\overline{P_1 Q} = |x_2 - x_1|$, as we saw in Chapter 3 when studying absolute value. Similarly, the segment $P_2 Q$ is vertical and so it can be thought of as a portion of a number line. Therefore, the distance $\overline{P_2 Q} = |y_2 - y_1|$. Now we apply the Pythagorean theorem to get

$$d^2 = |x_2 - x_1|^2 + |y_2 - y_1|^2$$

Figure 3

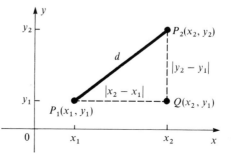

Since the square of a real number is always nonnegative, it follows that for any real number a, $|a|^2 = a^2$. So we can write

$$d^2 = (x_2 - x_1)^2 + (y_2 - y_1)^2$$

or

$$d = \sqrt{(x_2 - x_1)^2 + (y_2 - y_1)^2}$$

This is known as the **distance formula.** In applying this it makes no difference which point is labeled (x_1, y_1) and which point is labeled (x_2, y_2).

EXAMPLE 1   Find the distance between the points $(-1, 3)$ and $(7, 5)$.

Solution   We will let $(-1, 3)$ be (x_1, y_1) and $(7, 5)$ be (x_2, y_2). By the distance formula

$$d = \sqrt{[7 - (-1)]^2 + (5 - 3)^2} = \sqrt{8^2 + 2^2} = \sqrt{64 + 4}$$
$$= \sqrt{68} = \sqrt{4(17)} = 2\sqrt{17}$$

EXERCISE SET 2

A   In Problems 1 and 2 plot the given points.

1. **a.** $(2, 3)$ **b.** $(-4, -2)$ **c.** $(2, -3)$ **d.** $(0, 2)$ **e.** $(-2, 0)$
2. **a.** $(0, -4)$ **b.** $(-5, 2)$ **c.** $(-3, -4)$ **d.** $(4, -2)$ **e.** $(-3, 3)$

In Problems 3–15 a set of points in the plane is described. Illustrate each set by drawing a rectangular coordinate system and shading the set in question. When the set has a name, indicate what this is.

3. $\{(x, y): \ x \geq 0, \ y \geq 0\}$
4. $\{(x, y): \ x \leq 0, \ y \geq 0\}$
5. $\{(x, y): \ x \leq 0, \ y \leq 0\}$
6. $\{(x, y): \ x \geq 0, \ y \leq 0\}$
7. $\{(x, y): \ x < 0\}$
8. $\{(x, y): \ x = y\}$
9. $\{(x, y): \ x > y\}$
10. $\{(x, y): \ x = 0\}$
11. $\{(x, y): \ y = 0\}$
12. $\{(x, y): \ x = 0, \ y = 0\}$
13. $\{(x, y): \ x = 1\}$
14. $\{(x, y): \ y = -1\}$
15. $\{(x, y): \ y > 0\}$

In Problems 16–19 find the distance between the points.

16. **a.** $(-1, 3)$ and $(2, 5)$ **b.** $(4, 2)$ and $(-3, -4)$
17. **a.** $(0, 2)$ and $(2, -1)$ **b.** $(3, -2)$ and $(-3, 6)$
18. **a.** $(-2, -1)$ and $(3, 11)$ **b.** $(-3, 4)$ and $(5, 0)$
19. **a.** $(-9, 3)$ and $(6, -5)$ **b.** $(2, -3)$ and $(-1, -6)$

20. The distance between $(3, 4)$ and $(x, -2)$ is 10. Find x. (There are two solutions.)
21. The distance between $(-1, -2)$ and $(3, y)$ is $2\sqrt{5}$. Find y. (There are two solutions.)

B **22.** The converse of the Pythagorean theorem is also true, that is: If in a triangle with sides a, b, and c it is true that $c^2 = a^2 + b^2$, then the triangle is a right triangle. Use this to show that the triangle with vertices $(-2, 4)$, $(3, -6)$, and $(6, -2)$ is a right triangle.

23. An isosceles triangle is one in which two sides are equal. Show that the triangle with vertices $(-7, 5)$, $(-3, -2)$, and $(4, 2)$ is isosceles. Also show that it is a right triangle. (See Problem 22.)

24. Show that the triangle with vertices $(-4, -3)$, $(-3\sqrt{3}, 4\sqrt{3})$, and $(4, 3)$ is an equilateral triangle (all sides equal).

25. Show that the points $(-1, 2)$, $(2, 4)$, $(3, -4)$, and $(6, -2)$ are vertices of a rectangle. **Hint.** Use triangles.

26. Write and simplify an equation that specifies that the distance from (x, y) to $(3, 4)$ is 6. How would you describe the set of all points (x, y) satisfying this condition? Show the set graphically.

3 Graphs of Equations

Equations such as

$$y = 2x + 3 \qquad x^2 + y^2 = 4 \qquad y^2 = \frac{x-1}{x+3}$$

are said to be **equations in two variables.** In this section we will define what is meant by the graph of an equation of this type. In the equation $y = 2x + 3$ we may substitute any real number for x, but then the value of y is completely determined. In this case, then, we say that x is the **independent variable** and y is the **dependent variable.** Alternatively, we could solve this equation for x in terms of y, getting $x = \frac{1}{2}y - 3$, and then y would be the independent variable and x would be the dependent variable.

An ordered pair of numbers (a, b) is said **to satisfy an equation** if when a is substituted for x and b is substituted for y, **the equation becomes an identity.** For example, the equation $y = 2x + 3$ is satisfied by the pair $(1, 5)$, since when $x = 1$, we have $y = 2(1) + 3 = 5$. Similarly, this equation is satisfied by $(-2, -1)$, $(-1, 1)$, $(0, 3)$, $(\frac{1}{2}, 4)$, $(2, 7)$, and infinitely many other ordered pairs. Figure 4 shows these points, and they are seen to lie on a straight line.

Figure 4

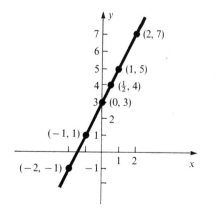

The graph of an equation is the collection of all points with coordinates x and y that satisfy the equation. Usually, it is impossible to locate (that is, to **plot**) every point with coordinates that satisfy an equation, because in most cases there are an infinite number of them. For the equations we will consider, a reasonably accurate graph can be drawn by plotting several points and joining these with a smooth curve. For many other equations, however, a more detailed analysis, such as would be studied in a calculus course, is needed. The next three examples illustrate how to draw graphs by plotting points.

EXAMPLE 2   Draw the graph of $y = 1/x^2$.

Solution   We make a table of values showing ordered pairs that satisfy the equation. We arbitrarily select several values of x and determine the corresponding y values. The graph is shown in Figure 5.

x	1	2	3	4	$\frac{1}{2}$	$\frac{1}{3}$	-1	-2	-3	-4	$-\frac{1}{2}$	$-\frac{1}{3}$
y	1	$\frac{1}{4}$	$\frac{1}{9}$	$\frac{1}{16}$	4	9	1	$\frac{1}{4}$	$\frac{1}{9}$	$\frac{1}{16}$	4	9

Figure 5

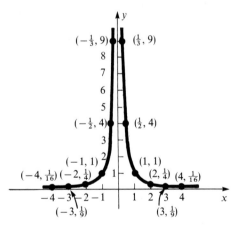

EXAMPLE 3   Draw the graph of $y = x^2 - 4$.

Solution

x	0	± 1	± 2	± 3
y	-4	-3	0	5

Here we have used the $\pm$ sign for x to shorten the table, since the value of y is the same when x is a given number or its negative. The graph is shown in Figure 6. (This is an example of a **parabola,** which we will study in Chapter 5.)

Figure 6

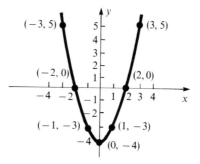

EXAMPLE 4   Draw the graph of $x^2 + y^2 = 36$.

Solution

x	0	± 1	± 2	± 3	± 4	± 5	± 6
y	± 6	$\pm\sqrt{35}$ $\approx \pm 5.92$	$\pm\sqrt{32}$ $\approx \pm 5.65$	$\pm\sqrt{27}$ $\approx \pm 5.20$	$\pm\sqrt{20}$ $\approx \pm 4.47$	$\pm\sqrt{11}$ $\approx \pm 3.32$	0

The graph is shown in Figure 7. (This is a **circle of radius 6,** as we shall show in Chapter 5.)

Figure 7

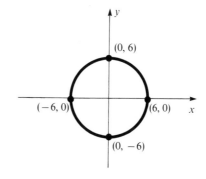

EXERCISE SET 3

In Problems 1–21 make a table of values and draw the graph.

A   **1.** **a.** $y = x + 1$ **b.** $y = x - 2$
     2. **a.** $y = 3x - 4$ **b.** $y = 3 - x$

3. **a.** $y = 2x$ **b.** $y = 3 + x$

4. **a.** $y = 2x - 5$ **b.** $x = 3 - y$

5. **a.** $x + y = 1$ **b.** $x - y = 2$

6. **a.** $2x + 3y = 4$ **b.** $3x - y = 2$

7. **a.** $5x - 3y = 4$ **b.** $4x + 5y + 7 = 0$

8. **a.** $y = -3x$ **b.** $x = 2y - 3$

9. **a.** $y = x$ **b.** $x + y = 0$

10. **a.** $x = 2$ **b.** $y = 3$

11. **a.** $x - 3 = 0$ **b.** $y + 1 = 0$

12. **a.** $y = \sqrt{x}$ **b.** $y = 1/x$

13. **a.** $y = x^2$ **b.** $y = 4 - x^2$

14. **a.** $y = |x|$ **b.** $y = |1 - x|$

B 15. **a.** $y^2 = x$ **b.** $x^2 + y^2 = 25$

16. **a.** $y = x^3/4$ **b.** $y = 2\sqrt[3]{x}$

17. **a.** $y = \dfrac{1}{1 + x^2}$ **b.** $y = \sqrt{36 - x^2}$

18. **a.** $y = x^2 - 2x$ **b.** $y = (x + 1)(x - 3)$

19. **a.** $y = x^3 - x$ **b.** $y = \dfrac{x - 1}{x}$

20. **a.** $y = |2x + 3|$ **b.** $y = 4 - |x|$

21. **a.** $|x + y| = 1$ **b.** $|x| + |y| = 1$

22. Coordinates are introduced into a town planner's map. One boundary of the town's limits corresponds to the graph of the equation $y = 2x + 1$, while the entrance to the fire station is at the point with coordinates $(3, 4)$. Find the formula, in terms of x, for the distance from the fire station entrance to each point on that boundary of the town.

23. A farmer estimates that if he plants x pounds of seed in a plot of land, the yield per pound of seed will be $50 - 2x$ pounds of produce. The total yield y from the plot is the number of pounds of seed planted multiplied by the yield per pound. Find the equation for y and graph that portion of the equation for which $x \geq 0$ and $y \geq 0$.

---

4 The Function Concept

The concept of a **function** occupies a central place in most branches of mathematics. A formal definition will be given below, but we will precede it with some examples.

To say that a certain thing is a function of another can be interpreted as meaning the first thing depends on the second. This terminology is often used in nonmathematical situations. For example, one might say "the size of the wheat crop is a function of the weather." Of course, it is a function of many other things as well. The mathematical use of the phrase "is a function of" is consistent with its everyday use, but the meaning is more precisely defined.

We might say that the circumference of a circle is a function of its radius. Given the length of the radius, the circumference can be determined; the circumference depends on the radius. In this case, we can state precisely the nature of the dependence by means of the equation $C = 2\pi r$, where C represents the circumference and r is the radius.

Or, consider a car traveling at a constant speed of 45 miles per hour. The distance d the car travels is a function of the time t elapsed. In this case, $d = 45t$.

As another example, if money is invested at compound interest, the accumulated amount of money A after t years is given by

$$A = P(1 + r)^t$$

where P is the original amount invested (the principal) and r is the annual interest rate. Suppose \$1,000 is invested at compound interest for 5 years. Then we can express A as a function of r by

$$A = 1,000(1 + r)^5$$

It is not always easy, and sometimes it is not possible, to express the dependence given by a functional relationship by means of an equation. Try, for example, to write an equation expressing precisely how the cost of mailing a first class letter is a function of its weight. It is easier in this case to state the relationship in words. Sometimes a table can be used to describe a function. For example, it is known that the number of deaths annually in the United States resulting from highway accidents is a function of the maximum speed limit. A table showing this relationship might look like the following:

Speed limit in miles per hour	50	55	60	65	70	Unlimited
Number of deaths annually	42,000	45,000	52,000	66,000	75,000	85,000

Whether given by an equation, stated in words, or shown by a table, there is always implicit in a functional relationship some sort of rule whereby the value of one quantity is determined by the value of another (or others). Moreover, each of these quantities assumes values ranging over some prescribed set. For example, in the equation relating the circumference of a circle to its radius, $C = 2\pi r$, the value of r may be chosen arbitrarily, but only from the set of positive real numbers. For each choice of r the value of C is uniquely determined, and the totality of values of C as r ranges over all positive real numbers also constitutes a set of positive numbers. In fact, it can be shown that C also assumes every positive real number value.

So the basic ingredients of a function are these: two sets and a rule relating the elements of one of the sets to elements in the other.

DEFINITION 1   Let A and B designate two nonempty sets. A function from A to B is a rule whereby each element in A is made to correspond to exactly one element in B.

The set A in this definition is called the **domain** of the function, and the set of elements of B that correspond, under the rule, to elements in A is called the **range** of the function. Note that the range may or may not be all of B. If the range is all of B, we say the function is from A **onto** B. The set B is sometimes called the **codomain** of the function. It is customary to designate functions by letters such as f, g, F, or ϕ. If f is a function from A to B, and x is an element of A, then the element of B that corresponds to x is called the **image of x under f**, and is designated $f(x)$, read "f of x." The set of all such images of elements of A is the range of f. If y designates an element in the range of f, and x is an element in A for which $f(x) = y$, then we say that x is a **preimage** of y. Depending on the nature of the function, an element of the range may have more than one preimage. This is in contrast to the fact that each element of the domain has exactly one image in the range.

The term *variable* is appropriate in speaking of functions. If x stands for an arbitrary value in the domain of a function f, and y designates the image of x under f—that is, $y = f(x)$—then x is the independent variable and y is the corresponding dependent variable. Any number in the domain may be substituted for the independent variable; the corresponding value for the dependent variable is then uniquely determined by means of the function.

To make these ideas more concrete, let us consider the function f with domain being the set of all real numbers, and value at x being $2x^2 + 3$. That is,

$$f(x) = 2x^2 + 3$$

This function is now completely defined. The set of images lies in R, and specifically the range of f is the subset of R consisting of those positive real numbers greater than or equal to 3:

$$\text{Range of } f = \{y: \quad y \geq 3\} \qquad \textbf{Why?}$$

We have, for example, $f(0) = 3$, since $f(0) = 2 \cdot (0)^2 + 3 = 2(0) + 3 = 3$. Also, $f(1) = 5$, $f(4) = 35$, and $f(-2) = 11$.

An equation such as $y = 2x^2 + 3$ implicitly defines a function, even though no letter is specified for the function. The equation is the rule, and the domain is assumed to be the largest subset of R such that for x in this subset, the corresponding value of y will be real. In this example, the domain, then, is all of R. Such an assumption about the domain of functions we use will always be made unless the domain is explicitly stated. In the function implied by the equation $y = \sqrt{x}$, then, we would assume the domain to be the set of all nonnegative real numbers.

To reiterate, we may write

$$f(x) = 2x^2 + 3$$

or

$$y = 2x^2 + 3$$

to define the same function. The first way has the advantage that it assigns a letter f to name the function, and images of various values of x are easily indicated, for example, $f(0) = 3$ and $f(1) = 5$. The second way has the advantage that a name, y, is assigned to the dependent variable. Both methods are useful and will be employed throughout the remainder of this text. It should be noted, too, that the particular letters used to designate the function or the variables are not important in themselves; the form is the important part. For example, the function g defined by

$$g(t) = 2t^2 + 3$$

is the very same function as f described previously, because both functions give the same result for each value of the independent variable. For this reason, we can write $f = g$. When the variables represent certain physical quantities, suggestive letters are often used, such as $v = f(t)$ for the velocity v in terms of time t.

EXAMPLE 5   Let $f(x) = \dfrac{x}{x + 1}$.   Find:

    a. $f(2)$ **b.** $f(0)$ **c.** $f(\tfrac{2}{3})$ **d.** $f(x - 1)$ **e.** $f(x + h)$

Solution **a.** $f(2) = \dfrac{2}{2 + 1} = \dfrac{2}{3}$ **b.** $f(0) = \dfrac{0}{0 + 1} = \dfrac{0}{1} = 0$

    c. $f(\tfrac{2}{3}) = \dfrac{\tfrac{2}{3}}{\tfrac{2}{3} + 1} = \dfrac{2}{2 + 3} = \dfrac{2}{5}$ **d.** $f(x - 1) = \dfrac{(x - 1)}{(x - 1) + 1} = \dfrac{x - 1}{x}$

    e. $f(x + h) = \dfrac{x + h}{x + h + 1}$

Note. Whatever occurs in the parentheses following f is substituted for x wherever x appears.

EXAMPLE 6   Let $g(x) = x^2 + 2x - 3$.   Find:

    a. $g(4)$ **b.** $g(-2)$ **c.** $g(a)$ **d.** $g(x + \Delta x)$

Solution **a.** $g(4) = 4^2 + 2(4) - 3 = 16 + 8 - 3 = 21$
    b. $g(-2) = (-2)^2 + 2(-2) - 3 = 4 - 4 - 3 = -3$
    c. $g(a) = a^2 + 2a - 3$
    d. $g(x + \Delta x) = (x + \Delta x)^2 + 2(x + \Delta x) - 3$
                $= x^2 + 2x(\Delta x) + (\Delta x)^2 + 2x + 2(\Delta x) - 3$

Note. The symbol Δx is read "delta x," and has historically been used in calculus. It is treated as a unit and does not mean the product of Δ and x.

EXAMPLE 7   Let $f(x) = \dfrac{1}{\sqrt{x}}$.   Find: $\dfrac{f(x+h) - f(x)}{h}$

Solution

$$\frac{f(x+h) - f(x)}{h} = \frac{\dfrac{1}{\sqrt{x+h}} - \dfrac{1}{\sqrt{x}}}{h} = \frac{\sqrt{x} - \sqrt{x+h}}{h\sqrt{x}\sqrt{x+h}}$$

In calculus, where this type of calculation is important, it is essential that the numerator be rationalized:

$$\frac{\sqrt{x} - \sqrt{x+h}}{h\sqrt{x}\sqrt{x+h}} \cdot \frac{\sqrt{x} + \sqrt{x+h}}{\sqrt{x} + \sqrt{x+h}} = \frac{x - (x+h)}{h\sqrt{x}\sqrt{x+h}\,(\sqrt{x} + \sqrt{x+h})}$$

$$= \frac{-h}{h\sqrt{x}\sqrt{x+h}\,(\sqrt{x} + \sqrt{x+h})}$$

$$= \frac{-1}{\sqrt{x}\sqrt{x+h}\,(\sqrt{x} + \sqrt{x+h})}$$

Sometimes the rule for a function is given by more than one expression, as the next example shows.

EXAMPLE 8   Let

$$f(x) = \begin{cases} -1 & \text{if} & x < 0 \\ x & \text{if} & 0 \le x < 2 \\ 1 - x^2 & \text{if} & x \ge 2 \end{cases}$$

Find:

a. $f(1)$ **b.** $f(2)$ **c.** $f(-2)$ **d.** $f(0)$ **e.** $f(3)$

Solution
a. When $0 \le x < 2$, $f(x) = x$. So $f(1) = 1$.
b. When $x \ge 2$, $f(x) = 1 - x^2$. So $f(2) = 1 - 4 = -3$.
c. When $x < 0$, $f(x) = -1$. So $f(-2) = -1$.
d. Since 0 is in the interval $0 \le x < 2$, for which $f(x) = x$, it follows that $f(0) = 0$.
e. When $x \ge 2$, $f(x) = 1 - x^2$. So $f(3) = 1 - 9 = -8$.

EXAMPLE 9   Express the area A of a circle as a function of its radius r.

Solution $A = \pi r^2$

EXAMPLE 10   A woman invests \$1,000 at 5% simple interest. Express the amount earned A as a function of time t in years.

Solution
$$A = (1{,}000)(0.05)t$$

EXAMPLE 11   A rectangle is inscribed in a circle of radius 2. Express the area of the rectangle as a function of its base.

Solution   Here it is helpful to draw a sketch, as shown in Figure 8. Since the radius of the circle is 2, its diameter is 4. Divide the rectangle into two triangles as shown by means of a diagonal. This diagonal is a diameter of the circle. Designate the base of the rectangle by b and its height by h. The area is $b \cdot h$, but to obtain this as a function of b only we need to express h in terms of b. This can be done by means of the Pythagorean theorem. For our triangle, this says that

$$b^2 + h^2 = 4^2$$

So $h^2 = 16 - b^2$, and $h = \sqrt{16 - b^2}$. Finally, then, the area A is given by

$$A = b\sqrt{16 - b^2}$$

Figure 8

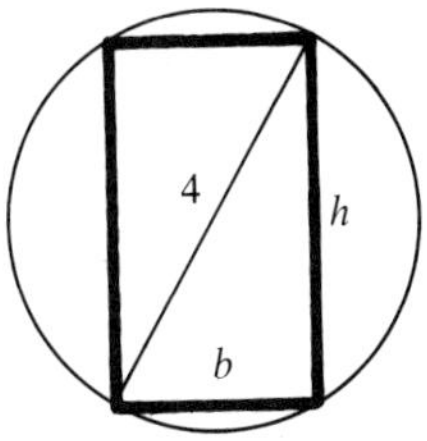

EXERCISE SET 4

A   In Problems 1–15 find the specified functional values for the given functions.

1.   For $f(x) = 2x - 1$,   find:
    a. $f(0)$ **b.** $f(1)$ **c.** $f(-2)$ **d.** $f(5)$

2.   For $h(t) = t^2 + 2t - 1$,   find:
    a. $h(-1)$ **b.** $h(2)$ **c.** $h(4)$ **d.** $h(-3)$

3.   For $F(x) = \dfrac{x+1}{x-1}$,   find:

    a. $F(2)$ **b.** $F(0)$ **c.** $F(-1)$ **d.** $F(-3)$

4.   For $g(x) = \sqrt{2x^2 - 3x - 5}$,   find:
    a. $g(3)$ **b.** $g(-2)$ **c.** $g(6)$ **d.** $g(-4)$

5.   For $f(x) = \sqrt{x-1}$,   find:
    a. $f(2)$ **b.** $f(5)$ **c.** $f(1)$ **d.** $f(2x)$

6.   For $f(t) = t^3 - 4t - 5$,   find:
    a. $f(2)$ **b.** $f(-1)$ **c.** $f(4)$ **d.** $f(-3)$

7. For $f(x) = \dfrac{2x - 3}{3x + 4}$, find:

    a. $f(-2)$ **b.** $f(\tfrac{1}{2})$ **c.** $f(\tfrac{5}{6})$ **d.** $f(-\tfrac{2}{3})$

8. For $g(x) = \dfrac{x - 1}{x + 2}$, find:

    a. $g(1)$ **b.** $g(-3)$ **c.** $g(0)$ **d.** $g\left(\dfrac{1}{x}\right)$

9. For $f(x) = 2x^2 - 4$, find:
    a. $f(1)$ **b.** $f(-2)$ **c.** $f(a)$ **d.** $f(x + 1)$ **e.** $f(x + h)$

10. For $f(t) = 1/t$, find:

    a. $f(2)$ **b.** $f(\tfrac{1}{2})$ **c.** $f\left(\dfrac{1}{a}\right)$ **d.** $\dfrac{f(t + h) - f(t)}{h}$

11. For $h(t) = \sqrt{1 - t^2}$, find:

    a. $h(1)$ **b.** $h(1 - t)$ **c.** $h\left(\dfrac{1}{t}\right)$ **d.** $h(t + \Delta t)$

12. For $f(x) = \begin{cases} x & \text{if } x \ge 0 \\ 1 - x & \text{if } x < 0 \end{cases}$ find:

    a. $f(0)$ **b.** $f(1)$ **c.** $f(-1)$ **d.** $f(2)$

13. For $g(x) = \begin{cases} x^2 & \text{if } x > 1 \\ 0 & \text{if } x = 1 \\ -1 & \text{if } x < 1 \end{cases}$ find:

    a. $g(2)$ **b.** $g(-2)$ **c.** $g(0)$ **d.** $g(1)$

14. For $f(x) = \begin{cases} x^2 - 4 & \text{if } x \ge 1 \\ \dfrac{1 - 2x}{3} & \text{if } x < 1 \end{cases}$ find:

    a. $f(5)$ **b.** $f(-2)$ **c.** $f(1)$

15. For $\phi(x) = \begin{cases} 1 & \text{if } x \text{ is rational} \\ 0 & \text{if } x \text{ is irrational} \end{cases}$ find:

    a. $\phi(1)$ **b.** $\phi(0)$ **c.** $\phi(\sqrt{2})$ **d.** $\phi(\pi)$

16. Let $F(x) = \sqrt{x}$. Show that

$$\frac{F(x + h) - F(x)}{h} = \frac{1}{\sqrt{x + h} + \sqrt{x}}$$

17. Let $\phi(x) = x/(x - 3)$. Find

$$\frac{\phi(4 + h) - \phi(4)}{h}$$

and simplify your result.

18. Find the domain of f if:
    a. $f(x) = \sqrt{2x - 4}$ **b.** $f(x) = \sqrt{x^2 - 2x - 3}$

19. Find the domain of g if

    a. $g(x) = \sqrt{\dfrac{x - 1}{x + 2}}$ **b.** $g(t) = \sqrt{t^2 - 4}$

20. Express the area of a triangle of base 10 as a function of its altitude h.
21. Express the perimeter of a rectangle of length 12 as a function of its width w.

22. Express the area of a circle as a function of its diameter d.

23. Express the circumference of a circle as a function of its diameter d.

24. The sum of $600 is invested at $5\frac{1}{2}\%$ simple interest for t years. Express the total interest earned as a function of t.

25. Express the amount of money accumulated from investing $10,000 at 7% compounded annually as a function of the number of years t.

26. Express the amount of money accumulated from investing $500 at compound interest (compounded annually) for 10 years as a function of the annual interest rate r.

27. A can is in the form of a right circular cylinder with the height twice the radius of the base. Express the following as functions of the radius of the base:
a. Volume   **b.** Lateral surface area   **c.** Total surface area
Hint. The volume of a right circular cylinder is the area of the base times the altitude, and the lateral surface area is the circumference of the base times the altitude.

28. Express the Celsius temperature ($°C$) as a function of the Fahrenheit temperature ($°F$).
Hint. Write $°F = a(°C) + b$ and find a and b from the fact that when $°C = 100$, $°F = 212$, and when $°C = 0$, $°F = 32$.

29. Let $f(x) = (x + 2)/(2x + 1)$. Show that if $x \neq 0$, then

$$f\left(\frac{1}{x}\right) = \frac{1}{f(x)}$$

30. Let $f(x) = 2x$. Show that for any real number k, $f(kx) = k \cdot f(x)$.

31. A function is said to be **even** if $f(-x) = f(x)$ for all x in the domain, and it is said to be **odd** if $f(-x) = -f(x)$ for all x in the domain. Determine in each of the following whether the function is even, odd, or neither:
a. $f(x) = x^2$ **b.** $f(x) = x^3$ **c.** $f(x) = x^2 + 2x$
d. $g(x) = x - 1$ **e.** $h(x) = 2x^2 - 3$

32. Follow the instructions in Problem 31 for the following:

a. $f(x) = \dfrac{x}{x^2 + 1}$ **b.** $g(x) = \dfrac{1}{x^3 - 2x}$ **c.** $h(t) = \dfrac{t + 2}{t - 1}$

d. $\phi(x) = \dfrac{x^2 - 1}{x^2 + 4}$ **e.** $F(t) = |t|$

33. A piece of manufacturing equipment is purchased for $200,000. For tax purposes the value is decreased by $40,000 times the number of years x since the equipment was purchased. Let $f(x)$ be the value of the equipment x years after purchase.
a. Write the rule for the function f.
b. Find the domain of f if the value of the equipment cannot be negative.

34. If a price of $$x$ per unit is charged for a certain product, the number of units sold will be $1,000 - 25x$. Let $f(x)$ be the revenue when the price is x, where revenue is the price per unit times the number of units sold.
a. Write the rule for the function f.
b. Find the domain of f if the revenue is to be nonnegative.

B   Find the domain of each of the functions in Problems 35 and 36.

35. $f(x) = \sqrt{\dfrac{x^2 - 2x - 8}{x + 5}}$ **36.** $g(t) = \sqrt{\dfrac{t + 4}{t^3 - t^2 - t + 1}}$

37. Express the area of an equilateral triangle as a function of one of its sides s.
Hint. Use the Pythagorean theorem.

38. An isosceles triangle of base x is inscribed in a circle of radius 2. Express the area of the triangle as a function of x.
Hint. Use the Pythagorean theorem.

39. A right circular cylinder is inscribed in a sphere of radius a. Express the volume of the cylinder as a function of its base radius r.

40. A rectangle is inscribed in an isosceles triangle of height 4 and base 6. Express the area of the rectangle as a function of its base.
    Hint. Use similar triangles.

41. A pyramid has a square base 4 feet on a side, and its height is 10 feet. Express the area of a cross-section, parallel to the base, that is h feet above the base as a function of h.

42. A storage tank is in the form of a cylinder x feet long and y feet in diameter, with hemispheres at both ends (see sketch).

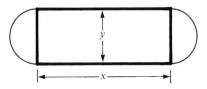

   a. Express the volume of the tank as a function of x and y.
   b. Express the surface area of the tank as a function of x and y.
   c. If $x = 4y$, express the volume and the surface area as functions of x only.

43. An organization issuing credit cards charges the card holder interest of $1\frac{1}{2}\%$ per month for the first $1,000 unpaid balance after 30 days and $\cdot 1\%$ a month on the unpaid balance above $1,000. Let $f(x)$ be the interest charged per month on an unpaid balance of $\$x$, and find the rule for f.

44. A company deducts 2% of gross salary from each employee's pay as the employee's contribution to the pension fund. However, no employee pays more than $400 into the fund in a single year. Let $f(x)$ be the contribution of an employee with gross salary of $\$x$ a year. Write the rule for f.

45. A right circular cone is inscribed in a sphere of radius a. Express the volume of the cone as a function of its base radius r.
    Hint. The volume of a cone is one-third the area of its base times its altitude.

46. Let f denote any function, and define g and h as follows:

$$g(x) = \frac{f(x) + f(-x)}{2} \qquad h(x) = \frac{f(x) - f(-x)}{2}$$

   a. Show that g is even and h is odd. (See Problem 31.)
   b. Show that $f(x) = g(x) + h(x)$. (This proves that every function can be expressed as the sum of an even function and an odd function.)

47. Prove that the range of the function $f(x) = 1 + \sqrt{x}$ is the set $\{y: \ y \geq 1\}$.
    Hint. Show first that the range is contained in the given set. Then show that for an arbitrary number k in that set, there is an x in the domain for which $f(x) = k$.

48. Prove that the range of the function $f(x) = 2x/(x^2 + 1)$ is the set $\{y: \ |y| \leq 1\}$.

5   One-to-One Functions

In the definition of a function it is required that each element in the domain correspond to an element in the range. However, it is entirely possible that two or more different elements in the domain correspond to the same element in the range. For example, let $f(x) = x^2$. We see that $f(2) = 4$ and also $f(-2) = 4$. So both 2 and -2 correspond to the same element, 4, in the range. In fact, if a is any nonzero number, $f(a) = a^2$ and $f(-a) = a^2$. Thus, with the exception of 0, each element of the range is the image of two distinct elements of the domain.

As an extreme example, consider the function g defined by $g(x) = 1$. Here, regardless of what element of the domain is selected, its image is 1; *all* elements in the domain have the same image.

In this section we consider a class of functions for which the situations described above cannot occur. These are the so-called **one-to-one functions** (also written "1–1"). In the next section we will see one of the important properties possessed by these functions.

DEFINITION 2 Let f be a function from A to B. Then f is said to be **one-to-one** provided that whenever x_1 and x_2 are in A, with $x_1 \neq x_2$, then $f(x_1) \neq f(x_2)$. Equivalently, f is one-to-one if whenever $f(x_1) = f(x_2)$, then $x_1 = x_2$.

The second criterion stated in the definition is often easier to use in testing whether a particular function is one-to-one. We illustrate with several examples.

EXAMPLE 12 Let $f(x) = 3x + 4$. Show that f is 1–1.

Solution Suppose $f(x_1) = f(x_2)$, that is, $3x_1 + 4 = 3x_2 + 4$. Then $3x_1 = 3x_2$, and hence $x_1 = x_2$. So f is 1–1.

EXAMPLE 13 Show that the function g defined by $g(x) = \sqrt{x}$ is 1–1 on its domain.

Solution Note that the domain consists of all nonnegative real numbers, that is, the set $\{x: \ x \geq 0\}$. Suppose $g(x_1) = g(x_2)$, that is, $\sqrt{x_1} = \sqrt{x_2}$. Squaring both sides yields immediately that $x_1 = x_2$. Thus, g is 1–1.

EXAMPLE 14 Let $h(x) = x^2$. Show that h is not 1–1 on all of R, but that it is 1–1 on the set $\{x: \ x \geq 0\}$.*

Solution When the domain is all of R, we can select any nonzero number and its negative to show that h is not 1–1. For example, $h(3) = 9$ and $h(-3) = 9$. So even though $3 \neq -3$, $h(3) = h(-3)$, which cannot happen with a 1–1 function.

In the second case, suppose $h(x_1) = h(x_2)$, so that $x_1^2 = x_2^2$. Then taking square roots, with both x_1 and x_2 restricted to nonnegative values, we must have $x_1 = x_2$. Thus, h is 1–1 on $\{x: \ x \geq 0\}$.

* Strictly speaking, we should consider these as two distinct functions, for even though the rules are the same, the domains are different. It is a convenience, however, to use the same letter to name both functions, and it is not likely to lead to difficulty as long as the proper domain is specified.

EXAMPLE 15   Show that the function ϕ defined by $\phi(t) = t^2 - 2t - 3$ is 1–1 on the domain $\{t: \ t \geq 1\}$.

Solution   If $\phi(t_1) = \phi(t_2)$, then $t_1^2 - 2t_1 - 3 = t_2^2 - 2t_2 - 3$. We rearrange and factor:

$$t_1^2 - t_2^2 - 2t_1 + 2t_2 = 0$$
$$(t_1 - t_2)(t_1 + t_2) - 2(t_1 - t_2) = 0$$
$$(t_1 - t_2)(t_1 + t_2 - 2) = 0$$

So either $t_1 = t_2$ or $t_1 + t_2 = 2$. But the only way the latter equation can be true when t_1 and t_2 are greater than or equal to 1 is for both t_1 and t_2 to equal 1. In both cases, then, $t_1 = t_2$, so that ϕ is 1–1 on the specified domain.

EXAMPLE 16   Determine whether the function $f(x) = x^3 - 1$ is 1–1 on R.

Solution   If $f(x_1) = f(x_2)$, then $x_1^3 - 1 = x_2^3 - 1$, and so $x_1^3 = x_2^3$, or $x_1^3 - x_2^3 = 0$. Factoring, we obtain

$$(x_1 - x_2)(x_1^2 + x_1 x_2 + x_2^2) = 0$$

from which it follows that either $x_1 = x_2$ or $x_1^2 + x_1 x_2 + x_2^2 = 0$. If we think of this second equation as being quadratic in x_1, the discriminant is $x_2^2 - 4x_2^2 = -3x_2^2$, which is negative unless $x_2 = 0$. So the solutions are imaginary if $x_2 \neq 0$. When $x_2 = 0$, the equation becomes

$$x_1^2 + x_1 \cdot 0 + 0^2 = 0$$
$$x_1^2 = 0$$

so that x_1 also is 0. Therefore, for real values of x_1 and x_2, we have in both cases that $x_1 = x_2$. So f is 1–1 on R.

EXERCISE SET 5

A   In Problems 1–12 show that the given functions are 1–1 on R.

1. $f(x) = 2x$
2. $f(x) = x + 4$
3. $g(x) = 3x - 2$
4. $h(x) = 2 - x$
5. $f(t) = 2t - 3$
6. $F(x) = 5x + 3$
7. $\phi(x) = \dfrac{3x}{4} + \dfrac{2}{3}$
8. $g(t) = t^3$
9. $f(x) = \dfrac{2(3 - x)}{5}$
10. $g(x) = \dfrac{1 - 2x}{3}$
11. $h(t) = \dfrac{2t}{3} - \dfrac{1}{4}$
12. $f(s) = \dfrac{4s - 5}{7}$

13. Let $\phi(x) = 1/x$. What are the domain and range of ϕ? Determine whether ϕ is 1–1.
14. What is the domain of the function $f(x) = \sqrt{x - 1}$? Is f a 1–1 function on this domain?

15. Let $\phi(x) = |x|$. Is ϕ a 1–1 function? Draw the graph of ϕ.

16. Let $h(x) = |x - 2|$. Is h a 1–1 function? Draw the graph of h.

17. Let $g(x) = 1/\sqrt{x}$. Show that g is a 1–1 function from R^+ to R^+. (Recall that R^+ means the set of all positive real numbers.)

18. Let h be the function from R to R defined by $h(x) = x^3 - x$. Show that h is not 1–1.
    Hint. Find two or more numbers that have 0 as their image.

19. Let $f(t) = t/(t + 1)$. What is the domain of f? Determine whether f is 1–1.

20. Let $f(t) = (2t - 1)/(2t + 1)$. What is the domain of f? Determine whether f is 1–1.

21. Determine whether the function ϕ defined by $\phi(t) = t^2 - 2t - 8$ is 1–1.

22. Let $g(x) = (1 - 2x)/(3x - 4)$. What is the domain of g? Determine whether g is 1–1.

23. Let $h(x) = 2 + \sqrt{x - 1}$. Find the domain of h, and prove that h is 1–1.

24. Show that if $f(x) = |3x - 5|$, then f is not 1–1 on R. Find a suitable restriction on the domain so that f will be 1–1.

B 25. **a.** Show that the function f defined by $f(x) = 1 - x^2$ is not 1–1.
    b. Restrict the domain of f in a suitable way so that it will be 1–1.

26. Let $g(x) = x^2 - 2x$. Show that g is not 1–1 on all of R but that it is 1–1 on the domain $\{x: \ x \geq 1\}$.

27. Show three distinct values of x that have the same image for the function $f(x) = x^3 - 4x^2 - x + 4$. What conclusion can you draw?

28. In Exercise Set 4 an even function was defined as one for which $f(-x) = f(x)$ and an odd function as one for which $f(-x) = -f(x)$, for all x in the domain. Show that an even function cannot be 1–1. Show by means of examples that an odd function may or may not be 1–1.

29. A function is said to be **increasing** if whenever $x_1 < x_2$, then $f(x_1) < f(x_2)$. Similarly, f is **decreasing** if whenever $x_1 < x_2$, then $f(x_1) > f(x_2)$. Show that if f is either increasing on its domain or decreasing on its domain, then f is 1–1.

30. A function f is said to be **periodic** if there exists a positive real number k such that $f(x + k) = f(x)$ for all x in the domain of f. Show that a periodic function cannot be 1–1.

31. To show that f is a function from A onto B, it is sufficient to show that whatever element y in B is chosen, there is some element x in A for which $f(x) = y$. Show that the function f for which $f(x) = \sqrt[3]{x}$ is a 1–1 function from R onto R.

32. Show that if $g(x) = (3 - 2x)/4$, then g is a 1–1 function from R onto R.

33. Let $f(x) = (x + 1)/(x - 1)$.
    a. Show that f is 1–1 on its domain.
    b. The set of images of f is contained in the set $\{y: \ y \geq 0\}$. Is f a 1–1 function from its domain onto this set?

34. The notation $[x]$ is defined to mean the greatest integer less than or equal to x. For example, $[2.3] = 2$, $[5\frac{3}{4}] = 5$, $[6] = 6$, and $[-1.5] = -2$. Show that the function f defined by $f(x) = [x]$ from R to R is neither 1–1 nor onto.

6 Composite Functions and Inverses

Functions are often compared to machines. For example, a value of x from the domain is fed into the machine. The machine (the function) then operates on x and out comes $f(x)$,* as shown in Figure 9. Now it frequently happens that having applied the function f to a number x and obtained $f(x)$ as an

---

* A calculator can be thought of as such a function machine.

Figure 9

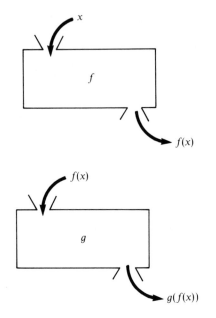

Figure 10

output, we then input $f(x)$ into a second function machine, say, g, as illustrated in Figure 10. The output of g would then be designated $g(f(x))$. When we first apply f and then g to the result, the overall effect on x is to produce $g(f(x))$. We could consider a new machine (a new function) that does this all at once, and this is usually designated $g \circ f$, as indicated in Figure 11. It is called the **composite** of g and f. For example, suppose that $f(x) = 2x^2 + 3$ and $g(x) = \sqrt{x}$. Then $g(f(x)) = \sqrt{2x^2 + 3}$.

Note. Observe that $g \circ f$ and $f \circ g$ need not be—and usually are not—equal. In this example, the function $f \circ g$ has the following effect on x:

$$(f \circ g)(x) = f(g(x)) = 2(\sqrt{x})^2 + 3 = 2x + 3$$

Care must be exercised in specifying the domain of a composite function. Since $(g \circ f)(x) = g(f(x))$, the (maximum) domain of $g \circ f$ consists of those values of x in the domain of f for which $f(x)$ lies in the domain of g; we cannot input $f(x)$ into the g function unless $f(x)$ lies in the domain of g. In our example above, where $f(x) = 2x^2 + 3$ and $g(x) = \sqrt{x}$, this was no problem, since the range of f consists of real numbers greater than or equal to 3, and these are

Figure 11

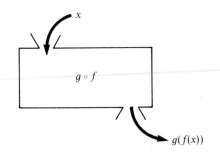

included in the domain of g. However, if we had been given $f(x) = x^2 - 4$ and $g(x) = \sqrt{x}$, then we would have $(g \circ f)(x) = \sqrt{x^2 - 4}$, so that the domain of $g \circ f$ would be restricted to those values of x for which $x^2 - 4 \geq 0$, or equivalently $(x - 2)(x + 2) \geq 0$. Using either of the techniques for solving quadratic inequalities given in Chapter 3, we find the solution set to be $\{x: \ x \geq 2\} \cup \{x: \ x \leq -2\}$. This, then, is the domain of $g \circ f$.

We now wish to introduce the concept of the **inverse** of a function. We will see that this has a direct relationship with composite functions. Roughly speaking, two functions are inverses of one another if each undoes what the other does. Thus, if $f(x) = 2x$ and $g(x) = x/2$, then f and g are inverses; f doubles any given number in its domain, whereas g halves numbers in its domain. So the effect of first applying f and then g, or vice versa, to a given number x is that there is no net change. This is typical of functions and their inverses.

DEFINITION 3   We say that f and g are inverses of one another provided that

$$\text{Range of } g = \text{Domain of } f$$
$$\text{Range of } f = \text{Domain of } g$$

and

$$(f \circ g)(x) = x$$
$$(g \circ f)(x) = x$$

EXAMPLE 17   Show that f and g, defined by $f(x) = 2x + 1$ and $g(x) = (x - 1)/2$, are inverses.

Solution   Both f and g have all of R as domain and range. Furthermore,

$$(f \circ g)(x) = f(g(x)) = 2\left(\frac{x - 1}{2}\right) + 1 = x$$

$$(g \circ f)(x) = g(f(x)) = \frac{(2x + 1) - 1}{2} = x$$

So f and g are inverses.

EXAMPLE 18   Show that if the domain of the function f defined by $f(x) = x^2$ is restricted to be $\{x: \ x \geq 0\}$, then if g is defined by $g(x) = \sqrt{x}$, f and g are inverses.

Solution   The range of f consists of all nonnegative numbers, which is precisely the domain of g, and the range of g consists of the set of nonnegative numbers, which by the given restriction is the domain of f. Also,

$$(f \circ g)(x) = f(g(x)) = (\sqrt{x})^2 = x$$
$$(g \circ f)(x) = g(f(x)) = \sqrt{x^2} = x$$

Note.   We have $\sqrt{x^2} = x$ since $x \geq 0$.

Figure 12

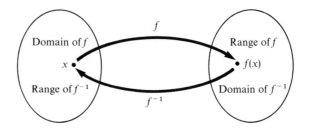

The symbol f^{-1} is customarily used to denote the inverse of f. So we always have

$$(f \circ f^{-1})(x) = (f^{-1} \circ f)(x) = x$$

We could say that $f \circ f^{-1}$ (or $f^{-1} \circ f$) is the **identity function,** because it leaves a number unchanged. Figure 12 shows how f and f^{-1} are related.

For the inverse of a function f to exist, it must be true that each element of the range of f is the image of precisely one element in the domain. If we begin with an x in the domain of f, its image is $f(x)$. Now if f^{-1} exists, it must make $f(x)$ correspond to x and to nothing else:

$$f^{-1}(f(x)) = x$$

But the requirement that each element in the range of f be the image of exactly one element in the domain is just the requirement that f be one-to-one.

Suppose now that f is 1–1. We will show that it has an inverse. All we need to do is to define f^{-1} as that function having domain equal to the range of f and for which

$$f^{-1}(y) = x$$

where $y = f(x)$. This is a valid definition, since each such y uniquely determines an x.

So we have the following result:

A function has an inverse if and only if it is 1–1.

Suppose we determine that a given function does have an inverse. How do we go about actually finding the inverse? Let

$$y = f(x)$$

Now apply f^{-1} to both sides:

$$f^{-1}(y) = f^{-1}(f(x)) = x$$

So

$$x = f^{-1}(y)$$

It appears, then, that to find an equation defining f^{-1}, we solve the equation $y = f(x)$ for x in terms of y. That is, we reverse the roles of dependent and independent variables. Consider, for example, the problem of finding the inverse of the function f defined by $f(x) = 2x + 3$. We can show that f is 1–1, so it has an inverse. Now set $y = 2x + 3$ and solve for x:

$$2x = y - 3$$

$$x = \frac{y - 3}{2}$$

So $f^{-1}(y) = (y - 3)/2$. This adequately defines the function f^{-1}; it is the function that subtracts 3 from the input value and then divides the result by 2. The particular letter used in the defining equation for a function is not important. We could just as well write

$$f^{-1}(t) = \frac{t - 3}{2} \quad \text{or} \quad f^{-1}(u) = \frac{u - 3}{2} \quad \text{or} \quad f^{-1}(x) = \frac{x - 3}{2}$$

Each of these in effect tells us what the function f^{-1} does. Since it is customary to use the letter x as the independent and y as the dependent variable in a functional relationship, this suggests the following procedure for finding the inverse of a 1–1 function f:

1. Set $y = f(x)$.
2. Solve the equation for x in terms of y.
3. Interchange x and y.

Again, consider the example $f(x) = 2x + 3$. We write

$$y = 2x + 3$$

$$x = \frac{y - 3}{2}$$

So,

$$y = \frac{x - 3}{2} = f^{-1}(x)$$

Two points should be emphasized regarding the procedure given above for finding the inverse. First, before attempting to apply the procedure it should be determined that the inverse actually exists. However, if this is not done in advance, the procedure itself will generally show this. For example, consider $f(x) = x^2 + 1$. We set $y = x^2 + 1$, solve for x:

$$x = \pm \sqrt{y - 1}$$

and interchange x and y:

$$y = \pm \sqrt{x - 1}$$

But the ambiguity of sign shows that no unique inverse exists.

Second, a major problem with the procedure is that it is not always possible to solve for x. For example, try to solve $y = x^5 + x - 1$ for x!

EXAMPLE 19   Show that if $f(x) = \sqrt{x - 1}$, then f has an inverse on the domain $\{x: \quad x \geq 1\}$, and find that inverse.

Solution   We show first that f is 1–1. Suppose $f(x_1) = f(x_2)$, so that $\sqrt{x_1 - 1} = \sqrt{x_2 - 1}$. It follows by squaring both sides that $x_1 - 1 = x_2 - 1$, or $x_1 = x_2$. Thus, f is 1–1 and hence has an inverse. Now set

$$y = \sqrt{x - 1}$$

Squaring and solving for x yields

$$y^2 = x - 1$$
$$x = y^2 + 1$$

Finally, the rule for f^{-1} is

$$y = x^2 + 1$$

that is,

$$f^{-1}(x) = x^2 + 1$$

But we must carefully state the domain of f^{-1}. It is precisely the range of f. Since $f(x) = \sqrt{x - 1}$, the range is contained in the set of nonnegative real numbers; in fact, this is the range of f. If we take any nonnegative real number k, we can find an x for which $f(x) = k$ as follows:

$$\sqrt{x - 1} = k$$
$$x - 1 = k^2$$
$$x = 1 + k^2$$

This value of x does work, since

$$f(x) = f(1 + k^2) = \sqrt{1 + k^2 - 1} = \sqrt{k^2} = k$$

Therefore, the domain of f^{-1} is the set of all nonnegative real numbers.

EXERCISE SET 6

A   In Problems 1–12 find $f \circ g$ and $g \circ f$.

1. $f(x) = x - 1$, $g(x) = 2x + 1$

2. $f(x) = 2x$, $g(x) = x + 3$

3. $f(x) = 3x - 5$, $g(x) = 2 - x$

4. $f(x) = 3x + 4$, $g(x) = \dfrac{2x - 1}{3}$

5. $f(x) = 2x^2$, $g(x) = x + 2$

6. $f(x) = \sqrt{x - 2}$, $g(x) = x^2 + 4$

7. $f(x) = \dfrac{1}{x + 1}$, $g(x) = \dfrac{1}{x - 3}$

8. $f(x) = x^3$, $g(x) = x^2$

9. $f(x) = x^2 - 2x + 1$, $g(x) = 2x - 3$

10. $f(x) = 1/x$, $g(x) = 3x + 4$

11. $f(x) = \dfrac{x}{x - 1}$, $g(x) = \dfrac{2}{x}$

12. $f(x) = 2$, $g(x) = x^2 - 4$

In Problem 13–15 find $f \circ g$ and $g \circ f$, and give the domain of each.

13. $f(x) = \sqrt{x - 1}$, $g(x) = x^2$

14. $f(x) = 2x^2 - 1$, $g(x) = \sqrt{x}$

15. $f(x) = \sqrt{x^2 - 1}$, $g(x) = 2x - 3$

16. Let $f(x) = x^2 - 1$. Find $(f \circ f)(x)$.

17. Let $f(x) = (2x - 3)/5$, $g(x) = x^2 - 3$, and $h(x) = (5x + 3)/2$. Find:
    a. $f(g(x))$ **b.** $g(f(x))$ **c.** $f(h(x))$ **d.** $h(f(x))$ **e.** $g(h(x))$

In Problems 18–32 show that an inverse exists, and find it.

18. $f(x) = x - 1$ **19.** $f(x) = 2x + 5$

20. $f(x) = 3x + 7$ **21.** $g(x) = \dfrac{x}{2} + 4$

22. $h(x) = \dfrac{2x - 5}{3}$ **23.** $G(x) = \dfrac{3x - 4}{7}$

24. $f(t) = \sqrt{t + 4}$ **25.** $g(x) = \dfrac{1}{x + 2}$

26. $F(x) = x^3$ **27.** $G(x) = \sqrt{x}$

28. $f(t) = \dfrac{1}{t}$ **29.** $h(t) = \dfrac{t}{t + 3}$

30. $g(t) = 1 + t^2$, $t \geq 0$ **31.** $F(z) = \sqrt{z^2 - 4}$, $z \geq 2$

32. $\phi(x) = \dfrac{x - 1}{x + 1}$

B   In Problems 33–36 show that an inverse does not exist. Then place a suitable restriction on the domain so that an inverse does exist, and find it.

33. $f(x) = x^2 - 4$ **34.** $g(x) = \dfrac{1}{x^2 + 1}$

35. $h(x) = x^2 - 2x$ **36.** $y = \sqrt{4 - x^2}$

37. Let $f(x) = (x + a)/(x + b)$, where $a \neq b$. Show that f has an inverse, and find it. What are the domain and range of f and of f^{-1}? Show that $f(f^{-1}(x)) = x$ and $f^{-1}(f(x)) = x$.

38. Let $f(x) = x^2 - 1$, $g(x) = 2x + 3$, and $h(x) = 1/(x + 1)$. Find $[f \circ (g \circ h)](x)$ and $[(f \circ g) \circ h](x)$. In general, if f, g, and h are arbitrary functions (with suitably restricted domains), what do you conjecture about $f \circ (g \circ h)$ and $(f \circ g) \circ h$?

39. If a satellite of mass m is at a distance $f(t)$ from the center of the earth at time t, then the force of gravity exerted by the earth on the satellite is $(h \circ f)(t)$, where $h(t) = km/t^2$, $k = gR$, g is the force of gravity at the surface of the earth, and R is the radius of the earth. If $f(t) = 6{,}400 + 4t - 0.0001t^2$, $g = 9.81$, and $R = 6{,}400$, find the rule for $h \circ f$.

7   Graphs of Functions

Let f be a function from A to B, where A and B are subsets of R. If x is an element in A, let y denote its image under f, that is, $y = f(x)$. Thus, f establishes a pairing of values (x, y) with x in A and y in B. The totality of such pairs as x varies over all of A can be plotted in a cartesian coordinate system, and the result is called the **graph of f**. If the rule for f is given in the form of an equation, the graph of f is identical to the graph of this equation as defined in Section 3. For example, the graph of the function f for which $f(x) = 2x + 3$ is the same as the graph of the equation $y = 2x + 3$. However, an important distinction must be made. Not all graphs of equations in two variables are graphs of

Figure 13

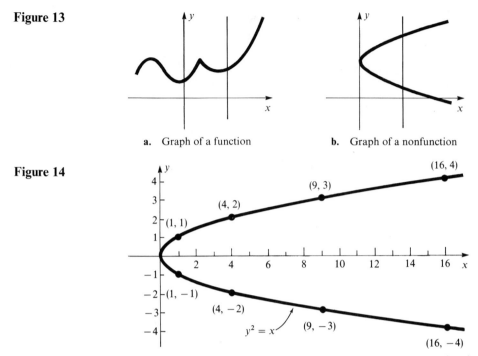

a. Graph of a function   **b.** Graph of a nonfunction

Figure 14

functions. This is because of the requirement that in the case of a function, each x in the domain determines exactly one value y in the range. **Geometrically, this amounts to the requirement that a vertical line through any point on the x axis strikes the graph of f at no more than one point** (Figure 13).

As a specific example of an equation that does not define y as a function of x, consider $y^2 = x$. If we solve for y, we get $y = \pm\sqrt{x}$, so that each $x > 0$ determines two values of y, which cannot happen in the case of a function. The graph of $y^2 = x$ (Figure 14) can be obtained by making a table of values, as follows:

x	1	4	9	16
y	±1	±2	±3	±4

It is possible to view the equation $y = \pm\sqrt{x}$ as defining two different functions, say,

$$y_1 = \sqrt{x} \quad \text{and} \quad y_2 = -\sqrt{x}$$

corresponding to the upper and lower halves, respectively, of Figure 14.* In a

---

* There are, in fact, various other ways to define functions from the equation, but the way given is the natural decomposition. We could, for example, let

$$y_1 = \begin{cases} \sqrt{x} & \text{if } x \text{ is rational and } \geq 0 \\ -\sqrt{x} & \text{if } x \text{ is irrational and } > 0 \end{cases} \qquad y_2 = \begin{cases} -\sqrt{x} & \text{if } x \text{ is rational and } \geq 0 \\ \sqrt{x} & \text{if } x \text{ is irrational and } > 0 \end{cases}$$

but it is unlikely that this would be very useful.

Figure 15

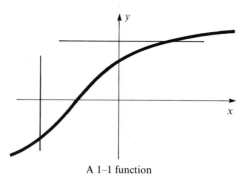

A 1–1 function

similar way, it is often possible to consider an equation whose graph is not a single function as defining more than one function.

A 1–1 function has the property that each point in the range corresponds to exactly one point in the domain. Geometrically, this means that each horizontal line intersects the graph at no more than one point. **Thus, for 1–1 functions both vertical and horizontal lines intersect the graph at no more than one point** (Figure 15).

EXAMPLE 20   Determine which of the graphs in Figure 16 are graphs of functions and which are not. For those that are graphs of functions, determine whether an inverse exists.

Figure 16

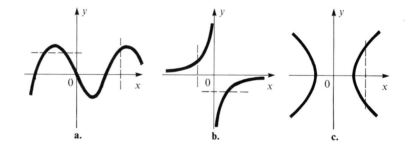

a.   b.   c.

Solution   **a.**   This a function since each vertical line cuts the curve in only one point. Some horizontal lines cut the curve in more than one point, so the function is not 1–1 and hence has no inverse.

b.   Each vertical line and each horizontal line cuts the curve at most in one point, so this is a function and it is 1–1. It does have an inverse.

c.   Some vertical lines cut the curve in two points, so this is not a function.

We can now see the geometrical relationship between a function and its inverse. Let f be a 1–1 function from A onto B, where A and B are subsets of R. In order to obtain the equation $y = f^{-1}(x)$, we solve the equation $y = f(x)$ for x and then interchange the roles of x and y. It follows that if (a, b) is a

Figure 17

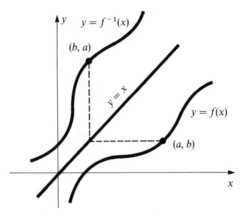

point on the graph of $y = f(x)$, then (b, a) is on the graph of $y = f^{-1}(x)$. The situation is shown in Figure 17. The graph of $y = f^{-1}(x)$ is the reflection of the graph of $y = f(x)$ in the line through the origin making a $45°$ angle with the x axis.* On this line the x coordinate and the y coordinate of any point are equal, and this holds true only for points on this line. Thus, the equation of the line is $y = x$. We say that the graphs of a 1–1 function and its inverse are **symmetric with respect to the line $y = x$.**

A function is said to be *increasing* on an interval if for any x_1 and x_2 on the interval, with $x_1 < x_2$, it is true that $f(x_1) < f(x_2)$. If when $x_1 < x_2$, it is true that $f(x_1) > f(x_2)$, then f is *decreasing* on the interval. Figure 18 illustrates increasing and decreasing functions.

Figure 18

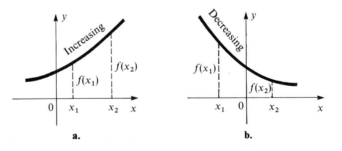

a.   b.

It is useful to be able to identify from the graph of a function the intervals on which it is increasing and those on which it is decreasing. The next example illustrates this.

EXAMPLE 21   Determine the intervals on which the function whose graph is shown in Figure 19 is increasing and decreasing.

---

* This presupposes that we have chosen the same scale on both the x and y axes.

Figure 19

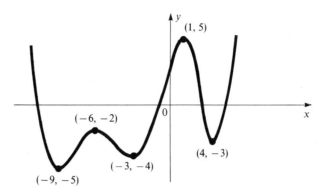

Solution   Increasing on the intervals $[-9, -6]$, $[-3, 1]$, and $[4, \infty)$.
Decreasing on the intervals $(-\infty, -9]$, $[-6, -3]$, and $[1, 4]$.

EXERCISE SET 7

A   Graph the functions in Problem 1–9.

1. **a.** $f(x) = x - 2$ **b.** $f(x) = 3x + 4$

2. **a.** $f(x) = 2x - 1$ **b.** $g(x) = 2 - 5x$

3. **a.** $f(x) = \dfrac{2x + 5}{3}$ **b.** $g(x) = \dfrac{4 - 3x}{5}$

4. **a.** $f(x) = x^2 - 2x$ **b.** $h(x) = x^3 - 1$

5. **a.** $g(x) = \sqrt{x - 1}$ **b.** $F(x) = \sqrt{4 - x}$

6. **a.** $f(x) = \dfrac{4}{x^2}$ **b.** $g(x) = \sqrt{4 - x^2}$

7. **a.** $f(x) = \dfrac{3}{2x}$ **b.** $h(x) = \dfrac{1}{x - 1}$

8. **a.** $f(x) = \begin{cases} x & \text{if } x \geq 0 \\ -1 & \text{if } x < 0 \end{cases}$ **b.** $g(x) = \begin{cases} x^2 & \text{if } x > 1 \\ 0 & \text{if } x = 1 \\ 2 - x & \text{if } x < 1 \end{cases}$

9. **a.** $F(x) = \begin{cases} 2x & \text{if } & x \geq 2 \\ 1 - x & \text{if } -1 \leq x < 2 \\ 2 & \text{if } & x < -1 \end{cases}$ **b.** $h(x) = \begin{cases} \sqrt{x - 4} & \text{if } x \geq 4 \\ 1 - \dfrac{x}{4} & \text{if } x < 4 \end{cases}$

In Problems 10 and 11 determine which of the graphs are graphs of functions and which are not. For those that are graphs of functions, state whether an inverse exists.

10. **a.**   **b.**   **c.**

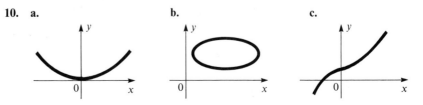

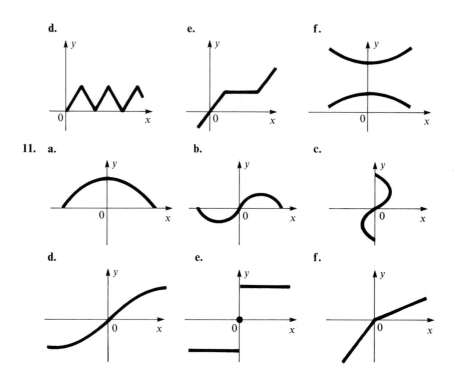

11.

In Problems 12–14 show that each function is 1–1. Graph each function and its inverse on the same set of axes.

12. **a.** $f(x) = \dfrac{x + 1}{3}$ **b.** $g(x) = x^2, \quad x \geq 0$

13. **a.** $f(x) = 2x - 5$ **b.** $g(x) = \sqrt{x - 2}$

14. **a.** $h(x) = \dfrac{3x + 4}{5}$ **b.** $\phi(x) = \sqrt{1 - x}$

In Problems 15 and 16 the graph of a function is given. On the same set of axes sketch the graph of the inverse.

15. **a.**   **b.**

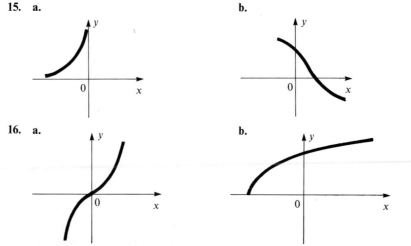

16. **a.**   **b.**

In Problems 17 and 18 determine the intervals on which the function whose graph is shown is increasing and those on which it is decreasing.

17.

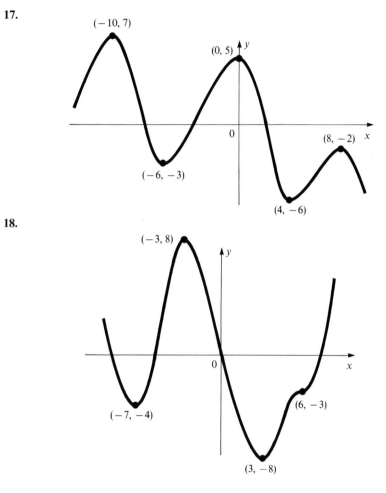

18.

B In Problems 19–22 exhibit two functions defined by the given equation and graph each function. State whether either function has an inverse.

19. $x^2 + y^2 = 4$ **20.** $y^2 = 1 - x$

21. $x = \sqrt{1 + y^2}$ **22.** $|x| + |y| = 1$

23. Graph the function f defined by

$$f(x) = \begin{cases} x^3 - 1 & \text{if} & x > 1 \\ 1 - x & \text{if} & 0 < x \leq 1 \\ 2 & \text{if} & x = 0 \\ 1 - x^2 & \text{if} & x < 0 \end{cases}$$

24. A model for the operation of a hydroelectric plant determines the function f below, which gives the amount of water that should be allowed to pass through the turbines if the amount of water in the reservoir is x. It is assumed that a steam generating plant supplying k units of energy is available to supplement the hydroelectric plant. If the total power required in a given period of time is A, then there are water levels a and b so that

$$f(x) = \begin{cases} x & \text{if} & 0 \le x < A - k \\ A - k & \text{if} & A - k \le x < a \\ \dfrac{k}{b-a}(x-b) + A & \text{if} & a \le x < b \\ A & \text{if} & x \ge b \end{cases}$$

Given that $A = 100$, $k = 70$, $a = 42$, and $b = 74$, graph the function f.

8 Variation

The following experiment demonstrates a recurrent type of functional relationship in physical science, as well as in the social and life sciences: An experiment in physics consists of attaching various weights to a spring and measuring the amount by which the spring is stretched; this is called the **elongation.** The object of the experiment is to determine a functional relationship between elongation and weight. Suppose that for a certain spring the measurements in Table 1 were found. From these data we would expect the pattern of each additional pound stretching the spring another 0.5 inch to continue, so long as the elastic limit of the spring is not exceeded. (At this limit the spring is stretched so much that it will not return to its original shape.)

Table 1

Weight (pounds)	Elongation (inches)
1	0.5
2	1.0
3	1.5
4	2.0

Let us designate the weight by w and the elongation by x. The functional relationship between w and x can be determined if we extend our table to include the ratio of w to x, as shown in Table 2. Since the ratio is constant for these four values, we assume it remains constant, so that for all x less than the elastic

Table 2

w	x	w/x
1	0.5	2
2	1.0	2
3	1.5	2
4	2.0	2

limit of the spring, $w/x = 2$, or $w = 2x$. This is an example of what is called **direct variation,** or **direct proportion.** The fact that w/x is constant is an expression of what is known as **Hooke's law.** The particular constant depends on the spring and is called the **spring constant.**

In other types of situations we may find that instead of the ratio of two quantities being constant, it is their product that is constant, and this leads to what is known as **inverse variation,** or **inverse proportion.** The following experiment in chemistry illustrates this: The experiment consists of measuring the pressure (p) of a gas occupying varying volumes (v) while held at constant temperature. The data are given in Table 3. Note that as the volume decreases, the pressure increases. It appears that for this gas at the given constant temperature, $pv = 6$, or $p = 6/v$.

Table 3

Pressure, p (millimeters of mercury)	Volume, v (cubic centimeters)	$p \cdot v$
2	3.00	6
3	2.00	6
5	1.20	6
8	0.75	6

The expressions *varies directly as* and *varies inversely as* are used to describe such situations as these two experiments illustrate. For example, Hooke's law says that the weight necessary to stretch a given spring (within its elastic limit) *varies directly* as the amount it is stretched beyond its natural length. As we saw, this means that $w = kx$. Also, we can say that for a gas at constant temperature the pressure *varies inversely as* the volume. As another example, Newton's law of gravitation states that the force of attraction between two bodies *varies jointly* as their masses and inversely as the square of the distance between them. Also, in biology we know from experiment that within certain limits the rate of growth of a bacteria culture is *proportional to* the number of bacteria present.

These concepts are made more precise by the following definition:

DEFINITION 4 The statement *y* **varies directly as** *x* means there is a real number $k \neq 0$ such that

$$y = kx$$

The statement *y* **varies inversely as** *x* means there is a real number $k \neq 0$ such that

$$y = \frac{k}{x}$$

The statement **w varies jointly as x and y** means there is a real number $k \neq 0$ such that

$$w = kxy$$

Instead of saying y varies directly as x we sometimes say y is **directly proportional** to x, and this is understood to mean the same thing. Similarly, inverse variation and joint variation can be stated using the language of proportionality. If the word *directly* is omitted, it is understood, but *inversely* and *jointly* must always be explicitly stated. There can be combined statements, such as "y varies jointly as x and the square of t and inversely as v^3," which means

$$y = \frac{kxt^2}{v^3}$$

The real number k in each case is called the **constant of variation,** or the **constant of proportionality.** Often, sufficient information is available to find k, in which case a completely defined functional relationship is determined and various values of the dependent variable can be found. The examples below should help to make these ideas clearer.

EXAMPLE 22   Translate each of the following into an appropriate mathematical statement:

a.   The temperature T of a gas in a container of fixed volume varies directly as the pressure p.

b.   The force F necessary to stop a body that is moving in a straight line varies directly as the square of the velocity v.

c.   The time t required for a satellite traveling in a circular orbit around the earth to complete its orbit is inversely proportional to its velocity v.

d.   The drag force F on an airplane varies jointly as the total cross-sectional area A and the square of the velocity v.

Solution   a. $T = kp$ b. $F = kv^2$ c. $t = \dfrac{k}{v}$ d. $F = kAv^2$

EXAMPLE 23   It is known that y varies directly as x and that when $x = 2$, then $y = 10$. Find y when $x = 7$.

Solution   The variational statement translates to $y = kx$. Now we know one pair of values satisfying this equation, namely $x = 2$, $y = 10$. So we substitute these to find k:

$$10 = k \cdot 2$$
$$k = 5$$

Therefore, the specific functional relationship is

$$y = 5x$$

From this we can find y for any other value of x. The problem asks for y when $x = 7$:

$$y = 5(7) = 35$$

Note that one set of values of the variables is sufficient to determine k, because k is constant. Having found it for one set of the variables, it is determined once and for all.

The preceding example is typical of most variation problems. There are four steps to their solution:

1. Translate the variational statement into a mathematical equation using Definition 4.
2. Substitute the given set of values for all the variables involved, and solve for k.
3. Substitute the value of k into the original equation. This gives an explicit functional relationship.
4. Solve for the unknown requested by substituting the corresponding known value (or values).

EXAMPLE 24   If w varies directly as t^2 and inversely as v, and $w = 15$ when $t = 2$ and $v = 3$, find w when $t = 6$ and $v = 8$.

Solution   We have

$$w = k \cdot \frac{t^2}{v}$$

$$15 = k \frac{(2)^2}{3}$$

$$15 = \frac{4k}{3}$$

$$4k = 45$$

$$k = \tfrac{45}{4}$$

So the original equation becomes

$$w = \frac{45}{4} \cdot \frac{t^2}{v}$$

Substituting $t = 6$ and $v = 8$, we get

$$w = \frac{45}{\overset{}{\underset{1}{4}}} \cdot \frac{\overset{9}{\cancel{36}}}{8} = \frac{405}{8}$$

EXAMPLE 25   The kinetic energy E of a moving object is jointly proportional to its weight w and the square of its velocity v. If an object weighing 32 pounds and moving with a velocity of 60 feet per second has a kinetic energy of 1,800 foot-pounds, find the kinetic energy of an object weighing 96 pounds and traveling at 20 feet per second.

Solution   The given statement translates to

$$E = kwv^2$$

We find k by substituting the known set of values:

$$1,800 = k \cdot (32)(60)^2$$
$$1,800 = k(32)(3,600)$$
$$k = \tfrac{1}{64}$$

So the equation becomes

$$E = \tfrac{1}{64}wv^2$$

Finally, we substitute $w = 96$ and $v = 20$:

$$E = \tfrac{1}{64} \cdot (96)(20)^2$$
$$= \tfrac{3}{2} \cdot 400 = 600 \text{ foot-pounds}$$

EXERCISE SET 8

A   In Problems 1–10 translate the given statement into a mathematical equation.

1. u varies directly as v
2. w varies inversely as t
3. z varies jointly as x and y
4. y is directly proportional to x^2
5. s is inversely proportional to $\sqrt{t}$
6. w varies directly as x and inversely as y
7. F is jointly proportional to m_1 and m_2 and inversely proportional to r^2
8. The force required to stretch a spring x units beyond its natural length is proportional to x.
9. The distance s traveled by a freely falling object varies directly as the square of the time t it has fallen.
10. The volume V occupied by a gas varies directly as the temperature T and inversely as the pressure P.

11. If y varies directly as x, and $y = 4$ when $x = 12$, then find y when $x = 20$.
12. If s varies inversely as r, and $s = 5$ when $r = 8$, then find s when $r = 3$.
13. If u varies jointly as x and y, and $u = 12$ when $x = 6$ and $y = 8$, then find u when $x = 4$ and $y = 12$.
14. If w varies directly as u and inversely as v, and $w = 6$ when $u = 4$ and $v = 2$, then find w when $u = 10$ and $v = 6$.
15. If x is jointly proportional to y and t^2, and $x = 108$ when $y = 4$ and $t = 3$, then find x when $y = 7$ and $t = 4$.

16. The volume of a sphere varies as the cube of its radius. A sphere of radius 3 has a volume of 36π. Find the constant of variation. Write the general formula for the volume of a sphere. What is the volume of a sphere of radius 5?

17. The weight of a body on or above the surface of the earth varies inversely as the square of the distance from the center of the earth. If a body weighs 50 pounds on the earth's surface, how much would it weigh 1,000 miles above the surface of the earth? (Assume the radius of the earth is 4,000 miles.)

18. A printing company has found that for orders between 5,000 and 50,000 the unit cost of printing a college catalogue is inversely proportional to the number printed. If the cost of printing 10,000 of these catalogues is 80¢ per copy, what is the unit cost of printing 25,600 copies?

19. According to Ohm's law the current I in a wire varies directly as the potential E and inversely as the resistance R of the wire. When $R = 20$ ohms and $E = 220$ volts, it is found that $I = 11$ amperes. Find I when $E = 110$ volts and $R = 5$ ohms.

20. The resistance R of a wire varies directly as its length l and inversely as the square of its diameter d. A certain type of wire 60 feet long and with diameter 0.02 inch has a resistance of 15 ohms. What would be the resistance of 90 feet of the same type of wire with cross-section 0.01 inch?

21. The volume of a right circular cone varies jointly as the height h and the square of the base radius r. A cone of base radius 2 and height 6 has volume 8π. Find the constant of variation, and write the general formula for the volume of any right circular cone. What is the volume of a cone of height 10 and base radius 6?

22. The force of the wind on a wall is jointly proportional to the area of the wall and the square of the wind velocity. When the wind is blowing at 20 miles per hour, the force on a wall having area 100 square feet is 180 pounds. Find the force on a wall of area 400 square feet caused by a wind blowing at 50 miles per hour.

23. Newton's law of cooling states that the rate r at which a body cools is proportional to the difference between the temperature T of the body and the temperature T_0 of the surrounding medium. A thermometer registering 70°F is taken outside where the temperature is 40°F, and at that instant the rate of cooling, r, is 18°F per minute. Find the rate of cooling when the thermometer reads 50°F.

24. The vibrating frequency (pitch) of a string varies directly as the square root of the tension in the string. If when the tension is 4 pounds, the frequency of vibration is 240 times per second, find the frequency of vibration when the tension is 16 pounds.

25. One mathematical model of the spread of an infectious disease in a community says that if x is the number of persons already infected and y is the number not infected, then the rate r at which additional persons become infected is jointly proportional to x and y. In a community of 500 persons, when 20 already have a certain infectious disease, the rate of spread is 2 additional cases per day. What will be the rate of spread when 80 persons in the community have the disease? (This model fails to take into consideration such things as quarantine and the duration of the disease.)

B 26. The intensity of light varies inversely as the square of the distance from the source. The intensity of a certain light is 200 candlepower at a point 8 feet from the source. At what distance from the source will the intensity be 100 candlepower?

27. Kepler's third law states that the square of the time it takes a planet to complete its orbit around the sun varies directly as the cube of its mean distance from the sun. The mean distance of Mars from the sun is approximately $1\frac{1}{2}$ times that of the earth. Find the approximate time (in "earth days") it takes Mars to complete its orbit.

28. The force of attraction between two bodies of masses m_1 and m_2, respectively, is jointly proportional to m_1 and m_2 and inversely proportional to the square of the distance d between them. What will be the effect on the force if m_1 is doubled, m_2 is tripled, and the distance d is cut in half?

29. The weight that can be safely supported by a wooden beam of rectangular cross-section varies jointly as the width and square of the depth of the cross-section and inversely as the length of the beam. A beam of length 12 feet with cross-section 4 inches wide by 8 inches deep will safely support 2,000 pounds. How much weight can be safely supported by a beam of the same material that is 8 feet long and has cross-section 2 inches wide by 4 inches deep?

30. The pressure P of a gas varies directly as its absolute temperature T and inversely as its volume V. Gas at pressure 200 pounds per square inch is at a temperature of 47°C and occupies a volume of 500 cubic feet. If the gas is allowed to expand so that its pressure is reduced to 20 pounds per square inch and its temperature is 17°C, find the volume it occupies.
    Note. **Absolute temperature** is measured through positive values from absolute 0, which is equivalent to -273°C, and one degree on the absolute scale equals one degree on the Celsius scale.

31. The mass of a spherical body varies jointly as its density and the cube of its radius. Find the ratio of the mass of Jupiter to that of the earth if the density of Jupiter is $\frac{5}{22}$ that of the earth and its radius is 11 times that of the earth.

32. The time of exposure necessary to photograph an object varies directly as the square of the distance d of the object from the light source and inversely as the intensity of illumination I. When the light source is 10 feet from the object, the correct exposure time for a certain type of film is $\frac{1}{30}$ second. For the same film, if the intensity of the light is doubled and the distance from the light to the object is cut in half (to 5 feet), what will be the correct exposure time?

Review Exercise Set

A In Problems 1–6 make a table of values and draw the graph.

1. **a.** $y = 2x - 1$ **b.** $2x + 3y = 4$
2. **a.** $x - 2y + 3 = 0$ **b.** $y - 3 = 0$
3. **a.** $5x - 2y = 4$ **b.** $x + 2 = 0$
4. **a.** $\dfrac{x + 2y}{4} = 3$ **b.** $\dfrac{y - 3}{2} = x$
5. **a.** $f(x) = 1 - x^2$ **b.** $g(x) = 1 - \sqrt{1 - x}$
6. **a.** $F(x) = \dfrac{1}{x + 2}$ **b.** $h(x) = \sqrt{x^2 + 4}$

7. Let $f(x) = x^2 - 1$. Find:
    a. $f(0)$ **b.** $f(-1)$ **c.** $f(4)$ **d.** $f(a)$ **e.** $f(x + h)$
8. Let $f(x) = x/(x - 2)$. Find:
    a. $f(1)$ **b.** $f(0)$ **c.** $f(3)$ **d.** $f(-5)$ **e.** $f(2.1)$
9. Let $g(t) = 2t/(t - 3)$. What is the domain of g? Find:
    a. $g(1)$ **b.** $g(-2)$ **c.** $g(1/t)$ **d.** $g(t + \Delta t)$
10. Let $g(x) = \sqrt{9 - x^2}$. What is the domain of g? Find:
    a. $g(0)$ **b.** $g(-3)$ **c.** $g(2)$ **d.** $g(\frac{12}{5})$ **e.** $g(-\frac{9}{5})$
11. Let

$$h(t) = \begin{cases} 2t^2 - 1 & \text{if} & t \geq 2 \\ 4 - t & \text{if} & 0 \leq t < 2 \\ 0 & \text{if} & t < 0 \end{cases}$$

Find:
a. $h(5)$ **b.** $h(-1)$ **c.** $h(0)$ **d.** $h(2)$ **e.** $h(1)$

12. Find the distance between each pair of points:
   a. $(-2, 1)$ and $(3, -4)$ **b.** $(0, 3)$ and $(4, 0)$
   c. $(4, -2)$ and $(-1, 10)$ **d.** $(6, 9)$ and $(-2, -6)$
   e. $(7, -5)$ and $(-2, 3)$

13. **a.** Find y so that the distance between $(-3, 2)$ and $(5, y)$ is 10. (There are two solutions.)
   b. Find x so that the distance between $(x, 2)$ and $(-7, -3)$ is 13. (There are two solutions.)

14. Let $f(t) = t - (1/t)$. Find:

   a. $f(1)$ **b.** $f\left(-\dfrac{1}{2}\right)$ **c.** $f\left(\dfrac{1}{t}\right)$ **d.** $\dfrac{1}{f(t)}$

15. Let $f(x) = 1/x^2$. Find $[f(x) - f(2)]/(x - 2)$ and simplify your result.

16. Let $f(x) = x/(x + 1)$. Find $[f(x) - f(-2)]/(x + 2)$ and simplify your result.

17. Let $F(x) = \sqrt{x - 1}$. Show that

$$\frac{F(x + h) - F(x)}{h} = \frac{1}{\sqrt{x + h - 1} + \sqrt{x - 1}}$$

18. Find the domain of each of the following:

   a. $f(x) = \sqrt{2x^2 - 3x - 5}$ **b.** $g(x) = \sqrt{\dfrac{x + 1}{x - 3}}$

19. Express the length of the diagonal of a square as a function of one of its sides s.
20. Express the diameter of a circle as a function of its area.
21. The sum of \$5,000 is invested in an account yielding 6% interest compounded annually. Express the total amount accumulated at the end of t years as a function of t.
22. Express the altitude h of an equilateral triangle as a function of one of its sides s.
23. The costs of renting a truck for 1 day are: \$25 rental fee, 20¢ for each mile driven, \$5.50 for insurance, 10¢ for each mile driven for gasoline. Let $C(x)$ be the total cost for driving x miles in a day. Write the rule for $C(x)$.

In Problems 24 and 25 show that each function is 1–1.

24. **a.** $f(x) = 3x - 7$ **b.** $f(x) = \dfrac{2x + 7}{3}$

25. **a.** $g(x) = 8 - 5x$ **b.** $h(x) = \dfrac{x + 1}{x}$

26. Show that the function f defined by $f(x) = 4 - x^2$ is not 1–1 on R. Find a suitable restriction on the domain of f so that it will be 1–1.

27. Determine which of the following functions are even, which are odd, and which are neither even nor odd (see Problem 31, Exercise Set 4):
   a. $f(x) = x^4 - 2x^2 + 1$ **b.** $g(x) = x(x^2 - 1)$ **c.** $h(t) = t^3 + 2t - 4$
   d. $F(x) = x - \dfrac{1}{x}$ **e.** $G(x) = \sqrt{\dfrac{4}{1 - x^2}}$

28. Show that the function $f(x) = |x - 1|$ is not 1–1 on R. Find a suitable restriction on the domain of f so that it will be 1–1.
29. Let $\phi(x) = (2x - 3)/5$ and $\psi(x) = (4 - 2x)/7$. Find $\phi \circ \psi$ and $\psi \circ \phi$.

30. Let $f(x) = x^2$ and $g(x) = \sqrt{x-4}$. Find $f \circ g$ and $g \circ f$, and determine the domain of each.

31. Let $f(t) = t^2 - 3$ and $g(t) = \sqrt{1-t}$. Find $f \circ g$ and $g \circ f$, and determine the domain of each.

32. Let $g(t) = 1/(t-1)$ and $h(t) = 1/(t+2)$. Find $g \circ h$ and $h \circ g$, and determine the domain of each.

33. At temperature x the thermal conductivity of a wall is given by $f(x) = k(1 + ax)$, where k and a are constants (k is the thermal conductivity at temperature $0°C$). Write the rule for the thermal resistance $g \circ f$, given that $g(x) = T/(Ax)$, where T is the thickness of the wall and A is its cross-sectional area.

In Problems 34–41 show that an inverse exists, and find it.

34. $f(x) = 4x - 5$

35. $f(x) = \dfrac{5x + 7}{3}$

36. $g(x) = 1 - \dfrac{2x}{3}$

37. $h(t) = \dfrac{2 - 3t}{5}$

38. $F(z) = 1 - \dfrac{2}{z}$

39. $g(x) = \dfrac{1}{\sqrt{x-2}}$

40. $f(t) = \dfrac{1}{\sqrt{t}}$

41. $h(x) = \dfrac{x+2}{x-3}$

In Problems 42 and 43 determine which graphs are graphs of functions and which are not. For those that are graphs of functions, state whether an inverse exists.

42.

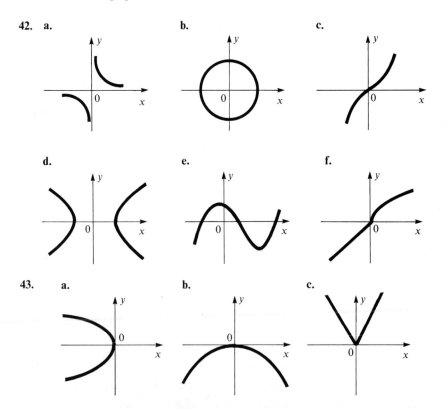

a.

b.

c.

d.

e.

f.

43.

a.

b.

c.

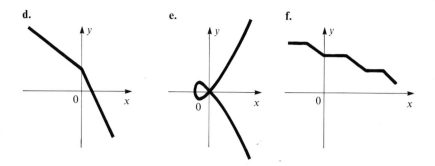

d.   e.   f.

In Problems 44 and 45 draw f and f^{-1} on the same set of axes.

44. $f(x) = \dfrac{3x - 5}{2}$ **45.** $f(x) = \sqrt{x + 1}$

46. Determine the intervals on which the function whose graph is shown is increasing and those on which it is decreasing.

a.

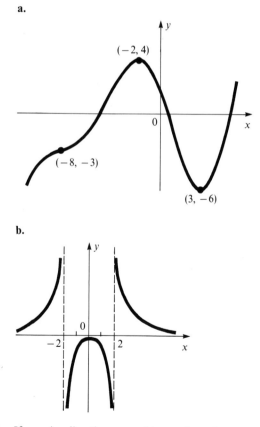

b.

47. If y varies directly as x and inversely as the square root of t, and $y = 5$ when $x = 3$ and $t = 4$, find y when $x = 8$ and $t = 9$.

48. If a body is dropped and falls freely for t seconds, the distance through which it falls varies directly as the square of t (neglecting air resistance). If after 2 seconds a body has fallen 64 feet, how far will it fall in 10 seconds?

49. The surface area S of a sphere is proportional to the square of its radius. If a sphere of radius 3 has a surface area of 36π, find a formula for the surface area of any sphere. What is the surface area of a sphere of diameter 10?

50. If the temperature is constant, the volume of a gas varies inversely as its pressure. A gas with a pressure of 150 pounds per square inch occupies a volume of 25 cubic feet. What volume will it occupy when the pressure is 45 pounds per square inch, assuming the temperature is held constant?

B 51. Make a table of values and draw the graph of each of the following:
    a. $y^2 = 18 - 2x^2$ **b.** $f(x) = 2 - x - x^2$

52. Find the domain of each of the following functions:
    a. $f(x) = \dfrac{2x + 3}{x^3 - 3x^2 - 4x + 12}$ **b.** $g(x) = \sqrt{\dfrac{4 - x^2}{x^3 + 4x^2}}$

53. Let
$$f(x) = \begin{cases} \sqrt{x^2 - 9} & \text{if} & x \geq 3 \\ 2 - x & \text{if} & 0 \leq x < 3 \\ -1 & \text{if} & x < 0 \end{cases}$$

Draw the graph of f and find:
    a. $f(5)$ **b.** $f(-1)$ **c.** $f(0)$ **d.** $f(3)$ **e.** $f(1)$

54. A rectangle of base b and altitude h is inscribed in an isosceles triangle of altitude 10 and base 6, as shown in the sketch. Express h as a function of b.

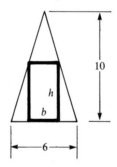

55. A circle is inscribed in an equilateral triangle. Express the area of the circle as a function of the length of a side of the triangle.

56. Show that the following four points are vertices of a rectangle: $(-1, 4)$, $(-7, -8)$, $(3, 2)$, $(-3, -10)$.

57. Let $f(x) = 4x - x^2$. Show that f is not 1–1 on R but that it is 1–1 on the domain $\{x: \ x \geq 2\}$. Find f^{-1} for this restricted domain.

58. Show that no inverse exists for the function defined by
$$f(x) = \frac{4}{(x - 2)^2}$$

but that by a suitable restriction on the domain an inverse does exist. Find this inverse.

59. Let $g(x) = x^2 - 4x - 5$. Show that g is not 1–1 on R but that it is 1–1 on the domain $\{x: \ x \leq 2\}$.

60. The strength of a beam of rectangular cross-section is jointly proportional to the width w and the square of the depth d of the cross-section and inversely proportional to the length l of the beam. What will be the effect on the strength if w and d are each doubled and l is cut in half?

5 Linear and Quadratic Functions

The importance of finding the equation of a line and recognizing linear equations is evident early in the study of calculus. A recurrent type of problem there involves finding the equations of a line tangent to a given curve and a line perpendicular to the tangent, called the *normal line*. The following problem is illustrative:*

Find an equation of the normal line to the curve $y = \sqrt{x - 3}$ which is parallel to the line $6x + 3y - 4 = 0$.

Solution Let l be the given line. To find the slope of l, we write its equation in the slope–intercept form, which is

$$y = -2x + \tfrac{4}{3}$$

Therefore, the slope of l is -2, and the slope of the desired normal line is also -2 because the two lines are parallel.

To find the slope of the tangent line to the given curve, we apply Definition 3.1 with $f(x) = \sqrt{x - 3}$, and we have . . .

$$m(x_1) = \frac{1}{2\sqrt{x_1 - 3}}$$

Because the normal line at a point is perpendicular to the tangent line at that point, the product of their slopes is -1. Hence, the slope of the normal line at (x_1, y_1) is given by $-2\sqrt{x_1 - 3}$. . . . Using the point–slope form of an equation we obtain

$$y - 1 = -2(x - 4)$$
$$2x + y - 9 = 0$$

In this chapter we will learn, among other things, what is meant by the slope of a line, the relationships between slopes of parallel and perpendicular lines, and the slope–intercept and the point–slope forms of the equation of a line.

159

1 Introduction

A function of the form

$$f(x) = a_n x^n + a_{n-1} x^{n-1} + \cdots + a_1 x + a_0 \qquad (a_n \neq 0)$$

where n is a nonnegative integer, is a **polynomial function** of degree n in the variable x. If $n = 1$, the function is **linear**; and if $n = 2$, it is **quadratic.** In this chapter we will study these two cases in some detail. The word *linear*, as you would expect, has some relationship to a straight line. In order to understand this more fully we begin by studying straight lines.

We know from plane geometry (and intuition) that a straight line is uniquely determined by two points. A line is also determined by one point and the "inclination," or **slope,** of the line. The next section makes this more precise.

2 The Equation of a Line

Suppose we are given a nonvertical straight line l and know any two points on the line, say, $P_1(x_1, y_1)$ and $P_2(x_2, y_2)$, as shown in Figure 1. Then we have the definition below.

Figure 1

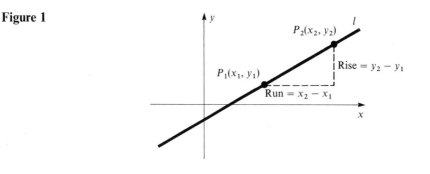

DEFINITION 1   The slope m of line l is defined as

$$m = \frac{y_2 - y_1}{x_2 - x_1}$$

(Can you see why we restricted l to be nonvertical?)

The slope is the vertical displacement from P_1 to P_2 divided by the horizontal displacement (sometimes described as the **rise** divided by the **run**). It does not matter which point is labeled P_1 and which is labeled P_2, since

$$\frac{y_2 - y_1}{x_2 - x_1} = \frac{y_1 - y_2}{x_1 - x_2} \qquad \textbf{Why?}$$

It can also be shown, using similar triangles, that it does not matter which particular two points on the line are chosen. The value found for the slope will always be the same.

EXAMPLE 1   A line passes through the points (1, 2) and (5, 4). Find the slope of the line.

Solution   We will let P_1 be the point (1, 2) and P_2 be (5, 4), but we could equally well reverse these. By Definition 1, we have

$$m = \frac{y_2 - y_1}{x_2 - x_1} = \frac{4 - 2}{5 - 1} = \frac{2}{4} = \frac{1}{2}$$

We can think of this as saying that for every 2 units we go horizontally to the right, the line rises 1 unit vertically, or for every 1 unit we go horizontally to the right, the line rises $\frac{1}{2}$ unit vertically. The graph is shown in Figure 2.

Figure 2

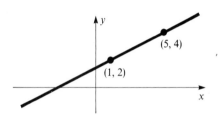

An examination of the definition of the slope of a line will show that lines that go upward to the right have positive slopes and those that go upward to the left have negative slopes (Figure 3). Also, a small numerical value for the slope indicates a line that is close to horizontal, whereas a large numerical value indicates a line that is nearly vertical. (What would a slope of 1 indicate? What about -1?) Horizontal lines have a slope of 0 (why?), and slope is not defined for vertical lines.

Figure 3

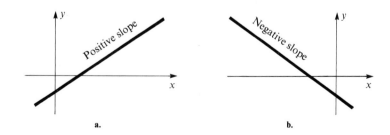

Let l be a nonvertical line passing through a given point $P_1(x_1, y_1)$ and having slope m. Any other point $P(x, y)$ will lie on l if and only if

$$\frac{y - y_1}{x - x_1} = m$$

If we clear this equation of fractions, we obtain

$$y - y_1 = m(x - x_1) \tag{1}$$

All points (including P_1) that lie on l have coordinates that satisfy this equation, and this is true for those points only. Thus, equation (1) is the equation of l. It is called the **point–slope form** of the equation of a line.

EXAMPLE 2   Find the equation of the line having slope $\frac{3}{4}$ and passing through the point $(-2, 5)$.

Solution   We simply substitute in equation (1), using $x_1 = -2$ and $y_1 = 5$:

$$y - 5 = \tfrac{3}{4}(x + 2)$$
$$4y - 20 = 3x + 6$$
$$3x - 4y + 26 = 0$$

Note.   In applying equation (1), x_1 and y_1 represent the coordinates of the given point and so will be specific numbers. On the other hand, x and y will remain in the equation as variables.

EXAMPLE 3   Find the equation of the line joining $(-6, -2)$ and $(2, 5)$.

Solution   First we must calculate the slope:

$$m = \frac{5 - (-2)}{2 - (-6)} = \frac{7}{8}$$

We may choose either point as (x_1, y_1) for the purpose of substituting in equation (1), so we select $(2, 5)$:

$$y - 5 = \tfrac{7}{8}(x - 2)$$
$$8y - 40 = 7x - 14$$
$$7x - 8y + 26 = 0$$

A special case of equation (1) results when the point (x_1, y_1) is taken as the point where the line crosses the y axis. The y coordinate of this point is called the **y intercept** and is usually denoted by b. Substituting $(0, b)$ for (x_1, y_1) in equation (1), we obtain

$$y - b = m(x - 0)$$

or

$$y = mx + b$$

This is called the **slope–intercept form** of the equation of a line (*intercept* is understood to mean the y intercept).

Note. The x coordinate of the point at which a line crosses the x axis is called its **x intercept** and is usually denoted by a. For a form of the equation of a line that uses both the x intercept and the y intercept, see Problem 27, Exercise Set 2.

EXAMPLE 4   Show that the equation $2x - 3y + 6 = 0$ is the equation of a line, and find the slope and y intercept of the line. Graph the line.

Solution   If we can put the equation in the form $y = mx + b$, we will know it is a line. This is done by solving for y:

$$3y = 2x + 6$$
$$y = \tfrac{2}{3}x + 2$$

Therefore, this is the equation of a line with slope $\tfrac{2}{3}$ and y intercept 2. We can use the slope to find another point on the line as follows: From $(0, 2)$ if we move 3 units in the x direction and 2 units in the y direction, we must come to another point on the line. This results in the point $(3, 4)$. The line can now be drawn as shown in Figure 4. An easier way to get a second point on the line is to find the x intercept by letting $y = 0$. This gives

$$0 = \tfrac{2}{3}x + 2$$
$$2x = -6$$
$$x = -3$$

So $(-3, 0)$ is another point on the line.

Figure 4

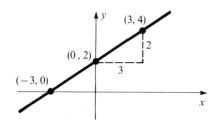

In either the point–slope form, $y - y_1 = m(x - x_1)$, or the slope–intercept form, $y = mx + b$, of the equation of a line, if the terms are rearranged, the result can be written in the form

$$Ax + By + C = 0 \qquad (2)$$

But do all equations of this form have straight lines as graphs? The answer is yes, as we now show.

We suppose that A and B are not both 0, because if they were, then C would also have to be 0, and the equation would read $0 = 0$, which is true but not of much interest. Now let us consider the case in which $B \neq 0$ (A may or may not be 0). We can solve for y to get

$$y = -\frac{A}{B}x + \left(-\frac{C}{B}\right)$$

This is in the form $y = mx + b$, with

$$m = -\frac{A}{B} \quad \text{and} \quad b = -\frac{C}{B} \tag{3}$$

So this represents a line with slope $-A/B$ and y intercept $-C/B$.

In the special case where $A = 0$ (but $B \neq 0$), we see from the above that

$$y = -\frac{C}{B}$$

or

$$y = b$$

Since this equation places no restriction on x, it follows that all points with coordinates of the form (x, b) satisfy the equation, and therefore the graph is a horizontal line. The slope is 0.

One final case needs to be considered, and that is the case in which $B = 0$ but $A \neq 0$. Equation (2) then has the form

$$Ax + C = 0$$

$$x = -\frac{C}{A}$$

If we let $a = -C/A$, this can be written

$$x = a$$

and this is the equation of a vertical line, since all points with coordinates of the form (a, y), where y can be anything whatsoever, satisfy the equation. Slope is not defined for a vertical line.

EXAMPLE 5   Draw the graphs of the lines:

a. $y = 3$ **b.** $x = -2$

Solution

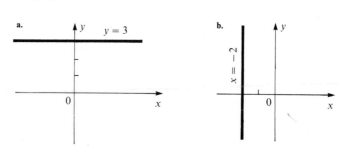

Note. A common mistake is to suppose that $y = 3$ means only the single point $(0, 3)$ or that $x = -2$ means just the point $(-2, 0)$. In the context of two dimensions both of these represent lines, not points.

Since equation (2) always represents a line, it is called the **general equation of a line.** It is also called the **general linear equation in two variables.**

EXERCISE SET 2

A 1. Find the slopes of the lines joining the pairs of points given. Sketch the lines.
   a. $(2, -1)$ and $(3, 4)$ **b.** $(-3, -2)$ and $(5, 4)$
   c. $(3, -4)$ and $(-2, 3)$ **d.** $(-5, 3)$ and $(-2, 0)$

2. The points $A(3, -2)$, $B(-6, -8)$, and $C(12, 4)$ all lie on a line. Calculate the slope of the line using three different pairs of points, and verify that the results are the same.

3. Find the slopes of the sides of the triangle with vertices at $A(3, 2)$, $B(-5, 4)$, and $C(-2, -3)$. Draw the triangle.

4. Find the slopes of the two diagonals of the quadrilateral with vertices at the points $(5, 6)$, $(-1, 3)$, $(-4, 0)$, and $(2, -2)$. Draw the figure.

5. What is the horizontal displacement between $(-3, -5)$ and $(2, -1)$? What is the vertical displacement?

6. In each of the following a line passes through the given point with the given slope. Find a second point on the line, and draw a sketch.
   a. $(-1, -3)$, slope $\frac{3}{4}$ **b.** $(2, -4)$, slope $-\frac{1}{2}$

7. The slope of the line passing through $(2, 3)$ and $(-4, y)$ is $-\frac{2}{3}$. Find y.

8. The slope of the line passing through $(x, 2)$ and $(6, 8)$ is $\frac{3}{5}$. Find x.

9. A quadrilateral has vertices $A(3, 2)$, $B(7, 5)$, $C(-1, 3)$, and $D(-5, 0)$. Find the slope of each of the sides. Does your result suggest anything special about this quadrilateral? Explain. Sketch the figure.

In Problems 10–18 find the equations of the lines satisfying the given conditions.

10. Slope $\frac{2}{5}$, passing through $(-2, -5)$ 11. Slope -2, passing through $(3, -4)$
12. Passing through $(3, 4)$ and $(-1, -2)$ 13. Passing through $(-1, 3)$ and $(2, -5)$
14. Passing through $(2, -3)$ and $(-6, 2)$ 15. Passing through $(-3, 0)$ and $(0, 4)$
16. Slope $-\frac{1}{2}$ and y intercept -3 17. Slope $\frac{2}{3}$ and y intercept 2
18. Slope 3 and y intercept $-\frac{1}{2}$

In Problems 19–24 find the slope m and the y intercept b. Draw the graph.

19. $y = 3x - 4$ 20. $3x + 5y - 2 = 0$
21. $2x + 6y = 7$ 22. $2y - 3x + 4 = 0$
23. $y = \dfrac{3x - 4}{5}$ 24. $\dfrac{4y - 5}{2} = x$

In Problems 25 and 26 write the equation of each horizontal or vertical line, as indicated.

25. **a.** y intercept 2, horizontal   **b.** x intercept 3, vertical
     c. Passing through $(-1, -3)$, horizontal   **d.** Passing through $(-1, -3)$, vertical

26. **a.** x intercept -4, vertical   **b.** y intercept -3, horizontal
    c. Passing through $(-3, 4)$, vertical   **d.** Passing through $(-3, 4)$, horizontal

27. Derive the **two-intercept form** of the equation of a line:

$$\frac{x}{a} + \frac{y}{b} = 1$$

Hint. You know two points on the line.

28. Use the result of Problem 27 to find the equations of the lines that have the given intercepts.
    a. x intercept 2, y intercept -3 **b.** x intercept $-\frac{3}{4}$, y intercept $\frac{5}{6}$

B   In Problems 29–31 show in two ways that the given points are **collinear** (that is, they lie on the same line).

29. $(2, 2), (-1, -7), (3, 5)$ 30. $(6, 2), (-2, -4), (2, -1)$
31. $(0, -3), (2, 1), (-1, -5)$

32. The line l_1 passes through the points $(2, 3)$ and $(-4, 6)$. Line l_2 passes through the point $(-1, -2)$ and has the same slope as l_1. Line l_3 passes through $(-1, -2)$, and its slope is the negative of the reciprocal of the slope of l_1. Find a second point on each of the lines l_2 and l_3, and sketch all three lines on the same set of axes. What relationship appears to exist among the three lines?

33. **a.** A line has slope $\frac{3}{2}$. Its equation can be written in the form $3x - 2y + C = 0$. Explain why this is so.
    b. If the line of part a passes through the point $(-5, 1)$, we can find C by substituting -5 for x and 1 for y in the equation $3x - 2y + C = 0$. Do this. After finding C, write the equation of the line.

34. Use the procedure of Problem 33 to find the equation of the line with slope $-\frac{3}{4}$ and passing through $(2, 3)$.

35. Find by two methods (see Problem 33 for one method) the equation of the line with slope $\frac{1}{2}$ and passing through $(3, -4)$.

36. What can you say about the relationship between the lines $3x - 4y + 7 = 0$ and $3x - 4y - 9 = 0$? Also compare $6x - 8y + 23 = 0$ with these two. How can you generalize your findings?

37. We may consider the equation $2x - 3y + C = 0$ as defining a **family of lines,** one line for each value of C. What can you conclude about each member of this family? What is the geometric significance of this?

38. A salesperson works for a fixed base weekly salary of $\$b$, plus a commission of $\$c$ for each unit sold. Write the weekly salary y as a function of the number x of units sold. Suppose the salesperson receives $\$300$ in a week in which 10 units were sold and $\$375$ in a week in which 15 units were sold. Find the person's base salary b and commission c. What will the person's salary be in a week in which 18 units are sold?

3   Parallel and Perpendicular Lines

It is intuitively obvious that two lines having the same slope are parallel, and conversely, if two lines are parallel, they have the same slope. This can be proved formally, but we will not do so here (the proof will be called for in Problem 24, Exercise Set 3).

But it is not obvious what the relationship is between slopes of perpendicular lines. If either line is horizontal, the other must be vertical; so we will consider only the case in which neither line is horizontal. Let $y = mx + b$ be the equation of such a line. We wish to find the slope of lines perpendicular to this one. First, we consider a line through the origin and parallel to the given line, since any line perpendicular to the original line will be perpendicular to all lines parallel to it, and vice versa (Figure 5). The equation of the line parallel to $y = mx + b$ and passing through the origin O is $y = mx$. We work with this line instead of the original, since it simplifies our computations. Designate this line by l_1.

Figure 5

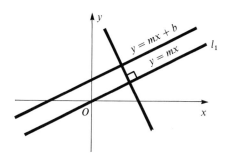

We find a second point on l_1 by letting $x = 1$. This gives $(1, m)$ as a point on the line. Let P be the point $(1, m)$ and let A be $(1, 0)$, as shown in Figure 6. Now consider a line through the origin and perpendicular to the line $y = mx$. Call this l_2. Draw the vertical line $x = -m$ and designate by B and Q the points of intersection with the x axis and l_2, respectively. The triangles OAP and QBO are congruent (since the sides are perpendicular and $\overline{OB} = \overline{AP}$). Thus, corresponding sides are equal in length. In particular, $\overline{BQ} = \overline{OA} = 1$. The coordinates of Q must therefore be $(-m, 1)$. Hence, the slope of l_2 is

$$\frac{1 - 0}{-m - 0} = -\frac{1}{m}$$

which is the result we were seeking.

Figure 6

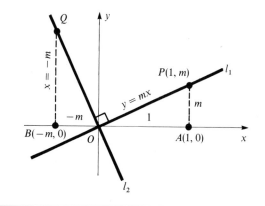

The slope of any line perpendicular to $y = mx + b$ is $-1/m$.

Figure 7

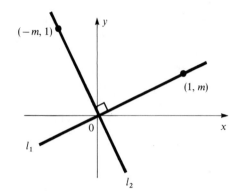

If the argument used above is reversed, we conclude that any line with slope $-1/m$ is perpendicular to any line with slope m.

Note. In the figures we have indicated that m is positive. The result is also true if m is negative, and the steps of the argument would be identical; only the figures would look different.

The number $-1/m$ is the **negative reciprocal** of m. Using this terminology, we can state our result as follows:

Perpendicular Lines

Two lines, neither of which is horizontal, are perpendicular if and only if their slopes are negative reciprocals of each other.

EXAMPLE 6   Find the slope of all lines perpendicular to the line $y = 2x + 3$.

Solution   Since the slope of the given line is 2, all lines perpendicular to it have slope $-\frac{1}{2}$.

Remark. The set of all lines satisfying a given condition is called a **family of lines.** So we could conclude in this case that the family of lines perpendicular to the given line is given by

$$y = -\tfrac{1}{2}x + C$$

where C can be any real number.

EXAMPLE 7   Find the equation of the line that passes through $(1, 3)$ and is perpendicular to the line $y = (2x/3) - 4$.

Solution   The line $y = (2x/3) - 4$ has slope $\frac{2}{3}$, so the line we are seeking has slope $-\frac{3}{2}$. Now we can use the point–slope form $y - y_1 = m(x - x_1)$ to find the desired equation:

$$y - 3 = -\tfrac{3}{2}(x - 1)$$
$$2y - 6 = -3x + 3$$
$$3x + 2y - 9 = 0$$

EXAMPLE 8   Find the equation of the family of all lines perpendicular to the line with equation $3x - 4y + 3 = 0$.

Solution   We know from equation (3) that the slope of the given line is $\tfrac{3}{4}$, so the slope of all lines perpendicular to it is $-\tfrac{4}{3}$. Since the slope of a line $Ax + By + C = 0$ is $-A/B$, we can let $A = 4$ and $B = 3$ to get

$$4x + 3y + C = 0$$

as the equation of the desired family. The constant C can take on any real value, and for each such choice, the line will be perpendicular to $3x - 4y + 3 = 0$.

If P and Q are two points on a curve, the line joining them is called a **secant** line for the curve. If P is held fixed and Q comes closer and closer to P, the secant lines through P and Q may approach some line as a limit; if so, this line is called the **tangent** to the curve at P. Figure 8 illustrates this.

Figure 8

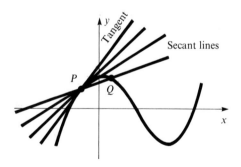

EXAMPLE 9   In calculus it is shown that the curve defined by $y = x^2 - 3x$ has a tangent at each point (x, y) whose slope is given by $m = 2x - 3$. Find the equation of the tangent line to this curve at the point for which $x = 2$.

Solution   When $x = 2$, $y = 2^2 - 3(2) = 4 - 6 = -2$. Also, the slope $m = 2(2) - 3 = 1$. So the equation is

$$y + 2 = 1(x - 2)$$
$$x - y - 4 = 0$$

The graph is shown in Figure 9.

Figure 9

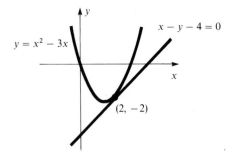

EXERCISE SET 3

A   In Problems 1–16 find the equations of the lines satisfying the given conditions.

  1. Passing through $(5, -3)$ and parallel to $3x - 4y = 12$
  2. Passing through $(0, 4)$ and parallel to $x - y = 4$
  3. Passing through $(-1, 2)$ and perpendicular to $y = 3x - 5$
  4. Passing through $(3, -1)$ and perpendicular to $2x - 3y - 4 = 0$
  5. Passing through $(0, 0)$ and parallel to $5x - 2y + 3 = 0$
  6. Passing through $(-2, -3)$ and parallel to $y = (x - 3)/4$
  7. Passing through $(6, 1)$ and perpendicular to $3y - 4x = 7$
  8. Passing through $(-2, 6)$ and perpendicular to $4x + 7y = 13$
  9. Parallel to the line joining $(3, 2)$ and $(-1, 5)$, passing through the point $(0, 2)$
  10. Perpendicular to the line joining $(-2, -3)$ and $(3, 5)$, passing through $(2, -4)$
  11. Passing through $(7, -3)$ and parallel to the line joining $(5, 4)$ and $(-1, -2)$
  12. Passing through $(3, 5)$ and perpendicular to the line joining $(-1, 3)$ and $(3, -5)$
  13. Parallel to $x + 2 = 0$ and passing through $(3, 5)$
  14. Parallel to $2y - 3 = 0$ and passing through $(-1, 4)$
  15. Perpendicular to $y = 4$ and passing through $(2, 3)$
  16. Perpendicular to $3x - 5 = 0$ and having y intercept 3
  17. Find the equation of the line perpendicular to $5x - 2y = 3$ and having the same y intercept as this line.
  18. Find the equation of the line parallel to $3x - 5y - 6 = 0$ and with x intercept that is the negative of the x intercept of this line.
  19. By using slopes show that the points $(2, 1)$, $(6, 9)$, and $(-2, 3)$ are vertices of a right triangle.
  20. Show that the points $(3, -1)$, $(5, 4)$, $(-5, 8)$, and $(-7, 3)$ are vertices of a rectangle.
  21. Show that the points $(2, 2)$, $(0, -1)$, $(-4, 1)$, and $(-2, 4)$ are the vertices of a parallelogram.
  22. Find the equation of the family of lines parallel to the line $5x - 4y = 6$. Find the equation of the member of this family that goes through the point $(4, 2)$.
  23. Find the equation of the family of lines perpendicular to the line $4x + 3y - 8 = 0$. Find the equation of the member of this family that has the same y intercept as that of the given line.

B   **24.** Let $y_1 = m_1 x + b_1$ and $y_2 = m_2 x + b_2$. Carry out the details of the argument outlined here to prove that the corresponding lines are parallel if and only if $m_1 = m_2$. We know that two lines are parallel if and only if they do not intersect. If $m_1 = m_2$, show that the lines do not intersect (unless they coincide). If $m_1 \neq m_2$, show that the lines do intersect.

25. A line parallel to $x - 2y - 4 = 0$ crosses the negative x axis and positive y axis so that the line and these axes form a right triangle of area 9. Find the equation of the line.

26. Prove that the points $(-2, 9)$, $(-4, -2)$, $(1, -12)$, and $(3, -1)$ are vertices of a rhombus (a parallelogram with all sides equal in length). Show that the diagonals are perpendicular.

27. Show that the midpoint of the line segment joining (x_1, y_1) and (x_2, y_2) is the point

$$\left(\frac{x_1 + x_2}{2}, \frac{y_1 + y_2}{2} \right)$$

Hint. Designate the unknown coordinates of the midpoint by (x, y), as shown in the figure:

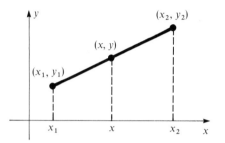

We must have $x - x_1 = x_2 - x$. (Why?) Solve this for x. Now do the same for y.

28. Using the result of Problem 27, find the equation of the perpendicular bisector of the line segment joining $(3, 4)$ and $(-1, 2)$.

29. Find the equations of the perpendicular bisectors of the sides of the triangle with vertices $(-5, 2)$, $(-7, -6)$, and $(3, -4)$.

30. Find the equations of the medians of the triangle whose vertices are $(3, 2)$, $(-5, -2)$, and $(7, -6)$. (See Problem 27.)

31. A physicist describes the results of an experiment on a plane coordinate system. Two magnets have centers at the points with coordinates $(0, -\frac{1}{2})$ and $(2, \frac{3}{2})$, respectively. The path of a particle in the experiment is the perpendicular bisector of the line segment joining the centers of the magnets. Find the equation of the path of the particle.

32. Prove that the lines joining the midpoints of the quadrilateral having vertices $(5, 10)$, $(-9, 2)$, $(-3, -8)$, and $(7, -4)$ form a parallelogram.
Note. It is an interesting fact that the lines joining the midpoints of *any* quadrilateral always form a parallelogram.

33. **a.** It can be shown by calculus that the slope of the tangent line to the curve $y = x^2 - 4$ at any point (x, y) is given by $m = 2x$. Find the equation of the tangent line at the point for which $x = 1$.
    b. The line perpendicular to the tangent line of a curve is called the **normal** line. Find the equation of the normal to the curve of part a at the given point. Draw the graph of $f(x) = x^2 - 4$, showing the tangent and normal at the given point.

34. The slope of the tangent line to the curve $y = x^3 - 2x^2 + 3$ at any point (x, y) can be shown by calculus to be given by the formula $m = 3x^2 - 4x$. Find the equation of the tangent line and of the normal line to this curve at the point for which $x = 2$. (See Problem 33.)

4   Linear Functions

In Section 1 we defined a linear function as a polynomial function of degree 1, that is, a function of the form $f(x) = a_1 x + a_0$, with $a_1 \neq 0$. The reason for calling this *linear* should now be clear. If we put $y = f(x)$, we get

$$y = a_1 x + a_0$$

which is of the form $y = mx + b$, where $m = a_1$ and $b = a_0$. It follows that the graph of every linear function is a straight line. The requirement that $a_1 \neq 0$ ensures that the line is not horizontal. If $a_1 = 0$, the function would be of the form $f(x) = a_0$, which is called a **constant function.** Its graph is the horizontal line having the equation $y = a_0$. So even though the graph is a line, the function is not called linear in this case.

We can now conclude that every linear equation in two variables,

$$Ax + By + C = 0$$

in which *neither A nor B is 0*, defines a linear function. The rule for the function is obtained by solving for y:

$$y = -\frac{A}{B} x + \left(-\frac{C}{B} \right)$$

or in functional notation,

$$f(x) = -\frac{A}{B} x + \left(-\frac{C}{B} \right)$$

Polynomial equations of the first degree in more than two variables are also called *linear*, although their graphs are no longer straight lines. For example, the equation $2x - 3y + 4z - 7 = 0$ is a linear equation in three variables. Graphing this equation requires three dimensions. Can you guess what sort of graph this equation would have? Equations such as this will be considered later, when we discuss systems of equations.

EXAMPLE 10   A clothing store owner is able to sell 60 jackets of a certain type per week if the price is $55 per jacket. He finds that if he reduces the price to $40, he is able to sell 90 of the jackets per week. Assume that the number $N(x)$ sold per week is a linear function of the price x, and express $N(x)$ in terms of x. How many of the jackets could he expect to sell if the price were $50?

Solution   We assume

$$N(x) = mx + b$$

From the given information $N(55) = 60$ and $N(40) = 90$. If we let $y = N(x)$, the points $(55, 60)$ and $(40, 90)$ are on the graph. So the slope m is

$$m = \frac{90 - 60}{40 - 55} = \frac{30}{-15} = -2$$

Therefore, $y = -2x + b$. To find b we can substitute either point. Putting $x = 40$ and $y = 90$ gives

$$90 = -2(40) + b$$
$$b = 170$$

Therefore, the functional relationship is

$$N(x) = -2x + 170$$

To find how many would be sold if the price were \$50, put $x = 50$:

$$N(50) = -2(50) + 170 = 70$$

So he could expect to sell 70 jackets at this price.

EXERCISE SET 4

A **1.** Let $f(x) = 3x - 2$. Identify and draw the graph of this function.

2. Let g be the linear function defined by $g(x) = (2 - x)/3$. What is the slope of the graph of g? Draw the graph.

3. Let $f(x) = (5x - 4)/3$. What are the slope and y intercept of the graph of f? Draw the graph.

4. Graph the constant function defined by $f(x) = 5$.

5. For a certain linear function f, we are given that $f(1) = 2$ and $f(-2) = 4$. Find $f(x)$.

6. For a certain linear function f, the slope of the graph is 2 and $f(0) = -5$. Find $f(x)$.

7. If g is a linear function with $g(0) = -1$ and $g(2) = 3$, find $g(x)$.

8. If $g(0) = \frac{2}{3}$, g is linear, and the slope of the graph of g is $-\frac{1}{2}$, find $g(x)$.

9. If h is linear, with $h(2) = 1$ and $h(-1) = 3$, find $h(x)$.

10. If f is a linear function with $f(0) = 4$ and $f(-2) = 0$, find $f(x)$.

11. For a function f it is known that $f(-2) = 3$ and $f(4) = 3$. Can f be linear? Explain. If f is a polynomial function of degree less than or equal to 1, find $f(x)$.

12. The function g is a polynomial function of degree less than 2, having the property that $g(0) = -5$ and $g(-3) = -5$. Find $g(x)$.

13. A car rental agency charges \$18 per day, plus 20¢ per mile for renting a certain model car. Express the daily cost of renting this type of car as a function of the number of miles driven. What is the cost for driving 375 miles?

14. If \$$P$ are invested at simple interest r, then the amount of money $A(t)$ accumulated at the end of t years is a linear function of t. Find this function. Find $A(10)$ if $P = $2,000$ and $r = 0.08$.

15. According to Hooke's law, the force $F(x)$ required to stretch a spring x inches beyond its natural length is a linear function of x. If for a certain spring it takes 40 pounds to stretch it 5 inches, find $F(x)$, and determine the force necessary to stretch the spring 8 inches.
Note. When $x = 0$, $F(x) = 0$.

16. The owner of a hamburger stand finds that she can sell an average of 200 hamburgers per day if she charges 95¢ per hamburger but only 170 hamburgers per day if she charges \$1.10 per hamburger. Assuming the relationship is linear, express the number $N(x)$ sold per day as a function of the price x. How many hamburgers could the owner expect to sell per day if she charged \$1.00 per hamburger?

17. Show that the composition of two linear functions is also a linear function.
Hint. Let $f(x) = a_1 x + a_0$ and $g(x) = b_1 x + b_0$, and consider $f \circ g$.

18. Let $f(x) = 2x - 3$ and $g(x) = (4 - x)/3$. Find $f \circ g$ and $g \circ f$. Give the slopes of the graphs of both $f \circ g$ and $g \circ f$, and draw their graphs.

B 19. Show that every linear function is 1–1 from R onto R.

20. What is the relationship between the slope of the graph of a linear function f and that of its inverse f^{-1}?

21. If f and g are linear functions with graphs that are perpendicular, show that the slopes of the graphs of $f \circ g$ and of $g \circ f$ are both -1.

22. A common practice in business and industry is for capital equipment to be **depreciated linearly** over a specified period of years. If we let $V(t)$ represent the value of the equipment t years after purchase, this means that $V(t) = at + b$, where a and b are constants determined by the following conditions: When $t = 0$, $V(t) = C$, the original cost. When $t = N$, $V(t) = 0$, where N is the number of years over which the equipment is being depreciated. Find the constants a and b in terms of C and N. What is the value after 8 years of an item originally costing \$10,000 which is to be depreciated linearly to 0 in 20 years?

23. In Problem 22, it is reasonable to assume that certain items will not depreciate to 0 but will have some residual value (at least as junk) after N years. If the residual value is R, find $V(t)$ in terms of C, N, and R. If the \$10,000 item in Problem 22 has a residual value of \$500 after 20 years, what is its value after 8 years?

24. A certain nonlinear function $L(x)$ is tabulated for x values given to two decimal places between $x = 1.00$ and $x = 9.99$. It is common practice to use **linear interpolation** to approximate $L(x)$ for x expressed to the nearest thousandth, that is, to assume $L(x)$ is linear between consecutive entries in the table. If $L(2.13) = 0.3284$ and $L(2.14) = 0.3304$, use linear interpolation to approximate $L(2.137)$.

25. A nonlinear function $C(x)$ is tabulated so that two consecutive entries in the table are $C(0.5992) = 0.8258$ and $C(0.6021) = 0.8241$. It is desired to find $C(0.6000)$. Use linear interpolation to approximate this. (See Problem 24.)

---

5 Quadratic Functions

As we indicated in Section 1, a polynomial function of degree 2 is called a *quadratic* function. So the general form is $f(x) = a_2x^2 + a_1x + a_0$, or as it is more customarily written,

$$f(x) = ax^2 + bx + c \qquad (a \neq 0) \qquad (4)$$

Let us consider several special cases and look at their graphs. The simplest case is $f(x) = x^2$. Denote the dependent variable by y, so that

$$y = x^2$$

The smallest value y can have is 0, which occurs when $x = 0$, because if $x \neq 0$, then $x^2 > 0$. Furthermore, a given value of x and its negative produce the same value of y, since $(-x)^2 = x^2$. Finally, we see that when $|x| > 1$, y increases more rapidly than x; for example, when x goes from 2 to 3, y goes from 4 to 9. With these facts, and by plotting a few points, we can draw a reasonably accurate graph, as shown in Figure 10. The graph is called a **parabola.** The low point on the parabola is called its **vertex,** and the y axis is the **axis** of the parabola.

The parabola is said to be **symmetric** to its axis because the part of the curve to the left of the axis is the mirror image of the part to the right.

Figure 10

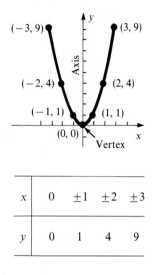

x	0	±1	±2	±3
y	0	1	4	9

We consider next the effect of a coefficient other than 1 on x^2:

$$y = ax^2$$

This causes every y coordinate to be multiplied by a. If $a > 0$, the general shape of the curve is the same as in Figure 10, but with the y coordinates stretched (if $a > 1$) or shortened (if $a < 1$). If $a < 0$, then each y value becomes negative, and the curve opens downward. In each case the curve is still a parabola with vertex at the origin and axis coinciding with the y axis. Figure 11 shows $y = 2x^2$, $y = \frac{1}{2}x^2$, and $y = -\frac{3}{2}x^2$ on the same set of axes.

Figure 11

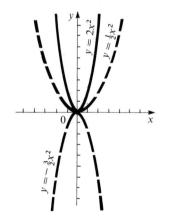

If a constant k is added, so that $y = ax^2 + k$, this has the effect of moving each y value upward (if $k > 0$) or downward (if $k < 0$). The vertex is shifted to the point $(0, k)$. Figure 12 shows the graph of $y = 2x^2 + 3$.

Figure 12

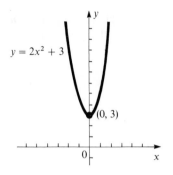

$$y = 2x^2 + 3$$

$(0, 3)$

Finally, we consider a function of the form

$$y = a(x - h)^2 + k$$

for example,

$$y = 2(x - 1)^2 + 3$$

In this case, y takes on its smallest value, 3, when $x = 1$, since for all other values of x, the term $2(x - 1)^2$ will be positive and so when added to 3 causes y to be greater than 3. Also, we note that if x is to the right or left of 1 by the same amount, the y value is the same. For example, if $x = 2$, then $y = 2(1)^2 + 3 = 5$, and when $x = 0$, then $y = 2(-1)^2 + 3 = 5$. We conclude that the curve is a parabola whose vertex is at the point $(1, 3)$ and whose axis is the line $x = 1$. In fact, this is the same parabola as in Figure 12 shifted one unit to the right, as shown in Figure 13.

Figure 13

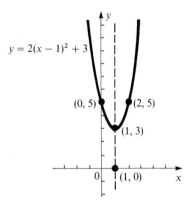

$$y = 2(x - 1)^2 + 3$$

$(0, 5)$ $(2, 5)$

$(1, 3)$

$(1, 0)$

Generalizing this, we can say that the graph of

$$y = a(x - h)^2 + k \tag{5}$$

is the same as the graph of

$$y = ax^2 + k$$

shifted h units horizontally. If $h > 0$, the shift is to the right, and if $h < 0$, it is to the left. In summary, we conclude that the graph of equation (5) is a parabola with the following properties:

---

1. Its vertex is at (h, k).
2. Its axis is the line $x = h$.
3. It opens upward if $a > 0$ and downward if $a < 0$.

Remark. If we write equation (5) in the form

$$y - k = a(x - h)^2$$

we can observe that shifting the parabola whose equation is $y = ax^2$ so that the vertex is at (h, k) results in x being replaced by $x - h$ and y being replaced by $y - k$. Such a shift is called a **translation,** and the result holds true for other curves as well. That is, **if the equation of a curve is known, and the curve is then translated h units horizontally and k units vertically, the new equation is obtained from the old by replacing x by $x - h$ and y by $y - k$.**

Now let us return to the general quadratic function (4). Let $y = f(x)$. We factor out the coefficient a from the first two terms and complete the square on x:

$$y = a\left(x^2 + \frac{b}{a}x\right) + c$$

$$= a\left(x^2 + \frac{b}{a}x + \frac{b^2}{4a^2}\right) + c - \frac{b^2}{4a}$$

$$= a\left(x + \frac{b}{2a}\right)^2 + \frac{4ac - b^2}{4a}$$

Note that we added $b^2/4a^2$ inside the parentheses (the square of one-half of the coefficient of x), which had the effect of adding $b^2/4a$ after multiplying by a, so we subtracted this same amount to balance what we added.

If we let

$$h = -\frac{b}{2a} \quad \text{and} \quad k = \frac{4ac - b^2}{4a} \tag{6}$$

we can finally write

$$y = a(x - h)^2 + k$$

Therefore, the graph of

$$y = ax^2 + bx + c$$

is a parabola with vertex at (h, k), where h and k are given by equations (6). Its axis is the line $x = h$, and the parabola opens upward if $a > 0$ and downward if $a < 0$.

Rather than memorize the values of h and k given in equations (6), it is probably better in each case to complete the square, as we do in the next example.

EXAMPLE 11   Find the vertex and the axis of the parabola with equation $y = x^2 - 2x + 3$.

Solution   We write

$$y = x^2 - 2x + 3$$
$$= (x^2 - 2x + 1) + 3 - 1$$
$$= (x - 1)^2 + 2$$

We added 1 inside the parentheses to complete the square and subtracted 1 outside the parentheses to balance this off. Comparing this with equation (5), we conclude that the vertex is $(1, 2)$ and the axis is line $x = 1$. The graph is shown in Figure 14.

Figure 14

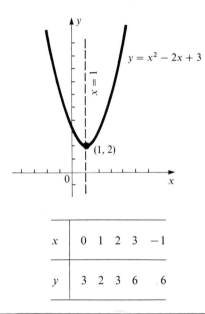

$y = x^2 - 2x + 3$

$(1, 2)$

x	0	1	2	3	-1
y	3	2	3	6	6

EXAMPLE 12   Find the largest value of the function $f(x) = 1 - 8x - 2x^2$ Draw the graph.

Solution   Let $y = f(x)$ and complete the square on x in order to put the equation in the form of equation (5):

$$y = -2x^2 - 8x + 1$$
$$= -2(x^2 + 4x) + 1$$
$$= -2(x^2 + 4x + 4) + 1 + 8$$
$$= -2(x + 2)^2 + 9$$

Notice that when we added 4 inside the parentheses to complete the square, we were actually adding -8 to the expression because of the factor -2, and so we had to balance this by adding $+8$.

The graph is a parabola, with vertex $(-2, 9)$, opening downward since the coefficient of the squared factor is negative. Its graph is shown in Figure 15. The highest point on the graph is the vertex $(-2, 9)$, so the largest value of y, that is $f(x)$, is 9.

Figure 15

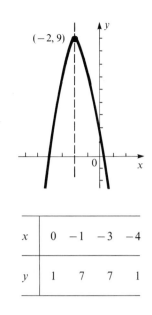

$(-2, 9)$

x	0	-1	-3	-4
y	1	7	7	1

Example 12 illustrates a technique for finding **maximum and minimum values** of quadratic functions. Since the graph is a parabola, its vertex is the lowest point on the curve if it opens upward and the highest point if it opens downward. When a curve attains such a highest or lowest point, this is called a **maximum point or minimum point,** respectively, of the curve, and the corresponding y coordinate is the **maximum or minimum value of the function.** More general techniques for finding maximum and minimum values of a much broader class of functions are studied in calculus. The next example illustrates an application of finding the maximum value of a function.

EXAMPLE 13

The average nightly attendance at a certain movie theater is 100 when the ticket price is $3.00. The manager finds that for each 10¢ reduction in price the average attendance goes up by 10 persons. If x represents the number of 10¢ reductions, the revenue $R(x)$ is given by

$$R(x) = 300 + 20x - x^2$$

Find the value of x that gives the maximum revenue, and determine the maximum revenue. What is the price per ticket and the average attendance for maximum revenue?

Solution

We want to maximize $R(x)$, so we look for the highest point on the graph of $y = R(x)$:

$$y = -x^2 + 20x + 300$$
$$= -(x^2 - 20x) + 300$$
$$= -(x^2 - 20x + 100) + 300 + 100$$
$$= -(x - 10)^2 + 400$$

The parabola opens downward, and so has its maximum point at its vertex (10, 400), as shown in Figure 16. Therefore, the maximum value of the revenue is $400, and this occurs when $x = 10$. So there should be 10 reductions in price of 10¢ each, making the cost of each ticket $2.00. Since the average attendance increases by 10 for each such reduction in price, the average attendance for maximum revenue is 200.

Figure 16

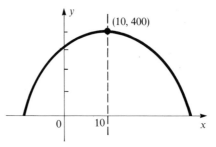

EXERCISE SET 5

A In Problems 1–20 find the vertex and axis of the parabola, and draw the graph.

1. $y = x^2/2$	**2.** $y = -2x^2$
3. $y = x^2 + 4$	**4.** $y = 3 - x^2$
5. $y = (x - 2)^2 + 1$	**6.** $y = -(x + 1)^2 + 2$
7. $y = 2(x - 3)^2 - 1$	**8.** $y = -\frac{3}{2}(x - 1)^2 - 2$
9. $y = x^2 + 4x + 5$	**10.** $y = 3 - 2x - x^2$
11. $y = 2x^2 - 4x + 7$	**12.** $y = 2x^2 + 3x - 1$
13. $y = 3x - x^2$	**14.** $y = 4 - 5x - 2x^2$
15. $y = 3x^2 - 2x + 4$	**16.** $y = 2x - 3x^2$
17. $y = \dfrac{x^2}{2} + x + 5$	**18.** $y = \dfrac{-3x^2 + 5}{2} - 3x$
19. $y = 2x(x - 3) - 5$	**20.** $y = x(1 - 2x) - \frac{3}{2}$

In Problems 21–28 find the maximum or minimum value of each function, and state the value of x that gives this maximum or minimum value.

21. $f(x) = 2x - x^2$	**22.** $f(x) = x^2 - 4x + 3$
23. $g(x) = 2x^2 - 6x + 3$	**24.** $h(x) = 5 - 3x - 2x^2$
25. $F(x) = 5x^2 - 2x - 3$	**26.** $G(x) = 4x^2 - 6x + 5$
27. $\phi(x) = 8x - 3x^2$	**28.** $\psi(x) = 6x(x - 2) + 25$

29. For a small manufacturing firm the cost in dollars of producing x items per day is given by the function

$$C(x) = x^2 - 120x + 4{,}000$$

How many items should be produced per day to minimize cost? What is the minimum cost?

30. A company that produces computer terminals analyzes its production and finds that the profit in dollars from selling x terminals per month is given by

$$P(x) = 160x - 0.1x^2 - 20{,}000$$

How many terminals should be sold per month to produce the maximum profit? What is the amount of the maximum profit?

31. An object is projected vertically upward from the ground at an initial velocity of 192 feet per second. If air resistance is neglected, its distance $s(t)$ in feet above the ground after t seconds is given by $s(t) = 192t - 16t^2$. Find the maximum height the object will rise.

32. An object is propelled vertically upward at time $t = 0$ in such a way that its height y above the ground at time t is given by

$$y = -16t^2 + \tfrac{247}{6}t + 4$$

Draw the graph of y. How long will the object remain in the air?

B **33.** Prove that if $f(x) = a(x - h)^2 + k$, then $f(h + t) = f(h - t)$.
Note. This proves that the parabola is symmetric to its axis, $x = h$.

34. A farmer wants to fence off a rectangular field that is bounded by a river on one side, so only three sides need to be fenced. He has 1,000 feet of fencing. What are the dimensions of the field if the area is to be a maximum?

35. Derive the formula for $R(x)$ in Example 13 from the information given there.

36. A restaurant has a fixed price of $10.00 for a complete dinner. The average number of customers per evening is 120. The owner estimates on the basis of prior experience that for each 50¢ increase in the price of the dinner, there will be 4 fewer customers per evening on the average. What should be charged for the dinner to obtain the maximum income?

37. A manufacturer sets the price of a certain item depending on the number sold per week, as follows:

Selling price (dollars) per item $= 100 - 0.05x$

where x is the number of items sold in a week. The total cost of producing x of these items in a week is given by

Cost (dollars) of producing x items $= 10{,}000 + 10x$

Express the profit $P(x)$ obtained by selling x items in a week as a function of x. Find the value of x that maximizes the profit. What should the selling price be for maximum profit?

6 Graphical Solution of Quadratic Inequalities

The values of x for which a function equals 0 are called the **zeros** of the function. These are the values of x for which the graph of the function crosses the x axis, that is, they are the x intercepts. For a quadratic function, $f(x) = ax^2 + bx + c$, finding the zeros amounts to the same thing as solving the quadratic equation $ax^2 + bx + c = 0$, and we know how to do this. If the polynomial is factorable with integral factors, the zeros are particularly easy to find and can be very helpful in obtaining a rapid sketch of the graph of the function. Sometimes, such a sketch is all that is needed. The next two examples show how *inequalities* can be solved using graphs. We refer to the technique as the **graphical solution of quadratic inequalities.**

EXAMPLE 14   Solve the inequality $x^2 - 12x < 0$ by graphical means.

Solution   We consider the function

$$y = x^2 - 12x = x(x - 12)$$

We know the graph is a parabola that opens upward, since the coefficient of x^2 is positive. From the factored form of the right-hand side, the zeros are seen to be $x = 0$ and $x = 12$. From this we sketch the parabola, as shown in Figure 17, without attempting to locate the vertex exactly. All we want to know are the x values for which y is negative, that is, for which the curve lies *below* the x axis. We can see from the figure that the answer is those values of x for which $0 < x < 12$. Or we could say the solution set is $\{x: \ 0 < x < 12\}$.

Figure 17

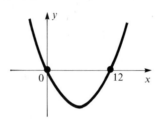

EXAMPLE 15   Solve by graphical means: $5 - 3x - 2x^2 \leq 0$

Solution   Let $y = 5 - 3x - 2x^2$. We will make a sketch of the graph and determine where $y \leq 0$. To find the zeros, we set $y = 0$ and factor:

$$5 - 3x - 2x^2 = 0$$
$$(5 + 2x)(1 - x) = 0$$

$5 + 2x = 0$	$1 - x = 0$
$x = -\frac{5}{2}$	$x = 1$

We know the graph is a parabola that opens downward, since the coefficient of x^2 is negative, and we draw it, as shown in Figure 18. So $y \leq 0$ when $x \geq 1$ or $x \leq -\frac{5}{2}$, that is, the solution set is $\{x: \ x \geq 1\} \cup \{x: \ x \leq -\frac{5}{2}\}$

Figure 18

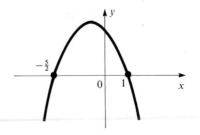

Note.   This graphical technique is not limited to solving quadratic inequalities. It will work whenever we can graph the function in question.

EXERCISE SET 6

A In Problems 1–15 use the graphical method to solve the inequalities.

1. $x^2 - 2x > 0$ **2.** $x^2 - 3x - 4 \leq 0$
3. $x^2 + 4x - 5 \leq 0$ **4.** $x^2 - x - 12 > 0$
5. $3x^2 - 5x - 2 < 0$ **6.** $6 - x - x^2 \geq 0$
7. $2x^2 + 5x - 12 \geq 0$ **8.** $6x^2 + 7x - 20 < 0$
9. $2x^2 - x > 3$ **10.** $x(2x + 1) \leq 15$
11. $2(1 - 3x^2) \geq x$ **12.** $3(x^2 - 4) \leq 5x$
13. $10x(x - 2) < 3(x - 4)$ **14.** $x(3x + 10) \geq 8$
15. $4x(3x - 2) > 15$

B Use graphical techniques to find the domains of the functions in Problems 16 and 17.

16. $f(x) = x + \sqrt{12x^2 + x - 20}$ **17.** $g(x) = \dfrac{2x - 1}{\sqrt{18 - 3x - 15x^2}}$

18. Solve the inequality below by graphical means.

$$x^3 - x^2 - 9x + 9 < 0$$

Hint. First factor the left-hand side to find the zeros of the function, and then plot a few points.

19. The revenue from the sale of x units of a certain commodity is 38 times the number of units sold. The cost of producing x units is $105 + 12x + x^2$. Write the profit function and find the range of values of x for which the profit is positive. What value of x produces the maximum profit, and what is that profit?

7 Other Second-Degree Equations; Conic Sections

We saw in Section 5 that the graph of an equation of the form

$$y = ax^2 + bx + c \qquad (a \neq 0) \tag{7}$$

is a parabola with a vertical axis, opening upward if $a > 0$ and downward if $a < 0$. By completing the square we found that we could write the equation in the form $y = a(x - b)^2 + k$. Now, for consistency with what follows, we choose to write this in the equivalent form

$$y - k = a(x - h)^2 \tag{8}$$

The vertex is (h, k), and the axis is the line $x = h$.

If in equation (7) we interchange the roles of x and y, we obtain

$$x = ay^2 + by + c \qquad (a \neq 0) \tag{9}$$

which is the equation of a parabola with horizontal axis, opening to the right if $a > 0$ and to the left if $a < 0$. Completing the square yields the following equation, which is analogous to equation (8):

$$x - h = a(y - k)^2 \tag{10}$$

Again the vertex is (h, k). The axis is the horizontal line $y = k$.

EXAMPLE 16   Sketch the graph of the equation $4y^2 - x - 8y + 2 = 0$.

Solution   First we solve for x, obtaining an equation in the form of equation (9):

$$x = 4y^2 - 8y + 2$$

We complete the square in y and write the equation in the form of equation (10):

$$x = 4(y^2 - 2y + 1) + 2 - 4$$
$$= 4(y - 1)^2 - 2$$
$$x + 2 = 4(y - 1)^2$$

So $h = -2$ and $k = 1$. These are the coordinates of the vertex. The axis is the line $y = 1$, and the parabola opens to the right, as shown in Figure 19.

Figure 19

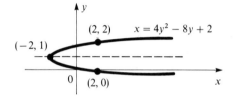

It should be emphasized that an equation of the form (9) does not define y as a function of x, since there are vertical lines that intersect the graph in more than one point.

The equations (7) and (9) each involves the square of *one* of the variables x or y, but not both. We now want to examine briefly some equations that involve the square of both x and y. The first of these is the equation of a **circle,** which we will derive.

A circle with radius r and center (h, k) is the set of all points whose distance from (h, k) equals r (Figure 20). A point (x, y) therefore lies on the circle if and only if

$$\sqrt{(x - h)^2 + (y - k)^2} = r$$

or equivalently,

$$(x - h)^2 + (y - k)^2 = r^2 \qquad (11)$$

Equation (11) is called the **standard form of the equation of a circle.** If the center is at the origin, it assumes the particularly simple form

$$x^2 + y^2 = r^2 \qquad (12)$$

Figure 20

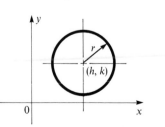

EXAMPLE 17   Identify and draw the graph of the equation $x^2 + y^2 - 6x + 4y + 9 = 0$.

Solution   We complete the square in both x and y to bring the equation to the standard form (11).

$$(x^2 - 6x + 9) + (y^2 + 4y + 4) = -9 + 9 + 4$$
$$(x - 3)^2 + (y + 2)^2 = 4$$

This is the equation of a circle with center $(3, -2)$ and radius 2, as shown in Figure 21.

Figure 21

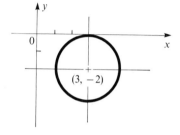

If we divide both sides of equation (12) by r^2, we get

$$\frac{x^2}{r^2} + \frac{y^2}{r^2} = 1$$

A closely related equation is of the form

$$\frac{x^2}{a^2} + \frac{y^2}{b^2} = 1 \qquad (13)$$

where a and b have different values. This is the equation of an **ellipse** with center at the origin. The x intercepts are $\pm a$, and the y intercepts are $\pm b$. If $a > b$, the ellipse is elongated on the x axis, and if $b > a$, it is elongated on the y axis. Figure 22 illustrates both cases. The line segments joining the intercepts

Figure 22

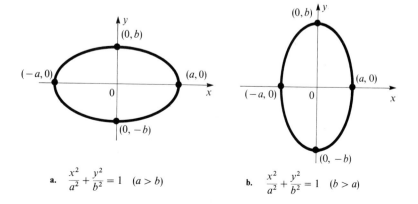

a. $\frac{x^2}{a^2} + \frac{y^2}{b^2} = 1$ $(a > b)$ b. $\frac{x^2}{a^2} + \frac{y^2}{b^2} = 1$ $(b > a)$

on the respective axes are called the **axes** of the ellipse. The longer axis is the **major axis** and the shorter one is the **minor axis.**

If the center of the ellipse is at the point (h, k), the standard equation of the ellipse is

$$\frac{(x - h)^2}{a^2} + \frac{(y - k)^2}{b^2} = 1 \tag{14}$$

EXAMPLE 18　　Identify and sketch the graph of the equation $4x^2 + 9y^2 - 36 = 0$.

Solution　　We write the equation in the form of equation (13) by adding 36 to both sides and then dividing both sides by 36.

$$\frac{x^2}{9} + \frac{y^2}{4} = 1$$

This is an ellipse with center at the origin and major axis along the x axis. Since $a^2 = 9$ and $b^2 = 4$, the x intercepts are ± 3 and the y intercepts are ± 2. These are the end points of the major and minor axes, respectively, as shown in Figure 23. We could find additional points on the graph by solving for y:

$$y = \pm \tfrac{2}{3}\sqrt{9 - x^2}$$

and substituting values for x. For a sketch, however, it is usually sufficient to use only the end points of the major and minor axes.

Figure 23

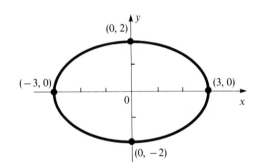

If instead of the sum of the squares as in equation (13), we have the difference

$$\frac{x^2}{a^2} - \frac{y^2}{b^2} = 1 \tag{15}$$

or

$$\frac{y^2}{b^2} - \frac{x^2}{a^2} = 1 \tag{16}$$

then the graph is that of a **hyperbola.** Figure 24 illustrates these two situations. The lines $y = bx/a$ and $y = -bx/a$ are called **asymptotes** of the hyperbola. The hyperbola approaches these lines more and more closely as the curve recedes indefinitely.

Figure 24

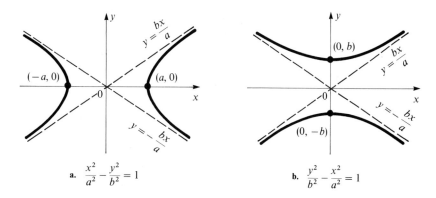

a. $\dfrac{x^2}{a^2} - \dfrac{y^2}{b^2} = 1$

b. $\dfrac{y^2}{b^2} - \dfrac{x^2}{a^2} = 1$

If the center is at (h, k), the two equations become

$$\frac{(x - h)^2}{a^2} - \frac{(y - k)^2}{b^2} = 1 \tag{17}$$

and

$$\frac{(y - k)^2}{b^2} - \frac{(x - h)^2}{a^2} = 1 \tag{18}$$

In this case, the asymptotes are the lines

$$y - k = \pm\frac{b}{a}(x - h)$$

EXAMPLE 19 Identify and sketch the graph of the equation $16x^2 - 9y^2 = 144$.

Solution We divide both sides by 144 to obtain

$$\frac{x^2}{9} - \frac{y^2}{16} = 1$$

So the graph is a hyperbola with center at the origin. The intercepts on the x axis are ± 3. There are no y intercepts; $y = 4x/3$ and $y = -4x/3$ are asymptotes.

Figure 25

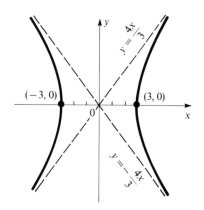

The circle, ellipse, parabola, and hyperbola are called **conic sections** (or simply **conics**) because they may be formed by the intersection of a plane with a right circular cone, as shown in Figure 26.

Figure 26

a. Plane perpendicular to axis of cone; intersection is a circle

b. Plane parallel to element of cone; intersection is a parabola

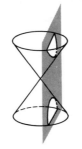

c. Plane parallel to axis of cone; intersection is a hyperbola

d. Any plane other than those in parts a, b, or c; intersection is an ellipse

EXERCISE SET 7

A   In Problems 1–16 identify the curve, and sketch its graph.

1. $4y^2 - x = 0$

2. $y^2 + 2x = 0$

3. $x^2 + y^2 - 9 = 0$

4. $x^2 + 4y^2 = 4$

5. $x^2 - y^2 = 1$

6. $4x^2 - 9y^2 + 1 = 0$

7. $4x^2 + 4y^2 = 25$

8. $9x^2 - 4y = 0$

9. $y^2 = \dfrac{16(9 - x^2)}{9}$

10. $y^2 = \dfrac{16(x^2 - 25)}{25}$

11. $x = 1 + y^2$

12. $(x - 1)^2 + (y - 2)^2 = 4$

13. $(x - 3) = -2(y + 2)^2$

14. $\dfrac{(x + 3)^2}{16} + \dfrac{(y - 2)^2}{25} = 1$

15. $\dfrac{(x + 2)^2}{4} - \dfrac{(y + 1)^2}{1} = 1$

16. $\dfrac{(y - 3)^2}{9} - \dfrac{(x + 2)^2}{16} = 1$

In Problems 17–22 graph the function by first squaring both sides and writing the result in a standard form. The graph must be appropriately restricted as dictated by the given equation.

17. $y = \sqrt{4 - x^2}$

18. $y = -\sqrt{1 - x}$

19. $y = \dfrac{2\sqrt{9 - x^2}}{3}$

20. $y = \dfrac{2\sqrt{9 + x^2}}{3}$

21. $f(x) = \sqrt{x + 2}$
Hint.   First let $y = f(x)$.

22. $g(x) = \sqrt{x^2 - 1}$

B   In Problems 23–30 write the equation in a standard form, identify, and sketch the graph.

23. $y^2 - 2x - 6y + 13 = 0$

24. $x^2 + y^2 + 2x - 4y + 1 = 0$

25. $4x^2 + 9y^2 - 16x - 20 = 0$

26. $16x^2 - 9y^2 - 32x - 54y + 79 = 0$

27. $2y^2 - 16y + x + 35 = 0$

28. $x^2 + y^2 - 3x + 5y - 21 = 0$

29. $4x^2 - y^2 + 8x - 4y - 4 = 0$

30. $25x^2 + 16y^2 + 100x - 32y - 284 = 0$

Review Exercise Set

A   **1.** Find the slopes of the lines joining the following pairs of points:
    a. $(-1, 3)$ and $(5, -4)$ **b.** $(-2, -6)$ and $(-4, 2)$

  2. In each of the following a line passes through the given point and has the given slope. Find a second point on the line.
    a. $(2, 5)$, slope $\frac{2}{3}$ **b.** $(-3, 2)$, slope $-\frac{5}{2}$

  3. Find the unknown if the line joining the two points has the given slope.
    a. $(5, -3)$ and $(7, y)$; slope 3 **b.** $(-4, 2)$ and $(x, -6)$; slope $-\frac{4}{3}$

In Problems 4–7 find the equations of the lines satisfying the given conditions.

  4. **a.** Slope $-\frac{1}{2}$, passing through $(-5, 3)$
    b. Passing through $(-1, -4)$ and $(2, -6)$

  5. **a.** Slope $\frac{3}{2}$, y intercept -2
    b. Passing through $(2, 7)$ and $(-1, 4)$

  6. **a.** Slope $\frac{3}{4}$, passing through $(-1, -5)$
    b. Passing through $(-2, 4)$ and $(5, 7)$

  7. **a.** y intercept 3, slope $\frac{3}{7}$
    b. Passing through $(5, -3)$ and $(-2, 6)$

In Problems 8 and 9 give the slope, x intercept, and y intercept, and draw the graphs.

  8. **a.** $5x - 4y + 20 = 0$ **b.** $3(y - 4) = 6 - 2x$

  9. **a.** $3(x - 4) = 2(y + 3)$ **b.** $\dfrac{2x - 3}{5} = \dfrac{1 - y}{3}$

In Problems 10–12 find the equations of the lines satisfying the given conditions.

10. **a.** Passing through $(3, 5)$ and parallel to $2x - 4y + 3 = 0$
    b. Passing through $(-1, 2)$ and perpendicular to $x + 5y - 6 = 0$

11. **a.** Perpendicular to the line joining $(2, 6)$ and $(-3, 2)$ and passing through $(1, -1)$
    b. Parallel to the line joining $(1, 4)$ and $(-3, 7)$ and passing through $(2, -5)$

12. **a.** Parallel to the line joining $(-1, 2)$ and $(5, 4)$ and passing through the origin
    b. Perpendicular to the line $3x + 2y - 4 = 0$ and having x intercept -1

13. For the triangle with vertices at $A(3, -2)$, $B(-1, 4)$, and $C(5, 6)$, find the following:
    a. The equation of the line through C and parallel to the side AB
    b. The equation of the altitude drawn from C

14. The line through $(-4, -1)$ and $(2, y)$ is parallel to the line $5x + 2y + 4 = 0$. Find y.

15. Find the equation of the perpendicular bisector of the line segment joining $(2, 7)$ and $(-4, -3)$. (See Problem 27, Exercise Set 3.)

16. **a.** Find the equation of the family of lines parallel to $3x - 5y + 2 = 0$. Find the equation of the member of this family that has y intercept 2.
    b. Find the equation of the family of lines perpendicular to $2x + 3y - 4 = 0$. Find the equation of the member of this family that has an x intercept at the same point as that of the given line.

17. **a.** Find the equation of the family of lines parallel to the line $8x + 3y + 7 = 0$. Find the equation of the member of this family that passes through the point $(2, -3)$.
    b. Find the equation of the family of lines perpendicular to the line $y = 3x - 4$. Find the equation of the member of this family that passes through the point $(-1, 5)$.

18. For a certain linear function f it is given that $f(2) = -1$ and $f(5) = 3$. Find $f(x)$ and draw the graph of f.

19. If g is a polynomial function of degree 1, and $g(0) = -3$ and $g(2) = -1$, find $g(x)$. What is the slope of the graph of g?

20. For a certain linear function f it is given that $f(-1) = 2$ and $f(3) = -4$. Find $f(x)$ and draw the graph of f.

21. Let $f(x) = (3x + 4)/5$ and $g(x) = 2 - 3x$. Find $f \circ g$ and $g \circ f$, and draw their graphs.

22. Let $g(x) = 4x - 5$ and $h(x) = (x - 2)/3$. Find $g \circ h$ and $h \circ g$, and draw their graphs.

23. The temperature of the air diminishes approximately linearly with altitude h, for altitudes of up to 6 miles, dropping about $2°F$ for each 1,000 feet of altitude. If T_0 is the temperature on the ground, write the temperature T as a linear function of the altitude h (feet). If $T_0 = 72°F$, find T at an altitude of 18,000 feet.

24. A bicycle shop owner sells an average of 20 ten-speed bicycles a month when the price is $175. When the price is reduced by 20%, sales go up by 50%. Assuming a linear relationship between price p and number of sales n, find the equation relating them. How many ten-speed bicycles should the shop owner expect to sell per month if the price is reduced by 10% from the original price?

In Problems 25 and 26 find the vertex and the axis of the parabola, and draw its graph.

25. **a.** $y = x^2 - 6x + 4$ **b.** $y = -2x^2 + 4x - 3$

26. **a.** $y = 4x - x^2$ **b.** $y = \dfrac{x^2}{2} + 2x + 3$

In Problems 27 and 28 find the maximum or minimum value of each function and state the value of x that gives this maximum or minimum value.

27. **a.** $f(x) = 3x^2 - 9x + 4$ **b.** $g(x) = 8 - 4x - x^2$
28. **a.** $F(x) = 2x^2 - 3x + 7$ **b.** $h(x) = 3x(2 - x) + 4$

29. The cost of producing x parts per month for a certain piece of equipment is given by the formula
$$C(x) = 2x^2 - 400x + 30,000$$
How many parts should be produced to minimize cost? What is the minimum cost?

30. The profit function $P(x)$ from producing and selling a certain type of pocket calculator is given by
$$P(x) = -0.02x^2 + 60x - 35,000$$
where x is the number produced and sold per week, and $P(x)$ is given in dollars. How many calculators should the manufacturer produce and sell each week to maximize profits? What is the maximum profit?

In Problems 31–33 solve the inequalities by the graphical method.

31. **a.** $x^2 - x - 2 < 0$ **b.** $3 - 2x - x^2 > 0$
32. **a.** $3x^2 - 5x - 2 \geq 0$ **b.** $2x^2 - 5x + 2 \leq 0$
33. **a.** $2(3x^2 + 2) \geq 11x$ **b.** $8 \geq x(2 + 3x)$

In Problems 34–38 identify the curve and draw its graph.

34. **a.** $x^2 + y^2 = 25$ **b.** $16x^2 + 25y^2 = 400$
35. **a.** $4x^2 - y^2 = 4$ **b.** $y^2 - x^2 = 1$
36. **a.** $y^2 + 2x = 0$ **b.** $x - 2 = \frac{1}{4}(y + 1)^2$
37. **a.** $(x - 3)^2 + (y + 3)^2 = 4$ **b.** $9(x - 1)^2 + 4(y + 2)^2 = 36$
38. **a.** $9(x - 3)^2 - 4(y + 1)^2 = 36$ **b.** $x = 3 - 4y - 2y^2$

B **39.** Show that the points $(-5, 2), (-3, -2), (6, 1)$, and $(4, 5)$ are the vertices of a parallelogram. Find the equations of the sides.

40. For the triangle with vertices at $(0, 5), (-4, -3)$, and $(6, 1)$ find the equations of the following:
  a. The altitudes   **b.** The medians
      (See Problem 27, Exercise Set 3.)

41. A space shuttle is placed in a circular orbit around the earth. When its engine is fired, it will move in a straight line tangent to the orbit. If with respect to a certain coordinate system for the plane of the orbit the center of the earth is at $(-1, 2)$ and the engine is fired when the spacecraft is at $(2, 3)$, what is the equation of the line along which the space shuttle will move?
  Note.   The tangent to the circle at a point is perpendicular to the radius drawn to that point.

42. By calculus it can be shown that the slope of the tangent line to the curve $y = x^4 - 2x^2 + 3x - 7$ at any point (x, y) is given by $m = 4x^3 - 4x + 3$. Find the equations of the tangent line and the normal line (the line perpendicular to the tangent) at the point for which $x = -1$.

43. In calculus it can be shown that the slope of the tangent line at any point (x, y) on the curve $y = 1/x$ is given by the formula $m = -1/x^2$. Find the equations of the tangent line and the normal line at the point for which $x = -1$. Draw the graph, showing the tangent and normal.

44. Use the graphical method to solve the inequality $x^3 + 4x^2 - x - 4 \geq 0$.

In Problems 45 and 46 write the equation in standard form, and identify and draw the graph.

45. **a.** $x^2 + y^2 - 2x + 4y + 1 = 0$ **b.** $x^2 + 4y^2 + 6x - 16y + 21 = 0$
46. **a.** $4x^2 - 9y^2 - 16x - 18y + 43 = 0$ **b.** $25x^2 + 16y^2 - 192y + 176 = 0$

47. Prove that the line $3x - 4y + 15 = 0$ is tangent to the circle $x^2 + y^2 - 4x + 2y - 20 = 0$ at the point $(-1, 3)$.

48. At a certain movie theater the price of admission is \$3.00, and the average daily attendance is 200. As an experiment the manager reduces the price by 5¢ and finds that the average attendance increases by 5 people per day. Assuming that for each further 5¢ reduction the average attendance would rise by 5, find the number of 5¢ reductions that would result in the maximum revenue.

49. A railroad company agrees to run an excursion train for a group under the following conditions: If 200 people or less go, the rate will be \$10 per ticket. The rate for each ticket will be reduced by 2¢ for each person in excess of 200 who go. What number of people will produce the maximum revenue for the railroad?

50. Find the dimensions of the rectangle of maximum area that can be inscribed in an isosceles triangle of altitude 8 and base 6.

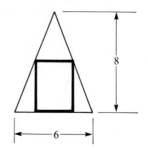

Hint. In a sketch use similar triangles to express the height of the rectangle in terms of its base.

6

Polynomial Functions of Higher Degree

While equations of order higher than 2 occur less frequently in calculus than linear and quadratic equations, they do occur, and it is important to learn techniques for solving them. As we shall show, the problem of solving a polynomial equation and the problem of factoring a polynomial are essentially the same, and the need to factor polynomials arises in various settings. The following example is an illustration:*

Evaluate

$$\int \frac{x^5 - x^4 - 3x + 5}{x^4 - 2x^3 + 2x^2 - 2x + 1} \, dx$$

Solution: The integrand is a fraction, but not a proper fraction. Hence, we divide first, obtaining

$$\frac{x^5 - x^4 - 3x + 5}{x^4 - 2x^3 + 2x^2 - 2x + 1} = x + 1 + \frac{-2x + 4}{x^4 - 2x^3 + 2x^2 - 2x + 1}$$

The denominator factors as follows:

$$x^4 - 2x^3 + 2x^2 - 2x + 1 = (x^2 + 1)(x - 1)^2$$

This is not the end of the calculus problem, but it is as far as we need to go to illustrate the point. The question is, how did the author determine the factors of the polynomial $x^4 - 2x^3 + 2x^2 - 2x + 1$? Certainly, none of the techniques of Chapter 2 will suffice. This is one of the things you will learn to do in this chapter.

---

* George B. Thomas, Jr., and Ross L. Finney, *Calculus and Analytic Geometry*, 5th ed. (Reading, Mass.: Addison-Wesley, 1979), p. 358. Reprinted by permission.

1 Historical Note

The problem of trying to find formulas analogous to the quadratic formula for the roots of polynomial equations of degree greater than 2 occupied mathematicians for centuries. For the cases $n = 3$ and $n = 4$, called the **cubic** and **quartic** equations, respectively, formulas were obtained around 1540 by the Italian mathematicians Tartaglia (cubic) and Ferrari (quartic). Attempts over the next two centuries to solve the general **quintic** (degree 5) equation all met with failure. Finally, in 1824, the brilliant Norwegian mathematician Niels Henrik Abel (1802–1829), at the age of 22, made the remarkable discovery that no such general formula for the solutions of polynomial equations of degree higher than 4 could be found. This settled the matter once and for all, but in a surprising way. It is indeed a curious fact that algebraic formulas for the solutions of degree n equations in terms of their coefficients exist for the cases $n = 1, 2, 3, 4$ but not for $n \geq 5$. It is not simply that no one has been able to find these formulas—it is *impossible* for anyone ever to find them because they do not exist. This is not to say that *no* equation of degree 5 or higher can be solved in terms of its coefficients; in some cases this can be done. The special case $ax^5 + b = 0$ can certainly be solved for one of its roots, namely, $x = \sqrt[5]{-b/a}$. Furthermore, we will show in this chapter a technique for finding the rational roots of polynomial equations (with integral coefficients) of arbitrary degree and also a way of approximating the irrational roots to any desired degree of accuracy. So the situation is not hopeless. As a matter of fact, the formulas developed by Tartaglia and Ferrari are of little practical importance. Cubic and quartic equations are typically solved not by their formulas but by the techniques to be presented here (or by numerical techniques with the aid of the computer). The quadratic formula, on the other hand, is used extensively because of its simplicity.

2 The Remainder Theorem and the Factor Theorem

Let $P_n(x)$ denote a polynomial function of degree n:

$$P_n(x) = a_n x^n + a_{n-1} x^{n-1} + \cdots + a_1 x + a_0$$

When this is set equal to 0, it becomes a polynomial equation, and we will be seeking ways to solve such equations for $n > 2$. Recall that we use the terms **root** and **solution** interchangeably, and if r is a root of the equation, then r is also called a **zero** of the polynomial, and conversely. For example, we may say that the roots of the polynomial equation $x^2 - x - 2 = 0$ are 2 and -1, or equivalently, that the zeros of the polynomial $x^2 - x - 2$ are 2 and -1. The zeros of a polynomial are the x intercepts of its graph.

If we divide $P_n(x)$ by $x - r$, we obtain a quotient of degree $n - 1$ and a constant term as remainder. For example, let us divide $2x^3 - 3x + 4$ by $x - 2$:

$$
\begin{array}{r}
2x^2 + 4x\ + 5 \\
x - 2\overline{)2x^3\qquad\ - 3x +\ 4} \\
\underline{2x^3 - 4x^2} \\
4x^2 - 3x \\
\underline{4x^2 - 8x} \\
5x +\ 4 \\
\underline{5x - 10} \\
14\quad \text{Remainder}
\end{array}
$$

The quotient $2x^2 + 4x + 5$ is one degree less than the dividend, and the remainder 14 is a constant. We could write in this case

$$
\frac{2x^3 - 3x + 4}{x - 2} = 2x^2 + 4x + 5 + \frac{14}{x - 2}
$$

and obtain from multiplication by $(x - 2)$

$$
2x^3 - 3x + 4 = (x - 2)(2x^2 + 4x + 5) + 14
$$

This is an identity in x, that is, it is true for all values of x. In a similar way, if we denote the quotient on dividing $x - r$ into $P_n(x)$ by $Q(x)$ and the remainder by R, we obtain the identity

$$
P_n(x) = (x - r)Q(x) + R \tag{1}
$$

Two results can now be obtained, and these are stated as theorems.

THE REMAINDER THEOREM

When the polynomial $P_n(x)$ is divided by $x - r$, the remainder is equal to $P_n(r)$.

Proof In equation (1) substitute $x = r$. We obtain

$$
\begin{aligned}
P_n(r) &= (r - r)Q(r) + R \\
&= 0 + R \\
&= R
\end{aligned}
$$

For example, if we let $P(x) = 2x^3 - 3x + 4$, we can conclude from the division problem above that $P(2) = 14$, since when we divided by $x - 2$, we obtained 14 as the remainder. We emphasize that to get $P(2)$ we divide by $x - 2$ and read off the *remainder* as the answer.

THE FACTOR THEOREM

The polynomial $x - r$ is a factor of the polynomial $P_n(x)$ if and only if r is a root of the polynomial equation $P_n(x) = 0$, or equivalently, if and only if r is a zero of the polynomial $P_n(x)$

For example, consider the polynomial $P(x) = x^3 - x^2 + x - 1$. By inspection we see that $x = 1$ is a root of the equation $P(x) = 0$. By the factor theorem it follows that $x - 1$ is a factor of $P(x)$. We can see this is true by factoring as follows:

$$x^3 - x^2 + x - 1 = x^2(x - 1) + (x - 1)$$
$$= (x^2 + 1)(x - 1)$$

Conversely, since $x - 1$ is a factor of $P(x)$, by putting $x = 1$, the polynomial equals 0.

These two theorems, although very simple, play a fundamental role in the solution of equations. The second theorem shows that the problem of finding roots of a polynomial equation and the problem of factoring a polynomial are essentially the same.

3   Synthetic Division

Since division will play a prominent role in the rest of this discussion, we will pause here to describe a shortened version of long division known as **synthetic division.** Consider a typical long division problem in which the divisor is of the special form $x - r$. For example, let us repeat the division of $2x^3 - 3x + 4$ by $x - 2$, shown in Section 2:

$$
\begin{array}{r}
2x^2 + 4x\ + 5 \\
x - 2{\overline{\smash{\big)}\,2x^3 \qquad\ - 3x +\ \ 4}} \\
\underline{(2x^3) - 4x^2} \\
4x^2 - 3x \\
\underline{(4x^2) - 8x} \\
5x + 4 \\
\underline{(5x) - 10} \\
14
\end{array}
$$

We will go through a series of steps that simplifies this procedure. First, there really is no need to write down the terms that have been circled above. We know that by subtraction the result will be 0. Leaving these out, we have

$$
\begin{array}{r}
2x^2\ + 4x\ + 5 \\
x - 2{\overline{\smash{\big)}\,②x^3 \qquad\ - 3x +\ \ 4}} \\
- 4x^2 \\
\underline{} \\
④x^2 - 3x \\
- 8x \\
\underline{} \\
⑤x +\ \ 4 \\
- 10 \\
\underline{} \\
14
\end{array}
$$

Next, note that the circled terms in the above display are the same as the coefficients in the quotient; this is because the coefficient of x in the divisor is unity. With this observation, then, we can leave out the x in the divisor. Also, we can dispense with the powers of x everywhere and let the position of the coefficient indicate the associated power of x. The display now takes the form

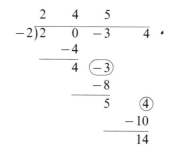

Note that we have written a 0 for the coefficient of the missing power. The first coefficient 2 in the quotient is the same as the first coefficient in the dividend. Each succeeding coefficient is obtained by multiplying the preceding coefficient by the -2 of the divisor and subtracting the result from the appropriate coefficient in the dividend. This can be simplified by changing the sign of the -2 in the divisor to $+2$ and *adding* rather than subtracting. Also, there is no need to repeat the circled terms above, because they occur in the dividend. This will enable us to compress the entire display as follows:

$$
\begin{array}{r}
2\quad 4\quad 5 \\
2)\overline{2\quad 0\ -3\quad 4} \\
4\quad 8\quad 10 \\
\hline
4\quad 5\quad 14
\end{array}
$$

Finally, if we bring down the initial coefficient 2 of the dividend to the bottom line, we see that the answer can be read from that line, so that we can omit the top line entirely. So we have

$$
\begin{array}{r|rrrr}
2 & 2 & 0 & -3 & 4 \\
 & \downarrow & 4 & 8 & 10 \\
\hline
 & 2 & 4 & 5 & 14
\end{array}
$$

Recall that the second line is obtained by multiplying the 2 in the divisor position by the numbers 2, 4, and 5, respectively, of the bottom line. Thus, $4 = 2 \cdot 2$, $8 = 2 \cdot 4$, and $10 = 2 \cdot 5$. The entries in the third line are obtained by adding the two entries above them. Finally, the answer is read, from right to left, as the remainder, constant term, coefficient of x, and coefficient of x^2.

Caution. Although 2 appears in the divisor position, the divisor is $x - 2$.

While the procedure was illustrated with a specific example, the method is general. Note carefully that the divisor must be of the form $x - r$, and r then is the number that appears in the position of the divisor. For example, if the divisor were $x + 2$, then this must be thought of as $x - (-2)$, so that -2 is in the divisor position. Remember, too, that since the positions of the coefficients indicate the powers of x, these coefficients must be arranged according to descending powers of x, and a 0 must be supplied as the coefficient of any missing power.

EXAMPLE 1   Divide $4x^4 - 15x^3 + 8x^2 - x + 7$ by $x - 3$.

Solution

$$
\begin{array}{r|rrrrr}
3 & 4 & -15 & 8 & -1 & 7 \\
  &   & 12 & -9 & -3 & -12 \\
\hline
  & 4 & -3 & -1 & -4 & -5
\end{array}
$$

Thus, the quotient is $4x^3 - 3x^2 - x - 4$, and the remainder is -5.

EXAMPLE 2   Divide $x^5 - 2x^4 + x$ by $x + 2$.

Solution

$$
\begin{array}{r|rrrrrr}
-2 & 1 & -2 & 0 & 0 & 1 & 0 \\
   &   & -2 & 8 & -16 & 32 & -66 \\
\hline
   & 1 & -4 & 8 & -16 & 33 & -66
\end{array}
$$

Thus, the quotient is $x^4 - 4x^3 + 8x^2 - 16x + 33$, and the remainder is -66.

Let us now consider the problem of evaluating a polynomial function $P_n(x)$ at a particular value of x, say $x = r$. There are two ways of doing this:

1. **The direct method:**   Substitute r for x wherever x appears.
2. **Use the remainder theorem:**   Divide $P_n(x)$ by $x - r$ using synthetic division; the answer sought is the remainder.

Consider, for example, the polynomial used previously, $P(x) = 2x^3 - 3x + 4$, and evaluate this by each method when $x = 2$. By the direct method,

$$P(2) = 2(2)^3 - 3(2) + 4 = 16 - 6 + 4 = 14$$

To apply method 2 we divide $P(x)$ by $x - 2$ using synthetic division and read off the remainder as the answer.

$$
\begin{array}{r|rrrr}
2 & 2 & 0 & -3 & 4 \\
  &   & 4 & 8 & 10 \\
\hline
  & 2 & 4 & 5 & \boxed{14 = P(2)}
\end{array}
$$

Remark.   Method 2 is especially well-suited to the use of a hand calculator. When the coefficients are messy (as is usually the case in practical applications), the use of a calculator is particularly helpful.

EXERCISE SET 3

A   In Problems 1–17 use synthetic division to divide the first polynomial by the second.

1. $x^2 - 2x - 4$ by $x - 1$
2. $3x^2 - 4x + 5$ by $x - 4$
3. $x^3 + 2x^2 - x + 3$ by $x - 2$
4. $2x^3 - 3x^2 + 4x - 5$ by $x + 1$
5. $x^3 + x^2 - 4$ by $x - 3$
6. $2x^3 - 5x^2 + 4$ by $x - 2$

7. $3x^3 - 4x + 5$ by $x + 2$
8. $2x^3 - 3x^2 + x - 7$ by $x - 3$
9. $2x^3 - 3x + 10$ by $x + 2$
   (What can you conclude?)
10. $5x^3 + 2x - 3$ by $x + 3$
11. $x^4 - 2x^3 - 25$ by $x - 3$
12. $x^4 - 3x^3 + 2x^2 - x - 3$ by $x - 1$
13. $x^5 - 2x^4 + 3$ by $x - 1$
14. $5x^3 - 4x^2 + x$ by $x + 4$
15. $3x^4 - 14x^3 - 6x^2 + 7x - 15$ by $x - 5$
16. $x^5 - 8x^3 - 6x^2 + 20$ by $x + 3$
17. $x^5 - 81$ by $x + 2$

In Problems 18–36 make use of synthetic division and the remainder theorem to find the indicated function values.

18. $P(x) = 3x^2 - 5x + 7$; find $P(4)$
19. $P(x) = x^3 - 2x^2 + 3x + 5$; find $P(2)$
20. $P(x) = x^3 + 7x^2 + 8x - 10$; find $P(-2)$.
21. $f(x) = 2x^3 - 10x - 8$; find $f(3)$
22. $P(x) = 2x^3 - 3x^2 + 4x - 7$; find $P(-3)$
23. $g(t) = t^4 - 2t^2 - 6$; find $g(-2)$
24. Let $f(x) = 3x^4 - 10x^3 - 24x - 16$; find $f(4)$. Also find $f(4)$ by substitution. Which method is easier?
25. Evaluate the polynomial $x^5 - 8x^3 + 10x + 12$ when $x = 2$.
26. $Q(t) = 4t^3 - 8t^2 - 9$; find $Q(-3)$
27. $P(x) = x^4 - 16$; find $P(3)$
28. $F(x) = 3x^4 - 8x^3 - 12x^2 + 5x - 9$; find $F(2)$
29. $P(y) = y^4 - 6y^2 + 3$; find $P(5)$
30. $P(x) = x^4 + 2x^3 - 3x^2 - 4x - 8$; find $P(2)$
31. $g(x) = 5x^4 - 20x^3 - 12x^2 + 10$; find $g(-2)$
32. $P(x) = x^5 - 2x^3 + 7$; find $P(3)$
33. $Q(s) = 3s^4 + 2s^3 - s^2 - 5s$; find $Q(-4)$
34. $P(x) = 3 - x - 2x^3 - x^4$; find $P(3)$
35. $Q(x) = x^5 + 12x^4 + 24x^3 - 10x^2 - 5$; find $Q(-3)$
36. $P(t) = 10 + x^2 - 3x^3 - 2x^4$; find $P(-2)$

B Use synthetic division to perform the divisions in Problems 37–40.

37. $(4x^4 - 3x^3 + x - 4) \div (x - \frac{1}{2})$
38. $(2x^4 + 3x^2 - 5x) \div (x + \frac{3}{2})$
39. $(x^3 - 5x^2 + 3x - 4) \div (2x - 1)$
   Hint. Write $(2x - 1) = 2(x - \frac{1}{2})$. Divide first by $x - \frac{1}{2}$, using synthetic division, and then divide the final quotient by 2.
40. $(2x^4 - 3x^3 + 2x - 5) \div (3x + 2)$ (See hint to Problem 39.)
41. Show by two methods that $x = -3$ is a zero of the polynomial $P(x) = 2x^4 - x^3 - 8x^2 + 34x - 15$.
42. Determine by two methods whether $x = 4$ is a root of the equation $3x^4 - 2x^3 - 50x^2 + 32x + 32 = 0$.
43. Show that $x + 6$ is a factor of $2x^4 + 11x^3 - 6x^2 + x + 6$ using synthetic division.
44. Let $P(x) = x^5 - 4x^4 - 5x^3 + 20x^2 + 4x - 16$. Use synthetic division and the remainder theorem to show that $P(4) = 0$. Show the complete factorization of $P(x)$.
45. Find $f(\frac{1}{2})$ using synthetic division if $f(x) = 4 - 2x + 3x^2 - x^4$.
46. Let $P(x) = 2x^4 - 3x^3 + 5x - 7$. Find $P(-\frac{3}{2})$ using synthetic division.
47. Let $Q(x) = 16x^4 - x^2 - 3x + 4$. Find $Q(\frac{3}{4})$ using synthetic division.
48. Show by synthetic division that $x = 4$ is a zero of the polynomial $x^3 - 6x^2 + 9x - 4$. Factor this polynomial completely. What are its other zeros?

49. Show by synthetic division that $x = -3$ is a root of the equation $x^3 + 2x^2 - 2x + 3 = 0$. What are the factors of the left-hand side? Are there any other real roots? Explain.

In Problems 50 and 51 find the specified function values using synthetic division and a hand calculator.

50. If $f(x) = 2.035x^3 - 1.768x^2 - 9.251x + 6.347$, find:
    a. $f(5.623)$ b. $f(-1.037)$
51. If $g(t) = 25.76 - 13.42t + 7.059t^3 - t^4$, find:
    a. $g(0.3254)$ b. $g(-2.167)$

4 The Fundamental Theorem of Algebra and Its Corollaries

As its name implies, the following theorem is fundamental to the theory of equations. Its proof is beyond the scope of this course, but two important corollaries will be proved.

THE FUNDAMENTAL THEOREM OF ALGEBRA	**Every polynomial equation $P_n(x) = 0$ has at least one root in the field of complex numbers.**

Our interest will center on polynomials with real, and usually integral, coefficients, but this theorem is valid when the coefficients are any complex numbers. The root that is asserted to exist, however, cannot be assumed to be real, even when the coefficients are. For example, the equation $x^2 + 4 = 0$ has the root $x = 2i$ (also $x = -2i$), which is not real, even though the coefficients are integers. As a corollary to the fundamental theorem, we have the following:

COROLLARY 1   **A polynomial of degree n can be factored into n (not necessarily distinct) linear factors over the complex number field.**

Proof   By the factor theorem, we know that $x - r$ is a factor of $P_n(x)$ if and only if $P_n(r) = 0$, that is, if and only if r is a root of the polynomial equation $P_n(x) = 0$. By the fundamental theorem, we know that $P_n(x)$ does have a root, say, r_1. Thus,

$$P_n(x) = (x - r_1)P_{n-1}(x)$$

where $P_{n-1}(x)$ is the quotient of $P_n(x)$ divided by $x - r$, and hence is of degree $n - 1$ and has the same leading coefficient, say, a_n, as $P_n(x)$. Further, $P_n(x) = 0$ if and only if

$$(x - r_1)P_{n-1}(x) = 0$$

So, applying the factor theorem and the fundamental theorem to $P_{n-1}(x)$, we know there is a root, say r_2, of $P_{n-1}(x) = 0$, so that

$$P_{n-1}(x) = (x - r_2)P_{n-2}(x)$$

and thus,

$$P_n(x) = (x - r_1)(x - r_2)P_{n-2}(x)$$

where $P_{n-2}(x)$ is of degree $n-2$ and has the same leading coefficient as $P_{n-1}(x)$ and hence as $P_n(x)$. We continue this process until arriving at the stage

$$P_n(x) = (x - r_1)(x - r_2) \cdot \cdots \cdot (x - r_{n-1})P_1(x)$$

where $P_1(x)$ is of degree 1 and has the same leading coefficient as $P_n(x)$. That is, $P_1(x) = a_n x + b = a_n[x + (b/a_n)]$. If we write $r_n = -b/a_n$, we have finally,

$$P_n(x) = a_n(x - r_1)(x - r_2) \cdot \cdots \cdot (x - r_n) \qquad (2)$$

and the proof is complete.

It must be emphasized that the factors in the above corollary may not all be distinct, which is equivalent to saying that the r_k's are not necessarily distinct. Each factor $(x - r_k)$ corresponds to a root of $P_n(x) = 0$, according to the factor theorem. We therefore have another corollary.

COROLLARY 2   **A polynomial equation of degree n has at most n roots.**

There will, in fact, be exactly n roots if all the r_k's are distinct. If, in the factorization given in equation (2), $(x - r_k)$ occurs as a factor m times, then r_k is said to be a **root of multiplicity m** of the equation $P_n(x) = 0$.

Suppose now that the coefficients in

$$P_n(x) = a_n x^n + a_{n-1} x^{n-1} + \cdots + a_1 x + a_0$$

are all real, and suppose $x = r$ is an imaginary root of the equation $P_n(x) = 0$. We will show that the conjugate $\bar{r}$ is also a root. For

$$\overline{P_n(r)} = \overline{a_n r^n + a_{n-1} r^{n-1} + \cdots + a_1 r + a_0}$$
$$= \bar{a}_n \bar{r}^n + \bar{a}_{n-1} \bar{r}^{n-1} + \cdots + \bar{a}_1 \bar{r} + \bar{a}_0$$

where we have used the properties of conjugates repeatedly. Now, since the coefficients are real, conjugation leaves them unaltered; that is,

$$\overline{P_n(r)} = a_n \bar{r}^n + a_{n-1} \bar{r}^{n-1} + \cdots + a_1 \bar{r} + a_0 = P_n(\bar{r})$$

Finally, since $P_n(r) = 0$ by hypothesis and 0 is a real number,

$$\overline{P_n(r)} = \bar{0} = 0 = P_n(\bar{r})$$

So $\bar{r}$ is a root of $P_n(x) = 0$ as asserted. This result shows that **in polynomial equations with real coefficients, imaginary roots always occur in conjugate pairs.** Thus, if $2 + 3i$ is a root of such an equation, then $2 - 3i$ is also.

EXERCISE SET 4

A   In Problems 1–5 verify the factorizations by making use of synthetic division.

1. $2x^3 - x^2 - 7x + 6 = 2(x - 1)(x + 2)(x - \frac{3}{2})$
2. $4x^3 - 21x^2 + 18x + 27 = 4(x + \frac{3}{4})(x - 3)^2$
3. $3x^3 + 4x^2 - 12x - 16 = 3(x + 2)(x - 2)(x + \frac{4}{3})$
4. $x^4 + 3x^3 - 6x^2 - 28x - 24 = (x - 3)(x + 2)^3$
5. $3x^4 - 2x^3 + 11x^2 - 8x - 4 = 3(x - 1)(x + \frac{1}{3})(x - 2i)(x + 2i)$

6. Construct a polynomial with integer coefficients whose only zeros are 2, -1, and $\frac{1}{2}$.
7. Find a polynomial equation with roots 3, 2, and -2.
8. Construct a polynomial with integer coefficients whose only zeros are -3, 4, and $\frac{3}{4}$.
9. Find a polynomial with 2 as a zero of multiplicity 3 and whose other zero is -1.
10. Find a polynomial equation with -3 as a root of multiplicity 2 and whose only other root is 4.
11. Show by direct substitution that both $1 + i$ and its conjugate are roots of the equation $5x^3 - 7x^2 + 4x + 6$.
12. Show by synthetic division that both $3i$ and $-3i$ are roots of the equation $x^4 - 5x^3 + 5x^2 - 45x - 36 = 0$.
13. Give an example of a quadratic equation with:
    a. No real roots   **b.** Two real roots   **c.** Exactly one real root ⌄

B 14. Show that every polynomial of *odd* degree with real coefficients has at least one real root.
15. Verify that $x = 2$ is a root of multiplicity 3 of the equation $x^4 - 4x^3 + 16x - 16 = 0$. What is the other root?
16. Find a polynomial whose only zeros are $2 + 3i$, $2 - 3i$, and 4.
17. Find a polynomial of fourth degree with integer coefficients two of whose zeros are $3 - 4i$ and $2 + i$.
18. Two of the roots of a certain fourth-degree polynomial equation with integer coefficients are $1 + 2i$ and $4 - 3i$. Find the equation.

5   Rational Roots

If the polynomial equation $P_n(x) = 0$ has rational numbers as coefficients, it is equivalent to one with integer coefficients. This can be seen by multiplying both sides of the equation by the LCD of the coefficients. For example, by multiplying both sides of

$$\tfrac{2}{3}x^3 - \tfrac{1}{4}x^2 + \tfrac{5}{6}x - 1 = 0$$

by 12, we obtain the equivalent equation

$$8x^3 - 3x^2 + 10x - 12 = 0$$

For this reason we will concentrate our attention on polynomial equations with integer coefficients. When the coefficients are irrational, the roots can be approximated to any desired degree of accuracy using numerical techniques. A hand calculator is very helpful in this situation.

Even when the coefficients are integers, the roots are not necessarily rational, or even real (for example, $x^2 + 1 = 0$ has imaginary roots). However, if there *are* rational roots, there is a procedure for finding them, which we now describe.

A rational number can always be written in the form p/q, where p and q are relatively prime integers (that is, they have in common no integer factor other than ± 1). Suppose p and q are relatively prime and $x = p/q$ is a root of the equation

$$P_n(x) = a_n x^n + a_{n-1} x^{n-1} + \cdots + a_1 x + a_0 = 0$$

where now it is assumed that all coefficients are integers. Then we have

$$a_n \frac{p^n}{q^n} + a_{n-1} \frac{p^{n-1}}{q^{n-1}} + \cdots + a_1 \frac{p}{q} + a_0 = 0 \tag{3}$$

We multiply both sides of (3) by q^n and obtain

$$a_n p^n + a_{n-1} p^{n-1} q + a_{n-2} p^{n-2} q^2 + \cdots + a_1 p q^{n-1} + a_0 q^n = 0 \tag{4}$$

We rewrite (4) in two equivalent forms. First, subtract $a_0 q^n$ from both sides and factor the common factor p from each term remaining on the left:

$$p(a_n p^{n-1} + a_{n-1} p^{n-2} + \cdots + a_1 q^{n-1}) = -a_0 q^n$$

From this we see that p is a factor of the left-hand side and hence also of the right-hand side. But p and q are relatively prime; hence, p is not a factor of q^n.* So p must divide a_0.

Now write equation (4) in another way—this time subtracting $a_n p^n$ from both sides and factoring q from what remains on the left:

$$q(a_{n-1} p^{n-1} + a_{n-2} p^{n-2} q + \cdots + a_0 q^{n-1}) = -a_n p^n$$

Reasoning as above, we see that q must divide a_n. These results are summarized as a theorem.

THEOREM

The only possible rational roots of the polynomial equation with integral coefficients

$$a_n x^n + a_{n-1} x^{n-1} + \cdots + a_0 = 0$$

are of the form p/q, where p is a factor of the constant term a_0 and q is a factor of the leading coefficient a_n.

---

* Although this really requires more proof, it is intuitively clear; we will not go into more detail here.

EXAMPLE 3   Find all roots of the equation $3x^3 - 8x^2 + 5x - 2 = 0$.

Solution   The possible numerators of rational roots are factors of 2, namely ± 1 and ± 2. The possible denominators are factors of 3, that is, 1 or 3. We need not consider negative denominators, since we get all possible signs by allowing the numerator only to vary. Thus, a listing of all possible rational roots is

$$\pm\tfrac{1}{3} \qquad \pm\tfrac{2}{3} \qquad \pm 1 \qquad \pm 2$$

We do not know whether any of these is actually a root, but at least the list of those we have to try is finite. There are various strategies for the order in which to test the possible rational roots. One way is to order them according to size and start with the smallest positive one and work up until (1) a root is found, (2) the list of positive possibilities is exhausted, or (3) it becomes evident that there is no need to proceed further. When a root r is encountered, we usually look for further roots in the so-called **depressed equation,** which is the quotient (on dividing by $x - r$) set equal to 0. If the depressed equation is quadratic, then the quadratic formula (or simple factorization) is used. If situation (2) or (3) above is encountered before finding all roots, then we can try the negative possibilities, working in reverse order.

For this problem the procedure is as follows:

$$
\begin{array}{r|rrrr}
\tfrac{1}{3} & 3 & -8 & 5 & -2 \\
  &   & 1 &   &   \\
\hline
  & 3 & -7 &   &
\end{array}
$$

This is not carried further since all remaining multiplications result in non-integers, and thus there is no possibility of getting a 0 remainder.

$$
\begin{array}{r|rrrr}
\tfrac{2}{3} & 3 & -8 & 5 & -2 \\
  &   & 2 & -4 &   \\
\hline
  & 3 & -6 & 1 &
\end{array}
$$

$$
\begin{array}{r|rrrr}
1 & 3 & -8 & 5 & -2 \\
  &   & 3 & -5 & 0 \\
\hline
  & 3 & -5 & 0 & -2
\end{array}
$$

$$
\begin{array}{r|rrrr}
2 & 3 & -8 & 5 & -2 \\
  &   & 6 & -4 & 2 \\
\hline
  & 3 & -2 & 1 & 0
\end{array}
$$

Now we have met with success. We have found that $x = 2$ is a root of the equation, since on division by $x - 2$ we have a remainder of 0. The polynomial can be factored as $(x - 2)(3x^2 - 2x + 1)$. Thus, all other roots are those of the depressed equation

$$3x^2 - 2x + 1 = 0$$

Since this is quadratic, we solve it by the quadratic formula:

$$x = \frac{2 \pm \sqrt{4 - 12}}{6} = \frac{1 \pm i\sqrt{2}}{3}$$

Note here, as we expected, the imaginary roots are conjugates. Now we have the complete solution set:

$$\left\{ 2, \ \frac{1 + i\sqrt{2}}{3}, \ \frac{1 - i\sqrt{2}}{3} \right\}$$

We also have the factorizations:

$$3x^3 - 8x^2 + 5x - 2 = (x - 2)(3x^2 - 2x + 1)$$

over the reals, and

$$3x^3 - 8x^2 + 5x - 2 = 3(x - 2)\left(x - \frac{1 + i\sqrt{2}}{3} \right)\left(x - \frac{1 - i\sqrt{2}}{3} \right)$$

over the complex numbers.

Before we go to another example of finding rational roots, we will illustrate a tabular arrangement which is convenient when we have several divisions to perform on the same polynomial. We illustrate with the polynomial

$$P(x) = 2x^3 - 5x^2 - 8x + 11$$

and we will find the value of $P(r)$ for various values of r, using synthetic division and the remainder theorem. Suppose, for example, that we want $P(2)$ and $P(-3)$. Instead of showing the synthetic divisions in the usual way, namely

$$
\begin{array}{r|rrrr}
2 & 2 & -5 & -8 & 11 \\
  &   & 4 & -2 & -20 \\
\hline
  & 2 & -1 & -10 & -9
\end{array}
$$

and

$$
\begin{array}{r|rrrr}
-3 & 2 & -5 & -8 & 11 \\
   &   & -6 & 33 & -75 \\
\hline
   & 2 & -11 & 25 & -64
\end{array}
$$

therefore obtaining $P(2) = -9$ and $P(-3) = -64$, we compress these into one table, as shown in Table 1. We have written the coefficients of $P(x)$ only once, and we have deleted the second line of the synthetic division process, carrying out the additions mentally. The numbers r whose function values we are seeking are listed in the first column, and the remainders, that is, the values of $P(r)$ corresponding to the value of r on the left, are listed in the right-hand column.

Table 1

r				$P(r)$
0	2	-5	-8	11
2	2	-1	-10	-9
-3	2	-11	25	-64

Table 2

r				$P(r)$
0	2	-5	-8	11
1	2	-3	-11	0
2	2	-1	-10	-9
3	2	1	-5	-4
4	2	3	4	27
5	2	5	17	96
-1	2	-7	-1	12
-2	2	-9	10	-9
-3	2	-11	25	-64

Now we extend the table, as shown in Table 2, and make some additional observations. Note first the sign change from $P(3)$ to $P(4)$. Now it can be proved that the graph of every polynomial function has no breaks in it (it is **continuous**). It follows that $P(x)$ must be 0 for some value of x between $x = 3$ and $x = 4$, since $P(x)$ is negative when $x = 3$ and positive when $x = 4$. Therefore, $P(x)$ has a zero between $x = 3$ and $x = 4$, or equivalently, a root of the equation $P(x) = 0$ occurs between $x = 3$ and $x = 4$. By the same sort of reasoning we can conclude that $P(x)$ has a zero between $x = -1$ and $x = -2$, since $P(-1)$ is positive, and $P(-2)$ is negative.

Our next observation is that 1 is a zero of $P(x)$. The line corresponding to $r = 1$ tells us that $P(x)$ can be factored as

$$P(x) = (x - 1)(2x^2 - 3x - 11)$$

So if we are interested in solving the equation $P(x) = 0$, we have

$$(x - 1)(2x^2 - 3x - 11) = 0$$

and thus either $x - 1 = 0$ or $2x^2 - 3x - 11 = 0$. So $x = 1$ is one root, and all other roots are contained in the equation $2x^2 - 3x - 11 = 0$. In this case the depressed equation is quadratic and can be solved by the quadratic formula:

$$x = \frac{3 \pm \sqrt{9 + 88}}{4} = \frac{3 \pm \sqrt{97}}{4} \approx \frac{3 \pm 9.85}{4}$$

So the two roots of the depressed equation are approximately 3.21 and -1.71, which agrees with our earlier conclusion that there were roots between 3 and 4 and between -1 and -2.

More generally, if we have found a root $x = r$ of a polynomial equation $P_n(x) = 0$, then the polynomial can be factored in the form

$$(x - r)P_{n-1}(r)$$

and the equation $P_{n-1}(x) = 0$ is the depressed equation corresponding to the root $x = r$. All other roots of the original equation are found by solving the depressed equation.

Returning now to Table 2, we want to make two more observations. Notice that all entries in the line corresponding to $r = 4$ are positive, as are the entries in the next line. We could have anticipated that the entries corresponding to $r = 5$ would also be positive, in fact, because they were positive when 4 was the multiplier, and so will be even larger when 5 is the multiplier. We can conclude from this that there are no roots of $P(x) = 0$ larger than $x = 4$. The number 4 is therefore an **upper bound** to the roots of $P(x) = 0$. Of course, since we have already actually found all roots in this particular problem, we get no new useful information by finding this upper bound, but we want to obtain a general result, which in some problems can be quite helpful.

Similarly, since the multiplier -2 is negative, and the signs of the entries in its row *alternate*, they will continue to alternate when a negative number greater in absolute value is used, as can be seen in the line corresponding to $r = -3$. Thus, we can conclude that -2 is a **lower bound** to the roots of $P(x)=0$.

More generally, **if the result of dividing a polynomial by $x - r$ using synthetic division yields all positive numbers when $r > 0$, then r is an upper bound to the roots of the equation $P_n(x) = 0$. If $r < 0$ and the signs alternate, then r is a lower bound to the roots.** (Note that for this test, 0 may be treated as either positive or negative.)

These concepts are employed in the next two examples.

EXAMPLE 4   Find all rational roots of $2x^4 - 7x^3 - 10x^2 + 33x + 18 = 0$.

Solution   The possible numerators are $\pm1, \pm2, \pm3, \pm6, \pm9, \pm18$ and possible denominators are 1, 2. Therefore, the possible rational roots are $\pm\frac{1}{2}, \pm1, \pm\frac{3}{2}, \pm2, \pm3, \pm\frac{9}{2}, \pm6, \pm9, \pm18$.

From Table 3, we see that $x = 3$ is a root. Now that a root has been found, we seek the remaining roots from the depressed equation

$$2x^3 - x^2 - 13x - 6 = 0$$

First, we observe that since the original polynomial can now be factored as

$$(x - 3)(2x^3 - x^2 - 13x - 6)$$

any root of the depressed equation is also a root of the original; so we know there is no need to try possible rational roots that have already been rejected. However, it is possible that $x = 3$ is a root of multiplicity 2 or greater, and so we test $x = 3$ in the depressed equation. This procedure is general; **when a root has been found, test this same number to see if it is a root of the depressed equation.**

Table 3

r					$P(r)$
0	2	-7	-10	33	18
$\frac{1}{2}$	2	-6	-13		
1	2	-5	-15	18	36
$\frac{3}{2}$	2	-4	-16	9	
2	2	-3	-16	1	20
3	2	-1	-13	-6	0

Table 4

r					P(r)
0	2	−7	−10	33	18
$\frac{1}{2}$	2	−6	−13		
1	2	−5	−15	18	36
$\frac{3}{2}$	2	−4	−16	9	
2	2	−3	−16	1	20
3	2	−1	−13	−6	0
3	2	5	2	0	

This is important, because otherwise multiple roots may not show up. Note, too, that the coefficients of the depressed equation are the last ones in Table 3, so we can simply continue the table, writing the remainder one column to the left. The entire table is repeated in Table 4 to illustrate this. And from Table 4, we see that $x = 3$ is a double root. We could continue testing the remaining possibilities in the new depressed equation

$$2x^2 + 5x + 2 = 0$$

but since this is quadratic, we can solve it by other means. In this case, it can be factored as

$$(2x + 1)(x + 2) = 0$$

So, $x = -\frac{1}{2}$ or $x = -1$. The complete solution set is

$$\{3, \quad -\tfrac{1}{2}, \quad -2\}$$

with 3 a root of multiplicity 2.

EXAMPLE 5 Find all rational roots of $x^5 - 8x^3 - 10x^2 + 12x - 20 = 0$.

Solution The possible rational roots are $\pm1, \pm2, \pm4, \pm5, \pm10, \pm20$.

Since we have tried all possible rational roots in Table 5 and none is actually a root, we conclude that there are no rational roots for this equation. We observe the sign change between $P(2)$ and $P(4)$, however, and conclude that

Table 5

	r						P(r)	
	0	1	0	−8	−10	12	−20	
	1	1	1	−7	−17	−5	−25	**Note**
	2	1	2	−4	−18	−24	−68 ⎫	sign
Upper bound →	4	1	4	8	22	100	380 ⎭	change
	−1	1	−1	−7	−3	15	−35	
	−2	1	−2	−4	−2	16	−52	
Lower bound →	−4	1	−4	8	−42	180	−740	

$P(x) = 0$ for some value of x between 2 and 4. We find $P(3)$ in order to narrow the bounds on the root:

$$\begin{array}{r|rrrrrr}
3 & 1 & 0 & -8 & -10 & 12 & -20 \\
  &   & 3 & 9 & 3 & -21 & -27 \\
\hline
  & 1 & 3 & 1 & -7 & -9 & -47
\end{array}$$

So $P(3) = -47$, which shows that the root lies between 3 and 4. In the next section we will see how we can obtain the root to any desired degree of accuracy.

We now summarize the procedure we have discussed for finding the rational roots of a polynomial equation $P_n(x) = 0$ having integral coefficients.

Step 1. List in order all numbers of the form $\pm p/q$, where p is a factor of the constant term of $P_n(x)$ and q is a factor of the leading coefficient.

Step 2. By synthetic division or otherwise, begin trying the positive numbers obtained in Step 1. Stop if (a) a root is found, (b) a bound on the roots is found, or (c) the possibilities are exhausted.

Step 3. If a root is found and the depressed equation is of degree 2, solve for all other roots (rational, irrational, and imaginary) by use of the quadratic formula or by factoring. If the depressed equation is of degree 3 or higher, continue testing in this equation all possibilities not previously rejected, beginning with the successful root just found. Use the same stopping criteria given in Step 2.

Step 4. When stopping criterion 2(b) or 2(c) is operative, begin testing negative possibilities in the current depressed equation until 2(a), 2(b), or 2(c) occurs.

Step 5. Repeat Step 3.

This procedure will always yield all rational roots, and in case the depressed equation at any stage is quadratic, all roots can be found. Even when it is not possible to find all roots, by observing changes in sign in the values of $P_n(x)$, we can roughly locate irrational roots.

EXERCISE SET 5

A In Problems 1 and 2 list all possible rational roots. (Do not test the possibilities.)
   1. a. $4x^4 - 8x^3 + 2x^2 - 3x + 3 = 0$ **b.** $6x^5 - 5x^3 + x^2 - 8x - 20 = 0$
   2. a. $15x^5 - 17x^3 + 5x^2 - 8x - 12 = 0$ **b.** $8x^4 + 12x^2 - 3x + 30 = 0$

In Problems 3 and 4 find upper and lower bounds for the roots. Indicate roots between consecutive integers when possible.

   3. $2x^3 + 3x^2 - 12x - 24 = 0$ **4.** $x^4 - 6x^3 - 5x^2 + 2x - 7 = 0$

In Problems 5–17 find all rational roots. When the depressed equation is quadratic, find all roots.

5. $x^3 - 3x - 2 = 0$ **6.** $x^3 - x^2 - 8x + 12 = 0$
7. $x^3 - 2x^2 - 6x - 8 = 0$ **8.** $2x^3 - 7x^2 + 9 = 0$
9. $4x^3 - 5x^2 + 10x + 12 = 0$ **10.** $3x^3 - 4x^2 - 17x + 6 = 0$
11. $2x^3 + 3x^2 - 14x - 15 = 0$ **12.** $8x^3 + 14x^2 - 33x - 9 = 0$
13. $x^4 - x^2 + 2x + 2 = 0$ **14.** $x^3 - 19x - 30 = 0$
15. $x^4 - 5x^3 + 20x - 16 = 0$ **16.** $x^4 - 2x^3 - 4x^2 - 8x - 32 = 0$
17. $x^4 - 6x^2 - 8x - 3 = 0$

In Problems 18–25 factor completely over the reals.

18. $3x^3 - x^2 - 8x - 4$ **19.** $x^3 - 13x + 12$
20. $6x^3 - x^2 - 27x - 20$ **21.** $x^4 - 2x^3 + x^2 - 8x - 12$
22. $5x^4 - 2x^3 - 35x^2 - 16x + 12$ **23.** $x^4 - 5x^2 - 10x - 6$
24. $x^4 + 2x^3 - 5x^2 - 18x - 36$ **25.** $4x^4 - 19x^2 + 3x + 18$

B   In Problems 26–28 find all roots.

26. $6x^3 + 5x^2 - 34x - 40 = 0$ **27.** $3x^3 + 22x^2 + 32x - 32 = 0$
28. $4x^5 - 18x^4 + 21x^3 - 14x^2 + 60x - 72 = 0$

29. Show that the equation $2x^4 + 13x^3 - 40x - 24 = 0$ has only one rational root. How many real roots does it have? Justify your answer.

6   Irrational Roots

The following example illustrates a technique for approximating irrational roots to any desired degree of accuracy. There are other methods requiring calculus which are more efficient in that they provide the same accuracy in fewer steps, but the method we present has the advantage of being straightforward and easily remembered. It is a definite advantage to use a calculator.

EXAMPLE 6   Approximate the real solution of the equation $x^3 - 3x^2 + 3x - 4 = 0$ to two decimal places of accuracy.

Solution   A quick check will show that there are no rational roots. By synthetic division we locate the root between consecutive integers (Table 6). If we denote the polynomial by $f(x)$, we see that $f(2) = -2$ and $f(3) = 5$, so there is a root between 2 and 3. Since 3 is an upper bound and 0 is a lower bound, there are no other real roots. On the interval $[2, 3]$ we approximate the graph of $y = f(x)$ by the line joining the two points $(2, -2)$ and $(3, 5)$, as shown in Figure 1. The

Table 6

Lower bound → 0	1	-3	3	-4
1	1	-2	1	-3
2	1	-1	1	-2
Upper bound → 3	1	0	3	5

Note sign change (for the last two values -2 and 5)

Figure 1

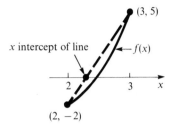

x intercept of this line is then an approximation of the zero of $f(x)$. This process is called **linear interpolation.** The slope of this line is 7, and its equation is

$$y + 2 = 7(x - 2)$$

or

$$y = 7x - 16$$

Setting $y = 0$, we find the x intercept to be $2\frac{2}{7}$, or approximately 2.3. So we try this as our next approximation to the root. We want to locate the root between consecutive tenths. Note from Table 7 that the approximation 2.3 found by linear interpolation was not quite as good an approximation as we might have hoped, since the actual root turns out to be between 2.4 and 2.5. This sometimes happens, but linear interpolation nevertheless provides a good starting point.

Now we use linear interpolation again. The line joining $(2.4, -0.26)$ and $(2.5, 0.38)$ has x intercept 2.44 (check this), which is our next approximation (Table 8). Then, in order to obtain two-place accuracy, we need to carry the process one more step and then round off. By linear interpolation we find the next approximation to be 2.442 (Table 9). The root is therefore between 2.442 and 2.443, so the answer correct to two decimal places is 2.44.

Table 7

	1	-3	3	-4	
2.3	1	-0.7	1.39	-0.80	
2.4	1	-0.6	1.56	-0.26	} **Note sign change**
2.5	1	-0.5	1.75	0.38	

Table 8

	1	-3	3	-4
2.44	1	-0.56	1.634	-0.0140
2.45	1	-0.55	1.652	0.0486

Table 9

	1	-3	3	-4
2.442	1	-0.558	1.6374	-0.00156
2.443	1	-0.557	1.6392	0.00469

We can summarize this procedure as follows. Suppose we have found that a root of $f(x) = 0$ lies between x_1 and x_2 (which may be consecutive integers, consecutive tenths, hundredths, etc.). The line joining $(x_1, f(x_1))$ and $(x_2, f(x_2))$ has the equation

$$y - f(x_1) = m(x - x_1)$$

where

$$m = \frac{f(x_2) - f(x_1)}{x_2 - x_1}$$

This line is used to approximate the actual graph of $y = f(x)$ on the interval $[x_1, x_2]$, and its x intercept provides an approximation to the root. This intercept is found by setting $y = 0$ in the equation of the line, which gives

$$x = x_1 - \frac{f(x_1)}{m} \tag{5}$$

The root is then approximated to any desired degree of accuracy by repeated application of the following two steps:

Step 1. If $f(x_1)$ and $f(x_2)$ are opposite in sign, where x_1 and x_2 are consecutive n-place decimals $(n = 0, 1, 2, \ldots)$, find x by equation (5) and round off to $n + 1$ decimal places.

Step 2. Test the value of x found in Step 1 and move forward or backward, as required, by consecutive $(n + 1)$-place decimals to locate new values of x_1 and x_2 for which $f(x_1)$ and $f(x_2)$ are opposite in sign.

The process is stopped when the root is located between consecutive decimals having *one more decimal place of accuracy than is desired*. The answer is found by rounding back one place.

We will do one more example, showing only the necessary steps. Since a calculator is being used for the calculations involving decimals, it is not necessary to show the intermediate steps in the synthetic division.

EXAMPLE 7   Find the real root of $x^3 - x^2 - 2x + 3 = 0$ correct to two decimal places.

Solution   From Table 10, the root is between -1.546 and -1.547, so the answer is -1.55, correct to two decimal places.

Table 10

	0	1	-1	-2	3	
	1	1	0	-2	1	
Upper bound →	2	1	1	0	3	
	-1	1	-2	0	3	$\leftarrow \left\{ m = 8, \quad x = -2 - \frac{-5}{8} = -1.4 \right.$
	-2	1	-3	4	-5	
	-1.4				1.096	
	-1.5				0.375	$\leftarrow \left\{ m = 8.31, \quad x = -1.6 - \frac{0.456}{8.31} = -1.54 \right.$
	-1.6				-0.456	
	-1.54				0.0561	$\leftarrow \left\{ m = 8.25, \quad x = -1.55 - \frac{-0.0264}{8.25} = -1.547 \right.$
	-1.55				-0.0264	
	-1.547				-0.001503	$\leftarrow$
	-1.546				0.00676	

Remark.   In solving an equation it is best to look for rational roots first. If any are found, go to the depressed equation. Only if there are no rational roots or if the depressed equation is of degree 3 or greater should the method of this section be used.

EXERCISE SET 6

A   In Problems 1–5 find the specified irrational root correct to two decimal places.

1.   The real root of $x^3 - x^2 + 2x - 3 = 0$
2.   The real root of $x^3 + 2x^2 + 3x - 10 = 0$
3.   The real root of $x^3 - x + 4 = 0$
4.   The negative root of $2x^3 - x^2 - 12x + 10 = 0$
5.   The largest positive root of $2x^4 + x^3 - 13x^2 - 4x + 12 = 0$

In Problems 6–15 find all rational and irrational roots, expressing irrational roots correct to two decimal places.

6. $x^4 + x^3 - 2x^2 - 7x - 5 = 0$ **7.** $x^4 - 4x^3 + 4x^2 + 8x - 16 = 0$
8. $2x^4 - 3x^3 - 3x^2 + 8x - 3 = 0$ **9.** $x^4 + x^3 - 7x^2 - 4x - 3 = 0$
10. $2x^4 + x^3 - 12x^2 + 27x - 27 = 0$

B   **11.** $x^3 + x^2 - 7x - 5 = 0$ **12.** $x^3 - 2.34x^2 - 5.16x - 4.87 = 0$
13. $x^4 - 2x^2 + 5x - 8 = 0$
14. $x^5 - 6x^4 + 7x^3 + 11x^2 - 16x - 4 = 0$
15. $2x^5 - 8x^4 - 11x^3 + 21x + 20 = 0$

7   Graphing Polynomial Functions

We can use the method of Section 5 to make a table of values to be used in graphing a polynomial function, $y = P(x)$. The zeros of the polynomial are particularly helpful, and if these are rational, we know how to find them. Real zeros that are irrational can at least be located between consecutive integers.

As we indicated earlier, the graph of a polynomial function is a smooth curve, with no breaks. The graph of a second-degree polynomial is a parabola, as we know. The graph of a third-degree polynomial typically has both a maximum point and a minimum point and looks like one of the curves shown in Figure 2. However, the "humps" may be straightened out, as in Figure 3.

Fourth-degree polynomials usually add one more maximum or minimum point, as shown in Figure 4. But again, some of the humps may not be present. The shape could be any of those shown in Figure 5.

This pattern continues. The number of maximum points and minimum points is at most 1 less than the degree of the polynomial.

The sign of the coefficient of the highest power of x tells us the sign of the function when x is very large in absolute value. For example, in

$$y = 2x^3 - 15x^2 - 30x - 100$$

Figure 2

Figure 3

Figure 4

Figure 5

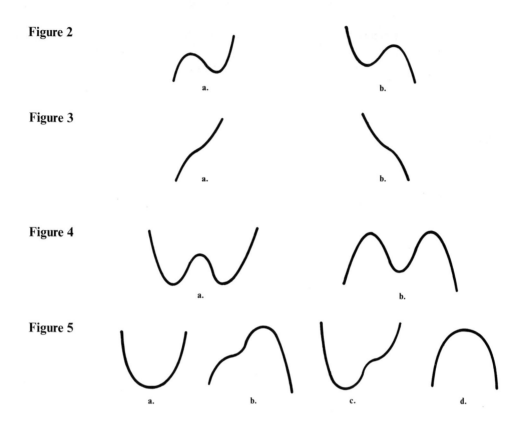

when x is very large, y will be positive, because the term $2x^3$ will dominate all other terms. Similarly, for x negative but large in absolute value, y will be negative. Thus, the general shape of the graph will be that of Figure 2a. By locating the real zeros and a few more points we can draw the graph with reasonable accuracy.

EXAMPLE 8   Graph the function $y = 2x^3 - 5x^2 - 8x + 11$.

Solution   This is the function we used as an illustration in Section 5. We can summarize the function values found in Table 2 as follows:

x	0	1	2	3	4	5	-1	-2	-3
y	11	0	-9	-4	27	96	12	-9	-64

Furthermore, we know that $x = 1$ is a zero of the polynomial, and that the other two zeros are between 3 and 4, and -1 and -2, respectively. In fact, we found these other two zeros to be approximately 3.21 and -1.71. The graph is shown in Figure 6. The exact location of the maximum and minimum points can be determined by the use of procedures studied in calculus.

Figure 6

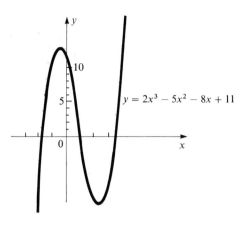

$$y = 2x^3 - 5x^2 - 8x + 11$$

EXAMPLE 9   Graph the function $y = x^4 - x^3 - 5x^2 + 12$.

Solution   We set up Table 11 using synthetic division. We find that 2 is a zero of the function, and we can write y in the factored form

$$y = (x - 2)(x^3 + x^2 - 3x - 6)$$

Table 11

	x					y
	0	1	-1	-5	0	12
	1	1	0	-5	-5	7
	2	1	1	-3	-6	0
Upper bound →	3	1	2	1	3	21
	-1	1	-2	-3	3	9
Lower bound →	-2	1	-3	1	-2	16

We test to see if 2 is also a zero of the second factor.

$$
\begin{array}{r|rrr}
2 & 1 & 1 & -3 & -6 \\
  &   & 2 & 6 & 6 \\
\hline
  & 1 & 3 & 3 & 0 \\
\end{array}
$$

This shows that 2 is a zero of the second factor and hence is a zero of multiplicity 2 of the original function. We can now factor y further as

$$y = (x - 2)^2(x^2 + 3x + 3)$$

Since the discriminant $b^2 - 4ac$ of the factor $x^2 + 3x + 3$ is $9 - 12 = -3$, it follows that there are no real zeros.

The graph is shown in Figure 7. The significance of the fact that 2 is a zero of multiplicity 2 is that the curve is *tangent to the x axis* at $x = 2$.

Figure 7

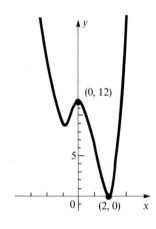

In Example 9 we indicated that the significance of the factor $(x - 2)^2$ occurring in the factored form of the polynomial was that the graph was tangent to the x axis at $x = 2$. **More generally, if a factor of the form $(x - r)^k$ occurs in the factored form of a polynomial, where $k \geq 2$, then the graph will be tangent to the x axis at $x = r$.** If k is even, then in the vicinity of $x = r$, the graph will have an appearance similar to one of the forms shown in Figure 8. If k is odd, the curve will look generally like one of the forms shown in Figure 9. The larger the exponent k is, the flatter the curve will be near the point of tangency.

Figure 8

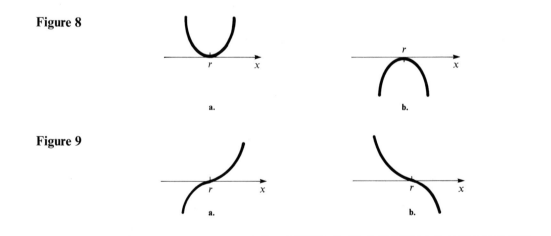

Figure 9

EXAMPLE 10   Graph the function $y = 4 + 3x^2 - x^4$.

Solution   The polynomial can be factored as

$$y = (4 - x^2)(1 + x^2) = (2 - x)(2 + x)(1 + x^2)$$

So the only real zeros are ± 2. Since the variable x in the polynomial appears only to even powers, it follows that substituting a given value for x or its negative will produce the same value of y. Since the function is a simple one,

Figure 10

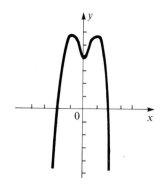

we will use direct substitution to make a table of values:

x	0	± 1	± 2	± 3
y	4	6	0	-50

EXERCISE SET 7

A In Problems 1–21 find the rational zeros, locate other real zeros between consecutive integers, make a table of values, and graph the function.

1. $y = x^3 - 1$

2. $y = x^3 - x^2 - 6x$

3. $y = x^3 - x^2 - 9x + 9$

4. $y = 1 - x^4$

5. $y = 4x - x^3$

6. $y = (x - 4)(x + 1)(x + 3)$

7. $y = (2x - 5)(x - 6)(x + 2)$

8. $y = (x - 1)(2x + 3)(x - 3)^2$

9. $y = (x + 1)^2(2x - 1)(x - 4)$

10. $y = \frac{1}{4}(x + 2)^3(x - 1)^2$

11. $y = x^3 - 3x + 2$

12. $y = 2x^3 - 7x^2 + 9$

13. $y = x^3 + x^2 - 10x - 12$

14. $y = x^4 - 10x^2 + 9$

15. $y = 8x^2 + 2x^3 - x^4$

16. $y = 12 - 3x + 4x^2 - x^3$

17. $y = x^4 + x^3 - 13x^2 - x + 12$

18. $y = 2x^4 - 5x^3 + 10x - 12$

19. $y = 2x^4 - 5x^3 + 10x - 12$

20. $y = x^5 - 4x^3$

21. $y = x^4 + 2x^3 - 12x^2 - 8x + 32$

In Problems 22–27 solve the inequalities by graphical means.

22. $(x - 3)(2x + 5)(x - 7)^3 \le 0$

23. $(x + 4)^2(3x - 4)(x - 5)(x + 6) > 0$

24. $x^3 - 3x^2 - x + 3 < 0$

25. $x^3 - 5x^2 + 2x + 8 \ge 0$

26. $x^3 - 2x^2 - 5x + 6 \le 0$

27. $x^4 - 9x^2 - 4x + 12 > 0$

Find the domain of each of the functions in Problems 28 and 29.

28. $f(x) = \sqrt{2x^3 - 9x^2 + x + 12}$

29. $g(x) = \dfrac{1}{\sqrt{x^4 - 9x^2 + 4x + 12}}$

B   In Problems 30 and 31 solve the inequalities by graphical means.

30. $2x^4 - 7x^3 - 26x^2 + 49x + 30 < 0$ **31.** $x(x^3 - 4x^2 + 3) \geq 14(1 + x)$

Graph the functions in Problems 32–36.

32. $y = x^4 - 3x^3 - 8x^2 + 10x + 12$ **33.** $y = x^4 - 2x^3 - 5x^2 + 6x$
34. $y = 3x^4 - 5x^3 - 8x^2 + 10x + 7$ **35.** $y = x^5 - x^4 - 15x^3 + x^2 + 38x + 24$
36. $y = 2x^5 - 5x^4 - 11x^3 + 23x^2 + 8x - 15$

8   Rational Functions

A function of the form

$$f(x) = \frac{P(x)}{Q(x)}$$

where $P(x)$ and $Q(x)$ are polynomials, is called a **rational function.** We will assume that $P(x)$ and $Q(x)$ have no nonconstant factor in common. The zeros of $f(x)$ are the same as those of $P(x)$, since if the numerator of a fraction is 0 and the denominator not 0, the fraction is 0. These zeros are the x intercepts of the graph of f. The y intercept, if any, is found by putting $x = 0$.

The zeros of $Q(x)$ are not in the domain of f, but the behavior of $f(x)$ for x near these zeros is particularly useful in drawing the graph of f. The next example illustrates how to analyze this behavior.

EXAMPLE 11   Examine the nature of the function below near $x = 2$:

$$f(x) = \frac{x + 1}{x - 2}$$

Solution   We set $y = f(x)$ and make a table of values for x close to 2.

x	1	1.5	1.9	1.99	3	2.5	2.1	2.01
y	-2	-5	-29	-299	4	7	31	301

As x approaches 2 from the left, y is negative but becomes arbitrarily large in absolute value. Similarly, when x approaches 2 from the right, y becomes arbitrarily large through positive values. The graph in the neighborhood of $x = 2$ is shown in Figure 11. The line $x = 2$ is called an **asymptote** to the curve, and the curve is said to be **asymptotic to this line.**

Figure 11

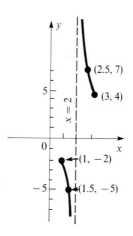

This example illustrates the general result that **if r is a zero of $Q(x)$, then the line $x = r$ is a vertical asymptote to the graph of f.**

Rational functions may also have horizontal asymptotes, and these can be determined by examing the behavior of $f(x)$ when x becomes arbitrarily large in absolute value. The next example illustrates a way of doing this.

EXAMPLE 12   Examine the behavior of the function

$$f(x) = \frac{2x^2 + x}{x^2 - 4}$$

as x becomes arbitrarily large in absolute value.

Solution   Let $y = f(x)$ and divide numerator and denominator by the highest power of x—in this case, x^2:

$$y = \frac{2x^2 + x}{x^2 - 4} = \frac{2 + (1/x)}{1 - (4/x^2)} \qquad (x \neq 0)$$

Now as x gets arbitrarily large in absolute value, the terms $1/x$ and $4/x^2$ both approach 0. So y approaches $\frac{2}{1} = 2$. For large values of $|x|$, the graph has the appearance shown in Figure 12. The line $y = 2$ is a horizontal asymptote.

Figure 12

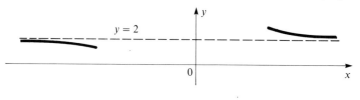

More generally, to find horizontal asymptotes, we divide numerator and denominator by the highest power of x that appears in either place and then see if y approaches some finite constant b as x becomes arbitrarily large in absolute value. If so, then the line $y = b$ is a horizontal asymptote.

Remark.   A rational function may have many vertical asymptotes (one for each zero of the denominator), but it can have at most one horizontal asymptote.

In the next example we make use of both intercepts and asymptotes to draw the graph.

EXAMPLE 13   Find the intercepts and asymptotes, and draw the graph of

$$y = \frac{2x - 3}{x^2 - 9}$$

Solution   The x intercept is $\frac{3}{2}$ and the y intercept is $\frac{1}{3}$. By factoring the denominator $x^2 - 9 = (x + 3)(x - 3)$, we see that the vertical asymptotes are $x = 3$ and $x = -3$. To determine whether a horizontal asymptote exists, we write

$$y = \frac{2x - 3}{x^2 - 9} = \frac{(2/x) - (3/x^2)}{1 - (9/x^2)} \qquad (x \neq 0)$$

As x increases without limit in absolute value, y approaches $\frac{0}{1} = 0$. So $y = 0$ is a horizontal asymptote. We now need only a few additional points:

x	1	2	4	-1	-2	-4
y	$\frac{1}{8}$	$-\frac{1}{5}$	$\frac{5}{7}$	$\frac{5}{8}$	$\frac{7}{5}$	$\frac{-11}{7}$

Figure 13

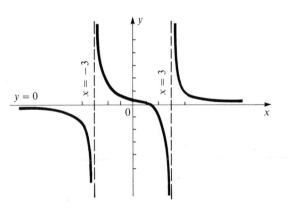

EXERCISE SET 8

A In Problems 1–6 find all vertical asymptotes.

1. **a.** $y = \dfrac{x+3}{x-4}$ **b.** $y = \dfrac{1}{x^2 - 1}$

2. **a.** $f(x) = \dfrac{x+2}{x^2 - 3x - 4}$ **b.** $g(x) = \dfrac{2x^2 - 1}{x^2 - 4x + 4}$

3. **a.** $y = \dfrac{x^3 + 4x - 1}{2x^2 - 3x - 5}$ **b.** $y = \dfrac{3x^2 + 4x + 6}{x^2 - 5x}$

4. **a.** $g(x) = \dfrac{x^2 + 1}{x^3 - 8}$ **b.** $h(x) = \dfrac{2x+3}{x^3 - x^2 - x + 1}$

5. **a.** $y = \dfrac{x^2 + 3x - 7}{x^3 - 4x^2 - 9x + 36}$ **b.** $y = \dfrac{1}{x^4 + x}$

6. **a.** $y = \dfrac{2x - 1}{x^3 - 3x - 2}$ **b.** $y = \dfrac{1 - x}{x^3 - 7x - 6}$

In Problems 7–12 find the horizontal asymptote or show that there is none.

7. **a.** $y = \dfrac{x+1}{x-2}$ **b.** $y = \dfrac{x-3}{2x+1}$

8. **a.** $f(x) = \dfrac{2x}{1 - x^2}$ **b.** $g(x) = \dfrac{x^2 - 1}{x^2 - 4}$

9. **a.** $y = \dfrac{2 - x^2}{x^2 + 2x + 1}$ **b.** $y = \dfrac{x^2 - 1}{x + 3}$

10. **a.** $y = \dfrac{2x^2 + 4}{x^3 + 1}$ **b.** $y = \dfrac{2x^2 + 5}{1 - x - 3x^2}$

11. **a.** $F(x) = \dfrac{x^2 + 4x - 5}{2x + 3}$ **b.** $G(x) = \dfrac{x^2}{x^3 - 1}$

12. **a.** $y = \dfrac{2x - x^2}{3x^2 + 4}$ **b.** $y = \dfrac{1 - x^3}{1 + x^3}$

In Problems 13–20 find intercepts, vertical and horizontal asymptotes, and draw the graph.

13. $f(x) = \dfrac{2x + 3}{x - 1}$ 14. $g(x) = \dfrac{2}{x + 2}$

15. $y = \dfrac{x - 1}{x^2 - 9}$ 16. $y = \dfrac{2x}{x^2 + 1}$

17. $y = \dfrac{x - 2}{x^2 - 2x - 3}$ 18. $y = \dfrac{x^2 - 1}{x^2 - 4}$

19. $y = \dfrac{x^2 - 2x - 8}{x^2 - 2x}$ 20. $y = \dfrac{x^2 - 3x}{x^2 - 4}$

B In Problems 21–23 find all vertical and horizontal asymptotes.

21. $y = \dfrac{1 - x - 2x^3}{x^3 - 3x^2 - 6x + 8}$ 22. $y = \dfrac{3x^3 + 4x - 7}{x^4 - 11x^2 + 18x - 8}$

23. $y = \dfrac{x^5 + 2}{x^4 - 4x^3 + 3x^2 + 4x - 4}$

In Problems 24–27 find intercepts, horizontal and vertical asymptotes, and draw the graph.

24. $y = \dfrac{x - 2}{x^2 + 4x}$

25. $f(x) = \dfrac{3 - 2x - x^2}{x^2 - x - 6}$

26. $y = \dfrac{2x + 5}{x^2 - 1}$

27. $y = \dfrac{x^2 - 1}{x + 2}$

28. Let $f(x) = P(x)/Q(x)$. If the degree of $P(x)$ is m and the degree of $Q(x)$ is n, prove that:
  a. If $m < n$, then $y = 0$ is a horizontal asymptote
  b. If $m = n$, there is a horizontal asymptote $y = b$, with $b \neq 0$. What can you conclude about the value of b?
  c. If $m > n$, there is no horizontal asymptote.

Review Exercise Set

A In Problems 1–3 use synthetic division to perform the indicated divisions.

1. **a.** $2x^3 + 3x^2 - 4x + 5$ by $x - 2$ **b.** $x^3 - 3x + 4$ by $x + 3$
2. **a.** $x^4 - 3x^3 + 5x - 7$ by $x - 3$ **b.** $3x^5 - 12x^3 + 10x^2 + 22x + 4$ by $x + 2$
(What can you conclude?)
3. **a.** $3x^4 - 2x^3 + x - 5$ by $x + 4$ **b.** $2x^5 - 10x^3 + 16$ by $x - 2$
(What can you conclude?)

4. **a.** Show by synthetic division that $x + 5$ is a factor of $x^5 + 4x^4 - 7x^3 - 9x^2 - 25$.
  b. Let $P(x) = 2x^4 - x^3 - 27x^2 + 36x$. Show that $P(3) = 0$ by synthetic division and give the complete factorization of $P(x)$.
5. Find a polynomial whose only zeros are $-2, \frac{3}{2}, 3i, -3i$.
6. Find a polynomial equation having 4 as a root of multiplicity 2 and whose only other root is -2.
7. Without testing for actual roots, list all possible rational roots of each of the following:
  a. $3x^5 - 2x^3 + 4x^2 - 8 = 0$ **b.** $10x^4 + 7x^3 - 3x^2 - 5x + 12 = 0$

In Problems 8–13 find all rational roots, and if the depressed equation is quadratic, find all roots.

8. $2x^3 - 7x^2 + 9 = 0$

9. $x^3 - 4x^2 - 3x + 18 = 0$

10. $x^4 - 8x^2 - 24x - 32 = 0$

11. $6x^4 + 5x^3 + 20x^2 + 20x - 16 = 0$

12. $2x^4 + 6x^3 + 3x^2 + 4x + 12 = 0$

13. $4x^4 - 19x^2 + 3x + 18 = 0$

In Problems 14–16 factor completely over the reals.

14. $x^3 + 9x^2 + 15x - 25$

15. $2x^3 - 9x^2 + 18x - 20$

16. $2x^4 - 6x^3 + 7x^2 - 9x + 6$

In Problems 17–21 draw the graph.

17. $y = 2x^3 - 3x^2 - 5x$

18. $y = 25 + 21x^2 - 4x^4$

19. $y = (x - 3)^2(2x + 1)(x + 4)$ **20.** $y = x^3 - 12x - 16$

21. $y = 2x^4 + x^3 - 19x^2 - 9x + 9$

In Problems 22–24 solve the inequalities by graphical means.

22. $12x - x^2 - x^3 \geq 0$ **23.** $x(x^2 - 3) < 2$ **24.** $x^3 > 7x - 6$

25. Find the domain of the function $f(x) = \sqrt{2x^3 - 5x^2 + 18}$.

In Problems 26–30 find intercepts and asymptotes, and draw the graph.

26. $f(x) = \dfrac{2x + 3}{x - 4}$ **27.** $g(x) = \dfrac{4x}{1 - x^2}$

28. $y = \dfrac{x^2 - 4}{x^2 - 1}$ **29.** $y = \dfrac{2x^2 - 3x - 2}{x^2 - 5x}$

30. $y = \dfrac{1 - x^2}{(x - 2)^2}$

31. Find the real root of $2x^3 - 10x + 9 = 0$ correct to two decimal places.

32. Find the largest positive root of $2x^4 + x^3 - 13x^2 - 4x + 12 = 0$ correct to two decimal places.

B   **33.** Construct a polynomial of fourth degree with integral coefficients three of whose zeros are $-2, \frac{5}{2},$ and $3 - 2i$.

In Problems 34 and 35 find all roots.

34. $6x^4 - 13x^3 + 2x^2 - 4x + 15 = 0$ **35.** $6x^4 + x^3 + 2x - 24 = 0$

Draw the graphs of the functions in Problems 36 and 37.

36. $y = (2x - 1)(x + 3)^3(x - 4)^2$ **37.** $P(x) = x^4 + 2x^3 + 7x - 6$

                                                           Discuss the nature of the zeros of this polynomial.

38. Solve the inequality $x^4 - 9x^2 + 4x + 12 > 0$.

39. Find the domain of the function

$$f(x) = \sqrt{\dfrac{x^3 - 4x^2}{x^3 - 7x + 6}}$$

40. Find all real roots of $x^3 - 3x^2 - 4x + 2 = 0$ correct to two decimal places.

41. Find the real root of $x^4 - 2.013x^3 + 0.025x^2 + 8.976x + 3.164 = 0$ correct to three decimal places.

In Problems 42–44 find the intercepts and asymptotes, and draw the graph.

42. $y = \dfrac{x^2}{x^2 - 2x - 8}$ **43.** $y = \dfrac{x^2 + x}{x - 1}$ **44.** $y = \dfrac{x^3 - 27}{x^4 - 16}$

7 Exponential and Logarithmic Functions

The term *exponential growth* is frequently seen or heard, and almost everyone has an intuitive understanding that it refers to very rapid growth. But it has a precise mathematical meaning that we will explore in this chapter. An example of exponential growth is the growth of a culture of bacteria. There is an analogous concept of the diminishing of some quantity, and this is called *exponential decay*. Radioactive substances exhibit exponential decay, for example. That is, they diminish at slower and slower rates. The following problem from calculus concerns exponential growth.* Involved in this are both exponential and logarithmic functions.

Example. Suppose that a culture of bacteria doubles its size from 0.15 grams to 0.3 grams during the seven-hour period from $t = 15$ hrs. to $t = 22$ hrs. What is its growth rate k?

Solution: Assuming normal exponential growth

$$y = ce^{kt}$$

our data are

$$0.15 = ce^{k \cdot 15} \qquad 0.30 = ce^{k \cdot 22}$$

Dividing the second equation by the first gives us

$$2 = e^{k(22-15)} = e^{k \cdot 7}$$

Solving this equation for k requires logarithms, which haven't come up yet but may be familiar to you from high school. In any event, if $y = e^x$, then x is called the natural logarithm of y, and is written

$$x = \log y$$

Thus, from $2 = e^{k \cdot 7}$ we get

$$k \cdot 7 = \log 2$$

$$k = \frac{\log 2}{7}$$

The formula $y = ce^{kt}$ for exponential growth is derived by calculus. Otherwise everything in this example will be accessible to you before this chapter is completed. In fact, you will be asked to do problems very similar to this one.

---

* Lynn Loomis, *Calculus* (Reading, Mass.: Addison-Wesley, 1974), p. 261. Reprinted by permission.

1   The Exponential Function

One of the very important functions that we have not yet studied is the **exponential function.** A typical example of such a function is given by

$$f(x) = 2^x$$

having as domain the set of all real numbers. We know how to evaluate this function for any integer x (positive, negative, or 0), and, in fact, for any rational number x. For example,

$$f(\tfrac{3}{2}) = 2^{3/2} = \sqrt{2^3} = \sqrt{2^2 \cdot 2} = 2\sqrt{2} \approx 2.828$$

(since $\sqrt{2} \approx 1.414$). How do we evaluate it for any irrational number x? For example, what is the meaning of $f(\pi) = 2^\pi$? Unfortunately, we cannot answer these questions in an entirely satisfactory way without more knowledge of limiting procedures. Nevertheless, it is not difficult to gain an intuitive feeling for the meaning of such expressions as 2^x, where x is an irrational number.

It can be shown that every irrational number can be approximated to any desired degree of accuracy by a rational number. This is a fundamental property of the real number system. Consider the irrational number we symbolize by π, for example. We can approximate it by 3.14, 3.142, 3.1416, and so on, and these numbers are rational, since they are terminating decimals. We *define* the number 2^π, then, as the number that is *approached* by $2^{3.14}$, $2^{3.142}$, $2^{3.1416}$, and so on. The fact that a unique number is approached more and more closely as the exponent approximates π with increasing accuracy, we will have to ask the reader to accept on faith at this stage.

More generally, let a be any positive number other than 1. (We exclude 1, since $1^x = 1$ for every x.) Now we define the function

$$f(x) = a^x \tag{1}$$

When x is rational, this is a well-defined number, and when x is irrational, we define a^x to mean the number approached by a^r as r takes on rational values that approach x arbitrarily closely. The function defined by equation (1) is called the **exponential function with base a.** The domain of the function consists of all real numbers, and the range is the set of all positive real numbers. While we cannot at this stage prove the latter assertion, it is intuitively reasonable that the range is at least a subset of R^+. Since roots, powers, and reciprocals of positive numbers are positive, then for a rational and positive x, say, $x = p/q$,

$$a^x = a^{p/q} = \sqrt[q]{a^p}$$

is positive, since $a > 0$. And since $a^{-x} = 1/a^x$, this too is positive. If x is irrational, then the value of a^x is approximated arbitrarily closely by numbers of the form a^r (with r rational), which is positive. It follows, then, that a^x also is positive. Finally, if $x = 0$, we know that $a^x = 1$.

It would lead to trouble to allow the base a to be either 0 or a negative number. If $a = 0$, then $a^x = 0$ for all $x > 0$, but a^x would be undefined for $x \le 0$. If $a < 0$, then a^x would lead to imaginary numbers for infinitely many values of x (for example, for $x = \tfrac{1}{2}, \tfrac{3}{4}$, and so on).

A table of values for $f(x) = 2^x$ and the graph of f (Figure 1) are shown. Notice that as x gets larger and larger positively, the function increases very

Figure 1

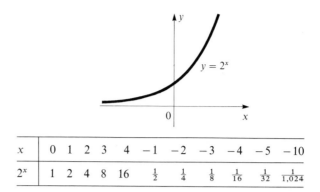

x	0	1	2	3	4	-1	-2	-3	-4	-5	-10
2^x	1	2	4	8	16	$\frac{1}{2}$	$\frac{1}{4}$	$\frac{1}{8}$	$\frac{1}{16}$	$\frac{1}{32}$	$\frac{1}{1,024}$

rapidly, and when x is negative and getting larger and larger in absolute value, the function gets closer and closer to 0, that is, the graph approaches the negative x axis more and more closely as we move to the left. The curve is said to become **asymptotic** to the negative x axis, and the x axis is said to be an **asymptote** to the curve.

The most common bases are those that are greater than 1. In fact, if $0 < a < 1$, then $a^x = (1/a)^{-x}$, and $1/a > 1$; so it really is sufficient to consider bases greater than 1. Figure 2 illustrates several exponential functions on the same coordinate system. The functions all have the following features in common.

Figure 2

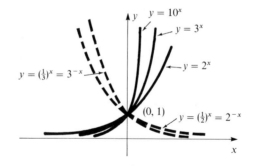

1. The y intercept is 1.
2. For $a > 1$, the negative x axis is an asymptote, and for $a < 1$, the positive x axis is an asymptote.
3. For $a > 1$, the curves all rise steadily as we move from left to right, and for $a < 1$, they all fall steadily from left to right.

Among all possible bases one stands out as being far and away the most important in advanced courses in mathematics and in applications to real-world phenomena. It is an irrational number that is universally designated by the letter e, and its value is approximately 2.71828. The following is one way of defining the number e: Let us consider the expression

$$\left(1 + \frac{1}{n}\right)^n$$

where n designates any positive integer. We evaluate this for a few values of n (using a calculator) in Table 1. This should be fairly convincing evidence (but

Table 1

n	$\left(1+\dfrac{1}{n}\right)^n$
1	2.00000
2	2.25000
3	2.37037
4	2.44141
5	2.48832
10	2.59374
100	2.70481
1,000	2.71692
10,000	2.71815
1,000,000	2.71828

it is not a proof) that as n becomes larger and larger, $[1 + (1/n)]^n$ approaches more and more closely some number whose first few digits are 2.718. We say that the **limit** of $[1 + (1/n)]^n$ as n becomes infinite is e, and write

$$e = \lim_{n \to \infty} \left(1 + \frac{1}{n}\right)^n \tag{2}$$

One way in which this number e arises in a practical situation concerns money invested at compound interest. If a certain principal amount P is invested, and interest is compounded annually at an interest rate of r per year (for example, if the interest rate is 5%, then $r = 0.05$), the amount A of money in the account at the end of t years can be shown to be

$$A = P(1 + r)^t$$

If the interest is compounded semiannually, the formula becomes

$$A = P\left(1 + \frac{r}{2}\right)^{2t}$$

and if quarterly,

$$A = P\left(1 + \frac{r}{4}\right)^{4t}$$

More generally, if interest is compounded k times a year, the amount after t years is

$$A = P\left(1 + \frac{r}{k}\right)^{kt} \tag{3}$$

For example, some banks make a big point of the fact that their interest is compounded daily. In this case equation (3) becomes

$$A = P\left(1 + \frac{r}{365}\right)^{365t}$$

Suppose that we wish to determine what the formula becomes when interest is compounded *continuously*. Some banks actually calculate their interest this way. What we must do is to let k become larger and larger in equation (3). That is, we want to find the limit of the expression on the right side of the equation as k gets arbitrarily large. In symbols, we want

$$\lim_{k \to \infty} P\left(1 + \frac{r}{k}\right)^{kt}$$

To find what this limit is, we manipulate the expression $[1 + (r/k)]^{kt}$ somewhat. First, let us make the substitution

$$n = \frac{k}{r}$$

Thus, $k = nr$. The substitution yields

$$\left(1 + \frac{r}{k}\right)^{kt} = \left(1 + \frac{1}{n}\right)^{nrt}$$

Now, as k gets larger and larger, so does n, and vice versa. So we can write the limit as*

$$\lim_{k \to \infty} P\left(1 + \frac{r}{k}\right)^{kt} = \lim_{n \to \infty} P\left(1 + \frac{1}{n}\right)^{nrt}$$

$$= P\left[\lim_{n \to \infty} \left(1 + \frac{1}{n}\right)^{n}\right]^{rt}$$

The expression in brackets is just the same as the right side of equation (2), and hence equals e. So, finally, our formula for the amount accumulated at the end of t years when the interest is compounded continuously is

$$A = Pe^{rt}$$

It should be emphasized that r is the annual interest rate, and t is the number of years.

EXAMPLE 1

If $1,000 is placed in a savings account yielding interest at 6%, find the amount accumulated after 5 years if interest is compounded:
a. Annually   b. Semiannually   c. Quarterly
d. Daily   e. Continuously
(Either a calculator or a handbook of tables is needed.)

Solution

a. $A = 1,000(1 + 0.06)^5 = 1,000(1.06)^5 = 1,000(1.33822) = \$1,338.22$

b. $A = 1,000\left(1 + \frac{0.06}{2}\right)^{5(2)} = 1,000(1.03)^{10} = 1,000(1.34392) = \$1,343.92$

c. $A = 1,000\left(1 + \frac{0.06}{4}\right)^{5(4)} = 1,000(1.015)^{20} = 1,000(1.34686) = \$1,346.86$

d. $A = 1,000\left(1 + \frac{0.06}{365}\right)^{5(365)} = 1,000(1.000164)^{1,825} = 1,000(1.34982)$

$= \$1,349.82$

e. $A = 1,000e^{(0.06)5} = 1,000(e^{0.3}) = 1,000(1.34986) = \$1,349.86$

---

* We use certain properties of limits here that are shown to be true in calculus.

The next example illustrates what is known as **exponential growth.** If y represents the size of a quantity at time t, then that quantity grows exponentially if it obeys the formula

$$y = Ce^{kt}$$

where C and k are positive constants. Under ideal conditions a culture of bacteria grows approximately exponentially for limited time intervals. Other applications of this law occur in chemistry, economics, and physics.

EXAMPLE 2   A culture of bacteria originally numbers 1,000, and after 2 hours the number has quadrupled. Assuming the law of exponential growth,

$$y = Ce^{kt}$$

find the number of bacteria after 6 hours.

Solution   We know that when $t = 0$, $y = 1,000$, so that

$$1,000 = C \cdot e^{k(0)}$$

giving $C = 1,000$. So our formula becomes

$$y = 1,000e^{kt} \tag{4}$$

We also know that when $t = 2$, $y = 4,000$, so that

$$4,000 = 1,000 \cdot e^{k(2)}$$

$$4 = e^{k(2)} \tag{5}$$

Instead of solving for k at this point, we solve for e^k. We take the square root of both sides of equation (5), or equivalently, raise both sides to the power $\frac{1}{2}$.

$$4^{1/2} = (e^{k(2)})^{1/2}$$

$$2 = e^k$$

Now we can substitute this in equation (4) to get

$$y = 1,000e^{kt} = 1,000(e^k)^t = 1,000 \cdot 2^t$$

This is the formula we need. It expresses the number y of bacteria in terms of the time t in hours. So when $t = 6$, we have

$$y = 1,000 \cdot 2^6 = 1,000 \cdot 64 = 64,000$$

In the case of radioactive substances, which exhibit **exponential decay,** a common measure of the rate of decay is the **half-life** of the substance. This is defined as the time required for the substance to diminish to one-half of the original amount. This is a property inherent in the particular substance and is independent of the amount present initially. Theoretically, a radioactive substance never decomposes completely, so it is not possible to give the "whole life." The basic formula (which is derived in calculus) for the amount y of a substance that decays exponentially is

$$y = Ce^{-kt}$$

where t represents time, and C and k are positive constants that are determined by the data in a particular instance.

EXAMPLE 3   A certain radioactive substance has a half-life of 50 years. How much of 100 grams will remain after 200 years?

Solution   The relevant data are:

> When $t = 0$, $y = 100$
>
> When $t = 50$, $y = 50$ Since half-life is 50 years
>
> When $t = 200$, $y = ?$

First we substitute $t = 0$ and $y = 100$ in the exponential decay equation, $y = Ce^{-kt}$:

$$100 = Ce^{-0} = C$$

So the constant C is 100. The formula for y is therefore

$$y = 100e^{-kt} \tag{6}$$

Now we substitute $t = 50$, $y = 50$:

$$50 = 100e^{-k(50)}$$

$$\tfrac{1}{2} = e^{-k(50)}$$

Again, we do not attempt to solve for the constant k but rather solve for e^{-k} by raising both sides of this equation to the power $\frac{1}{50}$. This gives

$$\left(\frac{1}{2}\right)^{1/50} = e^{-k}$$

Formula (6) now becomes

$$y = 100e^{-kt} = 100(e^{-k})^t = 100\left(\frac{1}{2}\right)^{t/50}$$

Finally, we let $t = 200$:

$$y = 100\left(\frac{1}{2}\right)^4 = 100\left(\frac{1}{16}\right) = \frac{25}{4} = 6.25$$

So after 200 years, 6.25 grams of the original 100 remain.

Finally, we list for reference the laws of exponents given in Chapter 1. Proofs of these for irrational exponents require greater sophistication than we wish to employ here, but the truth of these laws for rational exponents, together with the knowledge that a^x for any irrational x is approximated arbitrarily closely by numbers of the form a^r with r rational, should provide a degree of confidence in their truth in general. We assume $a > 0$ and $b > 0$.

Laws of Exponents

1. $a^0 = 1$

2. $a^{-x} = \dfrac{1}{a^x}$

3. $a^x \cdot a^y = a^{x+y}$

4. $\dfrac{a^x}{a^y} = a^{x-y}$

5. $(a^x)^y = a^{xy}$

6. $(ab)^x = a^x b^x$

EXERCISE SET 1

Note. In this and in other exercises in this chapter a calculator or tables will be needed for certain problems.

A In Problems 1 and 2 use a calculator to approximate the given numbers correct to four decimal places.

1. **a.** $2^{\sqrt{2}}$ **b.** 3^{π} **c.** $4^{-\sqrt{3}}$ **d.** e^2 **e.** e^{-1}
2. **a.** $(\sqrt{5})^{\sqrt{7}}$ **b.** e^{π} **c.** π^2 **d.** $2^{2/3}$ **e.** $e^{-1/2}$

In Problems 3–20 make a table of values and draw the graph.

3. $y = 4^x$ 4. $y = 4^{-x}$
5. $y = e^x$ 6. $y = e^{-x}$
7. $y = (\tfrac{2}{3})^x$ 8. $f(x) = 2^{x/2}$
9. $F(x) = 2^{x-1}$ 10. $g(x) = (\sqrt{3})^{-x}$ 11. $h(x) = 3^{(1-x)/2}$
12. $y = 2^{x^2}$ 13. $y = 1 - e^{-x}$ 14. $f(x) = 1 - 2^{-x}$
15. $g(x) = 2^{1-x^2}$ 16. $h(x) = 2^{|x|}$ 17. $f(x) = (\tfrac{3}{4})^{x-1}$
18. $F(x) = (\tfrac{3}{2})^{x/2}$ 19. $G(x) = (\tfrac{1}{2})^{2-x}$ 20. $y = 3^{|x-1|}$

21. The sum of $2,000 is placed in a savings account at $5\tfrac{1}{2}\%$ annual interest. Find the amount at the end of 4 years if interest is compounded:
    a. Annually **b.** Quarterly **c.** Continuously
22. A culture of bacteria originally numbers 500. After 2 hours there are 1,500 bacteria. Assuming exponential growth, find how many are present after 6 hours.
23. A radioactive substance has a half-life of 64 hours. If 200 grams of the substance are initially present, how much will remain after 4 days?
24. A culture of 100 bacteria doubles after 2 hours. How long will it take the number of bacteria to reach 3,200?
25. How long will it take 100 grams of a radioactive substance having a half-life of 40 years to be reduced to 12.5 grams?
26. One hundred kilograms of a certain radioactive substance decay to 40 kilograms after 10 years. Find how much will remain after 20 years.

27. In the initial stages a bacteria culture grows approximately exponentially. If the number doubles after 3 hours, what proportion of the original number will be present after 6 hours? After 12 hours?

28. Strontium-90 has a half-life of 25 years. If 200 kilograms were present originally, how much will remain after 10 years?

29. One bank offers 6% interest compounded quarterly. Another offers 6% compounded continuously. How much more income would result from depositing $1,000 in the second account for 5 years than in the first?

30. Determine which of the following is the better way to invest $500 for 2 years:
a. An account paying $7\frac{1}{2}\%$ compounded annually
b. An account paying $7\frac{1}{4}\%$ compounded continuously
If in part a interest were compounded quarterly, compare the result with part b.

31. The number N of items of a certain type of merchandise sold on a given day diminishes with time according to the formula

$$N = N_0 e^{-k(t-1)}$$

where N_0 is the number sold on the first day. If 20 of these items are sold on the first day and 10 on the fifth day, how many can be expected to be sold on the ninth day?

B In Problems 32–35 make a table of values and draw the graph.

32. $y = xe^{-x}$

33. $f(x) = \dfrac{e^x + e^{-x}}{2}$

34. $g(x) = \dfrac{e^x - e^{-x}}{2}$

35. $h(x) = \dfrac{e^x - e^{-x}}{e^x + e^{-x}}$

Note. The functions in Problems 33, 34, and 35 are called, respectively, **the hyperbolic cosine, the hyperbolic sine, and the hyperbolic tangent, written cosh x, sinh x, and tanh x.**

36. According to Newton's law of cooling, the rate at which a hot body cools is proportional to the difference in temperature between it and its surroundings. By calculus it can be shown that this results in the following formula for the temperature T of the body after time t:

$$T = T_0 + Ce^{-kt}$$

where T_0 is the temperature of the surrounding medium, and C and k are positive constants. A body is initially at 120°C and is placed in air at 20°C. After $\frac{1}{2}$ hour it has cooled to 80°C. Find its temperature after 1 hour.

37. A thermometer registering 20°C is placed in a freezer in which the temperature is $-10°$C. After 10 minutes the thermometer registers 5°C. What will it register after 30 minutes? (See Problem 36.)

38. A piece of metal is heated to 80°C and is then placed in the outside air at 20°C. After 15 minutes the temperature of the metal is 65°C. What will be its temperature after 15 more minutes? (See Problem 36.)

39. In 1930 the world population was approximately 2 billion and in 1960 approximately 3 billion. A crude approximation to the growth rate of the population is to assume it is exponential, that is, $y = Ce^{kt}$, where y represents the population t years after a given time. Using these data estimate the world population in the year 2000.

40. The air pressure p at an altitude h above the ground is approximately

$$p = p_0 e^{-kh}$$

where p_0 is the pressure at ground level. If the pressure at ground level is 15 pounds per square inch and at height 10,000 feet it is 10 pounds per square inch, find the approximate pressure at 20,000 feet.

41. The atmospheric pressure at sea level is approximately 15 pounds per square inch, and at Denver, which is 1 mile high, the pressure is approximately 12 pounds per square inch. What will be the approximate atmospheric pressure at the top of Vail Pass, which is 2 miles high? (See Problem 40.)

42. The intensity of light passing through a layer of translucent material decreases at a rate approximately proportional to the thickness of the material. It is shown in calculus that this results in the formula

$$I = I_0 e^{-kx}$$

where I_0 is the intensity with which the light first strikes the material and I is the intensity x units inside the material. If sunlight striking water is reduced to half of its original intensity at a depth of 10 feet, what will be the intensity (as a fractional part of the original intensity) at a depth of 30 feet?

---

2 Inverses of Exponential Functions

It is evident from its graph that the exponential function

$$f(x) = a^x \qquad (a > 0, \quad a \neq 1)$$

is one-to-one, since each horizontal line intersects the curve at most once. Thus, an inverse exists for this function. For historical reasons the name given to the inverse is the **logarithm function,** and its value is symbolized by $\log_a x$. This is read "log to the base a of x." So if $f(x) = a^x$, then

$$f^{-1}(x) = \log_a x$$

It is unfortunate that the name and symbol used here do not in any way suggest the relationship of the logarithm function to the exponential; this relationship simply must be learned.

Since for any function f having an inverse f^{-1} it is true that

$$f(f^{-1}(x)) = x \qquad \text{and} \qquad f^{-1}(f(x)) = x$$

we have that

$$a^{\log_a x} = x \tag{7}$$

and

$$\log_a a^x = x \tag{8}$$

Also, we know that a point (x_0, y_0) satisfies $y = f^{-1}(x)$ if and only if the point (y_0, x_0) satisfies $y = f(x)$; that is,

$$y_0 = f^{-1}(x_0)$$

if and only if

$$x_0 = f(y_0)$$

In terms of the logarithmic and exponential functions, this becomes (omitting the subscript)

$$y = \log_a x \qquad \text{if and only if} \qquad x = a^y \tag{9}$$

This is of fundamental importance and should be memorized. We refer to the left-hand equation in (9) as the **logarithmic form** and the right-hand equation as the **exponential form.** We can often deduce facts about logarithms by using an equivalent exponential form. For example, suppose we wish to obtain the value of $\log_2 \frac{1}{16}$. Let us denote this unknown value by y. Then

$$y = \log_2 \tfrac{1}{16}$$

and by (9) the equivalent exponential form is

$$\tfrac{1}{16} = 2^y$$

Since $\frac{1}{16} = (\frac{1}{2})^4 = 2^{-4}$, we have $2^{-4} = 2^y$, from which it follows that $y = -4$. Here are some other examples of this type.

EXAMPLE 4   Find the value of x if $x = \log_3 27$.

Solution   The exponential form is $3^x = 27$, and since $27 = 3^3$, we have $3^x = 3^3$, so that $x = 3$.

EXAMPLE 5   Find y if $6 = \log_2 y$.

Solution   In exponential form this is $2^6 = y$. So $y = 64$.

EXAMPLE 6   Solve for x: $3 = \log_x 125$

Solution   We have $x^3 = 125$, so that $x = \sqrt[3]{125} = 5$.

It helps in changing from logarithmic to exponential form to observe that the base in each case is the same. In the logarithmic form this appears as the subscript:

$$y = \log_@ x$$
$$\nwarrow \textbf{Base}$$

and in the exponential form it is the number that is raised to a power:

$$x = @^y$$
$$\nwarrow \textbf{Base}$$

Also, note that the exponent in the exponential form is the number y, that is, it is the value that the logarithm equals in the logarithmic form. This latter can be stated as follows:

> The logarithm to the base a of a number x is the exponent y of the power to which a must be raised to yield that number.

EXAMPLE 7   The formula for the amount y in grams present at time t in years of a radioactive substance is given by

$$y = Q_0 e^{-0.2t}$$

where Q_0 is the amount initially. Find how long it will be until only 10% of the original quantity is left.

Solution   We want to know the value of t for which $y = (0.1)Q_0$. Substituting this for y yields

$$(0.1)Q_0 = Q_0 e^{-0.2t}$$
$$0.1 = e^{-0.2t}$$

To solve for t we write this in logarithmic form:

$$\log_e(0.1) = -0.2t$$

So,

$$t = -\frac{\log_e(0.1)}{0.2} \approx -\frac{-2.30285}{0.2} \approx 11.513 \text{ years}$$

(We used a calculator here.)

EXAMPLE 8   A man deposits a sum of money in a savings account paying 5% interest compounded continuously. How long will it take for him to double his money?

Solution   Let P be the initial amount. Then, as we found in the previous section,

$$A = Pe^{0.05t}$$

is the formula for the amount after t years. We want to know t when $A = 2P$.

$$2P = Pe^{0.05t}$$
$$2 = e^{0.05t}$$

In logarithmic form this is

$$\log_e 2 = 0.05t$$

So,

$$t = \frac{\log_e 2}{0.05} \approx \frac{0.6931}{0.05} \approx 13.86 \text{ years}$$

EXAMPLE 9   It is found that in 2 years 10% of a radioactive substance disappears. Find the half-life of the substance.

Solution   Let Q_0 be the amount present initially. If y denotes the amount present at time t years later,

$$y = Ce^{-kt}$$

Substituting $y = Q_0$ when $t = 0$ yields $C = Q_0$. So

$$y = Q_0 e^{-kt}$$

We know that when $t = 2$, $y = (0.9)Q_0$. Thus,

$$(0.9)Q_0 = Q_0 e^{-k2}$$
$$0.9 = e^{-2k}$$

This time we will solve for k (rather than e^{-k}), using the logarithmic form:

$$\log_e 0.9 = -2k$$
$$k = -\tfrac{1}{2} \log_e 0.9 \approx -\tfrac{1}{2}(-0.10536) = 0.05268$$

So our formula for y becomes

$$y = Q_0 e^{-0.05268t}$$

Finally, we wish to determine the half-life, that is, the value of t when $y = Q_0/2 = 0.5Q_0$:

$$0.5Q_0 = Q_0 e^{-0.05268t}$$

After dividing both sides by Q_0, this can be written as

$$e^{-0.05268t} = 0.5$$

Again, we use the logarithmic form:

$$\log_e 0.5 = -0.05268t$$

So,

$$t = \frac{\log_e 0.5}{-0.05268} \approx \frac{-0.69315}{-0.05268} \approx 13.158 \text{ years}$$

EXAMPLE 10   The human ear can accommodate an enormous range of sound wave intensities, ranging from about 10^{-12} watt per square meter, taken to be the threshold of hearing, to about 1 watt per square meter, the threshold of pain for most people. Since the range is so great, it is customary to use a logarithmic scale to describe the relative loudness of a sound. The formula

$$\beta = 10 \log_{10} \frac{I}{I_0}$$

is used, where I_0 is the intensity of the threshold of hearing (10^{-12} watt per square meter) and I is the intensity of the sound in question. The number β is said to be the number of **decibels** of the sound. Find the decibel intensity of the noise of a heavy truck passing a pedestrian at the side of a road if the sound wave intensity I of the truck is 10^{-3} watt per square meter.

Solution   Since $I_0 = 10^{-12}$ and $I = 10^{-3}$, we have

$$\beta = 10 \log_{10} \frac{I}{I_0} = 10 \log_{10} \frac{10^{-3}}{10^{-12}}$$
$$= 10 \log_{10} 10^9$$
$$= 10(9) = 90$$

So the intensity is 90 decibels.

EXERCISE SET 2

A   Evaluate the expressions in Problems 1–11.

1.   **a.** $\log_2 2^3$ **b.** $2^{\log_2 3}$
2.   **a.** $\log_3 3^{10}$ **b.** $5^{\log_5 7}$
3.   **a.** $\log_b b^x$ **b.** $b^{\log_b x}$
4.   **a.** $\log_2 8$ **b.** $\log_3 \frac{1}{9}$
5.   **a.** $\log_{10} 10{,}000$ **b.** $\log_{10} 0.001$
6.   **a.** $\log_4 2$ **b.** $\log_2 0.125$
7.   **a.** $\log_4 64$ **b.** $\log_2 \frac{1}{16}$
8.   **a.** $\log_{10} 0.1$ **b.** $\log_3 81$
9.   **a.** $\log_{10} 100$ **b.** $\log_e(1/e^2)$
10. **a.** $e^{\log_e 4}$ **b.** $10^{\log_{10} 7}$
11. **a.** $\log_9 9^5$ **b.** $\log_6 6^{-5}$

Find the unknown in Problems 12–24.

12. **a.** $\log_4 x = 2$ **b.** $\log_x 16 = 4$
13. **a.** $\log_2 8 = y$ **b.** $\log_y 3 = -\frac{1}{2}$
14. **a.** $t = \log_2 \frac{1}{8}$ **b.** $\log_{10} x = -2$
15. **a.** $\log_{10} 1{,}000 = x$ **b.** $\log_{10} x = -3$
16. **a.** $\log_u 3 = \frac{1}{2}$ **b.** $\log_{1/2} u = -4$
17. **a.** $\log_4 1 = v$ **b.** $\log_v \frac{4}{9} = -\frac{2}{3}$
18. **a.** $\log_2 y = 6$ **b.** $\log_x 125 = 3$
19. **a.** $\log_4 0.25 = t$ **b.** $\log_4 x = \frac{1}{2}$
20. **a.** $\log_e 1 = y$ **b.** $\log_z 0.01 = -2$
21. **a.** $\log_9 u = -\frac{3}{2}$ **b.** $\log_8 4 = v$
22. **a.** $\log_x 16 = \frac{4}{3}$ **b.** $\log_{4/9} t = -\frac{3}{2}$
23. **a.** $\log_3(x - 1) = 2$ **b.** $\log_2 0.125 = y$
24. **a.** $w = \log_e(1/e)$ **b.** $\log_2 |w| = 3$

In Problems 25–28 write in logarithmic form.

25. **a.** $4^3 = 64$ **b.** $3^{-2} = \frac{1}{9}$
26. **a.** $10^4 = 10{,}000$ **b.** $10^{-3} = 0.001$
27. **a.** $2^8 = 256$ **b.** $2^{-3} = 0.125$
28. **a.** $r^s = t$ **b.** $z^x = m$

In Problems 29–32 write in exponential form.

29. **a.** $\log_2 16 = 4$ **b.** $\log_{10} 100 = 2$
30. **a.** $\log_e 1 = 0$ **b.** $\log_3 \frac{1}{9} = -2$
31. **a.** $\log_5 0.2 = -1$ **b.** $\log_8 4 = \frac{2}{3}$
32. **a.** $\log_a x = y$ **b.** $\log_x z = t$

In Problems 33–41 draw the graph of each function by first writing an equivalent exponential form and then making a table of values.

33. $y = \log_2 x$
34. $y = \log_3 x$
35. $y = \log_2(x + 1)$
36. $y = \log_3(1 - x)$
37. $y = \log_4(2x - 3)$
38. $y = \log_2(1 - 3x)$

39. $y = \log_2 x^2$

40. $y = \log_3 |x|$

41. $y = \log_e(-x)$

42. This problem is taken from an ARCO study guide for the Medical College Admission Test (MCAT):*

Which of the following equations is best represented by the graph?

(A) $x + y = a^b$ (B) $x - y = e^2$ (C) $ax^2 + by^2 = 0$

(D) $x + y = ab$ (E) $y = ax^b$

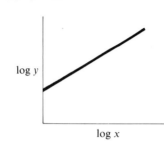

43. If \$10,000 is invested at 6% interest compounded continuously, in approximately how many years will the amount have grown to \$15,000?

44. In the early stages a bacterial culture grows approximately exponentially. Assuming this holds true for the time periods in question, if the quantity doubles in 2 hours, how long will it take to triple?

45. If \$5,000 is invested at $5\frac{1}{2}$% compounded continuously, approximately how long will it take for the amount to double?

46. A women has \$10,000 invested at 6% and her brother has \$9,000 invested at 7%. If in each account interest is compounded continuously, find how long it will take for the amount in the brother's account to equal the amount in his sister's.

47. If 100 grams of a radioactive material diminish to 80 grams in 2 years, find the half-life of the substance.

48. Use the information given in Example 10 to find the decibel level of sound at:

a. The threshold of hearing   **b.** The threshold of pain

49. At a distance of 10 meters from a rock band the intensity of sound is approximately 0.1 watt per square meter. Find the intensity in decibels.

50. At a distance of $\frac{1}{2}$ mile from the runway the noise level of a supersonic transport plane at takeoff is approximately 150 decibels. What is the sound wave intensity in watts per square meter?

51. Find the half-life of a radioactive substance that is reduced by 30% in 20 hours.

B   **52.** In a chemical reaction a substance is converted according to the formula

$$y = Ce^{-0.3t}$$

where y is the amount of the unconverted substance at time t minutes after the reaction began. If there are 20 grams of the substance initially, find how long it will take for only 4 grams of it to remain unconverted.

53. A body heated to 120°F is brought into a room in which the temperature is 70°F. After 15 minutes the temperature of the body is 100°F. How long will it take the temperature of the body to drop to 80°F? (See Problem 36, Exercise Set 1.)

54. If a body is brought into a warmer medium than the temperature of the body, a formula analogous to Newton's law of cooling holds true (see Problem 36, Exercise

---

* David R. Turner, *Medical College Admission Test*, 6th ed. (New York: ARCO Publishing, Inc., 1977), p. 242. Reprinted by permission.

Set 1), namely,

$$T = T_0 - Ce^{-kt}$$

where T_0 is the temperature of the surrounding medium and T is the temperature of the body after time t. An outside thermometer registers 5°C and is brought into a room where the temperature is 20°C. After 10 minutes the thermometer registers 12°C. How long will it take for it to register 18°C?

55.   Use the data given in Problem 42, Exercise Set 1, to find at what depth the intensity of illumination will be only $\frac{1}{10}$ the original.

56.   A thermometer registers -15°C in a freezer and is brought into a room in which the temperature is 25°C. After 5 minutes the thermometer registers 5°C. Find how long it will take for it to register 24°C. (See Problem 54.)

3   Properties of Logarithms

The domain of the exponential function

$$f(x) = a^x \qquad (a > 0)$$

consists of all real numbers, and its range is the set of all positive real numbers. It follows that for the inverse

$$f^{-1}(x) = \log_a x$$

the domain and range are interchanged:

Domain of $\log_a x$ = Set of all positive real numbers

Range of $\log_a x$ = Set of all real numbers

It should be emphasized that the logarithm function is defined *only for positive numbers*. This is also evident from the graph of $y = \log_a x$ (Figure 3), which is obtained by reflecting the graph of $y = a^x$ in the line $y = x$. (Alternately, one can obtain the graph of $y = \log_a x$ by plotting the graph of the equivalent equation $x = a^y$.)

The relations (7) and (8) of Section 2 enable us to exploit the properties of the exponential function to obtain corresponding properties of the logarithm function. In what follows we will repeatedly use the fact that if $a^x = a^y$, then $x = y$, which is a consequence of the fact that the exponential function is one-to-one.

By equation (7) we have that for any $x > 0$ and $y > 0$,

$$x = a^{\log_a x} \qquad \text{and} \qquad y = a^{\log_a y}$$

Figure 3

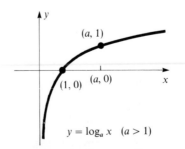

$$y = \log_a x \quad (a > 1)$$

Similarly,

$$xy = a^{\log_a xy}$$

Thus, we have the equality

$$a^{\log_a xy} = a^{\log_a x} \cdot a^{\log_a y}$$

But by the law of exponents for multiplication we can add the exponents on the right, and so obtain:

$$a^{\log_a xy} = a^{\log_a x + \log_a y}$$

The exponents must therefore be equal:

$$\log_a xy = \log_a x + \log_a y \tag{10}$$

Next consider the following two representations of x^p. By equation (7),

$$x^p = a^{\log_a x^p}$$

but also,

$$x^p = (a^{\log_a x})^p = a^{p \log_a x}$$

where the last equality is a consequence of one of the laws of exponents. The two expressions for x^p must be equal:

$$a^{\log_a x^p} = a^{p \log_a x}$$

which yields:

$$\log_a x^p = p \log_a x \tag{11}$$

Equations (10) and (11) can be used now to obtain a formula for the logarithm of a quotient. We have

$$\log_a \frac{x}{y} = \log_a\left(x \cdot \frac{1}{y}\right) = \log_a x + \log_a \frac{1}{y}$$

by equation (10). And by (11), with $p = -1$,

$$\log_a \frac{1}{y} = \log_a y^{-1} = -\log_a y$$

So finally,

$$\log_a \frac{x}{y} = \log_a x - \log_a y \tag{12}$$

The properties (10), (11), and (12) are basic and account for much of the utility of the logarithm function. These results are summarized in the box, where it is understood that $x > 0$ and $y > 0$.

Properties of Logarithms

1. $\log_a xy = \log_a x + \log_a y$

2. $\log_a \dfrac{x}{y} = \log_a x - \log_a y$

3. $\log_a x^p = p \log_a x$

Remark. The base e is the most important base for logarithms, and logarithms with this base are called **natural logarithms.** They also are sometimes called **Napierian** (after John Napier, a Scot who invented logarithms in the early seventeenth century). The symbol **ln** is often used instead of the more cumbersome $\log_e$. Also, many books, especially advanced books in mathematics, use "log" without any base indicated, with the understanding that this means the natural logarithm. So

$$\log_e x = \ln x = \log x$$

The last usage is not universal. In fact, in more elementary treatments "log" is usually understood to mean "$\log_{10}$," and this is called the **common logarithm.** Historically, common logarithms have been most used in computational work. Now, however, with the advent of hand calculators, they are seldom needed. In the exercises that follow we will often use "log" without explicitly indicating a base, since the results are valid regardless of what (admissible) base is being used. The properties of logarithms are used repeatedly. The following examples are illustrative. When no restriction on x is stated, we will understand that only values for which the logarithm is defined are allowed.

In Examples 11–14 we use the properties of logarithms to write the expression in a form that is free of logarithms of products, quotients, and powers.

EXAMPLE 11

$$\log x^2 = 2 \log x \qquad (x > 0)$$

If $x < 0$, we can write

$$\log x^2 = \log|x|^2 = 2 \log|x|$$

EXAMPLE 12

$$\log \frac{2x}{3x^2 + 1} = \log 2x - \log(3x^2 + 1)$$

$$= \log 2 + \log x - \log(3x^2 + 1)$$

A Word of Caution. The properties of logarithms enable us to simplify logarithms of products, quotients, and powers (and roots, since these can be treated as powers), but there is no way to simplify logarithms of sums or differences. Thus, $\log(x + y)$ *cannot* be changed to $\log x + \log y$, however tempting this might be. In fact, we know that $\log x + \log y$ is the same as $\log xy$, not $\log(x + y)$.

EXAMPLE 13

$$\log \sqrt{\frac{1 - x}{1 + x}} = \log \left(\frac{1 - x}{1 + x}\right)^{1/2}$$

$$= \frac{1}{2} \log \frac{1 - x}{1 + x} = \frac{1}{2}[\log(1 - x) - \log(1 + x)]$$

EXAMPLE 14

$$\log \frac{x^3 \sqrt{1 - x^2}}{(2x + 1)^{3/2}(x^4 + 1)} = \log x^3 \sqrt{1 - x^2} - \log(2x + 1)^{3/2}(x^4 + 1)$$

$$= \log x^3 + \log \sqrt{1 - x^2} - [\log(2x + 1)^{3/2} + \log(x^4 + 1)]$$

$$= 3 \log x + \tfrac{1}{2} \log(1 - x^2) - \tfrac{3}{2} \log(2x + 1) - \log(x^4 + 1)$$

We first treated this as a quotient. Then we worked with the products, and finally with the powers. Notice that in the second step we bracketed the two terms preceded by a minus sign. This procedure is highly recommended, for a common pitfall is to apply the minus to one term and not the other.

EXAMPLE 15   Solve for x: $\log x - \log(x - 2) = 2 \log 2$

Solution   By the properties of logarithms this can be written in the form

$$\log \frac{x}{x - 2} = \log 2^2$$

Since the logarithms are equal, so are the numbers (since the logarithm function is 1–1). Thus,

$$\frac{x}{x - 2} = 4$$

$$x = 4(x - 2)$$

$$x = 4x - 8$$

$$3x = 8$$

$$x = \tfrac{8}{3}$$

EXAMPLE 16   Solve for x: $\log_3(x + 1) + \log_3(x + 3) = 1$

Solution   We write the left-hand side as the logarithm of a product,

$$\log_3(x + 1)(x + 3) = 1$$

and then write the result in exponential form.

$$(x + 1)(x + 3) = 3^1$$

$$x^2 + 4x + 3 = 3$$

$$x^2 + 4x = 0$$

$$x(x + 4) = 0$$

$$x = 0 \quad | \quad x = -4$$

Upon checking in the original equation, we find that $x = 0$ works, since

$$\log_3(0 + 1) + \log_3(0 + 3) = \log_3 1 + \log_3 3 = 0 + 1 = 1$$

but $x = -4$ is not permissible, because this would lead to logarithms of negative numbers. The trouble was introduced when we wrote

$$\log_3(x + 1) + \log_3(x + 3) = \log_3(x + 1)(x + 3)$$

When $x = -4$, the left-hand side is not defined, but the product of the two negative numbers on the right is positive, so that the right-hand side is defined for $x = -4$.

EXAMPLE 17   Find the value of x in each of the following in terms of natural logarithms:
a. $2^x = 3$ **b.** $4^{1-x} = 3 \cdot 5^x$

Solution   **a.**   We take the natural logarithm of both sides, and make use of the properties of logarithms:

$$\ln 2^x = \ln 3$$
$$x \ln 2 = \ln 3$$
$$x = \frac{\ln 3}{\ln 2}$$

A Word of Caution.   Do not confuse $\ln 3/\ln 2$ with $\ln \frac{3}{2}$, which is equal to $\ln 3 - \ln 2$.

b.   Again, we equate the natural logarithm of the left side to that of the right:

$$\ln 4^{1-x} = \ln(3 \cdot 5^x)$$
$$(1 - x) \ln 4 = \ln 3 + \ln 5^x$$
$$\ln 4 - x \ln 4 = \ln 3 + x \ln 5$$
$$\ln 4 - \ln 3 = x(\ln 4 + \ln 5)$$
$$\ln \tfrac{4}{3} = x \ln(4 \cdot 5)$$
$$x = \frac{\ln \frac{4}{3}}{\ln 20}$$

Sometimes it is useful to change from one base for logarithms to another. Suppose we are given $\log_a x$ and want to find the value of this same expression involving logarithms to the base b. Let

$$y = \log_a x$$

We write this in exponential form and take the logarithm to the base b of both sides, making use of equation (11):

$$a^y = x$$
$$\log_b a^y = \log_b x$$
$$y \log_b a = \log_b x$$
$$y = \frac{\log_b x}{\log_b a}$$

Therefore, we have the change of base formula

$$\log_a x = \frac{\log_b x}{\log_b a} \tag{13}$$

If we put $x = b$, we obtain the interesting result

$$\log_a b = \frac{1}{\log_b a}$$

which can be used to give the following alternate form of equation (13):

$$\log_a x = (\log_a b)(\log_b x) \tag{14}$$

Another formula which is essentially a change of base formula for exponentials can be obtained as follows. By equation (7),

$$a^x = b^{\log_b a^x}$$

and since $\log_b a^x = x \log_b a$, this gives

$$a^x = b^{x \log_b a} \tag{15}$$

This formula is especially important in calculus when b is taken as e. Then we have

$$a^x = e^{x \ln a} \tag{16}$$

This shows that every exponential can be expressed as an exponential with base e.

EXERCISE SET 3

A   In Problems 1–26 use the properties of logarithms to write the expressions in a form that is free of logarithms of products, quotients, and powers.

1. **a.** $\log(x + 1)(x - 2)$ **b.** $\log \dfrac{x + 1}{x - 2}$

2. **a.** $\log(x + 4)^3$ **b.** $\log \sqrt{2x + 3}$

3. **a.** $\log(x - 2)^4$ **b.** $\log(2x + 1)^{3/2}$

4. **a.** $\log(x^2 + 2x)$ **b.** $\log \dfrac{x}{x + 1}$

5. **a.** $\log(x^2 - 2x - 3)$ **b.** $\log \dfrac{x^2 - 4}{x - 3}$

6. **a.** $\log \dfrac{x^3}{3x + 1}$ **b.** $\log \dfrac{x(x + 2)}{(x + 3)^2}$

7. **a.** $\log \sqrt[3]{(x + 1)^2}$ **b.** $\log \dfrac{1}{\sqrt{3x - 4}}$

8. **a.** $\log x^3 \sqrt{x + 1}$ **b.** $\log \dfrac{2x^2}{(x - 1)^3}$

9. **a.** $\log \sqrt{x^2 + y^2}$ **b.** $\log \sqrt{x^2 - y^2}$

10. $\log \dfrac{x^2}{(x - 1)(2x + 3)}$

11. $\log \dfrac{\sqrt[3]{1 - 2x}}{(x + 3)^2}$

12. $\log \dfrac{\sqrt{x+2}}{x^2 - 2x}$

13. $\log \dfrac{(2x-5)(3x+4)}{(x+2)(x-1)}$

14. $\log \dfrac{x^2 - x - 6}{x^2 - 1}$

15. $\log \sqrt{\dfrac{(x-1)^3}{x+2}}$

16. $\log \dfrac{2x\sqrt[3]{x-2}}{x^2}$

17. $\log \dfrac{3x^4}{\sqrt{(2x+1)^3}}$

18. $\log \sqrt{\dfrac{3x-7}{x^3}}$

19. $\log \dfrac{x^2\sqrt{1-3x}}{1+2x}$

20. $\log \dfrac{\sqrt{x^2-1}}{x^3 + 2x^2}$

21. $\log \dfrac{(2x+3)^{3/2}(x+1)^4}{3x\sqrt{1-x}}$

22. $\log \left(\dfrac{x^2 - 9}{x^2 - 2x - 5}\right)^{1/3}$

23. $\log \dfrac{3x(x^2+4)^2}{(x+2)\sqrt{2x+3}}$

24. $\log \dfrac{5x^4}{(x+3)^{2/3}(x-2)^3}$

25. $\log \dfrac{1}{\sqrt{2x^2 - x - 3}}$

26. $\log \dfrac{3x^2\sqrt{1-x}}{5(x+2)^3 \cdot \sqrt{2x+1}}$

Find the values of the expressions in Problems 27 and 28.

27. **a.** $2^{4 \log_2 3}$ **b.** $10^{2 \log_{10} 5}$

28. **a.** $e^{5 \ln 2}$ **b.** $e^{b \ln a}$

In Problems 29–34 use the properties of logarithms to combine each expression into a single term.

29. **a.** $\log x - \log y$ **b.** $2 \log x + 3 \log(x+1)$

30. **a.** $3 \log x - 2 \log y$ **b.** $\log 2 + \log x + \frac{1}{2} \log(x+1)$

31. **a.** $\frac{1}{2} \log(x-1) + \log 2 - 3 \log(x+2)$
    b. $3 \log(x+2) - \frac{1}{2} \log x - \frac{1}{2} \log(1-x)$

32. **a.** $\log 3 + \frac{2}{3} \log(x^2+4) - 2 \log(2x-3)$
    b. $2 \log(x+1) - 3 \log(x-1) - \frac{1}{2} \log(1-2x)$

33. **a.** $\frac{1}{2} \log(5x+3) - [2 \log(x+2) + \log(3x+1)]$
    b. $\log 2 + \log x - \frac{1}{2} \log(x+1) + \log C$

34. **a.** $\log(x^2+1) + \frac{2}{3} \log(x-3) - \log 2 - \frac{1}{2} \log(3x-4) + \log C$
    b. $\log 5 + 2 \log(6x+5) - \frac{1}{2}[\log(2x+1) + 3 \log(x-2)]$

In Problems 35–37 find x in terms of natural logarithms. (See Example 17.)

35. **a.** $4^x = 3$ **b.** $3^x = 5^{x-1}$

36. **a.** $2 \cdot 3^{-x} = 6^{2x}$ **b.** $3^{x+1} = e^{x/2}$

37. **a.** $e^{-x} = 2^{x-3}$ **b.** $3 \cdot 4^{2x} = 2 \cdot 3^{-x}$

In Problems 38–49 solve for x, and check your answers.

38. $\log 2 + \log x = \log 3 + \log(x-1)$

39. $2 \log x = \log 2 + \log(x+4)$

40. $2 \log x = \log(x+10) - \log 3$

41. $\log x + \log(2x-5) = \log 3$

42. $\log 3 + \log(x+2) = \log 2 + \log(3x-1)$

43. $\log x + \log(x - 4) = \log(x + 2) + \log(x - 5)$
44. $2 \log(x + 1) - \log(x + 4) = \log(x - 1)$
45. $\log(x - 1) - \log(x + 6) = \log(x - 2) - \log(x + 3)$
46. $\log x + \log(x - 2) = \log(x + 4)$
47. $\log_2 x + \log_2(x - 3) = 2$
48. $\log_2(x - 1) + \log_2(x - 2) = 1$
49. $\log_3 x + \log_3(x + 8) = 2$

In Problems 50–56 find y as a function of x.

50. $\log y = \log x - \log(x - 1) + \log C$
51. $2 \log y - 3 \log x = \log 4$
52. $\log y = 2 \log(x + 1) - \log x + \log C$
53. $\log y = 3 \log x + \log(2x - 3) - \frac{1}{2} \log(x + 2) + \log C$
54. $\ln y = x + \ln C$
55. $\ln(2y - 3) = 2 \ln(x - 1) + \ln C$
56. $\ln(y - 1) = 2x + \ln C$

57. Let $f(x) = \log_a x$ and $g(x) = \log_b x$. Verify each of the following:

    a. $f(x) = \dfrac{g(x)}{g(a)}$ **b.** $g(x) = \dfrac{f(x)}{f(b)}$ **c.** $f(b) \cdot g(a) = 1$

58. Explain the differences in meaning among the following:
    a. $\log x^2$ **b.** $(\log x)^2$ **c.** $\log(\log x)$
    Show by means of an example that these expressions yield different results.

In Problems 59–63 use a calculator to approximate the value of x to four significant figures.

59. **a.** $4^x = 3$ **b.** $3^{1-x} = 7$
60. **a.** $3^x = 2^{3/2}$ **b.** $5^{x/2} = 3^5$
61. **a.** $4 \cdot 2^{-x} = 3^{x+1}$ **b.** $2 \cdot 5^x = 3^{2x-4}$
62. **a.** $(2.034)^x = 12.06$ **b.** $(0.6305)^{-x} = 3.125$
63. **a.** $x = \log_2(3.104)$ **b.** $x = \dfrac{1}{\log_3 15.24}$

B   In Problems 64–69 solve for x and check your answers.

64. $2 \log_3(x - 2) - \log_3(x + 4) = 1$
65. $\log_e(2x - 1) + \log_e(x - 1) = 0$
66. $\log_4(x - 3) - \log_4(2x + 1) = -1$
67. $\log_{10} x + \log_{10} 2 = 1 - \log_{10}(x + 4)$
68. $\log_2 x + \log_2(x^2 - 3) = 1$
69. $\log_2(x + 3) - \log_2 3 = \log_2(3x + 4) - 3$

70. Show that $-\log(x - \sqrt{x^2 - 1}) = \log(x + \sqrt{x^2 - 1})$.

71. Show that $\log \dfrac{1}{\sqrt{x} - \sqrt{x - 1}} = \log(\sqrt{x} + \sqrt{x - 1})$.

In Problems 72–74 give the graphical solution of each system of inequalities.

72. $\begin{cases} x > 0 \\ y \le \ln x \\ x - y < 4 \end{cases}$ **73.** $\begin{cases} x < 1 \\ y + \ln(1 - x) \ge 0 \\ y - 2x \le 4 \end{cases}$ **74.** $\begin{cases} y \ge 1 + e^{-x} \\ y \le 1 + \ln x \\ 2x + 3y \le 12 \end{cases}$

75. Solve for x:

$$\frac{e^x - e^{-x}}{2} = 2$$

   Hint. By clearing of fractions and negative exponents, obtain the equation in the form $e^{2x} - 4e^x - 1 = 0$, and observe that this is quadratic in e^x.

76. Solve for x:

$$\frac{e^x + e^{-x}}{2} = 3$$

   (See hint for Problem 75.)

77. Solve for x:

$$x^{\log_2 x} = \frac{x^4}{8}$$

   Hint. Take logarithm to base 2 of both sides.

Review Exercise Set

A   In Problems 1 and 2 make a table of values and draw the graph of each function.

1. **a.** $y = (3/2)^x$ **b.** $y = 1 - 2^{-x}$
    c. $y = \log_2 x$ **d.** $y = \ln(1 - x)$
2. **a.** $f(x) = 2^{x/2} - 1$ **b.** $g(x) = \log_2(1 - 2x)$
    c. $h(x) = 3^{(x-1)/2}$ **d.** $F(x) = 2 - \log_e(x - 2)$

3. Write in exponential form:
    a. $\log_3 9 = 2$ **b.** $\log_{10} 0.001 = -3$ **c.** $\log_2 256 = 8$
    d. $v = \log_k y$ **e.** $z = \log_r s$
4. Write in logarithmic form:
    a. $4^{3/2} = 8$ **b.** $27^{-2/3} = \frac{1}{9}$ **c.** $10^3 = 1{,}000$
    d. $p = q^s$ **e.** $n = a^t$

In Problems 5–7 find the unknown.

5. **a.** $\log_2 x = 4$ **b.** $\log_y 16 = -2$ **c.** $\log_3 81 = z$
6. **a.** $y = \log_3 27$ **b.** $\log_x 0.0001 = -4$ **c.** $\log_4 z = -3$
7. **a.** $x = e^{\ln 3}$ **b.** $y = \log_5 5^2$ **c.** $\log_2(1 - x) = 3$

In Problems 8–10 write in a form that is free of logarithms of powers, roots, products, and quotients.

8. **a.** $\log \dfrac{2x^3}{\sqrt{x - 1}}$ **b.** $\log \dfrac{x(x - 2)}{2(x^2 + 1)^3}$

9. **a.** $\log \dfrac{x - 1}{(x + 2)^3}$ **b.** $\log \sqrt{\dfrac{2x - 1}{3x}}$

10. **a.** $\log \dfrac{x^2 - 9}{x^2 - 4x + 4}$ **b.** $\log \dfrac{4x^2 \sqrt{2x - 1}}{(x - 2)(x + 4)^{3/2}}$

Combine the expressions in Problems 11–13 into a single term.

11. **a.** $\frac{1}{2}\log(2x - 3) + \log x - 3\log 2 + \log C$
    b. $4\log x - \frac{1}{2}\log(x + 2) - \log(x - 1)$

12. **a.** $\log(3x + 4) - 2 \log x - \log(x + 2)$
 b. $3 \log(x + 2) + \frac{1}{2} \log(2x - 1) - \frac{3}{2} \log(2 - x)$
13. **a.** $\log 2 + 3 \log x - \frac{1}{3}[\log(x - 1) + \log(x + 2)]$
 b. $\frac{3}{2} \log(2x + 3) - \log 2 - 3 \log(x + 4)$

Evaluate the expressions in Problems 14 and 15.

14. **a.** $\log_{10} 10^{-12}$ **b.** $e^{\ln 5}$
 c. $2^{3 \log_2 5}$ **d.** $10^{2 \log_{10} 4}$
15. **a.** $\log_4 8$ **b.** $\log_2 0.125$
 c. $\log_3 \frac{1}{81}$ **d.** $\log_{10} 0.00001$

16. Solve for x in terms of natural logarithms.
 a. $3^x = 7$ **b.** $2^{x+1} = 3^{2x}$
 c. $4 \cdot 5^x = 3^{x-2}$ **d.** $3^x \cdot 2^{x+3} = 5^{2x}$
17. Solve the formula $y = C(1 - e^{-kt})$ for k.

In Problems 18–25 solve for x and check your answer.

18. $\log(3x + 2) - \log(2x - 1) = \log 2$
19. $\log x + \log(x - 2) = \log 3$
20. $\log x + \log(2x - 1) = \log 3$
21. $\log(x + 2) - \log(3x - 1) = \log(x - 1) - \log(x + 3)$
22. $\log_2(1 - x) + \log_2(4x + 1) = 0$
23. $\log_5(3x + 1) + \log_5(x + 1) = 1$
24. $\log_2(2x + 1) - \log_2(x - 1) = 2$
25. $\log_4(x - 3) + \log_4(2x + 1) = 1$

26. Solve for y as a function of x.
 a. $\ln(2y - 1) = 3x + \ln C$
 b. $\log y = 2 \log(x - 1) - \log(x + 1) + \log C$
27. A radioactive substance has a half-life of 6 years. If 20 pounds are present initially, how much will remain after 2 years?
28. A culture of bacteria increases from 1,000 to 5,000 in 8 hours. Assuming the law of exponential growth applies, how many will be present after 20 hours?
29. The population of a certain city in Florida increased by 40% between 1970 and 1978. If the population in 1978 was 210,000, what will be the population in the year 2000? (Assume the law of exponential growth holds.)
30. An object cools from 140°F to 120°F in 5 minutes when placed in outside air at 40°F. After how many more minutes will the temperature of the object be 70°F? (See Problem 36, Exercise Set 1.)
31. The sum of $5,000 is placed in a savings account paying 6% annual interest.
 a. How much money will there be after 5 years if interest is compounded quarterly? Continuously?
 b. If interest is compounded continuously, how long will it take for the amount to double?
32. At a distance of 1 meter the intensity of sound of normal conversation is about 10^{-6} watt per square meter. What is the intensity in decibels? (See Example 10.)
33. If atmospheric pressure at sea level is 14.7 pounds per square inch and at a height of 18,000 feet the pressure is half as great, find the pressure at a height of 10,000 feet, assuming pressure p obeys the law

$$p = p_0 e^{-kh}$$

where p_0 is the pressure at sea level and h is the height above sea level.

34. Find how many years are required for a principal of $10,000 to grow to $15,000 if it is invested at 6% compounded annually.

35. At what interest rate would an amount of money have to be invested in order to double after 10 years if interest is compounded continuously?

In Problems 36 and 37 use a calculator to approximate the unknown to four significant figures.

36. a. $5^{x-1} = 3$ **b.** $2^t = 3^{1-t}$

37. a. $4 \cdot 6^y = 7 \cdot 5^{y/2}$ **b.** $(3.025)^{2x-1} = (0.7134)^{-x}$

B In Problems 38 and 39 solve the systems of inequalities graphically.

38. $\begin{cases} x > 1 \\ y < \log_2(2x - 1) \\ 2x - 3y < 6 \end{cases}$ **39.** $\begin{cases} y - 1 \le \ln(x + 1) \\ y + 1 \ge e^{x-1} \end{cases}$

40. Solve for x:

$$\frac{e^x - e^{-x}}{e^x + e^{-x}} = \frac{1}{2}$$

(See Problem 75, Exercise Set 3.)

41. If sugar is placed in water, the amount undissolved after t seconds is given by the formula $A = Ce^{-kt}$. If 30 grams of sugar are placed in water and it takes 10 seconds for half of it to dissolve, how long will it take for all but 3 grams to dissolve?

42. If a body falls through the air and air resistance is taken into consideration, then by calculus it can be shown that the velocity after t seconds is given by

$$v = \frac{mg}{k} + \left(v_0 - \frac{mg}{k}\right)e^{-kt/m}$$

where m is the mass of the body, g is the acceleration due to gravity, v_0 is the initial velocity, and k is a positive constant. If $v_0 = 30$, $m = 2$, $g = 32$, and $k = 0.2$, find how long it will take to attain a velocity of 200.

43. When limitations on size are considered for the growth of a colony of bacteria (due to physical limitations or limitations on nutrients, for example) the growth is described reasonably accurately by what is known as the **law of logistic growth**:

$$Q = \frac{mQ_0}{Q_0 + (m - Q_0)e^{-kmt}}$$

where Q_0 is the initial quantity present, m is the maximum size, and k is a positive constant. If $Q_0 = 200$, $m = 1,000$, and $Q = 400$ when $t = \frac{1}{2}$, find k.

44. The so-called **learning curve** in educational psychology has an equation of the form $f(t) = C(1 - e^{-kt})$, where C and k are positive constants and t represents time. The value of $f(t)$ can be interpreted as the amount learned at time t. This also has applications in industry.

a. Draw the graph of f for $C = 10$ and $k = 0.1$.

b. Suppose that learning vocabulary in a foreign language follows the learning curve, and that for an average first-year college student $C = 20$, with time t measured in days. Suppose also that on the tenth day of study it has been found that first-year students can be expected to learn 15 new words. Find how many words a student would be expected to learn on the fifth day.

Cumulative Review Exercise Set II (Chapters 4–7)

1. Find the domain of each of the functions.

   a. $f(x) = \sqrt{\dfrac{2x-1}{x^2-4}}$ b. $g(x) = \sqrt{3x^3 - 5x^2 - 38x + 40}$

2. Find all zeros of the polynomial $P(x) = 6x^3 - 19x^2 - 26x + 24$. Factor the polynomial with integer coefficients.

3. Let

$$f(x) = \begin{cases} x^2 - 16 & \text{if} & x > 4 \\ \sqrt{x} & \text{if} & 0 < x \le 4 \\ -1 & \text{if} & x \le 0 \end{cases}$$

   Draw the graph of f. Find:
   a. $f(4)$ b. $f(0)$ c. $f(1)$ d. $f(5)$ e. $f(-1)$

4. a. Solve for x in terms of natural logarithms:

$$2^{(3-x)/2} = 4 \cdot 3^x$$

   b. Solve for x:

$$e^{3 \ln(x-1)} = \ln e^{(3-x)}$$

5. A piece of machinery valued at \$32,000 is estimated to have a residual value of \$2,000 after 20 years. If it is depreciated linearly, express its value after t years as a function of t. What will be its value after 16 years?

6. Let $g(x) = \frac{1}{2}(x^2 - 4x + 6)$. Draw the graph of g, and identify the curve. Is g 1–1 on R? If not, give a restriction on the domain so that on the restricted domain g is 1–1, and find g^{-1}. What are the domain and range of g^{-1}?

7. Let $f(x) = 1 + e^{2x}$ and $g(x) = \ln(x-1)$. Find $f \circ g$ and $g \circ f$, and draw their graphs. Give the domain of each.

8. a. Show that the function

$$f(x) = \frac{1-2x}{3}$$

   is 1–1 on R, and find f^{-1}.
   b. Prove that the function $g(x) = \sqrt{x-1}$ is 1–1 from the set $A = \{x: \ x \ge 1\}$ *onto* the set R^+.

9. Use graphical means to solve the inequalities.
   a. $x(4-x) \ge -12$ b. $x^3 - 12x + 16 > 0$

10. Find x and y intercepts, vertical and horizontal asymptotes, and sketch.

   a. $y = \dfrac{9-x^2}{x^2 + 2x - 8}$ b. $y = \dfrac{x+4}{x^2 - 4}$

11. A right circular cone has base radius r and altitude h. Express the area of a cross-section taken x units above the base as a function of x.

12. Sketch the graph of the polynomial function

$$y = (x-2)(x+1)^2(x-5)^3(x+4)$$

13. Let $f(x) = 1 + |x-1|$. Give the domain and range of f, and draw its graph. Is f 1–1? Explain why or why not.

14. A certain radioactive substance has a half-life of 8 years. If 200 grams are present initially, how much will remain at the end of 12 years? How long will it be until only 10% of the original amount remains?

15. Let $f(t) = \sqrt{t-1}$ and $g(t) = t^2 - 3$. Find $f \circ g$ and $g \circ f$, give the domain and range of each, and draw their graphs.

16. Find all roots of the equation

$$4x^4 - 16x^3 + 9x^2 + 5x + 25 = 0$$

Make a table of values using synthetic division, and draw the graph of the given polynomial.

17. Solve for x and check:

$$2 \log_2|x - 1| = 2 - \log_2(x + 2)$$

18. By calculus it can be shown that the slope of the tangent line to the curve $x^2 - y^2 = 9$ at any point (x, y) on the curve is given by $m = x/y$. Find the equations of the tangent and normal lines at each point on the curve with abscissa 5. Draw the curve, showing these tangent and normal lines.

19. Let $f(x) = 1/\sqrt{x}$. Find

$$\frac{f(x + h) - f(x)}{h}$$

and express the answer with a rational numerator.

20. **a.** Show that by taking logarithms of both sides of the equation $y = a^x$ and making the substitution $Y = \log y$, Y is a linear function of x.
b. Similarly, in $y = x^a$ take logarithms, and let $X = \log x$ as well as $Y = \log y$, and show that Y is a linear function of X.
c. For $y = 2^x$ graph $\log y$ versus x, and for $y = x^2$ graph $\log y$ versus $\log x$.

21. The formula $T = T_0 + Ce^{-kt}$ gives the temperature T of a body t units of time after being placed in a medium of temperature T_0. A body initially at 80°C is placed in outside air at 25°C, and after 15 minutes it has cooled to 60°C. What will its temperature be after 1 hour? How long will it take to reach a temperature of 30°C?

22. The volume of oil that flows through a pipeline per day is jointly proportional to the velocity of flow and the square of the diameter of the pipe. If approximately 3 million barrels of oil per day flow through a pipe which is 3 feet in diameter when the velocity is 10 feet per second, how many barrels would flow per day through a pipe 2 feet in diameter if the velocity is 15 feet per second?

23. A manufacturer of solar energy cells finds that the cost, in dollars, of producing x units of a certain type per month is given by $30x + 1,500$, and he sets the selling price per unit, depending on the number of units sold, as $120 - 0.1x$. How many units should be manufactured and sold per month to maximize profit? What is the maximum profit?

24. Sketch the graph of each of the following. Identify the graph in each case, and state whether the graph is that of a function of x.
a. $y^2 + 2x - 4y = 0$ **b.** $y = -\sqrt{9 - x^2}$
c. $x^2 + 2y^2 = 4$ **d.** $4x^2 - 9y^2 + 36 = 0$
e. $y = 2x^2 - 3x - 5$

25. The illumination given by a light varies directly as the candlepower of the light and inversely as the square of the distance from the source. Two lights are placed 15 meters apart. One has four times as much candlepower as the other. An object is to be placed on the line between the lights in such a way that it receives equal illumination from the two lights. How far from the stronger light should it be placed?

26. Prove that the points $(3, 2)$, $(-1, -1)$, and $(6, -2)$ are vertices of an isosceles right triangle. Find the equation of the altitude drawn from the right angle vertex to the hypotenuse.

27. The cost in hundreds of dollars for a company to produce x tons of a certain type of fertilizer per day is given by $C = 4x^2 - 20x + 29$. Find the number of tons that should be produced each day to minimize cost. What is the minimum cost?

28. Find the equation of the tangent line to the circle $x^2 + y^2 - 4x + 2y - 15 = 0$ at the point $(4, 3)$.

29. Draw the graph of each of the following:

    a. $f(x) = \ln(1 + x^2)$ **b.** $g(x) = 1 - e^{-|x|}$

    c. $y = 2^{(x-1)/2}$ **d.** $y = 1 + \log_2|x - 1|$

30. Let $f(x) = \ln(x^2 - x) - \ln(x^2 + 3) + \ln 2$.

    a. Find the domain of f.

    b. Find the zeros of f.

    c. Find all vertical asymptotes to the graph of f.

    d. Discuss the behavior of f as x gets large in absolute value. What can you conclude concerning horizontal asymptotes?

    e. Sketch the graph of f.

8

Introduction to Trigonometric Functions

Just as is true with algebra, trigonometry is an indispensable tool in the solution of certain calculus problems. Often the concepts of algebra, trigonometry, and calculus are so interrelated that it is difficult to separate one from another. The following simple example from a calculus textbook illustrates the merger of trigonometry and calculus:*

Example 24-4. A cameraman is televising a 100-yard dash from a position 10 yards from the track in line with the finish line (Fig. 24-2). When the runners are 10 yards from the finish line, his camera is turning at the rate of $\frac{3}{5}$ radians per second. How fast are the runners moving then?

Figure 24–2

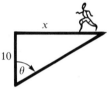

Solution. Figure 24-2 shows a runner x yards from the finish line, at which time the angle between the line joining the camera and the runners and the finish line extended to the camera is θ radians wide. We are told that $d\theta/dt = -\frac{3}{5}$ radians per second when $x = 10$, and we follow our outlined steps.

1. Figure 24-2 shows that $x = 10 \tan \theta$.
2. Therefore, according to the rule for differentiating $\tan \theta$ (Equation 8-10)

   (24-5)
   $$\frac{dx}{dt} = 10 \sec^2 \theta \frac{d\theta}{dt}$$

3. We are interested in the speed of the runners when $x = 10$. At that point $\theta = \frac{1}{4}\pi$, and hence $\sec \theta = \sqrt{2}$. So Equation 24-5 becomes

   $$\frac{dx}{dt} = 10(\sqrt{2})^2(-\tfrac{3}{5}) = -12$$

   and we see that our runners are approaching the finish line at the rate of 12 yards per second.

In this chapter we will learn the meaning of the term *radian*, why Step 1 of the example is true, and also where the information used in Step 3 comes from.

* Robert C. Fisher and Allen D. Ziebur, *Calculus and Analytic Geometry*, 3d ed. (Englewood Cliffs, N. J.: Prentice-Hall, 1975), p. 139. Reprinted by permission.

255

1   Introduction

The introduction of trigonometry as a separate area of study is generally believed to have been due to the Greek Hipparchus in the second century BC, but his works are lost. The first exposition of the subject still in existence comes from the work of Ptolemy, the Greek astronomer who lived in the second century AD. Rudiments of the subject are much older, however, and can be traced to Babylon and ancient Egypt. The principal use of trigonometry up until the fifteenth century was in astronomy, because it afforded a means of making indirect measurements. The name *trigonometry* was first used by Pitiscus (a German) in 1595 as the title of his book. The name is a combination of three Greek words meaning "three-angle measurement," or triangle measurement.

As the name implies, trigonometry originally meant a study of triangles, or more generally of angles of triangles, and its primary utility lay in the fact that by means of the trigonometric functions inaccessible quantities could be measured. In more modern times, while trigonometry still is of value for this purpose, it is the analytical aspects of the subject that make it indispensable in the study of many physical phenomena. We will indicate how it is useful in indirect measurement, but our emphasis will be on the analytical properties, since these are more important in calculus.

2   The Measurement of Angles

We all know what an angle is, yet it really is not very easy to define. When two **rays** emanate from the same point, we say they form an **angle** (Figure 1). We might define an angle, in fact, to be the set of points comprising two such rays. But there are some difficulties in this, since we want to distinguish between the interior angle shown in Figure 2a and the exterior angle shown in Figure 2b.

Figure 1   **Figure 2**

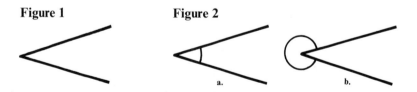

a.   b.

We usually get around this by drawing arcs indicating the chosen angle. It is sometimes useful to think of an angle as having been formed by two rays that were initially coincident, with one ray then rotated while the other remains fixed. This enables us to speak of **directed angles,** and they are indicated by adding an appropriate arrow to the arc. Figure 3 illustrates some possibilities.

Figure 3

a.   b.   c.   d.

An angle that is formed by a counterclockwise rotation is called **positive**; and by a clockwise rotation, **negative.**

The most familiar way of measuring the size of angles is by means of the **degree.*** If a circle is divided into 360 equal parts, then an angle formed by two rays from the center connecting two consecutive points is said to have a measure of one degree, indicated 1°. A **right angle** is an angle of 90°, and a **straight angle** is one of 180°. It is useful to observe properties of two other special angles, and for this purpose we make free use of some facts of geometry.

Consider an angle of 30° and form a triangle by dropping a perpendicular line from any point on either ray to the other, as in Figure 4. To facilitate discussion, we have labeled the vertices A, B, and C, and we will also speak of the angles A, B, C, meaning the interior angles formed by the sides meeting at these respective vertices. It is customary (but not mandatory) to use corresponding lowercase letters for the sides opposite the respective angles. Since the side BC was constructed perpendicular to AC in our figure, it follows (by the definition of perpendicularity) that C is a right angle. The side opposite C is called the **hypotenuse,** and the two shorter sides are often referred to as the **legs.** It is a fundamental property of triangles, proved in plane geometry, that the sum of the interior angles of any triangle is 180°. So angle B is 60°.

Figure 4

Figure 5

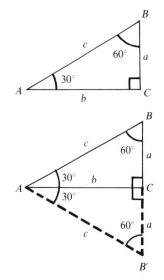

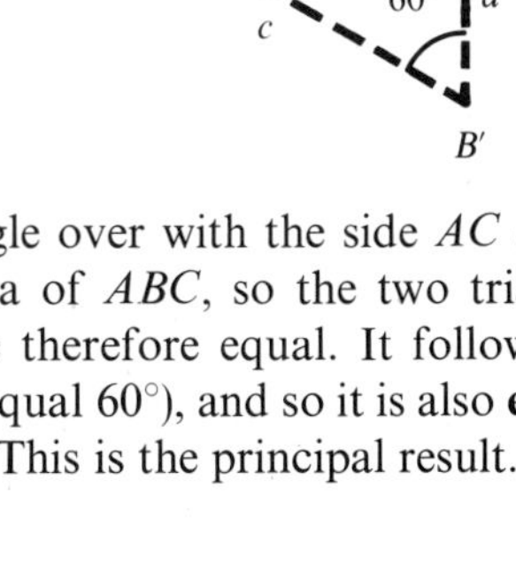

Now we flip this triangle over with the side AC as an axis (Figure 5). The triangle $AB'C$ is a replica of ABC, so the two triangles are congruent and corresponding angles are therefore equal. It follows that the triangle ABB' is **equiangular** (all angles equal 60°), and so it is also **equilateral** (all sides equal). Thus, $2a = c$, or $a = c/2$. This is the principal result. Returning to our original

---

* No one knows for sure the origin of the degree measurement of angles, but it can be traced back to the ancient Sumerians (4000–2000 BC). One theory is that 360 was used because of their erroneous calculation of 360 days in a year. They and the Babylonians (2000–600 BC) used 60 as a base in their arithmetic.

triangle, it says that the side opposite the 30° angle is one-half of the hypotenuse. This result should be committed to memory:

> In a **30°–60° right triangle,** the side opposite the 30° angle is equal to one-half of the hypotenuse.

There is another fact of fundamental importance about right triangles (that is, a triangle in which one angle is a right angle):

THE PYTHAGOREAN THEOREM

The sum of the squares of the lengths of the two legs of a right triangle equals the square of the length of the hypotenuse.

This is also proved in plane geometry. If we label a right triangle as before (but with angles A and B left unspecified, Figure 6), then we can write the Pythagorean theorem as

$$a^2 + b^2 = c^2$$

The converse of this theorem is also true; that is, if for a triangle having sides a, b, and c it is true that $a^2 + b^2 = c^2$, then the triangle is a right triangle with the 90° angle opposite side c.

Figure 6   **Figure 7**

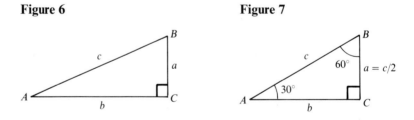

Returning to our 30°–60° triangle (Figure 7), we can now determine the side opposite the 60° angle. Since $a = c/2$ and $a^2 + b^2 = c^2$, we have $(c/2)^2 + b^2 = c^2$. So

$$b^2 = \frac{3c^2}{4}$$

$$b = \frac{\sqrt{3}}{2} c$$

So the side opposite the 60° angle is $\sqrt{3}/2$ times the hypotenuse.

Rather than memorizing this, it is probably easier just to use the Pythagorean theorem each time. For example, suppose we are given a 30°–60° right triangle with hypotenuse equal to 8 (Figure 8) and we wish to find the legs. We know that the side opposite the 30° angle is one-half the hypotenuse, so $a = 4$. By the Pythagorean theorem,

$$(4)^2 + b^2 = (8)^2$$
$$b^2 = 64 - 16 = 48$$
$$b = \sqrt{48} = \sqrt{16 \cdot 3} = 4\sqrt{3}$$

Figure 8

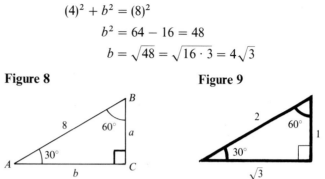

Figure 9

If the hypotenuse is 2 in a 30°–60° right triangle (Figure 9), we get a particularly easy combination to remember. It is useful to commit this to memory.

Another particularly easy right triangle to analyze is the isosceles right triangle. Since the legs are equal, the angles opposite them must be equal and hence each is equal to 45° (Figure 10). By the Pythagorean theorem,

$$c^2 = a^2 + b^2 = a^2 + a^2 = 2a^2$$

so,

$$c = \sqrt{2}a$$

Figure 10

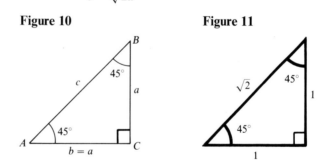

Figure 11

In particular, if $a = 1 = b$ (Figure 11), we have $c = \sqrt{2}$. This figure, too, should be memorized.

EXAMPLE 1   Find the unknown sides of the right triangle ABC for which:
a. $A = 60°$, $c = 10$ **b.** $B = 45°$, $a = 8$ **c.** $A = 30°$, $b = 12$

Solution   It is useful to sketch the triangle in each case.
a. Since $A = 60°$, B must be 30°. The side opposite the 30° angle is one-half the hypotenuse. So $b = 5$. Using the Pythagorean theorem,

$$a^2 + b^2 = c^2$$
$$a^2 + 25 = 100$$
$$a^2 = 75$$
$$a = 5\sqrt{3}$$

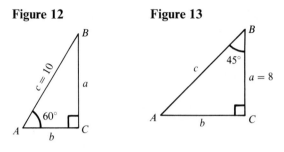

Figure 12   **Figure 13**

b. Since $B = 45°$, A also is $45°$, and the triangle is isosceles. So $b = 8$. By the Pythagorean theorem,

$$c^2 = a^2 + b^2 = 64 + 64 = 128$$
$$c = 8\sqrt{2}$$

c. The side a is equal to $c/2$. So by the Pythagorean theorem,

$$\left(\frac{c}{2}\right)^2 + b^2 = c^2$$
$$\frac{c^2}{4} + 144 = c^2$$
$$c^2 + 576 = 4c^2$$
$$3c^2 = 576$$
$$c^2 = \frac{576}{3} = 192$$
$$c = \sqrt{192} = \sqrt{3(64)} = 8\sqrt{3}$$
$$a = 4\sqrt{3}$$

Figure 14

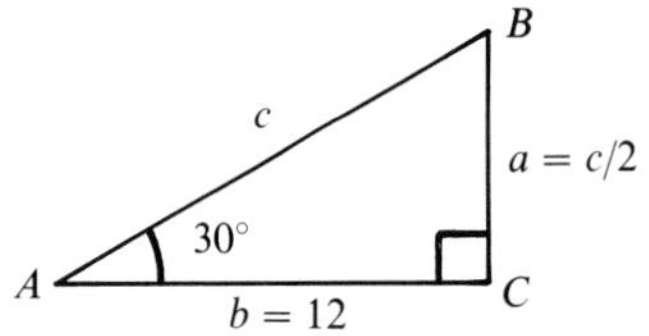

While the measurement of angles by degrees has the weight of history and the advantage of familiarity on its side, there is another method of measurement that, as will be seen later, is more natural and much more useful in the study of calculus. This is the **radian.** Consider an angle and construct any circle with its center at the vertex of the angle (Figure 15). Designate the angle by θ and the arc of the circle subtended by the angle by s. If r is the radius of the circle, then we define the radian measure of θ to be the ratio s/r. It would appear that this definition is dependent on which circle we use. But this is not the case, because if we take another circle of radius r' and subtended arc s', say, then since the two circular sectors are similar (just as if they were triangles), we have $s/r = s'/r'$. So we get the same number regardless of the radius. Often, the same

Figure 15

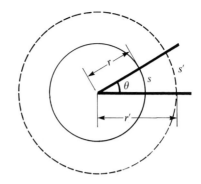

symbol is used for both the name of the angle and the radian measure of the angle. This usually causes no confusion. Thus, in this case, we write

$$\theta = \frac{s}{r} \qquad (1)$$

As a practical matter, it is useful to write this in the equivalent form

$$s = r\theta \qquad (2)$$

since angles and straight lines can usually be measured more easily than circular arcs. Formula (2) says that the length of arc subtended on a circle of radius r by an angle of θ *radians* (not degrees) equals the product $r\theta$. This should be remembered.

Another observation is that according to equation (1), the radian measure of θ is the ratio of two lengths, and hence the radian is dimensionless. In analyzing the dimensional properties of an expression, then, if an angle in radians appears in any way, the dimensionality is not affected.

According to equation (1), an angle of 1 radian subtends, on any circle with its center at the vertex, an arc equal in length to the radius (Figure 16). This

Figure 16

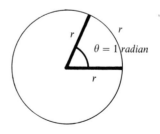

helps us to visualize the relative size of a radian. We know that the circumference of a circle is $2\pi r$. That is, we could mark off 2π arcs on the circle each equal in length to the radius (since $\pi \approx 3.1416$, we could mark off six full arcs equal to the radius and have a little more than a quarter of a radius left over). But for each arc on the circle of length r, there is an angle of 1 radian at the center. It follows then that there are 2π radians in one revolution. Since there are also 360° in one revolution, we have the following relationships:

$$2\pi \text{ radians} = 360°$$

$$\boldsymbol{\pi \textbf{ radians} = 180°}$$

$$1 \text{ radian} = \frac{180°}{\pi} \approx 57.3°$$

$$1 \text{ degree} = \frac{\pi}{180} \text{ radians}$$

Probably the easiest of these to remember is the second, π radians $= 180°$, and the others can be obtained from it. In the future, unless the degree symbol is used in speaking of the measure of an angle, we will automatically mean radian measure. Thus, if we write $\theta = \pi$, $\theta = \pi/6$, or even $\theta = 2$, we will mean π radians, $\pi/6$ radians, or 2 radians, respectively. We will almost always write radian measure as multiples or fractional parts of π. This is because these are particularly convenient angles. Since π radians $= 180°$, we have also that

$$\frac{\pi}{2} \text{ radians} = 90° \qquad \frac{\pi}{3} \text{ radians} = 60°$$

$$\frac{\pi}{4} \text{ radians} = 45° \qquad \frac{\pi}{6} \text{ radians} = 30°$$

But do not suppose that π must be used when giving the radian measure of an angle; angles of $\frac{1}{2}$ radian or 10 radians are perfectly valid.

EXAMPLE 2   **a.**   Convert $120°$ to radians.   **b.**   Convert $7\pi/6$ radians to degrees.

Solution   **a.**   Since $1° = \pi/180$ radians, we multiply 120 by $\pi/180$:

$$\overset{2}{\cancel{120}} \cdot \frac{\pi}{\underset{3}{\cancel{180}}} = \frac{2\pi}{3} \text{ radians}$$

b.   Since 1 radian $= 180°/\pi$, we multiply by $180/\pi$:

$$\frac{7\pi}{\cancel{6}} \cdot \frac{\overset{30}{\cancel{180}}}{\cancel{\pi}} = 210°$$

EXAMPLE 3   Find the arc length on a circle of radius 7 feet subtended by a central angle of $60°$.

Solution   We use $s = r\theta$, but first we must find θ in radians:

$$\theta = 60° \cdot \frac{\pi}{180°} = \frac{\pi}{3}$$

So,

$$s = 7 \cdot \frac{\pi}{3} = \frac{7\pi}{3} \text{ feet} \approx 7.33 \text{ feet}$$

EXAMPLE 4   A wheel of diameter 4 feet is rotating at 1,200 rpm. Find the linear velocity of a point on the rim.

Solution   The radius of the wheel is 2 feet. Since each revolution contains 2π radians, the wheel is turning at $2\pi(1,200) = 2,400\pi$ radians per minute $= 40\pi$ radians per second. This says that a central angle θ, as shown in Figure 17, is increasing by 40π radians each second. What we want to know is how fast the arc length s is increasing (since a point on the rim traces out such an arc). We know that

$$s = r\theta = 2\theta$$

Figure 17

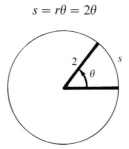

If in t seconds the point travels s feet around the arc, then $v = s/t$ is the average velocity for that interval of time. Similarly, θ/t is the average number of radians turned in t seconds. In this problem the number of radians turned per second is constant and equals 40π. Since $s = 2\theta$, we have

$$v = \frac{s}{t} = 2\frac{\theta}{t} = 2(40\pi) = 80\pi \text{ feet per second} = 251.33 \text{ feet per second}$$

EXERCISE SET 2

A   In Problems 1–20 find the unknown sides of the triangle ABC, with right angle at C.

1. $A = 30°$, $c = 5$
2. $B = 60°$, $a = 3$
3. $A = 60°$, $b = 10$
4. $A = 30°$, $b = 4\sqrt{3}$
5. $A = 30°$, $b = 6$
6. $B = 30°$, $a = 12$
7. $B = 45°$, $c = 4\sqrt{2}$
8. $A = 45°$, $c = 4$
9. $A = 60°$, $c = 5$
10. $B = 60°$, $b = 3$
11. $B = 45°$, $a = 3$
12. $A = 60°$, $a = 8$
13. $a = 2$, $b = 3$
14. $a = 5$, $c = 7$
15. $a = 3$, $b = 4$
16. $b = 12$, $c = 13$
17. $a = 7$, $c = 25$
18. $a = \sqrt{3}$, $b = 1$
        Also find angles A and B.
19. $a = 1$, $b = 2\sqrt{2}$
20. $a = 8$, $c = 17$

In Problems 21–23 the degree measure of an angle is given. Find the radian measure. (Leave answers in terms of π.)

21. **a.** $150°$ **b.** $225°$ **c.** $50°$ **d.** $600°$ **e.** $72°$
22. **a.** $144°$ **b.** $15°$ **c.** $270°$ **d.** $36°$ **e.** $330°$
23. **a.** $315°$ **b.** $-135°$ **c.** $240°$ **d.** $570°$ **e.** $80°$

In Problems 24–26 give the degree measure of the angles having the given radian measures.

24. **a.** $3\pi/4$ **b.** $5\pi/3$ **c.** 4π **d.** 2 **e.** $5\pi/2$
25. **a.** $5\pi/4$ **b.** $7\pi/3$ **c.** $8\pi/9$ **d.** $-7\pi/2$ **e.** $3\pi/10$
26. **a.** $5\pi/6$ **b.** $-4\pi/3$ **c.** $5\pi/9$ **d.** 3 **e.** $3\pi/2$

27. In each of the following, θ is a central angle in a circle of radius r, and the arc on the circle subtended by θ is s. Find the value (s, r, or θ) that is not given.

    a. $r = 3$, $\theta = 2$ **b.** $s = 4$, $\theta = 3$ **c.** $r = 5$, $s = 12$
    d. $r = 2$, $\theta = 30°$ **e.** $s = 8$, $\theta = 45°$

28. Find the arc length on the equator (in miles) subtended by an angle of $1°$ at the center of the earth (assume the radius of the earth is 3,960 miles).

29. What is the length of the arc swept out by the tip of the minute hand of a clock during the time interval from 6:00 to 6:20 if the minute hand is 4 inches long?

30. The diameter of the steering wheel of a car is 15 inches, and it has three equally spaced spokes. Find the distance on the steering wheel between consecutive spokes.

31. A wheel is revolving at an angular velocity of $5\pi/3$ radians per second. Find the number of revolutions per minute (rpm) through which the wheel is turning.

32. A flywheel is turning at 1,800 rpm. Through how many radians per second is it turning?

B 33. The earth completes one revolution about its axis in 24 hours. Find the velocity in miles per hour of a point on the equator (take $r = 3,960$ miles). What is the velocity in feet per second?

34. A train is traveling on a circular curve of $\frac{1}{2}$ mile radius at the rate of 30 miles per hour. Through what angle (in radians) will the train turn in 45 seconds? What is the angle in degrees?

35. A flywheel 4 feet in diameter is revolving at the rate of 50 rpm. Find the speed of a point on the rim in feet per second.

36. Assume the earth moves around the sun in a circular orbit with radius 93,000,000 miles. A complete revolution takes approximately 365 days. Find the approximate speed of the earth in its orbit in miles per hour.

Problem 37 and 38 are taken from an engineering textbook.*

37. Two points B and C lie on a radial line of a rotating disk. The points are 2 in. apart. $V_B = 700$ fpm [feet per minute] and $V_C = 880$ fpm. Find the radius of rotation for each of these points.

38. The tire of an automobile has an outside diameter of 27 in. If the rpm of the wheel is 700, determine the speed of the automobile (a) in miles per hour, (b) in feet per second, and (c) the angular speed of the wheel in radians per second.

39. By following the given steps, prove the Pythagorean theorem.
    a. Consider a right triangle with legs of lengths a and b and hypotenuse of length c. Construct a square of side $a + b$ as shown at the top of page 265, and draw four replicas (I, II, III, and IV) of the given triangle inside the square as shown.
    b. Show that the quadrilateral V is a square.
      Hint. Work with angles.
    c. Write the equation that puts into mathematical form the fact that the sum of the areas of I, II, III, IV, and V equals the area of the large square.
    d. By simplifying the equation in part c, draw the desired conclusion.

---

* George H. Martin, *Kinematics and Dynamics of Machines*, rev. printing (New York: McGraw-Hill Book Company, 1969), p. 43. Reprinted by permission.

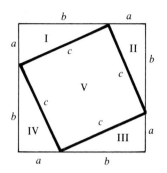

40. Follow the steps given to prove the converse of the Pythagorean theorem.
   a. Let ABC be a triangle in which $c^2 = a^2 + b^2$.
   b. Construct a right triangle $A'B'C'$ having legs a and b. Let the hypotenuse be of length c'. Use the Pythagorean theorem to express c' in terms of a and b.
   c. How can you conclude that $c' = c$?
   d. What does this say about triangle ABC compared with triangle $A'B'C'$?
   e. What do you conclude about angle C?

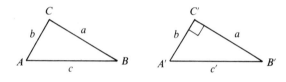

3 Right Triangle Trigonometry

We want to introduce next the trigonometric functions. We will do this in more than one way, since each way is useful. The first definition is the historical one, which still has great utility. Consider any right triangle and select either acute angle of that triangle. If the angle is called θ, then we define each function as shown in the box.

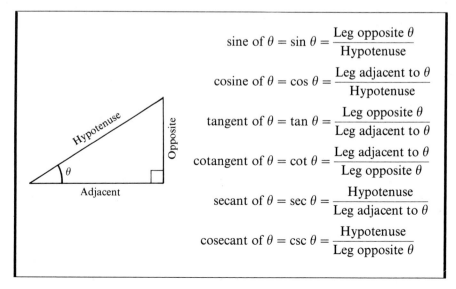

$$\text{sine of } \theta = \sin \theta = \frac{\text{Leg opposite } \theta}{\text{Hypotenuse}}$$

$$\text{cosine of } \theta = \cos \theta = \frac{\text{Leg adjacent to } \theta}{\text{Hypotenuse}}$$

$$\text{tangent of } \theta = \tan \theta = \frac{\text{Leg opposite } \theta}{\text{Leg adjacent to } \theta}$$

$$\text{cotangent of } \theta = \cot \theta = \frac{\text{Leg adjacent to } \theta}{\text{Leg opposite } \theta}$$

$$\text{secant of } \theta = \sec \theta = \frac{\text{Hypotenuse}}{\text{Leg adjacent to } \theta}$$

$$\text{cosecant of } \theta = \csc \theta = \frac{\text{Hypotenuse}}{\text{Leg opposite } \theta}$$

Note that we have departed from the standard functional notation. This is so that the traditional symbolism can be used. Note, too, that in these definitions the domain in each case consists of all acute angles, while the range is a subset of *R*.

Thus, in the triangle *ABC* (Figure 18), we have the following:

$$\sin A = \frac{a}{c} \qquad \csc A = \frac{c}{a}$$

$$\cos A = \frac{b}{c} \qquad \sec A = \frac{c}{b}$$

$$\tan A = \frac{a}{b} \qquad \cot A = \frac{b}{a}$$

Figure 18

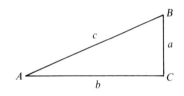

Also:

$$\sin B = \frac{b}{c} \qquad \csc B = \frac{c}{b}$$

$$\cos B = \frac{a}{c} \qquad \sec B = \frac{c}{a}$$

$$\tan B = \frac{b}{a} \qquad \cot B = \frac{a}{b}$$

The sine and cosine are called **cofunctions;** each is the confunction of the other. The same is true of the tangent and cotangent and the secant and cosecant. Also, each function of angle *A* is the cofunction of angle *B*, and conversely. The sum of angles *A* and *B* is 90°, and such angles are said to be **complementary. So, any function of an acute angle equals the cofunction of its complement.**

It is important that the functions be learned in terms of the words *opposite*, *adjacent*, and *hypotenuse*, rather than any particular set of letters, since the letters may change. For example, in Figure 19 the six functions of angle *R*

Figure 19

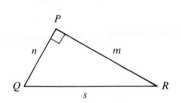

are as follows:

$$\sin R = \frac{n}{s}\frac{\text{Opposite}}{\text{Hypotenuse}} \qquad \csc R = \frac{s}{n}\frac{\text{Hypotenuse}}{\text{Opposite}}$$

$$\cos R = \frac{m}{s}\frac{\text{Adjacent}}{\text{Hypotenuse}} \qquad \sec R = \frac{s}{m}\frac{\text{Hypotenuse}}{\text{Adjacent}}$$

$$\tan R = \frac{n}{m}\frac{\text{Opposite}}{\text{Adjacent}} \qquad \cot R = \frac{m}{n}\frac{\text{Adjacent}}{\text{Opposite}}$$

It would appear that the definitions of the six trigonometric functions depend not only on the angle selected but on the particular right triangle that has the angle as one of its acute angles. However, the triangle is not important, as the following reasoning shows: Let θ denote any angle between $0°$ and $90°$ and consider any two right triangles formed by constructing perpendicular lines from one side of θ to the other, as shown in Figure 20. Since triangle ABC is similar to triangle $AB'C'$, it follows that corresponding sides are proportional. That is, $a/c = a'/c'$, $b/c = b'/c'$, and $a/b = a'/b'$. Thus, for example,

$$\sin \theta = \frac{a}{c} = \frac{a'}{c'}$$

Figure 20

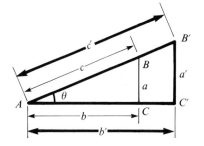

and similarly with all the functions of θ. It follows, then, that the functions of θ do not depend on which particular triangle is used to calculate the ratios. Thus, the trigonometric functions really depend only on the angle itself.

Extensive tables (such as those in the back of this book) of the six trigonometric functions of angles between $0°$ and $90°$ have been calculated. As you might guess, these calculations were not made by drawing right triangles and measuring sides. Much more sophisticated and accurate techniques were used. You will learn in the study of calculus about such techniques. But now that the tables are available, we can put them to good use. The modern scientific hand calculator, however, provides a more efficient and more accurate means of finding trigonometric functions and performing calculations with them. You should become familiar with the use of both tables and calculators.

Degrees are divided into minutes, with 60 minutes being equal to $1°$. Minutes are designated by the symbol $'$. Thus, $60' = 1°$. Minutes are further divided into seconds ($''$), with $60'' = 1'$. Tables customarily express angles either in radians or in degrees and minutes, such as 25°30'. Most scientific calculators also express angles in radians or in degrees, but fractions of degrees are given as decimals, rather than minutes. For example, on most calculators 25°30' would have to be expressed as 25.5°. In the examples and problems that follow, we will sometimes write angles using decimals and sometimes using minutes. You should be able to change from one form to the other as needed.

EXAMPLE 5   Solve the right triangle ABC shown in Figure 21, in which $A = 37°$ and $b = 17.3$.

Figure 21

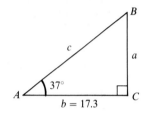

Solution   To solve a triangle means to find all missing parts. Since the acute angles are complementary, we get

$$B = 90° - 37° = 53°$$

To find a and c we select appropriate trigonometric functions, each involving only one unknown quantity. For a, we use

$$\tan A = \frac{a}{b}$$

$$a = b \tan A$$
$$= 17.3 \tan 37°$$
$$= 17.3(0.7536)$$
$$\approx 13.0$$

To get c we use

$$\cos A = \frac{b}{c}$$

$$c = \frac{b}{\cos A}$$
$$= \frac{17.3}{\cos 37°}$$
$$= \frac{17.3}{0.7986}$$
$$\approx 21.7$$

The next two examples illustrate how the trigonometric functions may be employed to calculate inaccessible distances.

EXAMPLE 6 Figure 22 shows a river and two points P and Q on opposite sides. It is desired to find the distance from P to Q.

Figure 22

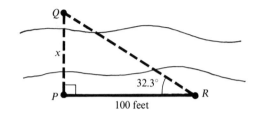

Solution By means of a transit at P, a line of sight perpendicular to PQ is determined, and a distance of 100 feet along this line from P to R is measured. At R the transit is again set up and the angle from RP to RQ is measured and found to be 32.3°. The calculation can now be done as follows:

$$\tan 32.3° = \frac{x}{100}$$
$$x = 100 \tan 32.3°$$
$$= 100(0.6322)$$
$$\approx 63.2 \text{ feet}$$

Remarks. We selected the tangent function because it involved the side opposite the known angle, namely, x, the side we were seeking, as well as the length of the adjacent side, 100, which was known. So the equation expressing the tangent of 32.3° has only one unknown in it. The cotangent could also have been used, but this would have involved a division instead of a multiplication (cot 32.3° = $100/x$, so $x = 100/\text{cot } 32.3°$). In selecting the function to use one is usually guided by the following considerations:

1. It should involve the unknown.
2. It should involve only known information—if possible.
3. In a choice between two functions satisfying considerations 1 and 2, normally one should choose the function that involves multiplication rather than division (although if a calculator is being used, this is of little consequence).

Also note that we rounded off the answer to three significant figures since the measurements involved are unlikely to be of greater accuracy than this.

EXAMPLE 7   Figure 23 shows measurements from which the height of a mountain is to be determined. Points A and B are assumed to be on the same level.

Figure 23

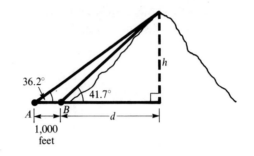

Solution   The **angles of elevation** of the top of the mountain are read at A and B and are as shown. We have introduced the auxiliary unknown d in the figure, since it will be involved in the functions we use. There are two right triangles, each of which involves the unknown height h. In each triangle, h is the side opposite the known angle, and we have some information about the adjacent side. We could therefore choose either the tangent or the cotangent to work with, because both involve h and d. By using both triangles, we will get two equations in these two unknowns. Our object will be to eliminate d between the two equations, thereby obtaining an equation in h alone. We choose the cotangent function, since this facilitates solving for d:

$$\cot 36.2° = \frac{1,000 + d}{h} \qquad \cot 41.7° = \frac{d}{h}$$

$$1,000 + d = h \cot 36.2° \qquad\qquad d = h \cot 41.7°$$

Now we substitute for d from the last equation into the second and then solve for h:

$$1,000 + h \cot 41.7° = h \cot 36.2°$$

$$h(\cot 36.2° - \cot 41.7°) = 1,000$$

$$h = \frac{1,000}{\cot 36.2° - \cot 41.7°}$$

The values in the denominator can be found either from a table or by a calculator. We obtain

$$h = \frac{1,000}{1.3663 - 1.1224}$$

$$= \frac{1,000}{0.2440}$$

$$\approx 4,099$$

So the mountain is approximately 4,100 feet above the surrounding countryside. If the elevation of points A and B above sea level is known, this can be added to give the height of the mountain above sea level.

EXERCISE SET 3

A In Problems 1–16 solve the triangles. The right angle in each case is C.

1. $A = 57.4°$, $b = 15$
2. $a = 32$, $b = 12$
3. $B = 67°$, $b = 13.5$
4. $A = 33°24'$, $c = 120$
5. $A = 49°$, $c = 24$
6. $B = 37°$, $c = 14$
7. $A = 32.5°$, $a = 24.7$
8. $B = 64.3°$, $a = 205$
9. $a = 15.35$, $c = 32.14$
10. $B = 62°18'$, $c = 104$
11. $B = 22°40'$, $b = 2.56$
12. $A = 43°30'$, $b = 10.5$
13. $A = 35.5°$, $a = 3.46$
14. $a = 32.5$, $b = 21.2$
15. $b = 25.4$, $c = 38.2$
16. $a = 5.1$, $c = 8.4$

17. In the triangle shown the right angle is at M. Write all six trigonometric functions of:
    a. Angle L **b.** Angle K

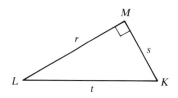

18. From a point on the ground 150 feet from the base of a building the angle of elevation of the top of the building is $68°$. How high is the building?
19. A rectangle is 14 inches long and 8 inches high. Find the angle between a diagonal and the base. What is the length of the diagonal?
20. From the top of a lighthouse the **angle of depression** (angle from the horizontal down to the line of sight) of a boat is $32.4°$. If the lighthouse is 80 feet high, how far is the boat from the base of the lighthouse?
21. A tower 102 feet high is on the bank of a river. From the top of the tower the angle of depression to a point on the opposite side of the river is $28°14'$. How long a cable would be required to reach from the top of the tower to the point on the opposite bank?
22. A guy wire for a telephone pole is anchored to the ground at a point 21 feet from the base of the pole, and it makes an angle of $64°$ with the horizontal. Find the height of the point where the guy wire is attached to the pole and the length of the guy wire.
23. A surveyor wants to find the distance between points A and B on opposite sides of a pond. With his transit at point A he determines a line of sight at right angles to AB and establishes point C along this line at a distance 150 feet from A. Then he measures the angle at C from AC to BC and finds it to be $62°10'$. How far is it from A to B?
24. Town A is 12 miles due west of town B, and town C is 15 miles due south of town B. How far is it from town A to town C? Assuming straight roads connect the three towns, what is the angle at A between the roads leading to B and C?

25. A bar with a pentagonal (five-sided) cross-section is to be milled from round stock. What diameter should the stock be if the dimension of each flat section is to be $2\frac{1}{2}$ inches? See the figure.

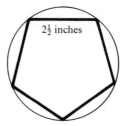

26. A 1,000 pound weight is resting on an inclined plane that makes an angle of 37° with the horizontal. Find the components of the weight parallel to and perpendicular to the surface of the plane. See the figure.

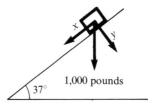

B 27. The accompanying sketch shows the angles of elevation of a balloon at a certain instant from points A and B, which are 1,200 feet apart. Find the height of the balloon at that instant.

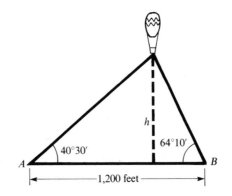

28. From a lighthouse 120 feet high, the angles of depression of two boats directly in line with the lighthouse are found to be 34°30′ and 27°40′. How far apart are the boats?

29. Find the length of 1° of longitude at the latitude of Chicago, 41°50′ (assume the radius of the earth is 3,960 miles).

30. At a point 250 feet from the base of a building, the angle of elevation of the top of the building is 27.5°, and the angle of elevation of the top of a statue on the top edge of the building is 30.2°. How high is the statue?

4 Trigonometric Functions of General Angles

Right triangle methods can also be used to solve oblique triangles (that is, triangles having no right angle), but before illustrating the techniques involved, it will be necessary to extend our definitions of the trigonometric functions to arbitrary angles, not just those between $0°$ and $90°$. To do this, let us return to the idea of an angle as having been formed by two rays, one of which has been rotated from an initial position coincident with the other. The fixed ray is called the **initial side** of the angle and the rotated ray is the **terminal side.** Given such an angle, we superimpose x and y axes so that the positive x axis coincides with the initial side and the vertex of the angle is at the origin. A typical situation is shown in Figure 24. Having done this, we select any point on the terminal side of the angle.* In general, let us denote this by (x, y). Let the distance along the terminal side of θ be designated by r (the **radius vector**).

Figure 24

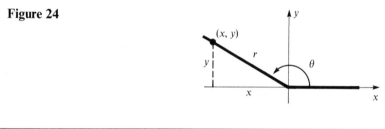

DEFINITION 1

$$\sin \theta = \frac{y}{r} \qquad \csc \theta = \frac{r}{y}$$

$$\cos \theta = \frac{x}{r} \qquad \sec \theta = \frac{r}{x}$$

$$\tan \theta = \frac{y}{x} \qquad \cot \theta = \frac{x}{y}$$

In this definition it will be seen that the domains of all functions except the sine and cosine are restricted. For example, the tangent and the secant are not defined for $90°$ or $270°$ or for angles **coterminal** (that is, having the same terminal side) with these, since for such angles $x = 0$ and the ratios y/x and r/x do not exist. Similarly, the cotangent and cosecant are not defined for $0°$ or $180°$ or for angles coterminal with these.

Notice that by dropping a perpendicular line from (x, y) to the x axis a triangle is formed, and the definitions above seem to be what we have previously used for the functions of the acute angle between the terminal side of θ and the x axis. There is one significant difference; in the present case, one or both

---

* That the choice of the particular point on the terminal side of θ is immaterial follows from the fact that because of similar triangles, the ratios given in the definitions that follow remain the same for all points (except the origin, which we exclude).

of the numbers x and y may be negative, whereas in our earlier definition all sides of the triangle were positive. We name the acute angle between the terminal side of θ and the x axis the **reference angle** for θ. We can therefore say that the functions of θ are *numerically* the same as those of its reference angle; only the correct sign must be affixed.

Let us consider some examples.

EXAMPLE 8   Find the six trigonometric functions of $150°$.

Solution   See Figure 25. The reference angle is $30°$, so we can choose $r = 2$ and have $y = 1$ and $x = -\sqrt{3}$. Notice that x is negative, since the terminal side lies in the second quadrant. We have

$$\sin 150° = \frac{1}{2} \qquad \csc 150° = 2$$

$$\cos 150° = -\frac{\sqrt{3}}{2} \qquad \sec 150° = -\frac{2}{\sqrt{3}}$$

$$\tan 150° = -\frac{1}{\sqrt{3}} \qquad \cot 150° = -\sqrt{3}$$

Observe that we can think in terms of opposite, adjacent, and hypotenuse, although these are not strictly applicable for the $150°$ angle, nor even for the $30°$ reference angle, since the functions of $30°$ would all be positive. Nevertheless, using this terminology is a good memory device.

Figure 25

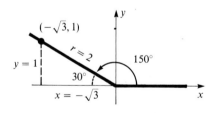

Whenever we are dealing with an angle having $30°$, $45°$, or $60°$ as a reference angle, we can find the functions readily by remembering the two basic triangles shown in Figure 26.

Figure 26

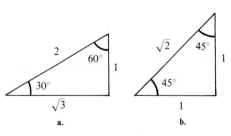

EXAMPLE 9   Find the six trigonometric functions of 225°.

Solution   See Figure 27. The reference angle is 45° and the terminal side is in the third quadrant. So we can use our 45° right triangle, noting that both x and y are negative. So,

$$\sin 225° = -\frac{1}{\sqrt{2}} \qquad \csc 225° = -\sqrt{2}$$

$$\cos 225° = -\frac{1}{\sqrt{2}} \qquad \sec 225° = -\sqrt{2}$$

$$\tan 225° = 1 \qquad \cot 225° = 1$$

Figure 27

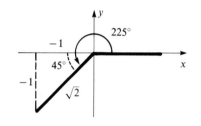

EXAMPLE 10   Find all trigonometric functions of 300°.

Solution   Our 30°–60° triangle occurs in the orientation shown in Figure 28, since 60° is the reference angle. Thus,

$$\sin 300° = -\frac{\sqrt{3}}{2} \qquad \csc 300° = -\frac{2}{\sqrt{3}}$$

$$\cos 300° = \frac{1}{2} \qquad \sec 300° = 2$$

$$\tan 300° = -\sqrt{3} \qquad \cot 300° = -\frac{1}{\sqrt{3}}$$

Figure 28

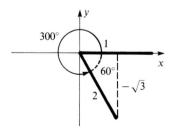

These examples illustrate how we can quickly find the functions of 30°, 45°, 60°, 120°, 135°, 150°, 210°, 225°, 240°, 300°, 315°, and 330°. We can, in fact, find functions of any other angles coterminal with any of these. For example, 510° is coterminal with 150°, and −120° is coterminal with 240°. Figure 29 illustrates this. All that really matters regarding the values of the trigonometric functions is the location of the terminal side, not how it got there. (But do not make the mistake of saying, for example, that 510° = 150°.)

Figure 29

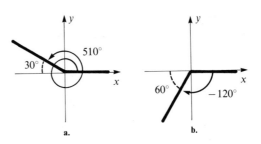

a.   b.

There are four more special angles for which the trigonometric functions are readily obtained: 0°, 90°, 180°, and 270° (and any angles coterminal with these). These are called **quadrantal angles.** For these angles we have no right triangles to work with, but Definition 1 can be applied, as shown in the boxes. Since the point on the terminal side is optional, we select one for which $r = 1$ in each case. Since x or y will be 0 for each of the angles, some of the functions will be undefined, as has been noted earlier.

$$\sin 0° = \frac{y}{r} = \frac{0}{1} = 0$$

$$\cos 0° = \frac{x}{r} = \frac{1}{1} = 1$$

$$\tan 0° = \frac{y}{x} = \frac{0}{1} = 0$$

$$\cot 0° = \frac{x}{y} = \frac{1}{0} \quad \text{undefined}$$

$$\sec 0° = \frac{r}{x} = \frac{1}{1} = 1$$

$$\csc 0° = \frac{r}{y} = \frac{1}{0} \quad \text{undefined}$$

0° (1, 0)

$x = 1$
$y = 0$
$r = 1$

$$\sin 90° = 1$$

$$\cos 90° = 0$$

$$\tan 90°, \quad \text{undefined}$$

$$\cot 90° = 0$$

$$\sec 90°, \quad \text{undefined}$$

$$\csc 90° = 1$$

(0, 1)

90°

$x = 0$
$y = 1$
$r = 1$

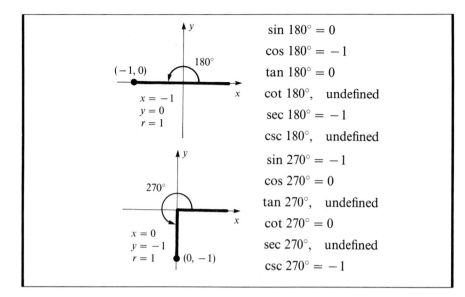

$$\sin 180° = 0$$
$$\cos 180° = -1$$
$$\tan 180° = 0$$
$$\cot 180°, \quad \text{undefined}$$
$$\sec 180° = -1$$
$$\csc 180°, \quad \text{undefined}$$

$$\sin 270° = -1$$
$$\cos 270° = 0$$
$$\tan 270°, \quad \text{undefined}$$
$$\cot 270° = 0$$
$$\sec 270°, \quad \text{undefined}$$
$$\csc 270° = -1$$

It is important to learn the special angles we have considered in terms of radians, so we list some corresponding values in Table 1.

Table 1

Degree measure	Radian measure	Degree measure	Radian measure
0°	0	180°	π
30°	$\pi/6$	210°	$7\pi/6$
45°	$\pi/4$	225°	$5\pi/4$
60°	$\pi/3$	240°	$4\pi/3$
90°	$\pi/2$	270°	$3\pi/2$
120°	$2\pi/3$	300°	$5\pi/3$
135°	$3\pi/4$	315°	$7\pi/4$
150°	$5\pi/6$	330°	$11\pi/6$

EXERCISE SET 4

A In Problems 1–14 find all six trigonometric functions of the angle given without using tables or a calculator.

1. $5\pi/4$ **2.** $330°$ **3.** $-4\pi/3$ **4.** $7\pi/6$

5. π **6.** $225°$ **7.** $270°$ **8.** $-5\pi/6$

9. $3\pi/4$ **10.** $150°$ **11.** $480°$ **12.** $10\pi/3$

13. 3π **14.** $600°$

In Problems 15–19 evaluate each function without tables or a calculator.

15. **a.** $\sin 120°$ **b.** $\cos 315°$ **c.** $\tan 210°$
  d. $\sec 240°$ **e.** $\csc 330°$

16. **a.** $\cos(3\pi/4)$ **b.** $\tan(2\pi/3)$ **c.** $\csc(5\pi/6)$
  d. $\sin(5\pi/4)$ **e.** $\cot(7\pi/6)$

17. **a.** $\tan 480°$ **b.** $\sec 900°$ **c.** $\sin(-270°)$
     d. $\cos(10\pi/3)$ **e.** $\cot(3\pi/2)$

18. **a.** $\sin(7\pi/6)$ **b.** $\cos 330°$ **c.** $\tan(7\pi/4)$
     d. $\sec 135°$ **e.** $\cot(3\pi/4)$

19. **a.** $\csc(11\pi/6)$ **b.** $\tan 600°$ **c.** $\sin(4\pi/3)$
     d. $\cos 225°$ **e.** $\sec(7\pi/6)$

20. Evaluate the expression below when $\theta = 5\pi/6$:

$$\frac{\sin \theta - \cos 2\theta}{\tan \theta \cot 2\theta}$$

21. Evaluate the expression below when $\theta = 2\pi/3$:

$$\left(\sec \theta - \csc \frac{\theta}{4}\right)\left(1 + \cos 3\theta\right)$$

B 22. Prove that $\cos n\pi = (-1)^n$, where n is any integer.

22.
23. Prove that

$$\sin \frac{(2n + 1)\pi}{2} = (-1)^n$$

where n is any integer.

24. For $0 \le \theta < \pi/2$, express all six trigonometric functions of each of the following in terms of θ:

a. $\pi - \theta$ **b.** $\pi + \theta$ **c.** $\dfrac{\pi}{2} + \theta$

d. $\dfrac{3\pi}{2} - \theta$ **e.** $2\pi - \theta$

Hint.   What is the reference angle in each case?

---

5   The Law of Sines

Consider an oblique triangle ABC, as shown in Figure 30. The altitude from C divides the triangle into two right triangles, as shown. From triangle ADC we obtain

$$\sin A = \frac{h}{b}$$

so that $h = b \sin A$. Similarly, from triangle BDC,

$$\sin B = \frac{h}{a}$$

so that $h = a \sin B$. The two expressions for h must be the same, giving $a \sin B = b \sin A$, or equivalently,

$$\frac{a}{\sin A} = \frac{b}{\sin B} \tag{3}$$

Figure 30

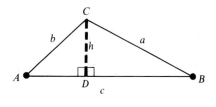

Figure 31

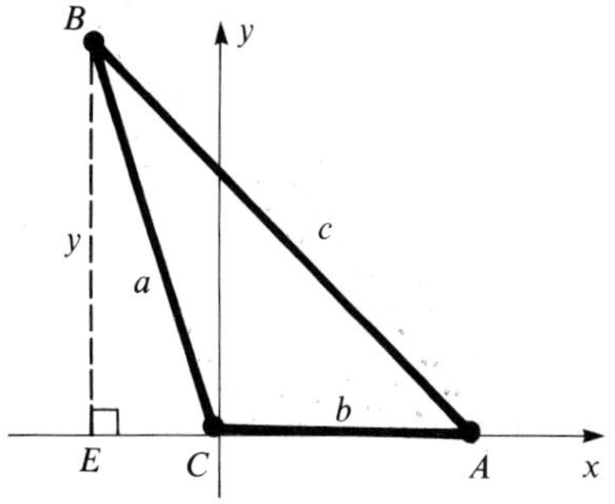

Next we orient the original triangle ABC with respect to a rectangular co-ordinate system so that C coincides with the origin and vertex A is on the positive x axis, as in Figure 31. Let the ordinate of B be designated by y. Then, since angle C is in standard position, we have that

$$\sin C = \frac{y}{a}$$

or $y = a \sin C$. If E is the foot of the perpendicular from B to the x axis, we can use the right triangle ABE to get

$$\sin A = \frac{y}{c}$$

so that $y = c \sin A$. Equating the two values of y gives $a \sin C = c \sin A$, or equivalently,

$$\frac{a}{\sin A} = \frac{c}{\sin C} \qquad (4)$$

Finally, combining equations (3) and (4), we get the **law of sines**:

The Law of Sines

$$\frac{a}{\sin A} = \frac{b}{\sin B} = \frac{c}{\sin C}$$

Although the triangle we used to derive this result had an obtuse angle (C), the derivation is valid also when all three angles are acute.

The law of sines can be employed to find the missing parts of a triangle in either of the following cases:

1. Two angles and one side are given
2. Two sides and the angle opposite one of them are given

In either case, an equation having only one unknown can be obtained from the law of sines. After solving for the unknown in the equation, all remaining unknowns can be found. The following examples illustrate typical situations.

EXAMPLE 11

In $\triangle ABC$, $A = 22°$, $B = 110°$, and $c = 13.4$. Find sides a and b and angle C.

Solution

Since the sum of the angles must be $180°$, we find that $C = 180° - 132° = 48°$. From the law of sines, we select $a/\sin A = c/\sin C$. Substituting known values, we obtain

$$\frac{a}{\sin 22°} = \frac{13.4}{\sin 48°}$$

so that

$$a = \frac{(13.4)(\sin 22°)}{\sin 48°} = \frac{(13.4)(0.3746)}{0.7431} = 6.75*$$

To find b we employ $b/\sin B = c/\sin C$. (We could have used $a/\sin A = b/\sin B$, but this involves the value of a just calculated, whereas the equation involving c relies primarily on original data. This helps to reduce an accumulation of errors.) The reference angle of $110°$ is $70°$. So $\sin 110° = \sin 70°$. Hence,

$$\frac{b}{\sin 70°} = \frac{13.4}{\sin 48°}$$

$$b = \frac{(13.4)(\sin 70°)}{\sin 48°} = \frac{(13.4)(0.9397)}{0.7431} = 16.9$$

The triangle is now solved.

Before proceeding to other examples, let us analyze what the possibilities are in case 2, in which we are given two sides and the angle opposite one of them. Suppose we are given a, b, and A. First, draw angle A with side b adjacent, thus determining vertex C (Figure 32). Now side a extends from C until it strikes the base. But it is quite evident that several possibilities exist, depending on the length of side a. A compass set at C with radius a clearly shows what can happen. If a exceeds the altitude from C but is shorter than b, then there are two distinct solutions, as Figure 33 shows. If a exceeds the altitude from C and also exceeds b, only one solution exists, as Figure 34 shows. If a equals the

---

* It is understood that all calculations such as this, in which tables (or a calculator) are used, result in answers that are approximations.

Figure 32   **Figure 33**

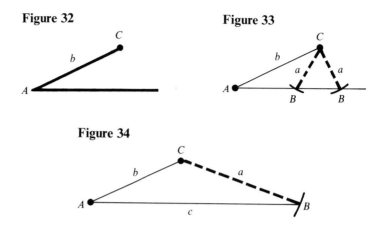

Figure 34

altitude from C, the triangle is a right triangle, and there is again just one solution, as shown in Figure 35. It should be noted that while this is a theoretical possibility, in actual practice it is extremely unlikely that this case will occur. Finally, if a is less than the altitude from C, there is no triangle at all, as shown in Figure 36.

Figure 35   **Figure 36**

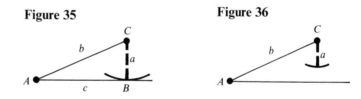

For obvious reasons, case 2 is known as the **ambiguous case**. Fortunately, it is not necessary to determine in advance which situation exists; after the law of sines is employed, the situation will become clear. It is necessary, however, to be alert and to interpret from the result whether one, two, or no solutions exist.

EXAMPLE 12   Solve the triangle for which $a = 125$, $b = 150$, and $A = 54°$.

Solution   From $a/\sin A = b/\sin B$ we get

$$\frac{125}{\sin 54°} = \frac{150}{\sin B}$$

so that

$$\sin B = \frac{(150)(\sin 54°)}{125} = \frac{(150)(0.8090)}{125} = 0.9708$$

From tables or a calculator we find that B is approximately $76°08'$. But the supplement of this angle is also a possibility. So we must consider an alternative value of B, $180° - 76°08' = 103°52'$, and then calculate the third angle in order to find out whether the second value of B is feasible. We have

$$C = 180° - (A + B) = 180° - (54° + 103°52')$$
$$= 180° - 157°52'$$
$$= 22°08'$$

Thus, two distinct solutions exist. Let us designate by B_1 the value $76°08'$ and by B_2 the value $103°52'$. Corresponding subscripts will be used for angle C and side c. Thus,

$$C_1 = 180° - (A + B_1) = 180° - (54° + 76°08')$$
$$= 180° - 130°08' = 49°52'$$

and $C_2 = 22°08'$, as we have already calculated.

We use $a/\sin A = c/\sin C$ to calculate side c for each case. So we have

$$\frac{125}{\sin 54°} = \frac{c_1}{\sin 49°52'} \quad \text{and} \quad \frac{125}{\sin 54°} = \frac{c_2}{\sin 22°08'}$$

These yield

$$c_1 = \frac{(125)(\sin 49°52')}{\sin 54°} = \frac{(125)(0.7646)}{0.8090} = 118$$

$$c_2 = \frac{(125)(\sin 22°08')}{\sin 54°} = \frac{(125)(0.3768)}{0.8090} = 58.2$$

The two solutions can be summarized as shown in Figure 37.

Figure 37

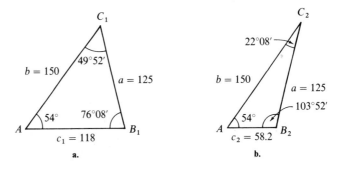

a.

b.

EXAMPLE 13   Solve the triangle for which $b = 13.2$, $c = 10.5$, and $B = 42°$.

Solution   From $b/\sin B = c/\sin C$ we find that

$$\sin C = \frac{c \sin B}{b} = \frac{(10.5)(\sin 42°)}{13.2} = 0.5323$$

so that C is either $32°10'$ or its supplement, $147°50'$. But if $C = 147°50'$, then $B + C = 189°50'$, which is impossible, since the sum of two angles of the triangle must be less than $180°$. Thus, only one solution exists.

Angle A is found to be

$$A = 180° - (B + C) = 180° - (42° + 32°10') = 105°50'$$

Also, $a/\sin A = b/\sin B$, so that

$$a = \frac{b \sin A}{\sin B} = \frac{(13.2)(\sin 105°50')}{\sin 42°}$$

$$= \frac{(13.2)(\sin 74°10')}{\sin 42°} \qquad \textbf{Why?}$$

$$= 19.0$$

The solution is shown in Figure 38.

Figure 38

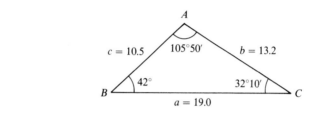

EXAMPLE 14   Find all unknown parts of the triangle for which $a = 23.5$, $c = 12.0$, and $C = 35°$.

Solution   We begin with $a/\sin A = c/\sin C$ and obtain

$$\sin A = \frac{a \sin C}{c} = \frac{(23.5)(\sin 35°)}{12.0} = 1.123$$

But this is an impossibility, since it is clear from the definition that the sine never exceeds 1. So there is no solution; that is, there is no triangle having a, c, and C as given.

EXERCISE SET 5

A   In Problems 1–15 use the law of sines to find all unknown angles and sides of the triangles determined by the data. In ambiguous cases determine whether there are one, two, or no solutions.

1. $a = 5$, $A = 40°$, $B = 20°$
2. $b = 37.4$, $A = 64.2°$, $C = 37.6°$
3. $b = 125$, $c = 85$, $C = 36°30'$
4. $a = 10.5$, $c = 12.8$, $A = 68°$
5. $A = 28°$, $a = 21$, $b = 10$
6. $C = 30°$, $b = 15.2$, $c = 7.6$
7. $B = 47.2°$, $C = 25.8°$, $c = 7.3$
8. $A = 63°$, $C = 72°$, $a = 102$
9. $A = 37°25'$, $B = 64°37'$, $b = 14.72$
10. $c = 3.5$, $B = 102°24'$, $C = 23°42'$
11. $B = 43°$, $b = 17.5$, $c = 10.2$
12. $C = 23.4°$, $a = 21.8$, $c = 15.2$
13. $B = 108°$, $b = 112$, $c = 73$
14. $A = 65.2°$, $a = 33.2$, $b = 68.7$
15. $C = 22°30'$, $a = 5.43$, $c = 4.87$

B   **16.** Points A and B are on opposite sides of a river, and it is desired to find the distance between them. Point C, on the same side of the river as A, is 35 feet from A. In the triangle ABC formed, the angle at C is found to be 42°30′ and the angle at A is found to be 103°24′. Find the approximate distance from A to B.

   17. The distance x across a gorge is to be determined in order that a cable car traversing it can be constructed. One side of the gorge is less steep than the other so that a distance along it can be measured as shown in the figure. The angles shown are also measured. Find the distance x.

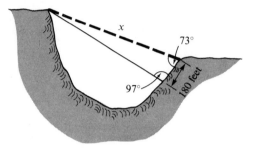

   18. Points A and B are on opposite sides of a swampy area. To find the distance between them a point C is located 254 feet from A and 197 feet from B. The angle at A from AB to AC is found to be 47.2°. Find the distance from A to B.

   19. A boat leaves point A at a heading of 48° east of due south and travels at an average speed of 15 knots (15 nautical miles per hour). A coast guard cutter at point B, which is 21 nautical miles due east of A wishes to intercept the boat. If the cutter can average 25 knots, what direction should it go in (assuming the cutter leaves at the same time)? When will the interception be made?

---

6   The Law of Cosines

When we are given two sides of an oblique triangle and the angle between them, or when we are given all three sides, the law of sines cannot be used, because there will always be two unknowns in any of the equations. In this section we derive a formula that will handle these situations.

Consider first a triangle as shown in Figure 39, with angle C acute. Construct the altitude AD and designate its length by h. Then, from $\triangle ADB$ and the Pythagorean theorem,

$$c^2 = h^2 + \overline{BD}^2 \tag{5}$$

From $\triangle ADC$ we have

$$b^2 = h^2 + \overline{CD}^2 \tag{6}$$

Figure 39

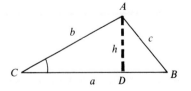

and since

$$\cos C = \frac{\overline{CD}}{b}$$

we obtain $\overline{CD} = b \cos C$. Since $\overline{BD} = a - \overline{CD}$, we can replace $\overline{BD}$ in equation (5) by $a - b \cos C$. Also, from equation (6), $h^2 = b^2 - \overline{CD}^2 = b^2 - (b \cos C)^2$. Making these substitutions in equation (5), we obtain

$$c^2 = b^2 - b^2 \cos^2 C + (a - b \cos C)^2$$
$$= b^2 - b^2 \cos^2 C + a^2 - 2ab \cos C + b^2 \cos^2 C$$
$$= a^2 + b^2 - 2ab \cos C$$

Figure 40

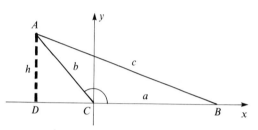

If C is obtuse, we introduce a coordinate system as in Figure 40, with C in standard position. As before, we have

$$c^2 = h^2 + \overline{BD}^2 \tag{7}$$

and

$$b^2 = h^2 + \overline{CD}^2 \tag{8}$$

Since the abscissa of A is negative, we have

$$\cos C = \frac{-\overline{CD}}{b}$$

or $\overline{CD} = -b \cos C$. Also,

$$\overline{BD} = a + \overline{CD} = a - b \cos C$$

Substituting these results into equations (7) and (8) and combining yields

$$c^2 = b^2 - (-b \cos C)^2 + (a - b \cos C)^2$$
$$= b^2 - b^2 \cos^2 C + a^2 - 2ab \cos C + b^2 \cos^2 C$$
$$= a^2 + b^2 - 2ab \cos C$$

which is the same as before.

By working from the other two vertices, in turn, analogous formulas are obtained. We have, then, the following three formulas:

$$a^2 = b^2 + c^2 - 2bc \cos A$$
$$b^2 = a^2 + c^2 - 2ac \cos B$$
$$c^2 = a^2 + b^2 - 2ab \cos C$$

These formulas are expressions of the **law of cosines.**

The Law of Cosines

The square of any side of a triangle equals the sum of the squares of the other two sides, minus twice the product of these two sides and the cosine of the angle between them.

If two sides and the angle between them are given, we select the form of the law of cosines that involves the given angle. Then the side opposite the angle can be calculated. To complete the solution of the triangle the law of sines is employed. For example, if a, c, and angle B are given, the formula

$$b^2 = a^2 + c^2 - 2ac \cos B$$

is employed to calculate b. Then either angle A or angle C can be found using the law of sines.

If all three sides of a triangle are given, any one of the forms of the law of cosines can be used to calculate one of the angles. Then another angle can be found either by another form of the law of cosines or by the law of sines. For example, suppose a, b, and c are given. Then we may use

$$a^2 = b^2 + c^2 - 2bc \cos A$$

to solve first for $\cos A$,

$$\cos A = \frac{b^2 + c^2 - a^2}{2bc}$$

and then find A from tables or a calculator.

EXAMPLE 15   In a triangle one angle is $120°$ and the two sides adjacent to this angle have lengths 3 and 5. Find all other parts of the triangle.

Solution   The missing side can be found without the use of tables. Let the given sides be $a = 3$, $b = 5$. Then $C = 120°$.

$$c^2 = a^2 + b^2 - 2ab \cos C$$
$$= 9 + 25 - 2(3)(5)(-\tfrac{1}{2})$$
$$= 9 + 25 + 15 = 49$$

so $c = 7$.

From the law of sines, $a/\sin A = c/\sin C$, we get

$$\sin A = \frac{a \sin C}{c} = \frac{3(\sqrt{3}/2)}{7} = \frac{3\sqrt{3}}{14} = 0.3712$$

$$A = 21°47'$$

Finally,

$$B = 180° - (A + C) = 180° - 141°47' = 38°13'$$

EXAMPLE 16   Find the angles in $\triangle ABC$ for which $a = 13$, $b = 21$, $c = 15$.

Solution   We have

$$\cos A = \frac{b^2 + c^2 - a^2}{2bc} = \frac{441 + 225 - 169}{2(21)(15)} = 0.7889$$

So $A = 37°55'$.

$$\cos B = \frac{a^2 + c^2 - b^2}{2ac} = \frac{169 + 225 - 441}{2(13)(15)} = -0.1205$$

and $B = 96°55'$. Thus, $C = 180° - (A + B) = 180° - 134°50' = 45°10'$.

EXERCISE SET 6

A   In Problems 1–10 find the remaining parts of the triangle ABC.

1. $A = 60°$, $b = 15$, $c = 10$
2. $B = 150°$, $a = 2.3$, $c = 1.2$
3. $C = 97°20'$, $a = 5$, $b = 11$
4. $a = 3$, $b = 4$, $c = 2$
5. $a = 12$, $b = 8$, $c = 15$
6. $A = 32.4°$, $b = 12.3$, $c = 15.8$
7. $B = 103.5°$, $a = 234$, $c = 160$
8. $C = 67°42'$, $a = 35.2$, $b = 42.3$
9. $a = 3.6$, $b = 4.7$, $c = 2.8$
10. $a = 21.5$, $b = 32.6$, $c = 50.7$

11. Two sides of a parallelogram are 8 inches and 10 inches, respectively, and the shorter diagonal is 12 inches. Find the length of the longer diagonal.
12. Two cars leave simultaneously from the same point, one going east at an average speed of 60 kilometers per hour and the other going southwest at an average speed of 80 kilometers per hour. How far apart are they after 2 hours?
13. Find the magnitude of the resultant of two forces of 25 pounds and 30 pounds acting on an object, if the angle between the forces is 23°.
    Hint.   The resultant force is the diagonal of a parallelogram formed by vectors representing the given forces and emanating from the same point.

B   14. A regular pentagon is inscribed in a circle of diameter 4 inches. Make use of the law of cosines to find the perimeter of the pentagon.
15. Airplane A leaves Chicago at 1:00 PM and flies at a heading 23° west of due south at an average speed of 350 miles per hour. Airplane B leaves from the same airport at 1:30 PM and flies 85° east of due north at an average speed of 400 miles per hour. How far apart are they at 2:00 PM?
16. A surveyor wishes to find the distance from point A to point B but cannot do so directly because there is a swamp between them. She sets up a transit at C and makes the following measurements: angle $C = 27°32'$, $\overline{CA} = 125.3$ feet, and $\overline{CB} = 117.5$ feet. Find $\overline{AB}$.
17. In order to determine the distance between two inaccessible points A and B on the opposite side of a river from an observer, two points C and D 50 feet apart are established and the angles are determined as shown in the sketch at the top of page 288. Find the distance between A and B.

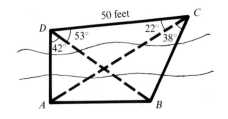

7   Areas of Triangles

The basic formula for the area of a triangle is well-known: Area $= \frac{1}{2}bh$, where b designates the base, and h is the altitude drawn from that base. When the triangle is a right triangle, either leg can be taken as the base and the other leg is then the altitude. For oblique triangles, the complication arises of how to find the altitude. The method of deriving the law of sines gives a clue as to how this can be done.

Suppose a triangle is oriented as shown in Figure 41. Since sin $A = h/b$, we have $h = b$ sin A. Since the base corresponding to this altitude is c,

$$\textbf{Area} = \tfrac{1}{2}bc \textbf{ sin } A \tag{9}$$

Notice that the formula involves two sides and the angle between them. By working from one of the other vertices, we obtain

$$\textbf{Area} = \tfrac{1}{2}ab \textbf{ sin } C = \tfrac{1}{2}ac \textbf{ sin } B$$

If enough information to solve the triangle is given, then the area can always be found in this way.

Figure 41

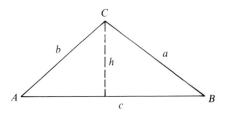

EXAMPLE 17   Find the area of $\triangle ABC$ for which $a = 10$, $b = 7$, and $C = 60°$.

Solution
$$\text{Area} = \frac{1}{2}ab \text{ sin } C = \frac{1}{2}(\overset{5}{\cancel{10}})(7)\left(\frac{\sqrt{3}}{2}\right) = \frac{35\sqrt{3}}{2} = 30.31$$

EXAMPLE 18   Find the area of a triangle having sides 5, 7, and 3.

Solution   Let $a = 5$, $b = 7$, and $c = 3$. We could work with any angle, but we choose B (a little scratch work will show this is the easiest one to find). By the law of cosines,

$$\cos B = \frac{a^2 + c^2 - b^2}{2ac} = \frac{25 + 9 - 49}{2(15)} = \frac{-15}{30} = -\frac{1}{2}$$

So $B = 120°$. (Why?) The area now can be found:

$$\text{Area} = \tfrac{1}{2}ac \sin B = \tfrac{1}{2}(5)(3) \sin 120°$$

$$= \frac{1}{2}(15)\left(\frac{\sqrt{3}}{2}\right) = \frac{15\sqrt{3}}{4} = 6.50$$

EXERCISE SET 7

A   Find the area of the triangle ABC in Problems 1–8.

1. $b = 36$, $c = 24$, $A = 30°$
2. $a = 7$, $b = 9$, $C = 23°$
3. $a = 1.2$, $b = 2.1$, $c = 2.8$
4. $a = 14$, $c = 10$, $B = 112°$
5. $b = 3.5$, $c = 2.8$, $A = 65.2°$
6. $a = 12.3$, $b = 18.2$, $C = 93°20'$
7. $a = 10$, $b = 6$, $c = 14$
8. $C = 102.5°$, $a = 24$, $b = 13$

B   9.   Show that equation (9) holds true when A is an obtuse angle.
10.   Derive the formula for the area of $\triangle ABC$,

$$\text{Area} = \frac{b^2 \sin A \sin C}{2 \sin B}$$

Hint.   Use equation (9) and the law of sines.
11.   Use the result of Problem 10 to find the area of the triangle in which $b = 8$, $B = 32°$, $C = 54°$.
12.   Two angles of a triangle are $47°$ and $56°$, respectively, and the side opposite the $47°$ angle is 13.4. Find the area of the triangle.
13.   Find the area of triangle ABC in which $A = 30°$, $a = 12$, and $b = 10$.
14.   Let the lengths of the two diagonals of an arbitrary quadrilateral be d_1 and d_2, and let α be their angle of intersection (either of the two angles). Prove that the area of the quadrilateral is given by $\tfrac{1}{2}d_1 d_2 \sin \alpha$.

Review Exercise Set

A   In Problems 1–14 solve the right triangles. Whenever possible solve without the use of a calculator or tables. The right angle is always at C.

1. $A = 60°$, $b = 5$
2. $B = 30°$, $c = 9$
3. $B = 60°$, $a = 4$
4. $A = 45°$, $a = 12$
5. $B = 45°$, $c = 10$
6. $A = 30°$, $c = 5$
7. $a = 7$, $c = 14$
8. $a = 8$, $b = 15$
9. $A = 33°$, $c = 13$
10. $a = 4$, $b = 7$
11. $b = 102$, $c = 151$
12. $A = 67°$, $a = 12.6$
13. $B = 26.2°$, $c = 33.5$
14. $B = 62.3°$, $a = 2.53$

15. Give the measure of the angle in radians:
    a. 240°   b. 315°   c. 15°   d. 540°   e. 160°

16. Give the measure of the angle in degrees:
    a. $5\pi/6$ b. $2\pi/9$ c. 3π d. 5   e. $7\pi/4$

17. a. What is the length of an arc on a circle of radius 20 subtended by à central angle of 36°?
    b. An arc of 8 feet is subtended by a central angle θ on a circle of radius 6 feet. Find θ in degrees.

18. A wheel 18 inches in diameter is turning at the rate of 600 rpm. What is the velocity in feet per second of a point on the extremity of the wheel?

In Problems 19 and 20 evaluate the function without using tables or a calculator.

19. a. $\cos(5\pi/3)$ b. $\sin 210°$ c. $\tan(-5\pi/6)$ d. $\cot 270°$
    e. $\sec(3\pi/4)$ f. $\csc 390°$ g. $\tan(5\pi/4)$ h. $\sin(-90°)$
    i. $\sec 480°$ j. $\cos(-7\pi/6)$

20. a. $\sin(3\pi/2)$ b. $\cos(11\pi/6)$ c. $\tan 225°$ d. $\sec(-3\pi)$
    e. $\cot(5\pi/3)$ f. $\sin 390°$ g. $\cos(4\pi/3)$ h. $\tan(-315°)$
    i. $\sec(5\pi/4)$ j. $\csc(7\pi/6)$

In Problems 21–31 solve the triangles.

21. $A = 34.3°$, $B = 26.8°$, $a = 13.4$
22. $B = 54°30'$, $C = 48°20'$, $a = 102.5$
23. $C = 32°$, $b = 15$, $c = 8$
24. $a = 4$, $b = 7$, $C = 110°$
25. $a = 22.3$, $b = 17.9$, $c = 10.5$
26. $A = 110°$, $C = 28°$, $c = 12$
27. $A = 63.2°$, $a = 14.3$, $b = 25.2$
28. $B = 36°20'$, $b = 34.5$, $c = 12.7$
29. $a = 8$, $b = 9$, $c = 13$
30. $a = 2.05$, $c = 3.72$, $B = 61°40'$
31. $A = 25.2°$, $c = 27.4$, $a = 15.1$

32. Find the area of each of the following triangles:
    a. $a = 32$, $b = 51$, $C = 25°$ b. $b = 1.2$, $c = 3.5$, $C = 102°$

33. A 40 foot high building is at the edge of a river. Directly opposite on the other edge an observer finds that the angle of elevation of the top of the building is $33°10'$. How wide is the river at that point?

34. The sides of a parallelogram are 6 and 10, and the longer diagonal is 14. Find the interior angles of the parallelogram.

35. A lot is in the shape of a triangle in which two sides are 100 feet and 120 feet, and the angle between them is 55°. Find the length of the third side and the area of the lot.

36. A portion of a modern metal sculpture consists of a large triangular plate mounted vertically, with one vertex at the top and with a horizontal base. The length of the base is 12 feet, and the base angles are 30° and 50°, respectively. Find the lengths of the other two sides of the triangle. What is its area?

B 37. An observer is 100 feet from the base of a building. She finds the angles of elevation to the bottom and top of a flagpole on the roof of the building to be 59.3° and 62.4°, respectively. How high is the flagpole?

38. Points A and B are on opposite sides of a pond, and it is desired to find the distance between them. Point C is located so that $\overline{AC} = 254$ feet and $\overline{BC} = 198$ feet. The angle at A from AB to AC is found to be $47°21'$. Find the distance from A to B.

39. At 4:00 PM an airplane leaves Chicago and heads due south at a cruising speed of 400 miles per hour. At 5:00 PM another plane leaves from the same airport and travels at a heading 40° west of due south, cruising at 500 miles per hour. How far apart are the two planes at 6:00 PM?

40. Two guy wires are attached on opposite sides of a pole. They make angles of 36° and 42° with the horizontal, and the points where they meet the ground are 50 feet apart. Find the length of each wire.

41. A ranger in a lookout tower spots a fire on a line 40° west of due north. A second ranger at a tower 10 miles to the west of the first tower also sees the fire and finds it to be on a line 35° east of due north. How far is the fire from each lookout tower?

9 Analytical Trigonometry

Relationships between the trigonometric functions, known as **trigonometric identities,** make possible the solution of many calculus problems. The use of such identities appears again and again in the part of calculus called *integration*. The following example from a calculus text illustrates this.*

Evaluate $\displaystyle\int \frac{dx}{(1 + x^2)^2}$.

... We use the substitution $x = \tan \theta$, which gives $dx = \sec^2 \theta \, d\theta$. This leads to

$$\int \frac{dx}{(1 + x^2)^2} = \int \frac{\sec^2 \theta \, d\theta}{(1 + \tan^2 \theta)^2} = \int \frac{\sec^2 \theta \, d\theta}{\sec^4 \theta} = \int \frac{d\theta}{\sec^2 \theta}$$

$$= \int \cos^2 \theta \, d\theta = \tfrac{1}{2} \int (1 + \cos 2\theta) \, d\theta = \tfrac{1}{2}\theta + \tfrac{1}{4} \sin 2\theta + C$$

Now, from the original substitution, we have

$$\theta = \text{Arctan } x$$

Figure 15.3

and, from Figure 15.3 or trigonometric identities,

$$\sin \theta = \frac{x}{\sqrt{1 + x^2}} \quad \text{and} \quad \cos \theta = \frac{1}{\sqrt{1 + x^2}}$$

Thus

$$\sin 2\theta = 2 \sin \theta \cos \theta = \frac{2x}{1 + x^2}$$

This gives

$$\int \frac{dx}{(1 + x^2)^2} = \tfrac{1}{2}\theta + \tfrac{1}{4} \sin 2\theta + C = \tfrac{1}{2} \text{Arctan } x + \frac{x}{2(1 + x^2)} + C$$

While this contains some mysterious symbols, it is loaded with notions that will be studied in this chapter. For example, the following relationships are employed that will be shown shortly:

$$1 + \tan^2 \theta = \sec^2 \theta \qquad \cos^2 \theta = \tfrac{1}{2}(1 + \cos 2\theta) \qquad \sin 2\theta = 2 \sin \theta \cos \theta$$

Also, the meaning of the notation "Arctan x" will be made clear in this chapter.

---

* Douglas F. Riddle, *Calculus and Analytic Geometry*, 3d ed. (Belmont, Ca.: Wadsworth Publishing Co., 1979), p. 468. Reprinted by permission.

1 Trigonometric Functions of Real Numbers

So far our definitions of the trigonometric functions have had as domain the set of all admissible angles. The values depend only on the angle, and not on whether the angle is measured in degrees or radians. The next stage of this course is an important one, although it is somewhat subtle. We wish to give meaning to such expressions as sin x, cos x, and tan x, where x stands for a *real number*.

Toward this end, let us recall that in the definition of the radian measure of an angle, we used an arbitrary circle with its center at the vertex of the angle. We find it useful in the present case to specify a circle of radius 1, with center at the origin, and we refer to this as the **unit circle** (Figure 1). If we are given an angle of θ radians, then the arc length subtended on a circle of radius r is $s = r\theta$, so if $r = 1$, the arc length s equals the angle θ. In other words, the linear measure (using the measurement units of r) of the arc length equals the angular measure of the central angle in radians. This relationship provides us with a way of defining the trigonometric functions of real numbers that is consistent with our previous definition.

Figure 1

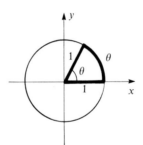

Figure 2

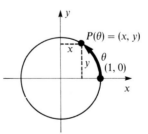

Given any real number θ (not at present to be thought of as the measure of an angle, but just a number), we mark off a distance of θ units on the unit circle, starting from (1, 0) and moving counterclockwise when θ is positive and clockwise when θ is negative. Let $P(\theta)$ be the point arrived at in this manner with coordinates (x, y), as shown in Figure 2.* With this established, we can define the following:

---

* $P(\theta)$ is thus a function of the real variable θ, whose range consists of ordered pairs (x, y) representing points on the circle.

DEFINITION 1

$$\sin \theta = y \qquad \csc \theta = \frac{1}{y}$$

$$\cos \theta = x \qquad \sec \theta = \frac{1}{x}$$

$$\tan \theta = \frac{y}{x} \qquad \cot \theta = \frac{x}{y}$$

Note that for the tangent and secant, θ cannot be any number for which $x = 0$; and for the cotangent and cosecant, θ cannot be any number for which $y = 0$.

The relationship between the functions defined in terms of numbers and those defined in terms of angles should now be apparent. We simply consider the angle with initial side the positive x axis and terminal side connecting the origin and $P(\theta)$, as shown in Figure 3. Then the radian measure of this angle is precisely θ, the same as the number of units of length of arc. Now by the definitions of the trigonometric functions of the *angle* θ given in Section 4 of Chapter 8, we have:

$$\sin \theta = \frac{y}{r} = \frac{y}{1} = y \qquad \csc \theta = \frac{r}{y} = \frac{1}{y}$$

$$\cos \theta = \frac{x}{r} = \frac{x}{1} = x \qquad \sec \theta = \frac{r}{x} = \frac{1}{x}$$

$$\tan \theta = \frac{y}{x} \qquad \cot \theta = \frac{x}{y}$$

Comparing these with Definition 1 for the functions of the *number* θ, we see that they are precisely the same.

Suppose, then, that we are given the expression "sin 2." How shall we interpret it? Does it mean the sine of the number 2 or the sine of the angle whose radian measure is 2? The answer is that it does not matter, because the values are the same. Conceptually these are quite different, but sin 2 where 2 is a number has the same value as sin 2 where 2 is the radian measure of an angle. More generally, any trigonometric function of a real number has the same value as the same trigonometric function of the angle of θ radians. Thus, insofar as

Figure 3

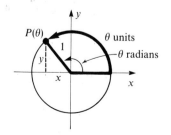

values of the trigonometric functions are concerned, it makes no difference whether we think in terms of an angle measured in radians or in terms of the real number that equals the number of radians in the angle.

EXAMPLE 1   The x coordinate of the point $P(\theta)$ on the unit circle is $-\frac{4}{5}$, and $P(\theta)$ is in the second quadrant. Find the y coordinate of $P(\theta)$ and write all six trigonometric functions of θ.

Solution   By the Pythagorean theorem we have

$$(-\tfrac{4}{5})^2 + y^2 = 1$$
$$y^2 = 1 - \tfrac{16}{25} = \tfrac{9}{25}$$
$$y = \tfrac{3}{5}$$

By Definition 1,

$$\sin \theta = y = \frac{3}{5} \qquad\qquad \csc \theta = \frac{1}{y} = \frac{5}{3}$$

$$\cos \theta = x = -\frac{4}{5} \qquad\qquad \sec \theta = \frac{1}{x} = -\frac{5}{4}$$

$$\tan \theta = \frac{y}{x} = -\frac{3}{4} \qquad\qquad \cot \theta = \frac{x}{y} = -\frac{4}{3}$$

EXAMPLE 2   Determine the value of θ, where $0 \le \theta < 2\pi$.
   a. $P(\theta) = (-\sqrt{3}/2, -\tfrac{1}{2})$ **b.** $P(\theta) = (-1/\sqrt{2}, 1/\sqrt{2})$

Solution   **a.**   We observe that the reference angle of the angle determined by the radius drawn to the given point is $30°$, or $\pi/6$ radians. So the angle itself is $7\pi/6$ radians. Therefore $\theta = 7\pi/6$.
   b.   The reference angle in this case is $45°$, or $\pi/4$ radians, and the angle is therefore $3\pi/4$ radians. So $\theta = 3\pi/4$.

Figure 4

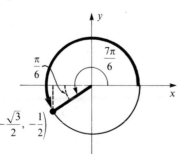

Figure 5

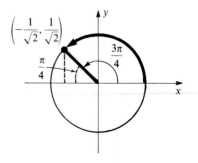

If you are using a calculator, you can find trigonometric functions of real numbers by considering the number as the radian measure of an angle, and this can be done regardless of the size of the number. For example, using a calculator we can read

$$\sin 2 = 0.9093 \qquad \cos 10 = -0.8391 \qquad \tan(-3.5) = -0.3746$$

and so on. However, to read these values from a table such as Table V at the end of this book requires some preliminary steps. This is because the table only gives functions of angles between 0° and 90°, or equivalently, between 0 and $\pi/2$ radians. For larger angles, we make use of reference angles. Since $\pi/2 \approx 1.5708$, we cannot find functions of numbers greater than this directly from the table. To get sin 2, for example, we proceed as follows: As an angle in radians 2 lies in the second quadrant (between $\pi/2 \approx 1.5708$ and $\pi \approx 3.1416$). We show this in Figure 6. The reference angle is therefore $\pi - 2 \approx 1.1416$. Since the sine is positive in the second quadrant, we have that sin 2 is approximately equal to sin 1.1416. Reading from Table V (and interpolating between values found there), we see that this is approximately 0.9093. We could also obtain cos 2 by observing that the cosine is negative in quadrant II. Then, again using the reference angle, we find $\cos 2 \approx -0.4161$. Thus, the coordinates of the point $P(2)$ on the unit circle are approximately $(-0.4161, 0.9093)$.

Figure 6

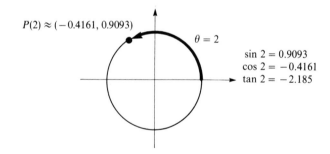

$P(2) \approx (-0.4161, 0.9093)$

$\theta = 2$

sin 2 = 0.9093
cos 2 = −0.4161
tan 2 = −2.185

To find the functions of 10 using Table V, we again must find the reference angle. Since 10 exceeds 2π, the first thing to do is to subtract 2π from 10, in order to obtain an angle in the first revolution that is coterminal with 10. This gives

$$10 - 2\pi \approx 10 - 6.2832 = 3.7168$$

This is seen to be in the third quadrant (Figure 7), and so the reference angle is

$$3.7168 - 3.1416 = 0.5752$$

From the table we can now obtain sin 0.5752 = 0.5440 and cos 0.5752 = 0.8391. Since both sine and cosine are negative in the third quadrant, we have finally that

$$\sin 10 \approx -0.5440 \qquad \text{and} \qquad \cos 10 \approx -0.8391$$

The advantage of a calculator for finding functions of real numbers should be evident.

Figure 7

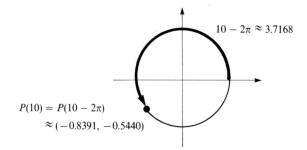

$P(10) = P(10 - 2\pi)$

$\approx (-0.8391, -0.5440)$

EXERCISE SET 1

A   In Problems 1–10 find all six trigonometric functions of θ.

1. $P(\theta) = (-\frac{1}{3}, -2\sqrt{2}/3)$ **2.** $P(\theta) = (-2/\sqrt{5}, 1/\sqrt{5})$
3. $P(\theta) = (\frac{5}{13}, -\frac{12}{13})$ **4.** $P(\theta) = (-0.6, 0.8)$
5.   The x coordinate of $P(\theta)$ is $\frac{1}{4}$, and $P(\theta)$ is in the fourth quadrant.
6.   The y coordinate of $P(\theta)$ is $\frac{4}{5}$, and $\pi/2 \leq \theta \leq \pi$.
7.   The abscissa of $P(\theta)$ is $-\frac{1}{2}$, and $\pi \leq \theta \leq 2\pi$. What is the value of θ?
8.   The ordinate of $P(\theta)$ is $\sqrt{3}/2$, and the abscissa of $P(\theta)$ is negative. What is the value of θ?
9.   The abscissa of $P(\theta)$ is $\frac{1}{3}$, and the ordinate is negative.
10.   The ordinate of $P(\theta)$ is $-\frac{5}{13}$, and $3\pi \leq \theta \leq 4\pi$.

In Problems 11 and 12 give the coordinates of the point $P(\theta)$ for the given values of θ.

11.   **a.** $P(\pi/2)$ **b.** $P(\pi)$ **c.** $P(3\pi/2)$
     d. $P(5\pi/6)$ **e.** $P(4\pi/3)$
12.   **a.** $P(5\pi/4)$ **b.** $P(8\pi/3)$ **c.** $P(-7\pi/4)$
     d. $P(5\pi)$ **e.** $P(31\pi/6)$

In Problems 13 and 14 indicate the approximate location of $P(\theta)$ on the unit circle for the given value of θ.

13.   **a.** $\theta = 0.5$ **b.** $\theta = -2$ **c.** $\theta = 4$ **d.** $\theta = -15$
14.   **a.** $\theta = 8.2$ **b.** $\theta = 12.75$ **c.** $\theta = -1.54$ **d.** $\theta = 4.72$

In Problems 15 and 16 give the value of θ if $0 \leq \theta < 2\pi$.

15.   **a.** $P(\theta) = (\sqrt{3}/2, \frac{1}{2})$ **b.** $P(\theta) = (1/\sqrt{2}, 1/\sqrt{2})$
     c. $P(\theta) = (\frac{1}{2}, \sqrt{3}/2)$ **d.** $P(\theta) = (0, 1)$
     e. $P(\theta) = (1, 0)$
16.   **a.** $P(\theta) = (-\frac{1}{2}, -\sqrt{3}/2)$ **b.** $P(\theta) = (0, -1)$
     c. $P(\theta) = (-\sqrt{3}/2, \frac{1}{2})$ **d.** $P(\theta) = (\sqrt{2}/2, -\sqrt{2}/2)$
     e. $P(\theta) = (-1, 0)$

In Problems 17 and 18 find $\sin \theta$, $\cos \theta$, and $\tan \theta$ using either a calculator or tables.

17.   **a.** $\theta = 0.3$ **b.** $\theta = 2.718$ **c.** $\theta = 14.03$ **d.** $\theta = -3.625$
18.   **a.** $\theta = 7.1432$ **b.** $\theta = 0.2967$ **c.** $\theta = -5.1324$ **d.** $\theta = 10.5146$

19. Find all values of θ between 0 and 2π for which:

    a. $\sin \theta = 1$ **b.** $\cos \theta = 1$ **c.** $\tan \theta = 1$

    d. $\sin \theta = 0$ **e.** $\cos \theta = 0$ **f.** $\tan \theta = 0$

    g. $\sin \theta = -1$ **h.** $\cos \theta = -1$ **i.** $\tan \theta = -1$

B   **20.** Use the unit circle to show that:

    a. $\sin(\pi - \theta) = \sin \theta$ **b.** $\cos(\pi - \theta) = -\cos \theta$

    c. $\sin(\pi + \theta) = -\sin \theta$ **d.** $\cos(\pi + \theta) = -\cos \theta$

    e. $\tan(\pi + \theta) = \tan \theta$

21. Show that for all integers n, each of the following is undefined for the stated values of θ:

    a. $\tan \theta$ for $\theta = \dfrac{2n+1}{2}\pi$ **b.** $\cot \theta$ for $\theta = n\pi$

    c. $\sec \theta$ for $\theta = \dfrac{2n+1}{2}\pi$ **d.** $\csc \theta$ for $\theta = n\pi$

22. Show that $|\sec \theta| \geq 1$ and $|\csc \theta| \geq 1$ for all θ in the domains of these functions.

23. Prove that the range of $\tan \theta$ is the set of all real numbers.

24. Let

$$S = \left\{\theta \in R: \quad \theta \neq \frac{(2n+1)\pi}{2}, n = 0, \pm 1, \pm 2, \ldots\right\}$$

$$T = \{\theta \in R: \quad \theta \neq n\pi, n = 0, \pm 1, \pm 2, \ldots\}$$

$$U = \{k \in R: \quad |k| \geq 1\}$$

Prove that if $f(\theta) = \sec \theta$ and $g(\theta) = \csc \theta$, then:

    a. f is a function from S onto U

    b. g is a function from T onto U

Is either f or g a 1–1 function? Explain.

---

2   Some Basic Trigonometric Identities

You may have already noticed the following relationships among the trigonometric functions:

$$\tan \theta = \frac{\sin \theta}{\cos \theta} \qquad\qquad \csc \theta = \frac{1}{\sin \theta}$$

$$\cot \theta = \frac{\cos \theta}{\sin \theta} \qquad\qquad \cot \theta = \frac{1}{\tan \theta}$$

$$\sec \theta = \frac{1}{\cos \theta}$$

These are seen to be true from Definition 1 of Chapter 8 or Definition 1 of this chapter. They are examples of **trigonometric identities**—so-called because they are true for all admissible values of θ. There are many other identities, and some are so basic that they should be committed to memory. These will be indicated as we proceed.

The Pythagorean theorem is the basis for another important set of identities. Using the theorem, we have, for (x, y) on the unit circle (see Figure 8),

$$x^2 + y^2 = 1$$

Figure 8

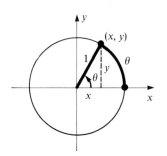

But $x = \cos \theta$ and $y = \sin \theta$. So

$$\sin^2 \theta + \cos^2 \theta = 1$$

If we divide both sides of the first equation by x^2 (assuming $x \neq 0$), we obtain

$$1 + \left(\frac{y}{x}\right)^2 = \left(\frac{1}{x}\right)^2$$

which, according to our definitions, is

$$1 + \tan^2 \theta = \sec^2 \theta$$

However, if we divide the first equation by y^2, we get

$$\left(\frac{x}{y}\right)^2 + 1 = \left(\frac{1}{y}\right)^2$$

or

$$\cot^2 \theta + 1 = \csc^2 \theta$$

We collect these results together:

$$\sin^2 \theta + \cos^2 \theta = 1$$
$$1 + \tan^2 \theta = \sec^2 \theta$$
$$1 + \cot^2 \theta = \csc^2 \theta$$

These are known as the **Pythagorean identities** and should be learned. Notice the similarity between the last two. In applying these, it may be necessary to write them in various equivalent forms, for example, $\cos^2 \theta = 1 - \sin^2 \theta$ or $\sec^2 \theta - 1 = \tan^2 \theta$, but it is probably best to concentrate on learning them in just one form, such as the one given. You can then mentally rewrite them in various ways.

Starting from any point on the unit circle, if we go 2π units along the circle either clockwise or counterclockwise, we wind up at the same point, since the circumference of the unit circle is 2π. Thus, for any θ, $P(\theta) = P(\theta + 2n\pi)$, where n is any integer, positive or negative. Since the values of the trigonometric functions depend only on the coordinates of $P(\theta)$, it follows that

$$\sin(\theta + 2n\pi) = \sin \theta$$
$$\cos(\theta + 2n\pi) = \cos \theta$$

and so on, for the remaining trigonometric functions. These can be expressed briefly by saying that each of the trigonometric functions is **periodic,** with **period 2π.*

Figure 9

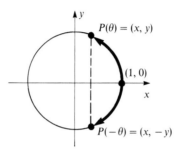

Consider next the relationships between the functions of θ and $-\theta$. By symmetry we see that if the coordinates of $P(\theta)$ are (x, y), then those of $P(-\theta)$ are $(x, -y)$ (Figure 9). The sine of $-\theta$ is by definition the y coordinate of $P(-\theta)$, but this is the negative of the y coordinate of θ. Thus, $\sin(-\theta) = -\sin \theta$. Since the x coordinates of $P(\theta)$ and $P(-\theta)$ are the same, it follows that $\cos(-\theta) = \cos \theta$. Similar relationships for the other functions can now be obtained:

$$\sin(-\theta) = -\sin \theta \qquad \csc(-\theta) = -\csc \theta$$
$$\cos(-\theta) = \cos \theta \qquad \sec(-\theta) = \sec \theta$$
$$\tan(-\theta) = -\tan \theta \qquad \cot(-\theta) = -\cot \theta$$

It is most important to learn the relationships $\sin(-\theta) = -\sin \theta$ and $\cos(-\theta) = \cos \theta$, since they occur more frequently than the others and since the others can be deduced easily from these two.

---

* In general, if $f(x + k) = f(x)$ for all x, then f is said to be *periodic with period k*. The smallest positive k for which this is true is called the **fundamental period.** For the sine, cosine, secant, and cosecant, the fundamental period is 2π, but for the tangent and cotangent, it is π.

EXAMPLE 3   Make use of the basic identities of this section to find the other five trigonometric functions of θ if $\tan \theta = -2$ and $\cos \theta < 0$.

Solution   First we observe that $\cot \theta = 1/\tan \theta = -\frac{1}{2}$. Next, since $\sec^2 \theta = 1 + \tan^2 \theta = 1 + (-2)^2 = 5$, we have $\sec \theta = \pm\sqrt{5}$. But $\cos \theta$ is negative, so that $\sec \theta$, which is $1/\cos \theta$, is also negative. Thus, $\sec \theta = -\sqrt{5}$. Therefore, $\cos \theta = -1/\sqrt{5}$.

Now we use $\sin^2 \theta = 1 - \cos^2 \theta = 1 - \frac{1}{5} = \frac{4}{5}$. Since $\tan \theta$ and $\cos \theta$ are both negative, and $\tan \theta = \sin \theta/\cos \theta$, it follows that $\sin \theta$ is positive. Thus, $\sin \theta = 2/\sqrt{5}$. Finally, $\csc \theta = 1/\sin \theta = \sqrt{5}/2$.

Note.   We could also have determined the signs of each of the functions by observing that for both $\tan \theta$, which is y/x, and $\cos \theta$, which is x, to be negative, we must have y positive, so that the point $P(\theta)$ is in the second quadrant, or equivalently, the angle of θ radians terminates in the second quadrant.

EXAMPLE 4   In each of the following use the basic identities to show that the first expression can be transformed into the second:

a. $\dfrac{\sin \theta}{\tan \theta}$, $\cos \theta$ **b.** $\dfrac{\cos \theta}{\tan \theta}$, $\csc \theta - \sin \theta$

Solution   **a.**   Since $\tan \theta = \sin \theta/\cos \theta$, we have

$$\frac{\sin \theta}{\tan \theta} = \frac{\sin \theta}{\dfrac{\sin \theta}{\cos \theta}} = \sin \theta \cdot \frac{\cos \theta}{\sin \theta} = \cos \theta$$

b.
$$\frac{\cos \theta}{\tan \theta} = \cos \theta \cdot \frac{\cos \theta}{\sin \theta} = \frac{\cos^2 \theta}{\sin \theta}$$

Now we use the identities $\cos^2 \theta = 1 - \sin^2 \theta$ and $\csc \theta = 1/\sin \theta$ to get

$$\frac{\cos^2 \theta}{\sin \theta} = \frac{1 - \sin^2 \theta}{\sin \theta} = \frac{1}{\sin \theta} - \frac{\sin^2 \theta}{\sin \theta} = \csc \theta - \sin \theta$$

EXERCISE SET 2

A   In Problems 1–8 find the remaining five trigonometric functions of θ, using the basic identities of this section.

1. $\sin \theta = -\frac{3}{5}$, $\tan \theta > 0$
2. $\sec \theta = \frac{13}{12}$, $\csc \theta < 0$
3. $\tan \theta = -\frac{4}{3}$, $\pi/2 < \theta < 3\pi/2$
4. $\cot \theta = \frac{5}{12}$, $\pi < \theta < 2\pi$
5. $\cos \theta = 2/\sqrt{5}$, $\cot \theta < 0$
6. $\csc \theta = -\frac{17}{5}$, $\tan \theta > 0$
7. $\sin \theta = \frac{1}{3}$, $5\pi/2 < \theta < 7\pi/2$
8. $\sec \theta = 3$, $3\pi < \theta < 4\pi$

In Problems 9–18 show that the first expression can be transformed into the second, using basic identities.

9. $\dfrac{\tan \theta}{\sin \theta}$, $\sec \theta$

10. $\dfrac{\cot \theta}{\csc \theta}$, $\cos \theta$

11. $\dfrac{\sin \theta}{\cot \theta}$, $\sec \theta - \cos \theta$

12. $\sec^2 \theta \sin^2 \theta$, $\sec^2 \theta - 1$

13. $\cot \theta \sec \theta$, $\csc \theta$

14. $\tan \theta \csc \theta$, $\sec \theta$

15. $\dfrac{\sec \theta}{\csc \theta}$, $\tan \theta$

16. $1 - \dfrac{\sin \theta}{\csc \theta}$, $\cos^2 \theta$

17. $\dfrac{1}{\sec^2 \theta} + \dfrac{1}{\csc^2 \theta}$, 1

18. $\sec^2 \theta - \sin^2 \theta \sec^2 \theta$, 1

B   19.   By making the substitution $t = \tan \theta$, where $-\pi/2 < \theta < \pi/2$, show that

$$\frac{\sqrt{1 + t^2}}{t} = \csc \theta$$

20.   By making the substitution $t = 2 \sin \theta$, where $-\pi/2 < \theta < \pi/2$, show that

$$\frac{t}{\sqrt{4 - t^2}} = \tan \theta$$

21.   By making the substitution $t = \frac{3}{2} \sec \theta$, where $0 < \theta < \pi/2$ if $t > 0$ and $\pi < \theta < 3\pi/2$ if $t < 0$, show that

$$\frac{\sqrt{4t^2 - 9}}{t} = 2 \sin \theta$$

22.   If $\tan \theta = t$, prove that

$$\sin \theta \cos \theta = \frac{t}{1 + t^2}$$

23.   If $\cot \theta = t$, prove that

$$\sec \theta \csc \theta = \frac{1 + t^2}{t}$$

---

3   The Addition Formulas for Sine and Cosine

We will develop in this section formulas for the sum and difference of two numbers (or of two angles). These are of fundamental importance in deriving other identities. We designate by α and β two numbers which we will initially restrict to be between 0 and 2π and for which $\alpha > \beta$. These restrictions will soon be removed.

In Figure 10 we show typical positions of $P(\alpha)$ and $P(\beta)$. Now the distance along the circle from $(1, 0)$ to $P(\alpha)$ is α, and the distance to $P(\beta)$ is β. So the arc length from $P(\beta)$ to $P(\alpha)$ is $\alpha - \beta$. If we think of sliding this latter arc along the circle until the point originally at $P(\beta)$ coincides with $(1, 0)$, then the distance $\alpha - \beta$ is in standard position and the end point of the arc is correctly labeled $P(\alpha - \beta)$. Since the x and y coordinates of a point $P(\theta)$ are $\cos \theta$ and $\sin \theta$, respectively, we can show the coordinates of the three points $P(\alpha)$, $P(\beta)$, and $P(\alpha - \beta)$ as in Figure 11. Now, since the arc from $(1, 0)$ to $P(\alpha - \beta)$ is equal in length to the arc from $P(\beta)$ to $P(\alpha)$, it follows that the chords connecting these respective pairs of points are also equal. We can get these lengths using the

Figure 10

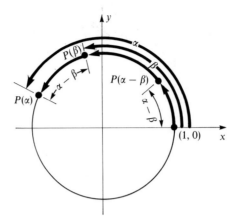

Figure 11

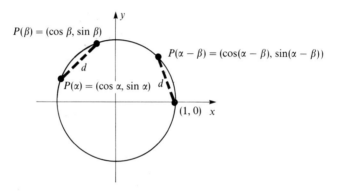

distance formula:

$$\text{Distance from } (1, 0) \text{ to } P(\alpha - \beta) = \sqrt{[\cos(\alpha - \beta) - 1]^2 + [\sin(\alpha - \beta) - 0]^2}$$

$$\text{Distance from } P(\beta) \text{ to } P(\alpha) = \sqrt{(\cos \alpha - \cos \beta)^2 + (\sin \alpha - \sin \beta)^2}$$

Since these distances are the same, their squares are the same; so we obtain the equation

$$[\cos(\alpha - \beta) - 1]^2 + [\sin(\alpha - \beta)]^2 = (\cos \alpha - \cos \beta)^2 + (\sin \alpha - \sin \beta)^2$$

On squaring and collecting terms, and making use several times of the identity $\sin^2 \theta + \cos^2 \theta = 1$, this gives

$$\cos^2(\alpha - \beta) - 2\cos(\alpha - \beta) + 1 + \sin^2(\alpha - \beta)$$

$$= \cos^2 \alpha - 2\cos \alpha \cos \beta + \cos^2 \beta + \sin^2 \alpha - 2\sin \alpha \sin \beta + \sin^2 \beta$$

$$2 - 2\cos(\alpha - \beta) = 2 - 2(\cos \alpha \cos \beta + \sin \alpha \sin \beta)$$

or finally,

$$\mathbf{cos(\alpha - \beta) = \cos \alpha \cos \beta + \sin \alpha \sin \beta} \tag{1}$$

We next remove the restrictions imposed on the sizes of α and β. First, suppose $\beta > \alpha$ and both α and β are between 0 and 2π. Then, since $\cos \theta = \cos(-\theta)$,

$$\cos(\alpha - \beta) = \cos[-(\alpha - \beta)] = \cos(\beta - \alpha)$$

and now equation (1) applies, with the roles of α and β reversed. So

$$\cos(\alpha - \beta) = \cos(\beta - \alpha)$$
$$= \cos\beta\cos\alpha + \sin\beta\sin\alpha$$
$$= \cos\alpha\cos\beta + \sin\alpha\sin\beta$$

that is, equation (1) is true regardless of whether α or β is larger, as long as both are between 0 and 2π. The formula is also true if $\alpha = \beta$, because then we have

$$\cos(\alpha - \beta) = \cos 0$$
$$= \cos\alpha\cos\alpha + \sin\alpha\sin\alpha$$
$$= \cos^2\alpha + \sin^2\alpha$$
$$= 1$$

which is true.

Now suppose α and β are not restricted to lie between 0 and 2π. Then we can always write

$$\alpha = \alpha_1 + 2n\pi \quad\text{and}\quad \beta = \beta_1 + 2m\pi$$

where α_1 and β_1 do lie between 0 and 2π and m and n are appropriately chosen integers. Because of periodicity we have that $\sin\alpha = \sin(\alpha_1 + 2n\pi) = \sin\alpha_1$, and similarly for β and β_1 with the cosine:

$$\cos(\alpha - \beta) = \cos(\alpha_1 + 2n\pi - \beta_1 - 2m\pi)$$
$$= \cos[(\alpha_1 - \beta_1) + (n - m)2\pi]$$
$$= \cos(\alpha_1 - \beta_1)$$
$$= \cos\alpha_1\cos\beta_1 + \sin\alpha_1\sin\beta_1$$
$$= \cos\alpha\cos\beta + \sin\alpha\sin\beta$$

The third step, namely, that

$$\cos[(\alpha_1 - \beta_1) + (n - m)2\pi] = \cos(\alpha_1 - \beta_1)$$

again follows from periodicity. This establishes equation (1) for *all* numbers α and β.

From equation (1) we are able to get a number of other results. Consider first $\cos(\alpha + \beta)$. We write $\alpha + \beta = \alpha - (-\beta)$ and apply (1), replacing β by $-\beta$:

$$\cos(\alpha + \beta) = \cos(\alpha - (-\beta)) = \cos\alpha\cos(-\beta) + \sin\alpha\sin(-\beta)$$

Since $\cos(-\beta) = \cos\beta$ and $\sin(-\beta) = -\sin\beta$,

$$\mathbf{\cos(\alpha + \beta) = \cos\alpha\cos\beta - \sin\alpha\sin\beta} \tag{2}$$

To get analogous formulas to equations (1) and (2) for the sine, we proceed as follows: In (1), replace α by $\pi/2$ and allow β to be arbitrary. Remember that $\sin\pi/2 = 1$ and $\cos\pi/2 = 0$. So,

$$\cos\left(\frac{\pi}{2} - \beta\right) = \cos\frac{\pi}{2}\cos\beta + \sin\frac{\pi}{2}\sin\beta$$
$$= 0 + \sin\beta$$
$$= \sin\beta$$

That is, for any number θ:

$$\cos\left(\frac{\pi}{2} - \theta\right) = \sin\theta \tag{3}$$

In particular, if we take $\theta = (\pi/2) - \alpha$, we obtain

$$\cos\left[\frac{\pi}{2} - \left(\frac{\pi}{2} - \alpha\right)\right] = \sin\left(\frac{\pi}{2} - \alpha\right)$$

or

$$\cos\alpha = \sin\left(\frac{\pi}{2} - \alpha\right) \tag{4}$$

for any number α. The letters θ and α used in equations (3) and (4) are independent of any relationships expressed between them in the derivation, that is, the formulas are self-contained. The letters are really just dummy variables, and any other letters would express the same relationships. So we could write, for example,

$$\cos\left(\frac{\pi}{2} - x\right) = \sin x$$

and

$$\sin\left(\frac{\pi}{2} - x\right) = \cos x$$

If in the first of these we take x as $\alpha + \beta$, we have, by equation (2),

$$\sin(\alpha + \beta) = \cos\left[\frac{\pi}{2} - (\alpha + \beta)\right] = \cos\left[\left(\frac{\pi}{2} - \alpha\right) - \beta\right]$$

$$= \cos\left(\frac{\pi}{2} - \alpha\right)\cos\beta + \sin\left(\frac{\pi}{2} - \alpha\right)\sin\beta$$

Now we use equations (3) and (4) again to get

$$\sin(\alpha + \beta) = \sin\alpha\cos\beta + \cos\alpha\sin\beta \tag{5}$$

This is valid for all numbers α and β. If we write $\alpha - \beta = \alpha + (-\beta)$, we can use equation (5) together with the facts that $\sin(-\theta) = -\sin\theta$ and $\cos(-\theta) = \cos\theta$ to get

$$\sin(\alpha - \beta) = \sin[\alpha + (-\beta)] = \sin\alpha\cos(-\beta) + \cos\alpha\sin(-\beta)$$

$$= \sin\alpha\cos\beta - \cos\alpha\sin\beta \tag{6}$$

We summarize formulas (1), (2), (5), and (6) in the box.

The Addition Formulas for Sine and Cosine

$$\sin(\alpha + \beta) = \sin\alpha\cos\beta + \cos\alpha\sin\beta$$

$$\sin(\alpha - \beta) = \sin\alpha\cos\beta - \cos\alpha\sin\beta$$

$$\cos(\alpha + \beta) = \cos\alpha\cos\beta - \sin\alpha\sin\beta$$

$$\cos(\alpha - \beta) = \cos\alpha\cos\beta + \sin\alpha\sin\beta$$

These are of fundamental importance and should be learned. Notice that for the sine, the sign between the terms on the right agrees with the sign between α and β on the left, whereas for the cosine these are reversed. So there really are just two basic patterns to learn.

EXAMPLE 5 Find the value of each of the following without the use of tables or a calculator:
a. $\sin(5\pi/12)$ **b.** $\sin(\pi/12)$ **c.** $\cos(13\pi/12)$ **d.** $\cos(-\pi/12)$

Solution **a.** We can write $5\pi/12$ as $2\pi/12 + 3\pi/12 = \pi/6 + \pi/4$. So

$$\sin \frac{5\pi}{12} = \sin\left(\frac{\pi}{6} + \frac{\pi}{4}\right)$$

$$= \sin\frac{\pi}{6}\cos\frac{\pi}{4} + \cos\frac{\pi}{6}\sin\frac{\pi}{4}$$

Now we know that we may treat $\pi/6$ and $\pi/4$ as if they are radian measures of angles. Thus,

$$\sin \frac{5\pi}{12} = \left(\frac{1}{2}\right)\left(\frac{1}{\sqrt{2}}\right) + \left(\frac{\sqrt{3}}{2}\right)\left(\frac{1}{\sqrt{2}}\right) = \frac{1 + \sqrt{3}}{2\sqrt{2}}$$

b. $$\sin \frac{\pi}{12} = \sin\left(\frac{3\pi}{12} - \frac{2\pi}{12}\right) = \sin\left(\frac{\pi}{4} - \frac{\pi}{6}\right)$$

$$= \sin\frac{\pi}{4}\cos\frac{\pi}{6} - \cos\frac{\pi}{4}\sin\frac{\pi}{6}$$

$$= \left(\frac{1}{\sqrt{2}}\right)\left(\frac{\sqrt{3}}{2}\right) - \left(\frac{1}{\sqrt{2}}\right)\left(\frac{1}{2}\right) = \frac{\sqrt{3} - 1}{2\sqrt{2}}$$

c. $$\cos \frac{13\pi}{12} = \cos\left(\frac{9\pi}{12} + \frac{4\pi}{12}\right) = \cos\left(\frac{3\pi}{4} + \frac{\pi}{3}\right)$$

$$= \cos\frac{3\pi}{4}\cos\frac{\pi}{3} - \sin\frac{3\pi}{4}\sin\frac{\pi}{3}$$

$$= \left(-\frac{1}{\sqrt{2}}\right)\left(\frac{1}{2}\right) - \left(\frac{1}{\sqrt{2}}\right)\left(\frac{\sqrt{3}}{2}\right) = -\frac{1 + \sqrt{3}}{2\sqrt{2}}$$

d. $\cos(-\pi/12) = \cos(\pi/12)$, since $\cos(-\theta) = \cos\theta$; so,

$$\cos\left(-\frac{\pi}{12}\right) = \cos\frac{\pi}{12} = \cos\left(\frac{3\pi}{12} - \frac{2\pi}{12}\right) = \cos\left(\frac{\pi}{4} - \frac{\pi}{6}\right)$$

$$= \cos\frac{\pi}{4}\cos\frac{\pi}{6} + \sin\frac{\pi}{4}\sin\frac{\pi}{6}$$

$$= \left(\frac{1}{\sqrt{2}}\right)\left(\frac{\sqrt{3}}{2}\right) + \left(\frac{1}{\sqrt{2}}\right)\left(\frac{1}{2}\right) = \frac{\sqrt{3} + 1}{2\sqrt{2}}$$

EXAMPLE 6   If $-\pi/2 \le x \le \pi/2$ and $0 \le y \le \pi$, $\sin x = \frac{3}{4}$, and $\cos y = -\frac{5}{6}$, find:
a. $\sin(x + y)$ b. $\cos(x - y)$

Solution   We conclude that x is between 0 and $\pi/2$, since if it were between 0 and $-\pi/2$, $\sin x$ would be negative. Similarly, we must have y between $\pi/2$ and π in order for $\cos y$ to be negative. By the Pythagorean identity, $\sin^2 x + \cos^2 x = 1$, we get

$$\cos^2 x = 1 - \sin^2 x$$

So

$$\cos x = \pm\sqrt{1 - \sin^2 x}$$

The positive sign must be chosen, since $0 \le x \le \pi/2$. So,

$$\cos x = \sqrt{1 - \left(\frac{3}{4}\right)^2} = \sqrt{1 - \frac{9}{16}} = \sqrt{\frac{7}{16}} = \frac{\sqrt{7}}{4}$$

Similarly,

$$\sin y = \sqrt{1 - \cos^2 y} = \sqrt{1 - \left(-\frac{5}{6}\right)^2} = \sqrt{1 - \frac{25}{36}} = \frac{\sqrt{11}}{6}$$

We now have all the necessary ingredients for obtaining the solutions.

a. $\sin(x + y) = \sin x \cos y + \cos x \sin y = \dfrac{3}{4}\left(-\dfrac{5}{6}\right) + \left(\dfrac{\sqrt{7}}{4}\right)\left(\dfrac{\sqrt{11}}{6}\right)$

$$= \frac{-15 + \sqrt{77}}{24}$$

b. $\cos(x - y) = \cos x \cos y + \sin x \sin y = \dfrac{\sqrt{7}}{4}\left(-\dfrac{5}{6}\right) + \dfrac{3}{4}\left(\dfrac{\sqrt{11}}{6}\right)$

$$= \frac{-5\sqrt{7} + 3\sqrt{11}}{24}$$

EXERCISE SET 3

A   In Problems 1–8 evaluate without using a calculator or tables.

1. $\sin(\alpha + \beta)$ and $\cos(\alpha + \beta)$,   where $\alpha = \pi/4$ and $\beta = \pi/3$
2. $\sin(\alpha - \beta)$ and $\cos(\alpha - \beta)$,   where $\alpha = \pi/4$ and $\beta = \pi/3$

3.   a. $\sin\left(\dfrac{\pi}{3} - \dfrac{3\pi}{4}\right)$ b. $\cos\left(\dfrac{5\pi}{6} + \dfrac{\pi}{4}\right)$

4.   a. $\sin\left(\dfrac{5\pi}{4} + \dfrac{11\pi}{6}\right)$ b. $\cos\left(\dfrac{4\pi}{3} - \dfrac{3\pi}{4}\right)$

5.   a. $\cos(7\pi/12)$ b. $\sin(17\pi/12)$
6.   a. $\sin(-\pi/12)$ b. $\cos(-5\pi/12)$
7.   a. $\sin 75°$ b. $\cos 15°$
8.   a. $\cos 255°$ b. $\sin 195°$

9. If $\sin \alpha = -\frac{3}{5}$, $\sin \beta = -\frac{5}{13}$, $P(\alpha)$ is in the third quadrant, and $P(\beta)$ is in the fourth quadrant, find the following:
   a. $\sin(\alpha - \beta)$ b. $\cos(\alpha + \beta)$

10. If $\cos \alpha = \frac{12}{13}$, $\cos \beta = -\frac{4}{5}$, $P(\alpha)$ is in the fourth quadrant, and $P(\beta)$ is in the second quadrant, find the following:
    a. $\sin(\alpha + \beta)$ b. $\cos(\alpha - \beta)$

11. If $\sin \alpha < 0$, $\cos \alpha = \frac{1}{3}$, $\cos \beta < 0$, and $\sin \beta = -\frac{2}{3}$, find the following:
    a. $\sin(\alpha + \beta)$ b. $\cos(\alpha - \beta)$

12. If $\tan x = -\frac{3}{4}$, $\csc x > 0$, $\sec y = \frac{13}{5}$, and $\cot y < 0$, find the following:
    a. $\sin(x - y)$ b. $\cos(x + y)$

13. If $\sin \alpha = -2/\sqrt{5}$, $\tan \alpha > 0$, $\cos \beta = -\frac{8}{17}$, and $\csc \beta > 0$, find the following:
    a. $\sin(\alpha - \beta)$ b. $\cos(\alpha + \beta)$

14. If $\cot A = -\frac{24}{7}$, $\sec A > 0$, $\tan B = \frac{4}{3}$, and $\sin B < 0$, find the following:
    a. $\sin(A + B)$ b. $\cos(A - B)$

15. If $\sin x = -\frac{3}{5}$, $\tan x > 0$, $\sec y = -\frac{13}{5}$, and $\cot y < 0$, find the following:
    a. $\csc(x + y)$ b. $\sec(x - y)$

Establish the formulas in Problems 16–20 by using addition formulas.

16. a. $\sin(\pi + \theta) = -\sin \theta$ b. $\cos(\pi + \theta) = -\cos \theta$

17. a. $\sin\left(\dfrac{\pi}{2} + \theta\right) = \cos \theta$ b. $\cos\left(\dfrac{\pi}{2} + \theta\right) = -\sin \theta$

18. a. $\sin\left(\dfrac{3\pi}{2} - \theta\right) = -\cos \theta$ b. $\cos\left(\dfrac{3\pi}{2} - \theta\right) = -\sin \theta$

19. a. $\sin(\pi - \theta) = \sin \theta$ b. $\cos(\pi - \theta) = -\cos \theta$

20. a. $\sin\left(\dfrac{\pi}{2} - \theta\right) = \cos \theta$ b. $\cos\left(\dfrac{\pi}{2} - \theta\right) = \sin \theta$

B 21. Prove that, in general, $\sin(\alpha + \beta) \neq \sin \alpha + \sin \beta$.

22. Find all values of x for which $0 \leq x < 2\pi$ and

$$\sin 5x \cos 4x = \cos 5x \sin 4x$$

23. Find all values of x for which $0 \leq x < 2\pi$ and

$$2 \cos 2x \cos x = 1 - 2 \sin 2x \sin x$$

24. Derive a formula for
    a. $\sin(\alpha + \beta + \gamma)$ b. $\cos(\alpha + \beta + \gamma)$
    Hint. First use the associative property for addition.

4 Double-Angle, Half-Angle, and Reduction Formulas

The importance of the addition formulas in Section 3 lies primarily in the fact that so many other identities can be derived from them. We will carry out the derivations for some of the most important of these.

The **double-angle formulas** are obtained from the addition formulas for sine and cosine of α and β by putting $\beta = \alpha$. If we denote the common value of α and β by θ, we obtain

$$\sin(\theta + \theta) = \sin \theta \cos \theta + \cos \theta \sin \theta$$
$$= 2 \sin \theta \cos \theta$$

So,

$$\sin 2\theta = 2 \sin \theta \cos \theta$$

Also,

$$\cos(\theta + \theta) = \cos \theta \cos \theta - \sin \theta \sin \theta$$
$$= \cos^2 \theta - \sin^2 \theta$$

So,

$$\cos 2\theta = \cos^2 \theta - \sin^2 \theta \tag{7}$$

Two other useful forms of $\cos 2\theta$ can be obtained by replacing, in turn, $\cos^2 \theta$ by $1 - \sin^2 \theta$ and $\sin^2 \theta$ by $1 - \cos^2 \theta$. This gives

$$\cos 2\theta = (1 - \sin^2 \theta) - \sin^2 \theta = 1 - 2 \sin^2 \theta$$

and

$$\cos 2\theta = \cos^2 \theta - (1 - \cos^2 \theta) = 2 \cos^2 \theta - 1$$

So we have

$$\cos 2\theta = 1 - 2 \sin^2 \theta \tag{8}$$

and

$$\cos 2\theta = 2 \cos^2 \theta - 1 \tag{9}$$

Whether to use equation (7), (8), or (9) depends on the objective. We will shortly see some examples where a particular form is clearly preferable.

If we solve equation (8) for $\sin^2 \theta$ and equation (9) for $\cos^2 \theta$, we obtain

$$\sin^2 \theta = \frac{1 - \cos 2\theta}{2}$$

and

$$\cos^2 \theta = \frac{1 + \cos 2\theta}{2}$$

These forms are employed extensively in calculus.

By replacing θ by $\alpha/2$ in the last two equations and then taking square roots, we get the **half-angle formulas:**

$$\sin \frac{\alpha}{2} = \pm \sqrt{\frac{1 - \cos \alpha}{2}}$$

$$\cos \frac{\alpha}{2} = \pm \sqrt{\frac{1 + \cos \alpha}{2}}$$

The ambiguity of sign has to be resolved in each particular instance according to the quadrant in which $\alpha/2$ lies.

From the addition formulas we can obtain a class of identities, sometimes called **reduction formulas,** of which equations (3) and (4) of Section 3 are special cases. Here are some others:

$$\sin(\pi + \theta) = \sin \pi \cos \theta + \cos \pi \sin \theta$$
$$= -\sin \theta$$
$$\sin(\pi - \theta) = \sin \pi \cos \theta - \cos \pi \sin \theta$$
$$= \sin \theta$$

Similarly,

$$\cos(\pi + \theta) = \cos \pi \cos \theta - \sin \pi \sin \theta$$
$$= -\cos \theta$$
$$\cos(\pi - \theta) = \cos \pi \cos \theta + \sin \pi \sin \theta$$
$$= -\cos \theta$$

In these, we have used the facts that $\sin \pi = 0$ and $\cos \pi = -1$. Since $\sin(3\pi/2) = -1$ and $\cos(3\pi/2) = 0$, we also have

$$\sin\left(\frac{3\pi}{2} + \theta\right) = \sin \frac{3\pi}{2} \cos \theta + \cos \frac{3\pi}{2} \sin \theta$$
$$= -\cos \theta$$
$$\sin\left(\frac{3\pi}{2} - \theta\right) = \sin \frac{3\pi}{2} \cos \theta - \cos \frac{3\pi}{2} \sin \theta$$
$$= -\cos \theta$$

Likewise, there are similar formulas for $\cos[(3\pi/2) \pm \theta]$.

Rather than continuing this approach, let us attempt to generalize. It appears that we should consider two categories:

1. Functions of $(X \pm \theta)$, where X is of the form $n\pi$ (that is, when considered as an angle in radians, X terminates on the x axis)
2. Functions of the form $(Y \pm \theta)$, where Y is an odd multiple of $\pi/2$ (that is, the angle of Y radians terminates on the y axis)

We give the following rule of thumb, which can be shown to be valid by means of the addition formulas:

Rule of Thumb for Reduction Problems

In case 1,

$$\sin(X \pm \theta) = \pm\sin \theta \quad \text{and} \quad \cos(X \pm \theta) = \pm\cos \theta$$

In case 2,

$$\sin(Y \pm \theta) = \pm\cos \theta \quad \text{and} \quad \cos(Y \pm \theta) = \pm\sin \theta$$

To determine the correct sign on the right-hand side, decide in which quadrant $X \pm \theta$ or $Y \pm \theta$ would lie if θ were acute, and use the sign of the given function on the left in that quadrant.

Let us illustrate with $\cos[(3\pi/2) + \theta]$. This is a case 2 problem, and so by our rule of thumb it reduces to $\pm\sin \theta$. But which sign should we choose? If θ were acute, $(3\pi/2) + \theta$ would lie in the fourth quadrant, and in the fourth quadrant the cosine is positive. Therefore, we choose the positive sign. So,

$$\cos\left(\frac{3\pi}{2} + \theta\right) = \sin \theta$$

Remark. It should be emphasized that in our rule of thumb we determine the sign by assuming θ is acute, but the result is valid for *all* values of θ. In Problem 23, Exercise Set 4, verification of this statement is asked for.

EXERCISE SET 4

A   In Problems 1–6 find $\sin 2\theta$ and $\cos 2\theta$.

1. $\sin \theta = \frac{4}{5}$ and θ terminates in quadrant II

2. $\cos \theta = -\frac{1}{3}$ and θ terminates in quadrant III

3. $\tan \theta = \frac{5}{12}$ and $\sin \theta < 0$ **4.** $\sec \theta = \sqrt{5}$ and $\csc \theta < 0$

5. $\cot \theta = -2$ and $0 \le \theta \le \pi$ **6.** $\csc \theta = \frac{17}{8}$ and $\frac{\pi}{2} \le \theta \le \frac{3\pi}{2}$

7. If $\sin \theta = x$ and $-\pi/2 \le \theta \le \pi/2$, find $\sin 2\theta$ and $\cos 2\theta$ in terms of x.

8. If $\cos \theta = x$ and $0 \le \theta \le \pi$, find $\sin 2\theta$ and $\cos 2\theta$ in terms of x.

In Problems 9–12 find $\sin \theta$ and $\cos \theta$.

9. $\cos 2\theta = \frac{1}{3}$ and $\pi \le 2\theta \le 2\pi$ **10.** $\sin 2\theta = -\frac{4}{5}$ and $\pi \le 2\theta \le 3\pi/2$

11. $\sec 2\theta = \frac{25}{7}$ and $-\pi \le 2\theta \le 0$ **12.** $\tan 2\theta = -\frac{12}{5}$ and $3\pi \le 2\theta \le 4\pi$

13. Use the half-angle formulas to find:
    a. $\sin(\pi/8)$ **b.** $\cos(\pi/12)$ **c.** $\sin 75°$ **d.** $\cos 67.5°$

In Problems 14–17 find $\sin(\alpha/2)$ and $\cos(\alpha/2)$.

14. $\cos \alpha = \frac{7}{25}, \quad 0 \le \alpha \le \pi$ **15.** $\sin \alpha = -\frac{12}{13}, \quad \pi \le \alpha \le 3\pi/2$

16. $\tan \alpha = -\frac{8}{17}, \quad 2\pi \le \alpha \le 3\pi$ **17.** $\sec \alpha = 3, \quad -\pi/2 \le \alpha \le 0$

18. If $P(\theta) = (1/\sqrt{5}, -2/\sqrt{5})$, find $P(2\theta)$.

19. If $0 \le 2\theta \le \pi$ and $P(2\theta) = (-\frac{7}{25}, \frac{24}{25})$, find $P(\theta)$.

In Problems 20–22 reduce to a function involving θ only.

20. a. $\sin(2\pi - \theta)$ **b.** $\cos\left(\dfrac{3\pi}{2} - \theta\right)$ **c.** $\sin\left(\dfrac{5\pi}{2} + \theta\right)$

    d. $\cos(-\pi + \theta)$ **e.** $\sin\left(-\dfrac{\pi}{2} - \theta\right)$

21. a. $\sin(\theta - \pi)$ **b.** $\cos\left(\dfrac{\pi}{2} + \theta\right)$ **c.** $\sin\left(-\dfrac{3\pi}{2} - \theta\right)$

    d. $\cos(\theta - 3\pi)$ **e.** $\sin\left(\dfrac{\pi}{2} + \theta\right)$

22. a. $\cos\left(\dfrac{3\pi}{2} + \theta\right)$ **b.** $\sin(\pi + \theta)$ **c.** $\cos(\theta - 2\pi)$

    d. $\sin\left(-\dfrac{\pi}{2} + \theta\right)$ **e.** $\cos\left(\theta - \dfrac{5\pi}{2}\right)$

B   **23.** Demonstrate that the rule of thumb for reduction problems is correct for each of the following cases:
    a. $X + \theta$, where X is coterminal with 0
    b. $X + \theta$, where X is coterminal with π

  c. $X - \theta$, where X is coterminal with 0
  d. $X - \theta$, where X is coterminal with π
  e. $Y + \theta$, where Y is coterminal with $\pi/2$
  f. $Y + \theta$, where Y is coterminal with $3\pi/2$
  g. $Y - \theta$, where Y is coterminal with $\pi/2$
  h. $Y - \theta$, where Y is coterminal with $3\pi/2$
24. If $\tan \theta = x$ and $-\pi/2 < \theta < \pi/2$, find $\sin 2\theta$ and $\cos 2\theta$ in terms of x.
25. If $\sec \theta = x$ and $0 \le \theta \le \pi$, find:
  a. $\sin 2\theta$ **b.** $\cos 2\theta$ **c.** $\sin \tfrac{1}{2}\theta$ **d.** $\cos \tfrac{1}{2}\theta$
26. By writing $3\theta = 2\theta + \theta$, find formulas for $\sin 3\theta$ and $\cos 3\theta$ in terms of functions of θ.
27. Derive the formula $\sin 4\theta = 4 \sin \theta \cos \theta - 8 \sin^3 \theta \cos \theta$.
28. Derive the formula $\cos 4\theta = 8 \cos^4 \theta - 8 \cos^2 \theta + 1$.
29. By calculating $\sin(\pi/12)$ in two ways, using the half-angle formulas and using the addition formulas, prove that

$$\sqrt{2 - \sqrt{3}} = \tfrac{1}{2}(\sqrt{6} - \sqrt{2})$$

5 Further Identities

To obtain addition formulas for the tangent, it is necessary only to use the fact that $\tan \theta = (\sin \theta)/(\cos \theta)$. Thus,

$$\tan(\alpha + \beta) = \frac{\sin(\alpha + \beta)}{\cos(\alpha + \beta)} = \frac{\sin \alpha \cos \beta + \cos \alpha \sin \beta}{\cos \alpha \cos \beta - \sin \alpha \sin \beta}$$

This formula can be improved by dividing the numerator and denominator by $\cos \alpha \cos \beta$:

$$\tan(\alpha + \beta) = \frac{\dfrac{\sin \alpha \cos \beta}{\cos \alpha \cos \beta} + \dfrac{\cos \alpha \sin \beta}{\cos \alpha \cos \beta}}{\dfrac{\cos \alpha \cos \beta}{\cos \alpha \cos \beta} - \dfrac{\sin \alpha \sin \beta}{\cos \alpha \cos \beta}}$$

$$= \frac{\tan \alpha + \tan \beta}{1 - \tan \alpha \tan \beta} \qquad (10)$$

And in an exactly similar way,

$$\tan(\alpha - \beta) = \frac{\tan \alpha - \tan \beta}{1 + \tan \alpha \tan \beta} \qquad (11)$$

  The double-angle formula for the tangent is obtained from equation (10) by taking $\alpha = \beta$. If we call this common value θ, we have

$$\tan(\theta + \theta) = \frac{\tan \theta + \tan \theta}{1 - \tan \theta \tan \theta}$$

So,

$$\tan 2\theta = \frac{2 \tan \theta}{1 - \tan^2 \theta}$$

We can derive half-angle formulas for the tangent as follows:

$$\tan \tfrac{1}{2}\alpha = \frac{\sin \tfrac{1}{2}\alpha}{\cos \tfrac{1}{2}\alpha} = \frac{\sin \tfrac{1}{2}\alpha}{\cos \tfrac{1}{2}\alpha} \cdot \frac{2 \cos \tfrac{1}{2}\alpha}{2 \cos \tfrac{1}{2}\alpha}$$

$$= \frac{2 \sin \tfrac{1}{2}\alpha \cos \tfrac{1}{2}\alpha}{2 \cos^2 \tfrac{1}{2}\alpha}$$

Since $2 \sin(\alpha/2) \cos(\alpha/2) = \sin 2(\alpha/2) = \sin \alpha$ and

$$2 \cos^2 \tfrac{1}{2}\alpha = 2 \left(\sqrt{\frac{1 + \cos \alpha}{2}} \right)^2 = 1 + \cos \alpha$$

we obtain

$$\tan \tfrac{1}{2}\alpha = \frac{\sin \alpha}{1 + \cos \alpha}$$

If in this derivation we had multiplied the numerator and denominator by $2 \sin(\alpha/2)$ instead of $2 \cos(\alpha/2)$, we would have obtained the equivalent formula

$$\tan \tfrac{1}{2}\alpha = \frac{1 - \cos \alpha}{\sin \alpha}$$

(See Problem 22, Exercise Set 5.)

We could obtain analogous formulas for the cotangent, secant and cosecant in a similar way, but these are so seldom used that we will not clutter up our already formidable list with them.

We do choose to list one more group of identities, which are known as the **sum and product formulas** for the sine and cosine. These can be obtained from the addition formulas (see Problems 20 and 21, Exercise Set 5). Although these are used less frequently than the other identities we have considered, they are indispensable at times.

Sum Formulas

$$\sin A + \sin B = 2 \sin \frac{A + B}{2} \cos \frac{A - B}{2}$$

$$\sin A - \sin B = 2 \cos \frac{A + B}{2} \sin \frac{A - B}{2}$$

$$\cos A + \cos B = 2 \cos \frac{A + B}{2} \cos \frac{A - B}{2}$$

$$\cos A - \cos B = -2 \sin \frac{A + B}{2} \sin \frac{A - B}{2}$$

Product Formulas

$$\sin \alpha \cos \beta = \tfrac{1}{2}[\sin(\alpha + \beta) + \sin(\alpha - \beta)]$$

$$\sin \alpha \sin \beta = \tfrac{1}{2}[\cos(\alpha - \beta) - \cos(\alpha + \beta)]$$

$$\cos \alpha \cos \beta = \tfrac{1}{2}[\cos(\alpha + \beta) + \cos(\alpha - \beta)]$$

EXERCISE SET 5

A 1. Use equations (10) and (11) to find:
   a. $\tan(5\pi/12)$ **b.** $\tan(\pi/12)$

   2. Find $\tan 2\theta$ if $P(\theta)$ is the point $(-\frac{1}{3}, 2\sqrt{2}/3)$.

   3. Find $\tan 2\theta$ if $\sin \theta = -\frac{3}{5}$ and $\cos \theta = -\frac{4}{5}$.

   4. Find $\tan(\alpha + \beta)$ if $\sin \alpha = \frac{4}{5}$, α is in the second quadrant, $\cos \beta = -\frac{5}{13}$, and β is in the third quadrant.

   5. Find $\tan(\alpha - \beta)$ if $\sec \alpha = 3$, $\csc \alpha < 0$, $\cot \beta = 2$, and $\cos \beta < 0$.

   6. If $\sec \alpha = \frac{13}{5}$, $\sin \alpha < 0$, $\csc \beta = \frac{17}{8}$, and $\cos \beta < 0$, find $\tan(\alpha + \beta)$ and $\tan(\alpha - \beta)$.

   7. For α and β as in Problem 6, find:
   a. $\tan 2\alpha$ **b.** $\tan 2\beta$ **c.** $\tan \frac{1}{2}\alpha$ **d.** $\tan \frac{1}{2}\beta$

   8. Find the value of each of the following without using tables or a calculator:
   a. $\tan 105°$ **b.** $\tan(\pi/12)$ **c.** $\tan(11\pi/12)$ **d.** $\tan 195°$
   e. $\tan(\pi/8)$

   9. Find $\tan 2\theta$ and $\tan \frac{1}{2}\theta$ if $\csc \theta = \frac{5}{3}$ and $\sec \theta < 0$.

   10. For the angles shown in the figures, find:

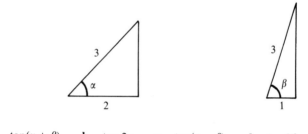

   a. $\tan(\alpha + \beta)$ **b.** $\tan 2\alpha$ **c.** $\tan(\alpha - \beta)$ **d.** $\tan 2\beta$
   e. $\tan \frac{1}{2}\alpha$ **f.** $\tan \frac{1}{2}\beta$

In Problems 11–14 evaluate by use of the sum and product formulas.

   11. **a.** $\sin(5\pi/12) + \sin(\pi/12)$ **b.** $\cos(7\pi/12) - \cos(\pi/12)$
   12. **a.** $\sin(5\pi/12)\cos(7\pi/12)$ **b.** $\cos(3\pi/8)\cos(\pi/8)$
   13. **a.** $\sin 105° - \sin 15°$ **b.** $\cos 165° + \cos 75°$
   14. **a.** $\sin 105° \sin 15°$ **b.** $\cos 165° \sin 75°$

   15. **a.** Write as a product: $\sin 5x + \sin 3x$
       b. Write as a sum: $\sin 5x \cos 3x$
   16. **a.** Write as a product: $\cos 7x - \cos 5x$
       b. Write as a sum: $\sin 7x \sin 5x$
   17. Derive the formula

$$\cot(\alpha + \beta) = \frac{\cot \alpha \cot \beta - 1}{\cot \alpha + \cot \beta}$$

   18. If $\cos \theta = x$ and $\sin \theta > 0$, find $\tan 2\theta$ and $\tan \frac{1}{2}\theta$ in terms of x.

B 19. Let X be an angle coterminal with 0 or π and let Y be an angle coterminal with $\pi/2$ or $3\pi/2$. Show that

$$\tan(X \pm \theta) = \pm\tan \theta \qquad \text{and} \qquad \tan(Y \pm \theta) = \pm\cot \theta$$

where the sign on the right in each equation is the sign of the tangent function in the quadrant in which $X \pm \theta$ and $Y \pm \theta$ would lie if θ were acute.

20. Use the addition formulas for the sine and cosine to derive the product formulas as follows:
   a. Add the formulas for $\sin(\alpha + \beta)$ and $\sin(\alpha - \beta)$ to get the formula for $\sin \alpha \cos \beta$.
   b. Add the formulas for $\cos(\alpha + \beta)$ and $\cos(\alpha - \beta)$ to get the formula for $\cos \alpha \cos \beta$.
   c. Subtract the formula for $\cos(\alpha + \beta)$ from that for $\cos(\alpha - \beta)$ to get the formula for $\sin \alpha \sin \beta$.

21. In the addition formulas for the sine and cosine make the following substitutions: $\alpha + \beta = A$ and $\alpha - \beta = B$. Solve these two equations simultaneously for α and β in terms of A and B, and obtain the sum formulas.

22. Derive the formula

$$\tan \frac{\alpha}{2} = \frac{1 - \cos \alpha}{\sin \alpha}$$

23. a. Let l be a nonvertical line and let α be the smallest positive angle from the positive x axis to l. Show that if m is the slope of l, then $m = \tan \alpha$.
   b. Let l_1 and l_2 be two nonvertical intersecting lines having slopes m_1 and m_2, respectively. Show that if θ is the angle from l_1 to l_2, then

$$\tan \theta = \frac{m_2 - m_1}{1 + m_1 m_2}$$

24. Use the formula in Problem 23, part b, to find the angle between the following pairs of lines:
   a. $3x - 2y + 7 = 0$ and $2x - 6y - 5 = 0$
   b. $2x - 3y + 4 = 0$ and $x + 2y - 6 = 0$

25. Find the interior angles of the triangle with vertices at $(3, 1)$, $(-1, -5)$, and $(-3, 3)$. (See Problem 23.)

---

6 Summary of Identities

We summarize below the identities given in Sections 2–5.

PYTHAGOREAN IDENTITIES $\sin^2 \theta + \cos^2 \theta = 1$ $1 + \tan^2 \theta = \sec^2 \theta$ $1 + \cot^2 \theta = \csc^2 \theta$

RECIPROCAL RELATIONS $\csc \theta = \dfrac{1}{\sin \theta}$ $\sec \theta = \dfrac{1}{\cos \theta}$ $\cot \theta = \dfrac{1}{\tan \theta}$

TANGENT AND COTANGENT IN TERMS OF SINE AND COSINE $\tan \theta = \dfrac{\sin \theta}{\cos \theta}$ $\cot \theta = \dfrac{\cos \theta}{\sin \theta}$

ADDITION FORMULAS	$\sin(\alpha \pm \beta) = \sin\alpha\cos\beta \pm \cos\alpha\sin\beta$
	$\cos(\alpha \pm \beta) = \cos\alpha\cos\beta \mp \sin\alpha\sin\beta$
	$\tan(\alpha \pm \beta) = \dfrac{\tan\alpha \pm \tan\beta}{1 \mp \tan\alpha\tan\beta}$

DOUBLE-ANGLE FORMULAS	$\sin 2\theta = 2\sin\theta\cos\theta$
	$\cos 2\theta = \cos^2\theta - \sin^2\theta = 2\cos^2\theta - 1 = 1 - 2\sin^2\theta$
	$\tan 2\theta = \dfrac{2\tan\theta}{1 - \tan^2\theta}$

SIN² AND COS² IN TERMS OF COSINE	$\sin^2\theta = \dfrac{1 - \cos 2\theta}{2} \qquad \cos^2\theta = \dfrac{1 + \cos 2\theta}{2}$

HALF-ANGLE FORMULAS	$\sin\dfrac{\alpha}{2} = \pm\sqrt{\dfrac{1 - \cos\alpha}{2}} \qquad \cos\dfrac{\alpha}{2} = \pm\sqrt{\dfrac{1 + \cos\alpha}{2}}$
	$\tan\dfrac{\alpha}{2} = \dfrac{\sin\alpha}{1 + \cos\alpha} = \dfrac{1 - \cos\alpha}{\sin\alpha}$

SUM FORMULAS	$\sin A + \sin B = 2\sin\dfrac{A + B}{2}\cos\dfrac{A - B}{2}$
	$\sin A - \sin B = 2\cos\dfrac{A + B}{2}\sin\dfrac{A - B}{2}$
	$\cos A + \cos B = 2\cos\dfrac{A + B}{2}\cos\dfrac{A - B}{2}$
	$\cos A - \cos B = -2\sin\dfrac{A + B}{2}\sin\dfrac{A - B}{2}$

PRODUCT FORMULAS	$\sin\alpha\cos\beta = \tfrac{1}{2}[\sin(\alpha + \beta) + \sin(\alpha - \beta)]$
	$\sin\alpha\sin\beta = \tfrac{1}{2}[\cos(\alpha - \beta) - \cos(\alpha + \beta)]$
	$\cos\alpha\cos\beta = \tfrac{1}{2}[\cos(\alpha + \beta) + \cos(\alpha - \beta)]$

REDUCTION FORMULAS

Let X be coterminal with 0 or π and let Y be coterminal with $\pi/2$ or $3\pi/2$. Then

$$\text{Any function of } X \pm \theta = \pm(\text{The } \textit{same} \text{ function of } \theta)$$
$$\text{Any function of } Y \pm \theta = \pm(\text{The } \textit{cofunction of } \theta)$$

The sign on the right-hand side is the same as the sign the function on the left has in the quadrant determined by $X \pm \theta$ or $Y \pm \theta$ (whichever occurs) when θ is acute.

Remarks

1. The particular letters used in these formulas are unimportant. The occurrence of α and β at some times, and θ or A and B at other times is only because of customary usage.
2. The formulas are often used in various equivalent forms. We have mentioned, for example, $\cos^2 \theta = 1 - \sin^2 \theta$ and $\sec^2 \theta - 1 = \tan^2 \theta$. We might also use $\sin \theta = 1/\csc \theta$ (although it is not too likely), or $\sin \theta \cos \theta = \frac{1}{2} \sin 2\theta$.
3. You should recognize the applicability of the formulas to expressions that are not precisely like those given. For example, we could use the double-angle formula for the sine to write

$$\sin 4x = 2 \sin 2x \cos 2x$$

Then, depending on what is desired, we could carry the right-hand side further by using the double-angle formulas once again. As another example, consider $\sin^4 t$. It may be (and often is) desirable to express this in terms of first powers of functions. We would proceed as follows:

$$\sin^4 t = (\sin^2 t)^2 = \left(\frac{1 - \cos 2t}{2}\right)^2$$
$$= \tfrac{1}{4}(1 - 2\cos 2t + \cos^2 2t)$$
$$= \frac{1}{4}\left(1 - 2\cos 2t + \frac{1 + \cos 4t}{2}\right)$$
$$= \frac{3}{8} - \frac{\cos 2t}{2} + \frac{\cos 4t}{8}$$

7 Proving Identities

By using the basic identities of Section 6, we can prove a multitude of others, but fortunately it is not particularly useful to try to memorize any of the results. By proving an identity, we mean to show that the given equation is true for all admissible values of the variable or variables involved. The general procedure is to work only on one side of the equation, and by use of the basic identities transform it so that in the final stage it is identical to the other side. Often, the more complicated side is the better to work with, since it offers more obvious possibilities for alteration, but there are times in calculus when it is better to change a simpler expression into a more complicated one. We illustrate the procedure by means of several examples.

EXAMPLE 7 Prove the identity: $\tan \theta + \cot \theta = \sec \theta \csc \theta$

Solution Perhaps it would be better to word the instructions, "Prove that the following is an identity," because we cannot prove it is true by assuming it is true, and this is an important point of logic that is often missed. For example, we cannot work on both sides of the equation, unless we verify that each step is reversible. A proper approach is to begin with the left-hand side and try to obtain the right-hand side:

$$\tan \theta + \cot \theta = \frac{\sin \theta}{\cos \theta} + \frac{\cos \theta}{\sin \theta}$$

$$= \frac{\sin^2 \theta + \cos^2 \theta}{\cos \theta \sin \theta}$$

$$= \frac{1}{\cos \theta \sin \theta}$$

$$= \frac{1}{\cos \theta} \cdot \frac{1}{\sin \theta}$$

$$= \sec \theta \csc \theta$$

Now the given equation is verified. It is true for all admissible values of θ, and in this case this means all values except 0, $\pi/2$, π, $3\pi/2$, and any angles coterminal with these, since at each of these values, two of the given functions are undefined. Usually, we will understand that such exceptions are necessary without mentioning them explicitly.

EXAMPLE 8 Prove the identity: $\dfrac{1}{\sec x + 1} = \cot x \csc x - \csc^2 x + 1$

Solution It would be possible to transform the right-hand side into the left, and this may even be easier, but we choose to work from the left-hand side, because in later applications the right-hand side will be seen to be the more desirable final form. We begin with a commonly used trick—multiplying the numerator and denominator by the same factor:

$$\frac{1}{\sec x + 1} = \frac{1}{\sec x + 1} \cdot \frac{\sec x - 1}{\sec x - 1} = \frac{\sec x - 1}{\sec^2 x - 1}$$

The object here is to bring $\sec^2 x - 1$ into the picture, because this is, by one of the Pythagorean identities, equal to $\tan^2 x$. So

$$\frac{1}{\sec x + 1} = \frac{\sec x - 1}{\tan^2 x} = \frac{\sec x}{\tan^2 x} - \frac{1}{\tan^2 x}$$

$$= \frac{1}{\cos x} \cdot \frac{\cos^2 x}{\sin^2 x} - \cot^2 x$$

$$= \frac{\cos x}{\sin^2 x} - (\csc^2 x - 1)$$

$$= \frac{\cos x}{\sin x} \cdot \frac{1}{\sin x} - \csc^2 x + 1$$

$$= \cot x \csc x - \csc^2 x + 1$$

EXAMPLE 9   Prove that: $\dfrac{2 \sin^3 x}{1 - \cos x} = 2 \sin x + \sin 2x$

Solution

$$\frac{2 \sin^3 x}{1 - \cos x} = \frac{2 \sin x(1 - \cos^2 x)}{1 - \cos x}$$

$$= \frac{2 \sin x(1 + \cos x)(1 - \cos x)}{1 - \cos x}$$

$$= 2 \sin x + 2 \sin x \cos x = 2 \sin x + \sin 2x$$

EXAMPLE 10   Prove that: $\dfrac{\sin 3x}{\sin x} + \dfrac{\cos 3x}{\cos x} = 4 \cos 2x$

Solution

$$\frac{\sin 3x}{\sin x} + \frac{\cos 3x}{\cos x} = \frac{\sin 3x \cos x + \cos 3x \sin x}{\sin x \cos x}$$

$$= \frac{\sin(3x + x)}{\sin x \cos x}$$

$$= \frac{\sin 4x}{\frac{1}{2} \sin 2x}$$

$$= \frac{2 \sin 2x \cos 2x}{\frac{1}{2} \sin 2x}$$

$$= 4 \cos 2x$$

A word is in order about the utility of proving identities. In calculus, as well as in more advanced courses, one is frequently confronted with an unwieldy expression involving trigonometric functions. Often, by use of the basic identities, these can be transformed into more manageable forms. So, in practice, one usually does not have a ready-made identity to verify, but rather an expression that is to be changed to some other, initially unknown form. It would be more accurate to describe this procedure as *deriving* an identity. The value of proving identities already provided lies in the fact that it gives a goal to shoot for in transforming an expression. Without this and with little or no experience in such activity, a student may go in circles or change to a less desirable form.

EXERCISE SET 7

Unless other instructions are given, prove that each of the following is an identity by transforming the left-hand side into the right-hand side.

A   **1.** $\dfrac{1 + \sin \theta}{\tan \theta} = \cos \theta + \cot \theta$

2. $(\tan x + \cot x)^2 = \sec^2 x + \csc^2 x$

3. $\sec x - \sin x \tan x = \cos x$

4. $\dfrac{\sec x + \tan x}{\sec^2 x} = \cos x(1 + \sin x)$

5. $(1 + \csc \theta)(\sec \theta - \tan \theta) = \cot \theta$

6. $\dfrac{1 - \cos x}{\sin x} + \dfrac{\sin x}{1 - \cos x} = 2 \csc x$

7. $\tan^2 x - \sin^2 x = \tan^2 x \sin^2 x$

8. $\dfrac{1}{1 - \sin x} + \dfrac{1}{1 + \sin x} = 2 \sec^2 x$

9. $\dfrac{\tan^2 x - 1}{\sin^2 x} = \sec^2 x - \csc^2 x$

10. $\sin \theta \tan \theta = \sec \theta - \cos \theta$

11. $\sin \theta + \cos \theta \cot \theta = \csc \theta$

12. $\sec^4 x = \tan^2 x \sec^2 x + \sec^2 x$

13. $\dfrac{\tan^2 \alpha - \sin^2 \alpha}{\sec^2 \alpha - 1} = \sin^2 \alpha$

14. $1 - \dfrac{\cos 2x - 1}{2 \cos^2 x} = \sec^2 x$

15. $\dfrac{1 + \tan x}{1 + \cot x} = \tan x$

16. $\sin^4 \theta \cot^2 \theta = \tfrac{1}{4} \sin^2 2\theta$

17. $\dfrac{\cos^2 \theta - \sin^2 \theta}{\sin \theta \cos \theta} = 2 \cot 2\theta$

18. $\dfrac{\sin x}{\cot x + \csc x} = 1 - \cos x$

19. $\dfrac{\sin^2 2x}{1 - \cos 2x} = 2 \cos^2 x$

20. $\dfrac{1}{1 - \sin \theta} = \sec^2 \theta + \tan \theta \sec \theta$

21. $(\sin x - \cos x)^2 = 1 - \sin 2x$

22. $\tan^4 x = \sec^2 x \tan^2 x - \sec^2 x + 1$

23. $\cos^4 x - \sin^4 x = \cos 2x$

24. $\dfrac{\tan x - \tan y}{\cot y - \cot x} = \tan x \tan y$

25. $\dfrac{1}{\sec x - \tan x} = \sec x + \tan x$

26. $\dfrac{\sin \theta}{1 - \cos \theta} = \csc \theta + \cot \theta$

27. $\dfrac{2}{1 + \cos 2x} = \sec^2 x$

28. $\dfrac{\sin 2\theta}{\sin^2 \theta - 1} + 3 \tan \theta = \tan \theta$

29. $\dfrac{1 - \tan^2 x}{1 + \tan^2 x} = \cos 2x$

30. $\tan \alpha + \cot \alpha = 2 \csc 2\alpha$

31. $\csc 2x - \cot 2x = \tan x$

32. $\dfrac{\cot \theta + \tan \theta}{\cot \theta - \tan \theta} = \sec 2\theta$

33. $\csc 2x + \cot 2x = \cot x$

34. $\dfrac{\tan^2 \theta}{1 + \tan^2 \theta} = \dfrac{1 - \cos 2\theta}{2}$

35. $\cos^4 x = \dfrac{3}{8} + \dfrac{\cos 2x}{2} + \dfrac{\cos 4x}{8}$

36. $\dfrac{\cos x + \sin x}{\cos x - \sin x} = \tan 2x + \sec 2x$

37. $\dfrac{\sin A - \sin B}{\cos A + \cos B} = \tan \tfrac{1}{2}(A - B)$

B **38.** $\sin^6 x = \dfrac{5}{16} - \dfrac{\cos 2x}{2} + \dfrac{3 \cos 4x}{16} + \dfrac{\sin^2 2x \cos 2x}{8}$

39. $\dfrac{1 - \cos 2nx}{\sin 2nx} = \tan nx$

40. $\dfrac{\sin 3\theta}{\sin \theta} - \dfrac{\cos 3\theta}{\cos \theta} = 2$

41. $\sin 3x + \sin x = 4 \sin x \cos^2 x$

42. $\dfrac{\sin(x + h) - \sin x}{h} = \dfrac{\sin(h/2)}{(h/2)} \cos\left(x + \dfrac{h}{2}\right)$

43. $\dfrac{\sin^3 x + \cos^3 x}{\sin x + \cos x} = 1 - \dfrac{1}{2} \sin 2x$

44. $\sqrt{\dfrac{1 - \cos \theta}{1 + \cos \theta}} = \dfrac{\sin \theta}{1 + \cos \theta}$ $(0 \le \theta < \pi)$

45. Make the substitution $x = \sin \theta$ $(-\pi/2 < \theta < \pi/2)$ in the expression $x^2/\sqrt{1 - x^2}$ and show that the result can be written in the form $\sec \theta - \cos \theta$.

46. Let $x = 2 \tan \theta$ $(-\pi/2 < \theta < \pi/2)$ in $x/\sqrt{4 + x^2}$, and show that the result is $\sin \theta$.

47. Substitute $x = 3 \sec \theta$, where θ lies either in the first or third quadrants, in $(x^2 - 9)^{3/2}/x$, and show that the result can be written in the form $9(\sec \theta \tan \theta - \sin \theta)$.

48. Substitute $x = \tfrac{3}{2} \tan \theta$ $(-\pi/2 < \theta < \pi/2)$ in $x/\sqrt{4x^2 + 9}$ and show that the result is $\tfrac{1}{2} \sin \theta$.

8 Trigonometric Equations

The solutions of most trigonometric equations cannot be obtained exactly, and we have to settle for approximate solutions obtained through some numerical procedure—usually with the aid of a calculator. There are, however, enough trigonometric equations for which elementary techniques will yield exact answers that some time devoted to these techniques is justified.

In order to make clear the sorts of solutions we are looking for, let us consider the very simple equation

$$2 \sin \theta - 1 = 0$$

The problem is to discover all real numbers θ for which this equation is true. We write the equation in the equivalent form

$$\sin \theta = \tfrac{1}{2}$$

and then rely on our knowledge of special angles to conclude that this is satisfied if and only if

$$\theta = \frac{\pi}{6} \quad \text{or} \quad \frac{5\pi}{6}$$

or any other number obtained from these by adding integral multiples of 2π to each. To see this, remember that the sine of a number θ is the y coordinate of the point on the unit circle that is θ units along the arc from $(1, 0)$. Thus, we are seeking those numbers θ for which the y coordinate of $P(\theta) = \frac{1}{2}$. There are only two points on the unit circle with y coordinate $\frac{1}{2}$, and our knowledge of the 30°–60° right triangle tells us that these points are $P(\pi/6)$ and $P(5\pi/6)$; see Figure 12. Since $P(\theta) = P(\theta + 2n\pi)$, we conclude that all solutions of the equation are given by $\pi/6 + 2n\pi$ or $5\pi/6 + 2n\pi$.

Figure 12

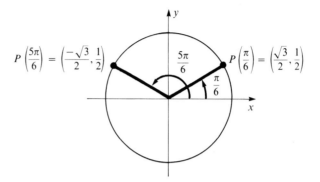

Usually it is sufficient to give only the **primary** solutions, that is, those lying between 0 and 2π. We would know then that all other solutions are obtainable from these by adding multiples of 2π.

We can usually condense the above reasoning as follows:

1. Determine the appropriate reference angle.
2. By the sign of the function locate all primary angles having this reference angle.
3. Write the answer as the radian measure of these angles.

In our example, since we know that $\sin 30° = \frac{1}{2}$, the reference angle is 30°. The sine is positive in quadrants I and II, so the angles are 30° and 150°. In radians these are $\pi/6$ and $5\pi/6$.

It might be useful to review at this time the functions of 0, 30°, 45°, 60°, and 90°:

$$\sin 30° = \sin \frac{\pi}{6} = \frac{1}{2} \qquad \sin 45° = \sin \frac{\pi}{4} = \frac{1}{\sqrt{2}} \qquad \sin 60° = \sin \frac{\pi}{3} = \frac{\sqrt{3}}{2}$$

$$\cos 30° = \cos \frac{\pi}{6} = \frac{\sqrt{3}}{2} \qquad \cos 45° = \cos \frac{\pi}{4} = \frac{1}{\sqrt{2}} \qquad \cos 60° = \cos \frac{\pi}{3} = \frac{1}{2}$$

$$\tan 30° = \tan \frac{\pi}{6} = \frac{1}{\sqrt{3}} \qquad \tan 45° = \tan \frac{\pi}{4} = 1 \qquad \tan 60° = \tan \frac{\pi}{3} = \sqrt{3}$$

$$\sin 0° = \sin 0 = 0 \qquad \sin 90° = \sin \frac{\pi}{2} = 1$$

$$\cos 0° = \cos 0 = 1 \qquad \cos 90° = \cos \frac{\pi}{2} = 0$$

$$\tan 0° = \tan 0 = 0 \qquad \tan 90° = \tan \frac{\pi}{2} \text{ is not defined}$$

The cotangent, secant, and cosecant are not listed here, because they occur less frequently and can always be obtained by the reciprocal relations. If you feel comfortable with radian measure by now, you can omit the degree measure entirely.

If an equation can be worked around to a form such as $\sin \theta = \pm a$, $\cos \theta = \pm b$, or $\tan \theta = \pm c$, for instance, where a and b are any of the numbers 0, 1, $\frac{1}{2}$, $1/\sqrt{2}$, or $\sqrt{3}/2$, and c is any of the numbers 0, 1, $1/\sqrt{3}$, or $\sqrt{3}$, then we can obtain all solutions in the manner outlined above. For other values of a, b, and c a calculator or tables can be used.

EXAMPLE 11   Find all values of θ, for which $0 \leq \theta < 2\pi$, satisfying $2 \cos^2 \theta - \cos \theta - 1 = 0$.

Solution   We treat this first as a quadratic equation in $\cos \theta$ and factor:

$$(2 \cos \theta + 1)(\cos \theta - 1) = 0$$

This is true if and only if $2 \cos \theta + 1 = 0$ or $\cos \theta = 1$, that is, if and only if

$$\cos \theta = -\tfrac{1}{2} \quad \text{or} \quad \cos \theta = 1$$

Since $\cos 60° = \frac{1}{2}$, the reference angle in the first of these is 60°. Since the cosine is negative in quadrants II and III, it follows that we are seeking the angles 120° and 240°. In radians these are $2\pi/3$ and $4\pi/3$. The only primary angle for which $\cos \theta = 1$ is $\theta = 0°$, or 0 radians. Thus, the primary solution set for the equation is $\{0, 2\pi/3, 4\pi/3\}$.

Remark.   In many equations involving trigonometric functions the variable must be treated as a real number, which is the reason for writing the above answers as the radian measures of the angles. Recall that the values of $\sin \theta$, $\cos \theta$, and the other trigonometric functions are unchanged whether we consider θ a real number or the radian measure of an angle. An example of a mixed algebraic and trigonometric equation will perhaps help to make this point clearer. The equation

$$2 \sin x - x = 0$$

is clearly satisfied when $x = 0$, but by trial and error, using a calculator (or by some more sophisticated numerical procedure), we can find that another solution is $x \approx 1.8955$. If we think of this x as the radian measure of an angle, then the degree measure of the angle is approximately 108.6°. Now it could *not* be true that $x = 108.6°$, because we would have

$$2 \sin 108.6° - 108.6° = 0$$

which is a totally meaningless statement.

EXAMPLE 12   Find all primary solutions of $\sin 2x + \cos x = 0$.

Solution   This time we will omit most discussion and simply go through the steps that would be expected in doing the problem.

$$\sin 2x + \cos x = 0$$
$$2 \sin x \cos x + \cos x = 0$$
$$\cos x(2 \sin x + 1) = 0$$

$$\cos x = 0 \qquad \bigg| \qquad 2 \sin x = -1$$
$$x = \frac{\pi}{2}, \frac{3\pi}{2} \qquad \bigg| \qquad \sin x = -\frac{1}{2}$$
$$\qquad \qquad \qquad \bigg| \qquad x = \frac{7\pi}{6}, \frac{11\pi}{6}$$

The solution set is $\{\pi/2, 7\pi/6, 3\pi/2, 11\pi/6\}$.

EXAMPLE 13   Find all primary solutions of $\tan \alpha - 3 \cot \alpha = 0$.

Solution
$$\tan \alpha - 3 \cot \alpha = 0$$
$$\tan \alpha - \frac{3}{\tan \alpha} = 0$$
$$\tan^2 \alpha - 3 = 0$$
$$\tan \alpha = \pm\sqrt{3}$$

The reference angle is $60°$, or $\pi/3$ radians. Therefore,

$$\alpha = \frac{\pi}{3}, \frac{2\pi}{3}, \frac{4\pi}{3}, \frac{5\pi}{3}$$

Since we multiplied by an unknown expression, namely, $\tan \alpha$, it is essential that we check to see that this was not 0 for the values of α found, which is clearly the case here, since $\tan \alpha = \pm\sqrt{3} \neq 0$. Therefore the solution set is $\{\pi/3, 2\pi/3, 4\pi/3, 5\pi/3\}$.

EXAMPLE 14   Find all primary solutions of $\sqrt{3} \cos x = 2 + \sin x$.

Solution   The fact that $\cos^2 x = 1 - \sin^2 x$ suggests that if we square both sides of the given equation, it will be easier to work with.

$$3 \cos^2 x = 4 + 4 \sin x + \sin^2 x$$
$$3(1 - \sin^2 x) = 4 + 4 \sin x + \sin^2 x$$
$$4 \sin^2 x + 4 \sin x + 1 = 0$$
$$(2 \sin x + 1)^2 = 0$$
$$\sin x = -\frac{1}{2}$$
$$x = \frac{7\pi}{6}, \frac{11\pi}{6}$$

We again must check our answers, because squaring does not necessarily lead to an equivalent equation; it may introduce extraneous roots. So we check in the original equation.

When $x = 7\pi/6$, we get

$$\sqrt{3}\left(-\frac{\sqrt{3}}{2}\right) \overset{?}{=} 2 + \left(-\frac{1}{2}\right)$$

$$-\frac{3}{2} \neq \frac{3}{2}$$

So $7\pi/6$ is not a solution.

When $x = 11\pi/6$,

$$\sqrt{3}\left(\frac{\sqrt{3}}{2}\right) \overset{?}{=} 2 + \left(-\frac{1}{2}\right)$$

$$\frac{3}{2} = \frac{3}{2}$$

So $x = 11\pi/6$ is the only primary solution.

EXAMPLE 15   Find all primary solutions of $2 \sin 3\theta - \sqrt{3} \tan 3\theta = 0$.

Solution $2 \sin 3\theta - \dfrac{\sqrt{3} \sin 3\theta}{\cos 3\theta} = 0$

$\sin 3\theta(2 \cos 3\theta - \sqrt{3}) = 0$ **We multiplied by cos 3θ.**

$\sin 3\theta = 0 \quad | \quad \cos 3\theta = \dfrac{\sqrt{3}}{2}$ **So we did not multiply by 0.**

Now we want all values of θ lying between 0 and 2π. We must therefore find all values of 3θ lying between 0 and 6π. In general, if $n\theta$ is involved, in order to find all values of θ between 0 and 2π, we find all values of $n\theta$ between 0 and $2n\pi$, and then divide by n. We have from $\sin 3\theta = 0$ that

$$3\theta = 0, \quad \pi, \quad 2\pi, \quad 3\pi, \quad 4\pi, \quad 5\pi$$

and

$$\theta = 0, \quad \frac{\pi}{3}, \quad \frac{2\pi}{3}, \quad \pi, \quad \frac{4\pi}{3}, \quad \frac{5\pi}{3}$$

From $\cos 3\theta = \sqrt{3}/2$, we have

$$3\theta = \frac{\pi}{6}, \quad \frac{11\pi}{6}, \quad \frac{13\pi}{6}, \quad \frac{23\pi}{6}, \quad \frac{25\pi}{6}, \quad \frac{35\pi}{6}$$

$$\theta = \frac{\pi}{18}, \quad \frac{11\pi}{18}, \quad \frac{13\pi}{18}, \quad \frac{23\pi}{18}, \quad \frac{25\pi}{18}, \quad \frac{35\pi}{18}$$

In both cases we found the first two values of 3θ, then added 2π once, and then 2π again. The complete solution set is

$$\left\{0, \frac{\pi}{18}, \frac{\pi}{3}, \frac{11\pi}{18}, \frac{2\pi}{3}, \frac{13\pi}{18}, \pi, \frac{23\pi}{18}, \frac{4\pi}{3}, \frac{25\pi}{18}, \frac{5\pi}{3}, \frac{35\pi}{18}\right\}$$

EXAMPLE 16   Find all real solutions to $\tan x - \cot x = 2$.

Solution

$$\tan x - \cot x = 2$$

$$\frac{\sin x}{\cos x} - \frac{\cos x}{\sin x} = 2$$

$$\sin^2 x - \cos^2 x = 2 \sin x \cos x$$

$$-\cos 2x = \sin 2x$$

$$\tan 2x = -1$$

$$2x = \frac{3\pi}{4}, \frac{7\pi}{4}, \frac{11\pi}{4}, \frac{15\pi}{4} \qquad \textbf{We "went around" twice.}$$

$$x = \frac{3\pi}{8}, \frac{7\pi}{8}, \frac{11\pi}{8}, \frac{15\pi}{8}$$

At none of these values is $\sin x$, $\cos x$, or $\cos 2x$ equal to 0, so all our operations are legal. We have found all primary solutions. To get *all* solutions, we add arbitrary multiples of 2π to each of these. So the complete solution set is

$$\left\{\frac{3\pi}{8}, \frac{7\pi}{8}, \frac{11\pi}{8}, \frac{15\pi}{8}, \frac{19\pi}{8}, \frac{23\pi}{8}, \frac{27\pi}{8}, \frac{31\pi}{8}, \frac{35\pi}{8}, \frac{39\pi}{8}, \frac{43\pi}{8}, \frac{47\pi}{8}, \ldots\right\}$$

or we could write

$$\left\{\frac{3\pi}{8} + 2n\pi, \ \frac{7\pi}{8} + 2n\pi, \ \frac{11\pi}{8} + 2n\pi, \ \frac{15\pi}{8} + 2n\pi, \ n = 0, \pm 1, \pm 2, \ldots\right\}$$

We could continue with examples, each possessing its own idiosyncracies, but these illustrate the main techniques. It might be useful to summarize some of the points that the examples were meant to bring out:

1. Use algebraic techniques such as factoring, multiplying by the LCD, and squaring in conjunction with basic trigonometric identities to simplify so as to obtain one or more elementary equations of the form $\sin x = \pm a$, $\cos x = \pm b$, and so on.
2. Find the radian measure of all primary angles (that is, $0 \le x < 2\pi$) satisfying the elementary equation in Step 1. If *all* solutions are desired, add arbitrary multiples of 2π to each of these to obtain the complete solution set.
3. If in arriving at the elementary equations both sides of an equation were squared, or if both sides were multiplied by or divided by any unknown expression, it is necessary to check the answers. (In the case of multiplying

or dividing by an unknown, it is sufficient to check that the multiplier or divisor was not 0.)

4. If the final elementary equation(s) is of the form $\sin nx = \pm a$, $\cos nx \pm b$, and so on, then to get all primary values of x, find all values of nx between 0 and $2n\pi$, and divide by n. This amounts to finding the angles in the first revolution and then adding 2π a total of $n - 1$ separate times and finally dividing everything by n.

EXERCISE SET 8

A   Find all primary solutions unless otherwise specified.

1. $2 \cos x = \sqrt{3}$
2. $\cot x + 1 = 0$
3. $\sec x = 2$
4. $\sin^2 x = 1$ Give all solutions.
5. $\cos x - 2 \cos^2 x = 0$
6. $\sin 2\theta - \cos \theta = 0$
7. $\cos 2x + \cos x = 0$
8. $2 \tan \psi - \tan \psi \sec \psi = 0$
9. $\sin^2 t - \cos^2 t = 1$
10. $\sec^2 \theta = 2 \tan \theta$
11. $2 \cos x - \cot x = 0$
12. $2 \cos^2 x + \sin x - 1 = 0$
13. $4 \tan^2 \alpha = 3 \sec^2 \alpha$
14. $\sin^2 x = 2(\cos x - 1)$
15. $\dfrac{1 - \cos x}{\sin x} = \sin x$
16. $2 \cos^2 2\theta - 3 \cos 2\theta + 1 = 0$
17. $\sin 6\theta + \sin 3\theta = 0$
18. $\sin^2 4\theta - \sin 4\theta - 2 = 0$
19. $2(\sin^2 2\theta - \cos^2 2\theta) = \sqrt{3}$
20. $2 \cos x - \cot x \csc x = 0$

B   21. $\sin \theta + \cos \theta = 1$ Give all solutions.
22. $\cos 2x \tan 3x + \sqrt{3} \cos 2x = 0$
23. $\dfrac{\sin x}{1 - \cos x} = \dfrac{\cos x}{1 + \sin x}$
24. $\tan^3 x + \tan^2 x - 3 \tan x - 3 = 0$
25. $8 \sin^4 \theta - 2 \sin^2 \theta - 3 = 0$
26. $2 \sin^3 x + 3 \sin^2 x - 1 = 0$

---

9   Graphs of the Trigonometric Functions

To obtain the graphs of $y = \sin x$ and $y = \cos x$ we make use of the definitions of the sine and cosine as the ordinate and abscissa, respectively, of a point on the unit circle (Figure 13). Notice that we have used x to represent arc length on the circle from $(1, 0)$ to $P(x)$ rather than the customary θ. We have used capital letters to label the axes. Table 1 indicates how $\sin x$ and $\cos x$ vary as x varies from 0 to 2π. These facts can be verified by visualizing the point $P(x)$ moving around the circle in a counterclockwise direction, starting at $(1, 0)$. Furthermore, it should be clear from considering Figure 13 that the ordinate and abscissa of $P(x)$ (that is, $\sin x$ and $\cos x$) vary in a uniform way, with no breaks or sudden changes, as $P(x)$ moves around the circle.

This information, together with our knowledge of $\sin x$ and $\cos x$ for $x = \pi/6$, $\pi/4$, $\pi/3$ and related values in the other quadrants, enables us to draw the graphs with reasonable accuracy. Since $\sin x$ and $\cos x$ each has period 2π, the graphs repeat every 2π units. So once we know the graphs in the interval

Figure 13

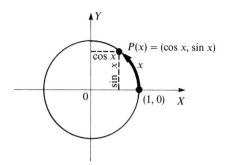

Table 1

As x goes from	sin x goes from	cos x goes from
0 to $\pi/2$	0 to 1	1 to 0
$\pi/2$ to π	1 to 0	0 to -1
π to $3\pi/2$	0 to -1	-1 to 0
$3\pi/2$ to 2π	-1 to 0	0 to 1

Figure 14

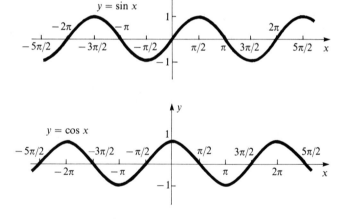

Figure 15

from 0 to 2π, we can extend them indefinitely in either direction. The graphs are shown in Figures 14 and 15.

The maximum height attained by the sine curve and by the cosine curve is called the **amplitude.** So, for $y = \sin x$ and $y = \cos x$ the amplitude is 1.

To obtain the graph of $y = \tan x$ we again refer to Figure 13 and use the fact that $\tan x = \sin x/\cos x$. Therefore, we consider the ratio of the ordinate of $P(x)$ to its abscissa as $P(x)$ moves around the unit circle.

When $x = 0$, $\tan x = 0/1 = 0$, and as x increases toward $\pi/2$, the ratio of the ordinate of $P(x)$ to its abscissa steadily increases, taking on the value 1 at $x = \pi/4$. For x near $\pi/2$ the ratio becomes very large and gets larger and larger without limit as x approaches $\pi/2$. At $x = \pi/2$ the ratio is not defined. For x slightly greater than $\pi/2$ the ratio is negative, since the ordinate is positive and the abscissa is negative, but its absolute value is large. At $x = 3\pi/4$ the ratio is -1, and at π, it is back to 0. Since

$$\tan(x + \pi) = \frac{\tan x + \tan \pi}{1 - \tan x \tan \pi} = \frac{\tan x + 0}{1 - (\tan x) \cdot 0} = \tan x$$

it follows that $\tan x$ has period π. So we can draw its graph between 0 and π and duplicate this in each succeeding interval of length π. Similarly, we can extend it to the left. With this information and our knowledge of $\tan x$ for special values of x such as $\pi/6$ and $\pi/3$, we can draw the graph, as in Figure 16. The vertical lines at odd multiples of $\pi/2$ are asymptotes.

Figure 16

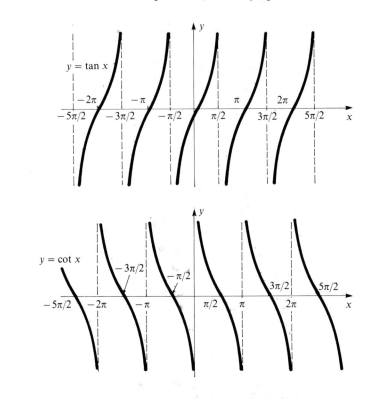

Figure 17

The graph of $y = \cot x$ can be similarly obtained and we show its graph in Figure 17.

The secant of x is defined as the reciprocal of the X coordinate of $P(x)$, that is, as the reciprocal of $\cos x$. Since $|\cos x| \le 1$, it follows that $|\sec x| \ge 1$. The signs of $\sec x$ and $\cos x$ are the same. When $\cos x = 1$, $\sec x = 1$, and when $\cos x = 0$, $\sec x$ is undefined. This latter occurs at $x = \pi/2$, $3\pi/2$, $-\pi/2$, $5\pi/2$, and so on. As x gets close to $\pi/2$, $\cos x$ gets close to 0; so $\sec x$ gets arbitrarily large. Using this analysis, together with the values of $\sec x$ for $x = \pi/6, \pi/4, \pi/3$, and so on, we sketch the graph in Figure 18. The graph of $y = \csc x$ can be similarly obtained (Figure 19).

Both $\sec x$ and $\csc x$ have a fundamental period of 2π. No amplitude is defined for them. They each have asymptotes; for the secant they are the lines $x = \pi/2$, $-\pi/2$, $3\pi/2$, and so on, and for the cosecant the lines $x = 0$, π, $-\pi$, 2π, and so on.

Figure 18

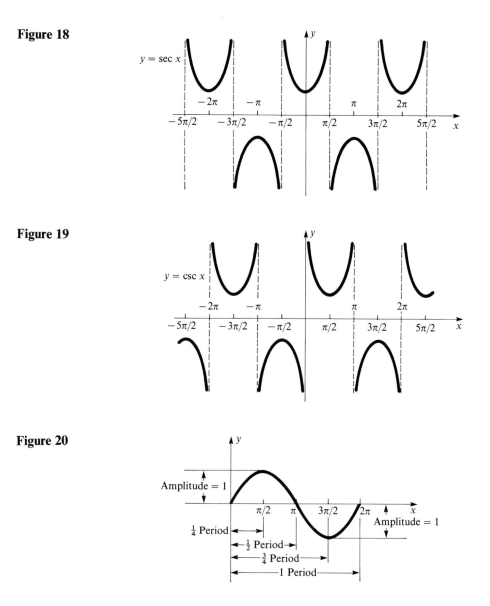

Figure 19

Figure 20

What we have shown are the basic graphs of the trigonometric functions. You should become especially familiar with the first three. The crucial facts about the sine curve are illustrated in Figure 20. This, together with the knowledge of the periodicity, enables us to obtain a rapid sketch of the curve.

For the basic sine curve $y = \sin x$, the fundamental period is 2π and the amplitude is 1. Now we want to consider the effect of introducing positive constants a and b:

$$y = a \sin bx$$

We will take one at a time. First consider $y = a \sin x$. This has the effect of multiplying every y value of the basic curve by a. So when the basic curve is at its maximum height of 1, the new curve will be a units high. Thus, the ampli-

Figure 21

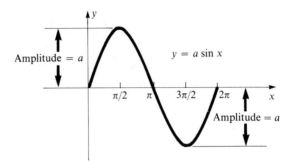

tude becomes a, and the period remains unchanged, as shown in Figure 21. We have exhibited the curve for one period only (also called one **cycle**), since the extension is obvious.

Next, let us consider $y = \sin bx$. We know that the sine curve completes one cycle when bx goes from 0 to 2π. But then x goes from 0 to $2\pi/b$. So, we conclude that the period of $\sin bx$ is $2\pi/b$. This can also be seen in another way. To say that $\sin x$ has period 2π means that $\sin(x + 2\pi) = \sin x$. Thus, $\sin(bx + 2\pi) = \sin bx$. But $\sin(bx + 2\pi) = \sin b[x + (2\pi/b)]$. So,

$$\sin b\left(x + \frac{2\pi}{b}\right) = \sin bx$$

Thus, when we add $2\pi/b$ to any x, we get the same value as $\sin bx$. That is, $\sin bx$ has period $2\pi/b$. The effect of the multiplier b in this position, then, is to alter the period; it is shortened if $b > 1$ and lengthened if $b < 1$ (Figure 22).

Figure 22

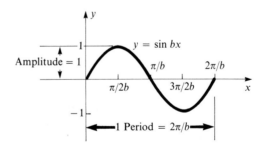

Remark. When the variable x represents time, the period is the time required to complete one cycle. The reciprocal of this, called the **frequency,** gives the number of cycles (or fraction of a cycle) completed per unit of time. Thus,

$$\text{Frequency} = \frac{1}{\text{Period}} = \frac{b}{2\pi}$$

This is a widely used concept in electronics.

Now we can put these two changes together to get the graph of $y = a \sin bx$, as shown in Figure 23. If, in addition, a constant k is added to the right-hand side, giving $y = a \sin bx + k$, this has the effect of shifting the entire graph of $y = a \sin bx$ vertically by k units—upward if $k > 0$ and downward if $k < 0$.

We consider next an equation of the form

$$y = a \sin b(x - \alpha) \tag{12}$$

Figure 23

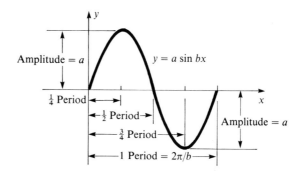

The sine will complete one cycle when $b(x - \alpha)$ goes from 0 to 2π. Setting $b(x - \alpha)$ equal to 0 gives

$$b(x - \alpha) = 0$$
$$x - \alpha = 0$$
$$x = \alpha$$

and setting $b(x - \alpha)$ equal to 2π gives

$$b(x - \alpha) = 2\pi$$
$$x - \alpha = \frac{2\pi}{b}$$
$$x = \frac{2\pi}{b} + \alpha$$

So a complete cycle occurs in the interval from α to $(2\pi/b) + \alpha$. This is a distance of $2\pi/b$, so the period of $2\pi/b$ remains unchanged, but the curve is shifted α units horizontally. If $\alpha > 0$, the shift is to the right, and if $\alpha < 0$, it is to the left. We call α the **phase shift,** and say that the curve is $|\alpha|$ units out of phase with the curve $y = a \sin bx$. This is illustrated in Figure 24.

Finally, we note the effect on the graph of $y = a \sin bx$ if either a or b is negative. If $a < 0$, every y value is the negative of what it would have been if a were positive. For example, in $y = -2 \sin x$, every y value is the negative of the corresponding y value in $y = 2 \sin x$. So the effect is to flip the graph of $y = 2 \sin x$ about the x axis. If $b < 0$, we make use of $\sin(-\theta) = -\sin \theta$.

Figure 24

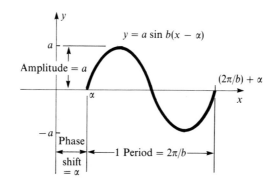

For example, we would write $y = \sin(-2x)$ as $y = -\sin 2x$, and proceed as above.

A similar analysis to that for the sine holds for the cosine function (see Figure 25). Note, however, for the cosine that $\cos(-\theta) = \cos\theta$, and so the graph of $y = \cos(-2x)$ would be identical to the graph of $y = \cos 2x$. With appropriate modifications because of the different period (for the tangent and cotangent) and lack of amplitude, the other functions, too could be analyzed in an analogous way. You will be asked to consider certain of these in Exercise Set 9.

Figure 25

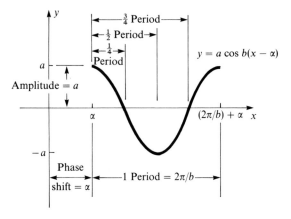

EXAMPLE 17   Sketch the graph of $y = 3 \sin \frac{1}{2}x$.

Solution   The amplitude is 3 and the period is $2\pi \div \frac{1}{2} = 4\pi$ (Figure 26).

Figure 26

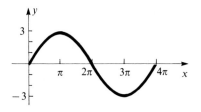

EXAMPLE 18   Sketch the graph of $y = 2 \sin(3x - \pi)$.

Solution   We first factor out the 3 to put this in the form of equation (12):

$$y = 2 \sin 3\left(x - \frac{\pi}{3}\right)$$

This is therefore a sine curve with amplitude 2, period $2\pi/3$, and phase shift $\pi/3$ units to the right (Figure 27).

Figure 27

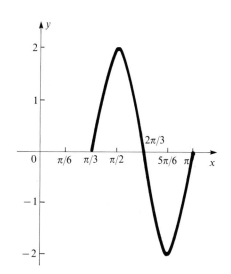

EXAMPLE 19   Sketch the graph of $y = \cos[2x + (\pi/3)]$.

Solution   We first write the equation in a form analogous to equation (12):

$$y = \cos 2\left(x + \frac{\pi}{6}\right) = \cos 2\left[x - \left(-\frac{\pi}{6}\right)\right]$$

So the phase shift α is $-\pi/6$. The curve is therefore shifted $\pi/6$ units to the left. The amplitude is 1, and the period is $2\pi/2 = \pi$ (Figure 28).

Figure 28

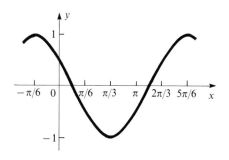

EXAMPLE 20   Sketch $y = 2 \tan \frac{1}{2}x$.

Solution   There is no amplitude defined for the tangent curve, but the effect of the coefficient 2 is to multiply all y values of the basic curve by 2; in particular, when the basic curve is 1 unit high, the new curve will be 2 units high. Since the fundamental period of the basic tangent curve is π, the period of this curve is $\pi \div \frac{1}{2} = 2\pi$. We show two complete cycles in Figure 29.

Figure 29

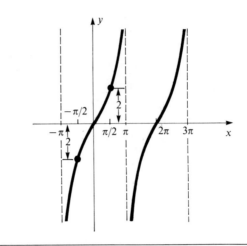

EXERCISE SET 9

In Problems 1–25 sketch one cycle of each of the curves. Give the period and, where appropriate, the amplitude and phase shift.

A **1.** $y = \sin 2x$ 　　　　　　　　　　**2.** $y = \cos 3x$

3. $y = 2 \sin x$ 　　　　　　　　　　**4.** $y = 2 \sin 3x$

5. $y = 3 \cos(\pi x/2)$ 　　　　　　　**6.** $y = 4 \sin(\pi x/3)$

7. $y = 2 \sin(-3x)$ 　　　　　　　　**8.** $y = 3 \cos(-\pi x)$

9. $y = -\sin(x/2)$ 　　　　　　　　**10.** $y = -2 \cos x$

11. $y = \sin 2x + 1$ 　　　　　　　　**12.** $y = 2 \cos x - 3$

13. $y = 2 \sin 3x - 1$ 　　　　　　　**14.** $y = 3 \cos \pi x + 2$

15. $y = \frac{1}{2} \tan \pi x$ 　　　　　　　　**16.** $y = 2 \cot(x/2)$

17. $y = \sin\left(x - \dfrac{\pi}{3}\right)$ 　　　　　　**18.** $y = \cos\left(x - \dfrac{\pi}{4}\right)$

19. $y = \sin 2\left(x + \dfrac{\pi}{4}\right)$ 　　　　　**20.** $y = \cos 3\left(x + \dfrac{\pi}{3}\right)$

B **21.** $y = 2 \sin(3x + 2)$ 　　　　　　**22.** $y = 3 \cos(3 - 2x)$

23. $y = 1 - 2 \sin\left(\pi x - \dfrac{\pi}{4}\right)$ 　　　**24.** $y = \frac{1}{2} \tan\left(2x - \dfrac{\pi}{2}\right)$

25. $y = 2 + 2 \cos\left(3x - \dfrac{\pi}{2}\right)$

26. **a.** By multiplying and dividing by $\sqrt{a^2 + b^2}$ show that $a \sin x + b \cos x$ can be written in the form

$$\sqrt{a^2 + b^2} \sin(x + \theta)$$

where θ is as shown in the accompanying sketch.

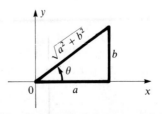

    b. Use the technique of part a to analyze and sketch the graph of

$$y = \sin x + \sqrt{3} \cos x$$

    What is the value of θ in this case?

27. By the technique of Problem 26, part a, write each of the following as a sine function. Determine the amplitude, period, and phase shift.

    a. $y = \sin x - \cos x$ **b.** $y = 3 \cos x - 4 \sin x$

28. Disucss the effect on the basic secant curve of introducing positive constants a and b to obtain $y = a \sec bx$.

29. Sketch $y = 2 \sec(x/3)$.

10 The Inverse Trigonometric Functions

Let us look again at the graph of $y = \sin x$ (Figure 30). It is immediately evident that this is not a one-to-one function, so it has no inverse. However, by a suitable restriction on the domain of the sine, an inverse can be found. The standard choice is to restrict x so that $-\pi/2 \le x \le \pi/2$. The sine curve with this domain will be called the **principal part** of the sine curve.

Figure 30

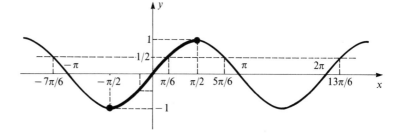

If x is so restricted, then for each y such that $-1 \le y \le 1$, there is exactly one x such that $y = \sin x$. For example, if we take $y = \frac{1}{2}$, we get $x = \pi/6$. On the other hand, if $y = -\frac{1}{2}$, then $x = -\pi/6$. The equation $y = \sin x$ expresses y in terms of x; we would like to solve this equation for x in terms of y. Unfortunately, we as yet have no way of doing this. With the aid of the graph (or tables or a calculator), we can find x for any given y, but we have no simple equation that expresses x in terms of y. What we do is invent a symbolism,

$$x = \sin^{-1} y$$

which is read "x is the inverse sine of y." Actually, we should probably read it as the "*principal* inverse sine of y," since it is the principal part of the sine curve that is used in finding x. Unless otherwise stated, we will in the future understand that $\sin^{-1} y$ means the principal value.

There is another symbol in wide use that means the same thing as $\sin^{-1} y$; it is **arcsin y,** read "arcsine of y." Its origin probably lies in the length of arc on a unit circle used in the definition of the trigonometric functions. Thus, arcsin $\frac{1}{2}$ might be interpreted as "the length of arc on the unit circle for which the sine is $\frac{1}{2}$." We know the length in this case is $\pi/6$, so

$$\sin \frac{\pi}{6} = \frac{1}{2} \quad \text{and} \quad \frac{\pi}{6} = \arcsin \frac{1}{2}$$

are two ways of viewing the same fact. The first says that the sine of the number $\pi/6$, that is, the measure of the arc on the unit circle from $(1, 0)$, is $\frac{1}{2}$. The second says that $\pi/6$ is the number, that is, the measure of arc length on the unit circle from $(1, 0)$, whose sine is $\frac{1}{2}$.

In what follows we will use both notations for the inverse sine (as well as inverses of the other trigonometric functions), since both are in wide use. Let us consider some examples.

EXAMPLE 21   Find the value of:
  a. $\sin^{-1} 1$ **b.** $\sin^{-1}(-\frac{1}{2})$ **c.** $\sin^{-1} 0$

Solution   **a.** $\sin^{-1} 1 = \pi/2$, since $\pi/2$ is that value of x on the principal part of the siné curve whose sine is 1

  b.
$$\sin^{-1}(-\tfrac{1}{2}) = -\pi/6$$

It is important here to note that the answer is not $11\pi/6$, though an angle of $11\pi/6$ radians and an angle of $-\pi/6$ radians are coterminal. The distinction is most clearly seen by looking at the graph of the sine curve. On the x axis, $-\pi/6$ certainly is not the same point as $11\pi/6$, and even though the sine of each is $-\frac{1}{2}$, we choose $-\pi/6$ since it falls within the restricted range on x that defines the principal part of the sine curve.

  c.
$$\sin^{-1} 0 = 0$$

EXAMPLE 22   Evaluate:

  a. $\arcsin \dfrac{\sqrt{3}}{2}$ **b.** $\arcsin \left(-\dfrac{1}{\sqrt{2}}\right)$ **c.** $\arcsin(-1)$

Solution   **a.** $\arcsin \dfrac{\sqrt{3}}{2} = \dfrac{\pi}{3}$ **b.** $\arcsin \left(-\dfrac{1}{\sqrt{2}}\right) = -\dfrac{\pi}{4}$ **c.** $\arcsin(-1) = -\dfrac{\pi}{2}$

To plot the graph of the inverse of the principal part of the sine curve, we follow our usual procedure with inverses of solving for x and then interchanging x and y, giving

$$y = \sin^{-1} x$$

Now, since the roles of x and y have been switched, we must restrict x so that $-1 \le x \le 1$ and y will fall in the range $-\pi/2 \le y \le \pi/2$. The graph is just the reflection of the principal part of the sine curve in a 45° line through the origin (Figure 31). Any value of x between -1 and 1 corresponds to a unique value

Figure 31

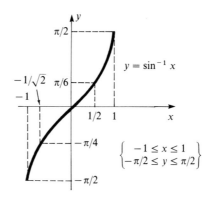

of y between $-\pi/2$ and $\pi/2$. When x is positive, y is positive, and when x is negative, y is negative. Notice that $y = \sin^{-1} x$ if and only if $\sin y = x$ and $-\pi/2 \leq y \leq \pi/2$. So when we wish to evaluate $\sin^{-1} x$ for a particular x, we ask what number (or what angle in radians) between $-\pi/2$ and $\pi/2$ has x as its sine.

We consider next the inverse of the tangent function, since it has much in common with the sine. The principal part (or **principal branch**) of the graph of $y = \tan x$ is that portion between $-\pi/2$ and $\pi/2$ (Figure 32). On this portion, if any value of y is specified, a unique value of x is determined. Solving $y = \tan x$ for x leads to either

$$x = \tan^{-1} y \qquad \text{or} \qquad x = \arctan y$$

Figure 32

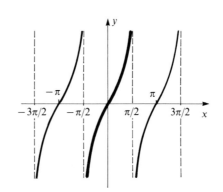

Again, we interchange the roles of x and y, and write

$$y = \tan^{-1} x \qquad \text{or} \qquad y = \arctan x$$

and restrict y such that $-\pi/2 < y < \pi/2$. The graph of this function is shown in Figure 33. With our knowledge of special angles, we conclude, for example, that

$$\tan^{-1} 1 = \frac{\pi}{4} \qquad\qquad \tan^{-1} \sqrt{3} = \frac{\pi}{3}$$

$$\tan^{-1}(-1) = -\frac{\pi}{4} \qquad\qquad \tan^{-1} 0 = 0$$

Figure 33

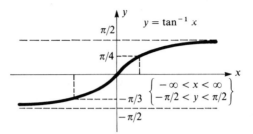

A consideration of the graph of the cosine function will make it clear that some different portion will have to be used as the principal part (Figure 34), because when x lies between $-\pi/2$ and $\pi/2$, y is always positive and hence does not assume all of its possible values. Furthermore, for a given y there is generally not a unique x determined. We choose instead the portion of the curve lying between 0 and π as the principal part. Then, for each y such that $-1 \le y \le 1$, there is a unique x for which $y = \cos x$. Again, we express the dependence of x on y by writing

$$x = \cos^{-1} y \qquad \text{or} \qquad x = \arccos y$$

Interchanging the roles of x and y, we obtain the graph of $y = \cos^{-1} x$ (Figure 35). When x is positive, y is positive and between 0 and $\pi/2$; when x is negative, y is positive and between $\pi/2$ and π.

Figure 34

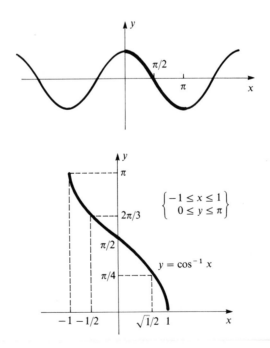

Figure 35

The inverse cotangent is seldom used, but is defined in much the same way as the inverse cosine. Its graph is shown in Figure 36.

The secant and cosecant present certain problems, and there is no general agreement on what parts of the curves to invert, that is, which portions to define

Figure 36

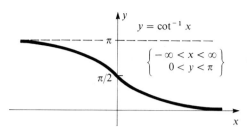

as the principal branches. Fortunately, this lack of agreement is not serious. The inverse cosecant is seldom used, and since it can always be circumvented, we will omit it entirely. The inverse secant is sufficiently useful to warrant consideration. For purposes of later use in calculus, there is some justification for the choice of principal parts made below. Let us consider first the graph of $y = \sec x$ (Figure 37). No single branch of the graph is suitable for defining the inverse function. While the reason is not now apparent, we select the highlighted portions in Figure 37 to define the inverse function. Thus, when $y \geq 1$, we take $0 \leq x < \pi/2$, and when $y \leq -1$, we take $\pi \leq x < 3\pi/2$. When we interchange the roles of x and y, we obtain the graph shown in Figure 38.

Figure 37

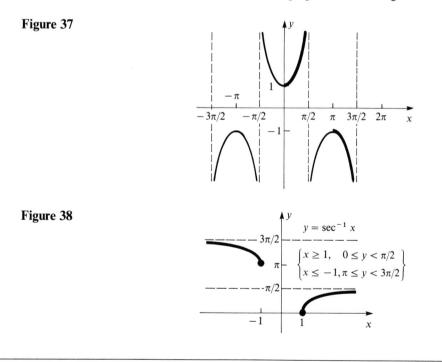

Figure 38

EXERCISE SET 10

In Problems 1–32 give the value.

A **1.** **a.** $\sin^{-1}(-\sqrt{3}/2)$ **b.** $\arccos(-\sqrt{3}/2)$
  2. **a.** $\tan^{-1}(-1)$ **b.** $\arcsec(-2)$
  3. **a.** $\arccot(-1)$ **b.** $\sin^{-1} 1$

4. **a.** $\arccos 1$ **b.** $\tan^{-1} 1$

5. **a.** $\text{arcsec } 1$ **b.** $\sin^{-1}(-\frac{1}{2})$

6. **a.** $\arctan 0$ **b.** $\cos^{-1}(-1)$

7. **a.** $\arccos 0$ **b.** $\sec^{-1}(-2/\sqrt{3})$

8. **a.** $\cos \alpha$ if $\alpha = \sin^{-1}(-\frac{2}{3})$ **b.** $\tan \theta$ if $\theta = \arccos(-\frac{1}{4})$

9. **a.** $\sin x$ if $x = \sec^{-1}(-3)$ **b.** $\sec \theta$ if $\theta = \tan^{-1} \frac{3}{4}$

10. **a.** $\sin[\cos^{-1}(-\frac{1}{3})]$ **b.** $\sec[\sin^{-1}(-\frac{1}{4})]$

11. **a.** $\tan(\arccos \frac{3}{5})$ **b.** $\cos[\tan^{-1}(-\frac{4}{3})]$

12. **a.** $\sin(\arctan \frac{7}{12})$ **b.** $\csc[\arccos(-\frac{2}{3})]$

13. **a.** $\cos[\sin^{-1}(-\frac{1}{3})]$ **b.** $\cot[\cos^{-1}(-\frac{7}{25})]$

14. $\sin(\alpha - \beta)$ if $\alpha = \sin^{-1} \frac{3}{5}$ and $\beta = \cos^{-1}(-\frac{5}{13})$

15. $\cos(\alpha + \beta)$ if $\alpha = \arctan(-\frac{1}{2})$ and $\beta = \sec^{-1} \frac{5}{3}$

16. $\tan(\alpha + \beta)$ if $\alpha = \sin^{-1}(-\frac{4}{5})$ and $\beta = \cos^{-1}(-\frac{12}{13})$

17. $\sin 2\theta$ if $\theta = \cos^{-1}(-\frac{3}{5})$ 18. $\cos 2\theta$ if $\theta = \sin^{-1}(-\frac{1}{3})$

19. $\tan 2\theta$ if $\theta = \sin^{-1}(-\frac{5}{13})$ 20. $\cos[2 \cos^{-1}(-\frac{2}{3})]$

21. $\tan(2 \tan^{-1} \frac{1}{2})$ 22. $\sin[2 \sin^{-1}(-\frac{3}{5})]$

23. $\cos[2 \sin^{-1}(-\frac{2}{3})]$ 24. $\sin(\frac{1}{2} \cos^{-1} \frac{7}{25})$

25. $\cos[\frac{1}{2} \arcsin(-\frac{3}{5})]$ 26. $\tan \frac{1}{2}[\tan^{-1}(-\frac{12}{5})]$

27. $\sin \theta$ and $\cos \theta$ if $2\theta = \tan^{-1}(-\frac{24}{7})$

28. $\sin(\alpha/2)$, $\cos(\alpha/2)$, and $\tan(\alpha/2)$ if $\alpha = \cos^{-1} \frac{3}{5}$

29. $\sin(\sin^{-1} \frac{12}{13} + \cos^{-1} \frac{4}{5})$ 30. $\cos[\cos^{-1} \frac{8}{17} - \sin^{-1}(-\frac{7}{25})]$

31. $\tan(\tan^{-1} \frac{2}{3} + \tan^{-1} \frac{1}{4})$

B   32. $\tan^{-1} \frac{2}{3} - \tan^{-1}(-\frac{1}{5})$
  Hint. Call this $\alpha - \beta$ and find $\tan(\alpha - \beta)$.

33. If $\theta = \arcsin x$, show that $x/\sqrt{1 - x^2} = \tan \theta$.

34. If $\theta = \arctan x$, show that $\sqrt{1 + x^2}/x^2 = \csc \theta \cot \theta$.

35. Show that the expression $x^2/3\sqrt{x^2 + 9}$ changes to $\sec \theta - \cos \theta$ when the substitution $\theta = \tan^{-1}(x/3)$ is made.

36. Show that $\sqrt{x^2 - 4}/x = \sin \theta$ under the substitution $\theta = \sec^{-1}(x/2)$.

---

11   Trigonometric Form of Complex Numbers

An interesting and useful relationship exists between complex numbers and trigonometric functions. Let $z = a + bi$ be any complex number. We associate with this point (a, b) in the plane. As in Figure 39, let r denote the distance from the origin to the point (a, b), and let θ denote the angle from the positive x axis to the line from the origin to the point (a, b). Then we have

$$r = \sqrt{a^2 + b^2}$$
$$a = r \cos \theta \tag{13}$$
$$b = r \sin \theta$$

The complex number $z = a + bi$ can therefore be written in the form

$$z = r(\cos \theta + i \sin \theta) \tag{14}$$

Figure 39

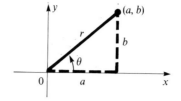

This is called the **trigonometric form** (or **polar form**) of the complex number z. The form $a + bi$ is referred to as the **rectangular form.** It should be noted that the trigonometric form is not unique, because if any integral multiple of 2π is added to θ, the value of z is unchanged. For example, if

$$z = 2\left(\cos\frac{\pi}{3} + i\sin\frac{\pi}{3}\right)$$

then also

$$z = 2\left(\cos\frac{7\pi}{3} + i\sin\frac{7\pi}{3}\right) = 2\left(\cos\frac{13\pi}{3} + i\sin\frac{13\pi}{3}\right)$$

and so on, where integral multiples of 2π are added to the angle. Normally, we choose θ to be between 0 and 2π, but there are exceptions.

The number r is called the **modulus** of z and is often denoted by $|z|$. Recall that for real number x, the absolute value $|x|$ can be interpreted geometrically as the distance on the number line between the origin and the point representing x. Similarly, $|z|$ represents the distance in the plane between the origin and the point representing z. In fact, $|z|$ is sometimes referred to as the absolute value of z. The angle θ is called the **argument** of z.

EXAMPLE 23   Find the trigonometric form of the complex number $z = \sqrt{3} + i$. What are the modulus and argument of z?

Solution   The number $\sqrt{3} + i$ is in rectangular form $a + bi$, with $a = \sqrt{3}$ and $b = 1$. So $r = \sqrt{3 + 1} = 2$, and $\theta = \pi/6$ (Figure 40). Therefore, by equation (14),

$$z = 2\left(\cos\frac{\pi}{6} + i\sin\frac{\pi}{6}\right)$$

The modulus of z is 2 and its (primary) argument is $\pi/6$.

Figure 40

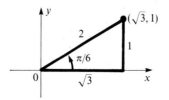

One of the advantages of the trigonometric form of complex numbers is that products, quotients, powers, and roots are particularly easy to calculate with numbers in this form. Consider first the product of two such numbers, say $z_1 = r_1(\cos\theta_1 + i\sin\theta_1)$ and $z_2 = r_2(\cos\theta_2 + i\sin\theta_2)$. On multiplication, making use of the fact that $i^2 = -1$ and arranging terms, we have

$$z_1 z_2 = r_1 r_2[(\cos\theta_1\cos\theta_2 - \sin\theta_1\sin\theta_2) + i(\sin\theta_1\cos\theta_2 + \cos\theta_1\sin\theta_2)]$$

By the addition formulas for the cosine and the sine, this can be written in the form

$$z_1 z_2 = r_1 r_2[\cos(\theta_1 + \theta_2) + i\sin(\theta_1 + \theta_2)] \qquad (15)$$

So the modulus of the product is the product of the moduli, and the argument of the product is the sum of the arguments of the two numbers.

EXAMPLE 24   Find the product of the two complex numbers

$$z_1 = 2\left(\cos\frac{\pi}{6} + i\sin\frac{\pi}{6}\right) \quad \text{and} \quad z_2 = 3\left(\cos\frac{\pi}{3} + i\sin\frac{\pi}{3}\right)$$

Solution   By equation (15), the product is

$$z_1 z_2 = 2\cdot 3\left[\cos\left(\frac{\pi}{6} + \frac{\pi}{3}\right) + i\sin\left(\frac{\pi}{6} + \frac{\pi}{3}\right)\right]$$

$$= 6\left(\cos\frac{\pi}{2} + i\sin\frac{\pi}{2}\right)$$

This can be put in rectangular form by evaluating the trigonometric functions:

$$z_1 z_2 = 6(0 + i\cdot 1) = 0 + 6i = 6i$$

In a similar way, we can prove that if $z_2 \neq 0$, then

$$\frac{z_1}{z_2} = \frac{r_1}{r_2}[\cos(\theta_1 - \theta_2) + i\sin(\theta_1 - \theta_2)] \qquad (16)$$

EXAMPLE 25   Find the quotient z_1/z_2 if

$$z_1 = 4\left(\cos\frac{5\pi}{6} + i\sin\frac{5\pi}{6}\right) \quad \text{and} \quad z_2 = 2\left(\cos\frac{\pi}{2} + i\sin\frac{\pi}{2}\right)$$

Solution

$$\frac{z_1}{z_2} = \frac{4}{2}\left[\cos\left(\frac{5\pi}{6} - \frac{\pi}{2}\right) + i\sin\left(\frac{5\pi}{6} - \frac{\pi}{2}\right)\right]$$

$$= 2\left(\cos\frac{\pi}{3} + i\sin\frac{\pi}{3}\right)$$

In rectangular form this is

$$\frac{z_1}{z_2} = 2\left(\frac{1}{2} + i\frac{\sqrt{3}}{2}\right) = 1 + i\sqrt{3}$$

It is in raising to a power that the trigonometric form has the greatest advantage. We will state the following important result due to the French mathematician Abraham De Moivre (1667–1754):

DE MOIVRE'S THEOREM

Let $z = r(\cos\theta + i\sin\theta)$ be the trigonometric form of any complex number. Then for any natural number n

$$z^n = r^n(\cos n\theta + i\sin n\theta)$$

Thus, the modulus of z^n is the nth power of the modulus of z, and the argument is n times the argument of z. That the result is plausible follows from a consideration of a few powers of z. By equation (15),

$$z^2 = r \cdot r[\cos(\theta + \theta) + i\sin(\theta + \theta)]$$
$$= r^2(\cos 2\theta + i\sin 2\theta)$$

Similarly, applying equation (15) again,

$$z^3 = z^2 \cdot z = r^2 \cdot r[\cos(2\theta + \theta) + i\sin(2\theta + \theta)]$$
$$= r^3(\cos 3\theta + i\sin 3\theta)$$

This process could be continued, and for any given power of z, De Moivre's theorem would be confirmed. However, this does not constitute a proof of the theorem. It can be proved using a technique called *mathematical induction*, which we will study later in this book.

EXAMPLE 26   Expand $(1 + i)^6$.

Solution   With the aid of a sketch (Figure 41) we determine that $r = \sqrt{2}$ and $\theta = \pi/4$. Writing $z = 1 + i$, we have

$$z^6 = (1 + i)^6 = \left[\sqrt{2}\left(\cos\frac{\pi}{4} + i\sin\frac{\pi}{4}\right)\right]^6$$
$$= (\sqrt{2})^6\left(\cos\frac{6\pi}{4} + i\sin\frac{6\pi}{4}\right)$$
$$= 8\left(\cos\frac{3\pi}{2} + i\sin\frac{3\pi}{2}\right)$$

In rectangular form the answer is

$$(1 + i)^6 = 8[0 + i(-1)] = -8i$$

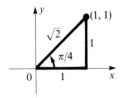

Remark. You might wish to compare this solution with the work involved in doing this problem by expanding by the binomial theorem and simplifying the result. You will find that the method we used is much simpler.

Finally, we consider the problem of taking roots of complex numbers. Again let $z = r(\cos\theta + i\sin\theta)$, and suppose we wish to find $\sqrt[n]{z}$. We first recall that if any integral multiple of 2π is added to θ, the same value of z results. That is, for any integer k,

$$z = r[\cos(\theta + 2k\pi) + i\sin(\theta + 2k\pi)]$$

Making use of De Moivre's theorem we can prove the result

$$z^{1/n} = r^{1/n}\left[\cos\left(\frac{\theta + 2k\pi}{n}\right) + i\sin\left(\frac{\theta + 2k\pi}{n}\right)\right] \tag{17}$$

Furthermore, there are exactly n distinct roots that can be obtained by taking $k = 0, 1, 2, \ldots, n - 1$. These n roots are equally spaced on a circle of radius $r^{1/n}$.

Note. In working with the trigonometric form of complex numbers, the angle θ may be expressed in either degrees or radians. If degrees are used, then equation (17) must be written

$$z^{1/n} = r^{1/n}\left[\cos\left(\frac{\theta + k \cdot 360°}{n}\right) + i\sin\left(\frac{\theta + k \cdot 360°}{n}\right)\right]$$

EXAMPLE 27   Find all cube roots of -8.

Solution   Let $z = -8 = -8 + 0i$. Then $r = 8$ and $\theta = \pi$ (Figure 42). So by equation (17),

$$\sqrt[3]{z} = z^{1/3} = 8^{1/3}\left[\cos\left(\frac{\pi + 2k\pi}{3}\right) + i\sin\left(\frac{\pi + 2k\pi}{3}\right)\right]$$

and the three distinct roots correspond to $k = 0$, $k = 1$, and $k = 2$. Let ζ_0, ζ_1, and ζ_2 be the roots corresponding to these values of k. Then

$$\zeta_0 = 2\left(\cos\frac{\pi}{3} + i\sin\frac{\pi}{3}\right) = 2\left(\frac{1}{2} + i\frac{\sqrt{3}}{2}\right) = 1 + i\sqrt{3}$$

$$\zeta_1 = 2\left(\cos\frac{\pi + 2\pi}{3} + i\sin\frac{\pi + 2\pi}{3}\right) = 2(\cos\pi + i\sin\pi) = -2$$

$$\zeta_2 = 2\left(\cos\frac{\pi + 4\pi}{3} + i\sin\frac{\pi + 4\pi}{3}\right) = 2\left(\cos\frac{5\pi}{3} + i\sin\frac{5\pi}{3}\right)$$

$$= 2\left(\frac{1}{2} - \frac{i\sqrt{3}}{2}\right) = 1 - i\sqrt{3}$$

These roots are shown graphically in Figure 43.

Figure 42

Figure 43

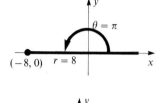

EXERCISE SET 11

A In Problems 1–10 find the trigonometric form of the complex number. Draw a sketch in each case.

1. $1 - i$

2. $2 + 2i$

3. $\sqrt{3} - i$

4. $1 + i\sqrt{3}$

5. -4

6. $-8i$

7. $-2 - 2i\sqrt{3}$

8. $4\sqrt{2} - 4i\sqrt{2}$

9. 5

10. $i - 1$

In Problems 11–20 find the rectangular form of the complex number.

11. $2\left(\cos\frac{4\pi}{3} + i\sin\frac{4\pi}{3}\right)$

12. $4\left(\cos\frac{3\pi}{4} + i\sin\frac{3\pi}{4}\right)$

13. $5\left(\cos\frac{3\pi}{2} + i\sin\frac{3\pi}{2}\right)$

14. $7(\cos\pi + i\sin\pi)$

15. $6\left(\cos\frac{11\pi}{6} + i\sin\frac{11\pi}{6}\right)$

16. $2\left(\cos\frac{5\pi}{3} + i\sin\frac{5\pi}{3}\right)$

17. $\sqrt{2}\left(\cos\frac{7\pi}{4} + i\sin\frac{7\pi}{4}\right)$

18. $5(\cos 0 + i\sin 0)$

19. $3(\cos 150° + i\sin 150°)$

20. $8(\cos 315° + i\sin 315°)$

In Problems 21–27 find $z_1 \cdot z_2$ and z_1/z_2 using equations (15) and (16). When possible without using tables or a calculator, write answers in rectangular form.

21. $z_1 = 2\left(\cos\frac{\pi}{4} + i\sin\frac{\pi}{4}\right)$, $z_2 = 3\left(\cos\frac{5\pi}{4} + i\sin\frac{5\pi}{4}\right)$

22. $z_1 = 8(\cos \pi + i \sin \pi), \quad z_2 = 4\left(\cos \dfrac{2\pi}{3} + i \sin \dfrac{2\pi}{3}\right)$

23. $z_1 = 4\left(\cos \dfrac{\pi}{2} + i \sin \dfrac{\pi}{2}\right), \quad z_2 = 2\left(\cos \dfrac{3\pi}{4} + i \sin \dfrac{3\pi}{4}\right)$

24. $z_1 = \sqrt{2} + i\sqrt{2}, \quad z_2 = 4\sqrt{2} - 4i\sqrt{2}$

25. $z_1 = \sqrt{3} - i, \quad z_2 = 2 + 2i\sqrt{3}$

26. $z_1 = 3(\cos 70° + i \sin 70°), \quad z_2 = 5(\cos 50° + i \sin 50°)$

27. $z_1 = 4(\cos 100° + i \sin 100°), \quad z_2 = 6(\cos 70° + i \sin 70°)$

In Problems 28–34 find the indicated power using De Moivre's theorem. Express answers in rectangular form.

28. $\left[2\left(\cos \dfrac{\pi}{6} + i \sin \dfrac{\pi}{6}\right)\right]^4$

29. $\left[3\left(\cos \dfrac{3\pi}{4} + i \sin \dfrac{3\pi}{4}\right)\right]^6$

30. $[2(\cos 240° + i \sin 240°)]^5$

31. $(\sqrt{3} - i)^8$

32. $(1 - i)^{10}$

33. $(2 - 2i\sqrt{3})^4$

34. $(-\sqrt{2} - i\sqrt{2})^5$

B In Problems 35–40 find all the roots indicated. Express answers in rectangular form whenever this can be done without tables or a calculator.

35. Cube roots of $8i$

36. Fourth roots of $-\dfrac{1}{2} - \dfrac{i\sqrt{3}}{2}$

37. Sixth roots of 1

38. Fourth roots of $-8 + 8i\sqrt{3}$

39. Square roots of $-16i$

40. Fifth roots of $-16\sqrt{3} + 16i$

In Problems 41–44 find the complete solution set.

41. $x^6 + 64 = 0$

42. $x^4 + 256 = 0$

43. $x^6 = 64$

44. $x^3 + 27i = 0$

45. Prove equation (17).
Hint. First write $z = r[\cos(\theta + 2k\pi) + i \sin(\theta + 2k\pi)]$, and let $\zeta = \rho(\cos \phi + i \sin \phi)$ be an unknown nth root of z. By De Moivre's theorem determine ρ and ϕ such that $\zeta^n = z$.

Problem 46 is taken from an engineering textbook.* The notation $\underline{/\theta}$ means $\cos \theta + i \sin \theta$. Use a hand calculator to evaluate.

46. Express the result of each of these complex number manipulations in polar form, using six significant figures just for the pure joy of calculating:
(a) $(5 - 3\underline{/21.6°})(4.11\underline{/-161.9°} - 2\underline{/18.7613°});$ (b) $1/(1\underline{/189.2°} - 1\underline{/-15.6°});$
(c) $2.18\underline{/57.9°} + 4.11\underline{/-161.9°} + 1.89\underline{/-49.2°}$

---

* William H. Hayt, Jr., and Jack E. Kemmerly, *Engineering Circuit Analysis*, 3d ed. (New York: McGraw-Hill Book Company, 1978), p. 755. Reprinted by permission.

12 A Useful Inequality

In this section we will derive an inequality that plays an important role in the calculus of trigonometric functions. In Figure 44 the shaded area is called a **sector** of the circle. We begin by finding a formula for the area of a sector of a circle. The area depends on the radius r and the angle θ. As usual, we use θ both to name the angle and to designate its measure in radians. We take it as known that the area of the entire circle (actually, we should say the area within the circle) is πr^2. The area of the sector is a fractional part of this, but what fractional part? It is just the ratio of θ to 2π, since θ is the measure of the central angle of the sector and 2π is the measure of the entire revolution. So we want to multiply the fraction $\theta/2\pi$ by the area of the circle:

$$\frac{\theta}{2\pi} \cdot \pi r^2 = \frac{1}{2}r^2\theta$$

Area of a sector $= \dfrac{1}{2}r^2\theta$

Figure 44

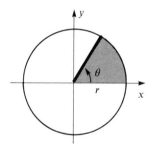

Let us now consider a unit circle and an angle θ between 0 and $\pi/2$ (Figure 45). The angle of θ radians subtends an arc length of θ units. The point $P(\theta)$ has coordinates $(\cos\theta, \sin\theta)$. So the length of the vertical line segment from $P(\theta)$ to the x axis is $\sin\theta$. If θ is small, it would appear that this vertical segment and the arc from $(1, 0)$ to $P(\theta)$ are very nearly equal, that is, $\sin\theta$ and θ are

Figure 45

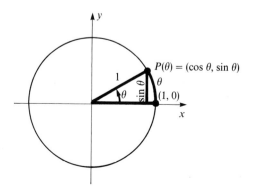

$P(\theta) = (\cos\theta, \sin\theta)$

Figure 46

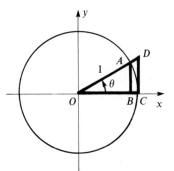

almost equal. We want to consider the ratio $(\sin\theta)/\theta$, since this plays an important role in calculus. From the observation above, it seems reasonable to expect that when θ is small, this ratio is almost 1. Of course, we cannot allow θ to be 0, because then the ratio becomes $0/0$, which is meaningless. To get a more precise estimate on the size of $(\sin\theta)/\theta$, we consider the areas shown in Figure 46. Let

$$A_1 = \text{Area of triangle } OAB$$

$$A_2 = \text{Area of sector } OAC$$

$$A_3 = \text{Area of triangle } ODC$$

Then it is evident from the geometry that $A_1 < A_2 < A_3$. The point A is what we have customarily called $P(\theta)$ and it is θ units along the arc of the circle from C, that is, arc $AC = \theta$. Also, by definition, the x and y coordinates of A are $\cos\theta$ and $\sin\theta$, respectively. Thus, $\overline{OB} = \cos\theta$ and $\overline{AB} = \sin\theta$. The area A_1 can now be calculated:

$$A_1 = \tfrac{1}{2}(\overline{OB})(\overline{AB}) = \tfrac{1}{2}\sin\theta\cos\theta$$

We have already found that

$$A_2 = \tfrac{1}{2}(1)^2 \cdot \theta = \tfrac{1}{2}\theta$$

To find A_3, note that $\overline{OC}$ of triangle OAC is 1 unit in length, since it is the radius of the circle. From that triangle we know that

$$\tan\theta = \frac{\overline{CD}}{\overline{OC}} = \frac{\overline{CD}}{1}$$

So $\overline{CD} = \tan\theta$. The area A_3 is therefore

$$A_3 = \frac{1}{2}(\overline{OC})(\overline{CD}) = \frac{1}{2}(1)(\tan\theta) = \frac{1}{2}\tan\theta = \frac{1}{2}\left(\frac{\sin\theta}{\cos\theta}\right)$$

Now the inequality $A_1 < A_2 < A_3$ can be written

$$\frac{1}{2}\sin\theta\cos\theta < \frac{1}{2}\theta < \frac{1}{2}\left(\frac{\sin\theta}{\cos\theta}\right)$$

For $0 < \theta < \pi/2$, $\sin\theta > 0$, so we can divide by $\tfrac{1}{2}\sin\theta$ to obtain

$$\cos\theta < \frac{\theta}{\sin\theta} < \frac{1}{\cos\theta}$$

and again using the fact that all terms are positive, we can invert each term, reversing inequality signs (since $a < b$ implies $1/a > 1/b$ for positive a and b). This gives

$$\frac{1}{\cos \theta} > \frac{\sin \theta}{\theta} > \cos \theta$$

or equivalently,

$$\cos \theta < \frac{\sin \theta}{\theta} < \frac{1}{\cos \theta} \tag{18}$$

and this is true for all θ where $0 < \theta < \pi/2$.

If θ lies in the range $-\pi/2 < \theta < 0$, we can arrive at inequality (18) again, although it is necessary to be careful with negative signs. We will not present the details. For all values of θ, then, where $-\pi/2 < \theta < \pi/2$, and $\theta \neq 0$, we have

$$\cos \theta < \frac{\sin \theta}{\theta} < \frac{1}{\cos \theta}$$

We know that $\cos 0 = 1$, and for θ very close to 0, $\cos \theta$ will be very close to, but less than, 1. Our inequality tells us, then, that for small values of θ, positive or negative, $(\sin \theta)/\theta$ is hemmed in between two numbers—one slightly less than 1 and the other slightly greater than 1. Furthermore, the closer θ is to 0, the closer both $\cos \theta$ and $1/(\cos \theta)$ are to 1. Without defining precisely what we mean by a limiting process, we can nevertheless say that $(\sin \theta)/\theta$ will be as close to 1 as we wish for all values of θ in a sufficiently small deleted neighborhood of 0. In calculus, this will be indicated by the symbolism

$$\lim_{\theta \to 0} \frac{\sin \theta}{\theta} = 1$$

which is read "the limit of $(\sin \theta)/\theta$ as θ approaches 0 is 1."

EXERCISE SET 12

A **1.** Make a table of values (using a calculator) and draw the graph of

$$y = \frac{\sin x}{x}$$

for $-\pi/2 \leq x \leq \pi/2$, $x \neq 0$. Plot several points near $x = 0$. Does your result tend to confirm the limit found in this section?

2. Show that

$$\lim_{\theta \to 0} \frac{\tan \theta}{\theta} = 1$$

Hint. $\dfrac{\tan \theta}{\theta} = \dfrac{\sin \theta}{\theta} \cdot \dfrac{1}{\cos \theta}$

3. By multiplying the numerator and denominator of $(1 - \cos \theta)/\theta$ by $1 + \cos \theta$, show that

$$\lim_{\theta \to 0} \frac{1 - \cos \theta}{\theta} = 0$$

In Problems 4–9 find the limits indicated.

4. $\lim\limits_{\theta \to 0} \dfrac{\sin^2 \theta}{\theta}$

5. $\lim\limits_{x \to 0} \dfrac{\sin 2x}{x}$

6. $\lim\limits_{t \to 0} t \cot t$

7. $\lim\limits_{x \to 0} \dfrac{\cos 2x - 1}{x^2}$

8. $\lim\limits_{x \to 0} \dfrac{1}{x^2(1 + \cot^2 x)}$

9. $\lim\limits_{\theta \to 0} \dfrac{1 - \cos 2\theta}{\theta \sin 2\theta}$

B   **10.**   If $\alpha > 0$, show that

$$\lim_{\alpha \to 0} \frac{\sqrt{2(1 - \cos \alpha)}}{\alpha} = 1$$

Hint.   Use a half-angle formula.

11.   Show that

$$\lim_{x \to \infty} x \sin \frac{1}{x} = 1$$

Hint.   Let $\theta = 1/x$.

12.   Show that inequality (18) remains valid for $-\pi/2 < \theta < 0$.

Review Exercise Set

A   **1.**   Evaluate each of the following without the use of tables or a calculator:
   a. $\tan(7\pi/12)$ **b.** $\sin(23\pi/12)$ **c.** $\cos(\pi/8)$ **d.** $\cos(13\pi/12)$

2.   Reduce each of the following to a function involving θ only:

   a. $\sin\left(\dfrac{3\pi}{2} - \theta\right)$ **b.** $\tan(\theta + \pi)$ **c.** $\sec(4\pi - \theta)$

   d. $\cos\left(\theta - \dfrac{\pi}{2}\right)$ **e.** $\csc\left(\dfrac{9\pi}{2} + \theta\right)$

3.   Evaluate:
   a. $\arccos(-\frac{1}{2})$ **b.** $\tan^{-1}(-\sqrt{3})$ **c.** $\arcsin(-\sqrt{3}/2)$
   d. $\cos^{-1}(-1)$ **e.** $\tan^{-1}(-1)$

4.   If $\sec \theta = \frac{5}{4}$ and $\sin \theta < 0$, $0 \le \theta \le 2\pi$, find:
   a. $\cos 2\theta$ **b.** $\tan 2\theta$ **c.** $\cos \frac{1}{2}\theta$

5.   If $\sin \theta = \frac{4}{5}$ and $\tan \theta < 0$, $0 \le \theta \le 2\pi$, find:
   a. $\sin 2\theta$ **b.** $\cos 2\theta$ **c.** $\sin \frac{1}{2}\theta$

6.   If $\alpha = \arcsin(-\frac{3}{5})$ and $\beta = \arccos(-\frac{5}{13})$, find:
   a. $\sin(\alpha + \beta)$ **b.** $\cos(\alpha - \beta)$ **c.** $\tan(\alpha + \beta)$

7.   If $\alpha = \tan^{-1}(-\frac{4}{3})$ and $\beta = \cos^{-1}(-\frac{8}{17})$, find:
   a. $\sin(\alpha - \beta)$ **b.** $\cos(\alpha + \beta)$ **c.** $\tan(\alpha - \beta)$

8.   If $\tan 2\theta = -\frac{24}{7}$ and $\pi/2 < 2\theta < \pi$, find $\sin \theta$ and $\cos \theta$.

9.   If $\tan \theta = \frac{15}{8}$ and $\pi < \theta < 3\pi/2$, find $\sin(\theta/2)$, $\cos(\theta/2)$, and $\tan(\theta/2)$.

10.   Evaluate:
   a. $\cos[\arcsin(-\frac{2}{3})]$ **b.** $\sin[2 \sin^{-1}(-\frac{5}{13})]$

11.   Evaluate:
   a. $\tan[2 \cos^{-1}(-\frac{4}{5})]$ **b.** $\cos(\frac{1}{2} \tan^{-1} 4\sqrt{3})$

In Problems 12–25 prove that the given equations are identities.

12. $\csc\theta - \cos\theta\cot\theta = \sin\theta$

13. $\dfrac{2}{\tan\theta + \cot\theta} = \sin 2\theta$

14. $\dfrac{\cos\theta + \cot\theta}{1 + \csc\theta} = \cos\theta$

15. $\dfrac{\cos\theta}{\sec\theta - \tan\theta} = 1 + \sin\theta$

16. $\dfrac{\cot\theta - \tan\theta}{\cot\theta + \tan\theta} = \cos 2\theta$

17. $\sin\theta\tan\theta + \cos\theta = \sec\theta$

18. $\dfrac{\tan\theta}{\sec\theta + 1} + \dfrac{\sec\theta - 1}{\tan\theta} = \dfrac{2\tan\theta}{\sec\theta + 1}$

19. $\tan\theta(\cos 2\theta + 1) = \sin 2\theta$

20. $\sin\theta + \cos\theta\cot\theta = \csc\theta$

21. $\sin 2\theta + (\cos\theta - \sin\theta)^2 = 1$

22. $1 + \dfrac{\tan^2\theta}{\sec\theta + 1} = \sec\theta$

23. $\cot\theta - \tan\theta = 2\cot 2\theta$

24. $\dfrac{1 + \tan^2\theta}{1 - \tan^2\theta} = \sec 2\theta$

25. $\dfrac{\sin 2\theta}{1 - \cos 2\theta} = \cot\theta$

In Problems 26–31 find all primary solutions.

26. $\sin\theta - 2\sin^2\theta = 0$

27. $3\tan^3\theta - \tan\theta = 0$

28. $\sin 2\theta + \cos\theta = 0$

29. $\cos 2\theta - \sin\theta = 0$

30. $2\cos^2 3\theta - 3\cos 3\theta - 2 = 0$

31. $2\cos^2\theta = 1 - \sin\theta$

In Problems 32–34 sketch the graphs.

32. a. $y = 2\sin x + 1$ **b.** $y = 3\cos 2x$

33. a. $y = 2\tan\dfrac{\pi}{4}x$ **b.** $y = 2\cos\dfrac{\pi}{2}x + 3$

34. a. $y = \sin\left(3x + \dfrac{\pi}{2}\right)$ **b.** $y = 2\cos\left(\pi x - \dfrac{\pi}{3}\right)$

35. If

$$z_1 = 2\left(\cos\frac{5\pi}{6} + i\sin\frac{5\pi}{6}\right) \quad \text{and} \quad z_2 = 3\left(\cos\frac{4\pi}{3} + i\sin\frac{4\pi}{3}\right)$$

find $z_1 \cdot z_2$ and z_1/z_2. Express answers in rectangular form.

In Problems 36 and 37 expand using De Moivre's theorem.

36. $(1 - i)^8$

37. $(2 + 2i\sqrt{3})^4$

In Problems 38 and 39 evaluate the limits.

38. $\displaystyle\lim_{x\to 0}\frac{x\sin x}{1 - \cos 2x}$

39. $\displaystyle\lim_{\theta\to 0}\frac{\sec^2\theta - 1}{\theta^2}$

B 40. Evaluate:

    a. $\cos\left[\tan^{-1}\frac{5}{12} - \sin^{-1}\left(-\frac{8}{17}\right)\right]$ **b.** $\tan^{-1}\frac{1}{3} + \tan^{-1}(-2)$

In Problems 41–46 prove the identities.

41. $\dfrac{\sin \theta + \cos 2\theta - 1}{\cos \theta - \sin 2\theta} = \tan \theta$

42. $\sin^2 \theta + \dfrac{2 - \tan^2 \theta}{\sec^2 \theta} - 1 = \cos 2\theta$

43. $\dfrac{\sec \theta - 1}{\sec \theta + 1} = 2 \csc^2 \theta - 2 \cot \theta \csc \theta - 1$

44. $\sec \theta \csc \theta - 2 \cos \theta \csc \theta = \tan \theta - \cot \theta$

45. $\dfrac{\sin 8\theta - \sin 6\theta}{\cos 8\theta + \cos 6\theta} = \tan \theta$

46. $\dfrac{\cos \theta}{1 - \sin \theta} = \tan \theta + \sec \theta$

In Problems 47–50 find all primary solutions.

47. $\tan^2 2\theta + 3 \sec 2\theta + 3 = 0$

48. $\cos 2\theta \sec \theta + \sec \theta = 1$

49. $\tan 2\theta = 2 \cos \theta$

50. $1 + \sin \theta = \cos \theta$
    Hint. Square both sides.

51. Draw the graph of:

   a. $y = 3 - 2 \cos\left(\dfrac{\pi}{3} - 2x\right)$ **b.** $y = \sqrt{3} \cos x - \sin x$

52. Find all eighth roots of 1.

53. Find all cube roots of $-64i$.

54. Find the complete solution set to the equation $x^6 + 1 = 0$.

55. Show that if $x < 0$, then

$$\lim_{x \to 0} \frac{\sqrt{1 - \cos 2x}}{x} = -\sqrt{2}$$

Cumulative Review Exercise Set III (Chapters 8 and 9)

1. In each of the following a right triangle ABC is given, with right angle at C. Find all missing sides and angles without the use of a calculator or tables.
   - **a.** $A = 30°$, $b = 12$
   - **b.** $B = 60°$, $a = 4$
   - **c.** $A = 45°$, $c = 16$
   - **d.** $a = 10$, $c = 20$
   - **e.** $b = 6\sqrt{2}$, $c = 12$

2. Without using a calculator or tables find all six trigonometric functions of each of the following. Leave answers in exact form.
   - **a.** $4\pi/3$
   - **b.** $-3\pi/4$
   - **c.** $17\pi/6$
   - **d.** $75°$
   - **e.** $990°$
   - **f.** $17\pi/12$
   - **g.** $-22.5°$
   - **h.** $(2n - 1)\pi$, n an integer

3. Evaluate without a calculator or tables:
   - **a.** $\sin[2 \sin^{-1}(-\frac{3}{5})]$
   - **b.** $\cos[\cos^{-1}(-\frac{5}{13}) + \tan^{-1}(-\frac{4}{3})]$
   - **c.** $\tan(\frac{1}{2})\theta$, where $\sec\theta = \frac{3}{2}$ and $\sin\theta < 0$
   - **d.** $\csc\left(\dfrac{\pi}{2} - \theta\right)$, where $\cot\theta = \frac{1}{2}$ and $\cos\theta < 0$

4. Prove the identities:
   - **a.** $\dfrac{\sec\theta - \cos\theta + \tan\theta}{\tan\theta + \sec\theta} = \sin\theta$
   - **b.** $\dfrac{(\sin\theta - \cos\theta)^2}{\cos 2\theta} = \dfrac{1 - \tan\theta}{1 + \tan\theta}$

5. Let
$$z_1 = 4\left(\cos\frac{4\pi}{3} + i\sin\frac{4\pi}{3}\right) \quad \text{and} \quad z_2 = 2\left(\cos\frac{\pi}{6} + i\sin\frac{\pi}{6}\right)$$
   Find:
   - **a.** $z_1 z_2$
   - **b.** z_1/z_2
   - **c.** z_2^4
   - **d.** The square roots of z_1

   Express answers in rectangular form.

6. Points A and B are on level ground on a line with the base of a tower. From the top of the tower the angles of depression of A and B, respectively, are $24.6°$ and $31.2°$. If points A and B are known to be 20 meters apart, find the height of the tower.

7. Evaluate without using a calculator or tables:
   - **a.** $\cos[2 \sin^{-1}(-\frac{1}{3})]$
   - **b.** $\sin[2 \arctan(-3)]$
   - **c.** $\tan\frac{1}{2}[\cos^{-1}(-\frac{7}{25})]$
   - **d.** $\cos\frac{1}{2}\alpha$, where $\csc\alpha = -\frac{5}{4}$ and $\pi \le \alpha \le 3\pi/2$

8. Find all primary solutions:
   - **a.** $2\sin x + \sqrt{3}\tan x = 0$
   - **b.** $2\sin^2\frac{1}{2}x - 2\cos^2 x = 1$

9. A chord on a circle of radius 6 inches is 10 inches long. Find:
   - **a.** The central angle formed by the radii to the end points of the chord
   - **b.** The arc length cut off by the chord
   - **c.** The area of the sector formed by this arc and the radii
   - **d.** The area of the triangle formed by the chord and the radii

10. In each of the following, triangle ABC is a right triangle, with right angle at C. Find all unknown sides and angles.
    - **a.** $A = 16.3°$, $b = 4.68$
    - **b.** $B = 52°40'$, $c = 171.5$
    - **c.** $a = 13.7$, $b = 21.3$
    - **d.** $b = 50$, $c = 130$

11. **a.** A curve on a railroad track is in the form of a circular arc which is 1.5 kilometers long. If the central angle corresponding to this arc is $72°$, find the radius of the circle.
    - **b.** A girl is riding a bicycle with 28 inch diameter tires which are rotating at the rate of 200 revolutions per minute. Approximately what is the speed of the bicycle in miles per hour?

12. Find all unknown sides and angles of the triangles ABC. Also, find their areas.
    - **a.** $A = 32°$, $B = 47°$, $c = 25.3$
    - **b.** $B = 28.3°$, $b = 23.5$, $c = 39.2$
    - **c.** $A = 98°24'$, $b = 124.3$, $c = 89.7$
    - **d.** $a = 32.5$, $b = 26.8$, $c = 18.6$

13. Prove the identities:

   a. $\dfrac{\sin \theta}{\sec \theta - 1} + \dfrac{\sin \theta}{\sec \theta + 1} = 2 \cot \theta$

   b. $\dfrac{\sin 2x \cos x}{(1 + \cos 2x) \cos^2(x/2)} = 2 \tan(x/2)$

14. Find all primary solutions:
   a. $\sin x + \sin 2x = 0$
   b. $\cos 4x = \cos 2x$

15. Find all solutions to the equation $x^3 + 64i = 0$.

16. Evaluate:
   a. $\sin[\cos^{-1}(-\tfrac{4}{5}) - \tan^{-1}(-\tfrac{12}{5})]$
   b. $\arctan(-2) + \arctan \tfrac{1}{2}$
   c. $\cos(\alpha + 2\beta)$, where $\sec \alpha = \tfrac{13}{5}$, $\csc \alpha < 0$, $\cot \beta = 2$, $\sec \beta < 0$

17. Draw the graphs:

   a. $y = 1 + 2 \sin\left(\dfrac{x}{2} - \dfrac{\pi}{3}\right)$
   b. $y = \tfrac{1}{2} \cos\left(\pi x + \dfrac{\pi}{4}\right)$

18. An airplane leaves an airport at 2:00 PM and heads 32° east of due south at an average cruising speed of 240 miles per hour. Another plane leaves the same airport at 2:15 PM at a heading 25° east of due north and cruises at an average speed of 400 miles per hour. If they fly at the same altitude, how far apart are they at 3:30 PM?

19. Prove the identities:

   a. $\dfrac{2 \cos^3 \theta}{1 + \sin \theta} + \sin 2\theta = 2 \cos \theta$
   b. $\tan\left(\dfrac{\alpha}{2} - \dfrac{\pi}{4}\right) = \dfrac{\sin \alpha - 1}{\cos \alpha}$

20. Prove the identities:

   a. $\cot \theta \sin 2\theta - \cos 2\theta = 1$
   b. $\dfrac{\sin \theta}{1 - \cos \theta} - \tan \dfrac{\theta}{2} = 2 \cot \theta$

21. Draw the graphs:
   a. $y = \tfrac{1}{3} \tan(\pi x/2)$
   b. $y = \tfrac{1}{2} \sin^{-1} 2x + \pi/3$

22. Observer A is stationed at a point due west of a monument, and measures the angle of elevation of the top of the monument to be 45°. Observer B is 100 feet due south of observer A, and B measures the angle of elevation of the top of the monument to be 30°. Without using a calculator or tables, find the height of the monument.

23. Find the complete solution set of each of the following:

   a. $\sqrt{3} \sin x = \cos x - 1$
   b. $\dfrac{\cos 3x}{\sin x} + \dfrac{\sin 3x}{\cos x} = \csc 2x$

24. An offshore oil drilling rig in the Gulf of Mexico is viewed from points A and B on the shore, with the point on the shore nearest the rig lying between A and B. In the triangle formed by the rig and the points A and B, the angle at A is 76°25′, and the angle at B is 68°54′. If A and B are 6 kilometers apart, find the distance from each of these points to the drilling rig. How far is the rig from the nearest point on shore?

10 Systems of Equations and Inequalities

There are many occasions in calculus when finding the simultaneous solution of systems of two or more equations is a necessary part of a larger problem. The equations may be linear or nonlinear. They may even involve transcendental functions. The following example illustrates how a linear system arises in one aspect of calculus:*

If $f(x, y, z) = 4x^2 + y^2 + 5z^2$, find the point on the plane $2x + 3y + 4z = 12$ at which $f(x, y, z)$ has its least value.

Solution: We wish to find the minimum value of $f(x, y, z)$ subject to the constraint $g(x, y, z) = 2x + 3y + 4z - 12 = 0$. If . . . we let

$$w = 4x^2 + y^2 + 5z^2 + \lambda(2x + 3y + 4z - 12)$$

then . . .

$$w_x = 8x + 2\lambda = 0$$
$$w_y = 2y + 3\lambda = 0$$
$$w_z = 10z + 4\lambda = 0$$
$$w_\lambda = 2x + 3y + 4z - 12 = 0$$

The first three equations give us

$$\lambda = -4x = -\tfrac{2}{3}y = -\tfrac{5}{2}z$$

Consequently, for a local extremum we need

$$y = 6x \quad \text{and} \quad z = \tfrac{8}{5}x$$

Substituting in the equation $w_\lambda = 0$ we obtain

$$2x + 18x + \tfrac{32}{5}x - 12 = 0$$

or $x = \tfrac{5}{11}$. Hence $y = 6(\tfrac{5}{11}) = \tfrac{30}{11}$ and $z = (\tfrac{8}{5})(\tfrac{5}{11}) = \tfrac{8}{11}$. It follows that the minimum value occurs at the point $(\tfrac{5}{11}, \tfrac{30}{11}, \tfrac{8}{11})$.

The main thing to observe is that an essential part of this problem was the solution of a system of four linear equations in four unknowns. There are different ways to solve such systems, and in this chapter we present the most important of these.

---

* Earl W. Swokowski, *Calculus with Analytic Geometry*, 2d ed. (Boston: Prindle, Weber, & Schmidt, 1979), pp. 813–814. Reprinted by permission.

1 Linear Systems

We saw in Chapter 5 that the general form of the equation of a line is

$$Ax + By + C = 0$$

This is also called a **general linear equation in two variables.** The word *linear* continues to apply, however, regardless of the number of variables as long as the equation is of degree 1. Thus, $2x - 3y + 4z = 0$ and $5x_1 - 2x_2 + 3x_3 - 4x_4 + 8x_5 = 2$ are linear, whereas $x^2 - 2x + 3y = 4$, $xy - 3x + 4y - 7$, and $\sqrt{x_1} + x_2 = 4$ are not. Techniques for solving **systems of linear equations** form the basis of a large part of what is called **linear algebra.** A complete treatment of linear systems is not appropriate in this course, but a review of techniques studied in high school algebra, with an introduction to more advanced techniques, is in order.

We consider first **two linear equations in two variables.** For example,

$$\begin{cases} 2x - 3y = 4 \\ 5x + 4y = 7 \end{cases}$$

Our object is to find all pairs (x, y) that simultaneously satisfy both equations. The collection of all such pairs is the solution to the system. In this context we will refer to x and y as the unknowns, since we are seeking to determine values for them.

Since two linear equations correspond geometrically to two lines, we can reason in advance that one of three situations will occur:

1. The lines intersect in one and only one point, and its coordinates constitute the unique solution. The equations are said to be **consistent.**
2. The lines are parallel and hence do not intersect, so there is no simultaneous solution. The equations are said to be **inconsistent.**
3. The lines coincide and hence intersect at infinitely many points, so there are infinitely many solutions. The equations are said to be linearly **dependent.**

Let us consider the example given above.

EXAMPLE 1   Solve the system: $\begin{cases} 2x - 3y = 4 \\ 5x + 4y = 7 \end{cases}$

Solution   By appropriate multiplications and additions or subtractions, we seek to eliminate one of the unknowns. One way is to multiply the top equation by 4 and the bottom equation by 3 and then add the resulting equations, thus eliminating y:

$$\begin{array}{r} 8x - 12y = 16 \\ 15x + 12y = 21 \\ \hline 23x \qquad\;\; = 37 \\ x = \tfrac{37}{23} \end{array}$$

This is the x coordinate of a point on both lines; we can therefore find the y coordinate by substituting into either equation. By substituting in the first equation, we have

$$2(\tfrac{37}{23}) - 3y = 4$$
$$-3y = \tfrac{18}{23}$$
$$y = \tfrac{-6}{23}$$

So this problem illustrates a consistent system. There is a unique solution $(\tfrac{37}{23}, \tfrac{-6}{23})$. The situation is graphed in Figure 1.

Figure 1

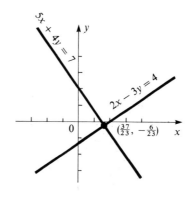

The technique used to solve the system in Example 1 is known as **elimination.** We eliminated y from the two equations and obtained an equation in x alone. Then we were able to solve for x, and y was found by substituting this value of x into one of the original equations. In general, to eliminate an unknown we use one of two methods: elimination by addition or subtraction, or elimination by substitution.

To eliminate an unknown from two equations by addition or subtraction, the coefficients of that unknown must be the same in absolute value. If they are not the same, then we make them the same by multiplying the equations by appropriate numbers. In Example 1 we multiplied both sides of the first equation by 4 and both sides of the second equation by 3. When the coefficients are the same in absolute value, we *add* if the coefficients are opposite in sign (as in Example 1) and *subtract* if they are like in sign.

To eliminate an unknown from two equations by substitution, we solve one of the equations for that unknown in terms of the other (or others) and substitute this into the other equation. Consider, for example, the system

$$\begin{cases} 3x - 4y = 7 \\ 2x - \ y = 3 \end{cases}$$

We can solve the second equation for y to get $y = 2x - 3$ and then substitute $2x - 3$ in place of y in the first equation:

$$3x - 4(2x - 3) = 7$$
$$3x - 8x + 12 = 7$$
$$-5x = -5$$
$$x = 1$$

Having found x, we obtain y from $y = 2x - 3$. This gives

$$y = 2(1) - 3$$
$$= -1$$

So this is a consistent system with unique solution $(1, -1)$.

The next two examples illustrate inconsistent and dependent systems, respectively.

EXAMPLE 2   Solve the system: $\begin{cases} 2x - 3y = 4 \\ 4x - 6y = 15 \end{cases}$

Solution   After multiplying both sides of the top equation by 2 and subtracting, we get $0 = -7$, which is clearly impossible. There is no simultaneous solution; that is, the lines are parallel. In fact, we could have seen in advance that the two lines had the same slope. The situation is graphed in Figure 2.

Figure 2

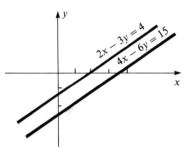

EXAMPLE 3   Solve the system: $\begin{cases} 2x - 3y = 4 \\ 4x - 6y = 8 \end{cases}$

Solution   This system differs from the one in Example 2 only in the fact that the entire second equation is obtainable from the first by multiplying both sides by 2. But this means that the second equation is really equivalent to the first and so has the same graph. Attempting to solve simultaneously would yield $0 = 0$. This is an example of a dependent system. We could say that the simultaneous solutions are the pairs (x, y) corresponding to all points on the line given by either equation. From the first equation we can solve for y to get

$$y = \frac{2x - 4}{3}$$

Now let x take on an arbitrary value, say, $x = c$; then $y = (2c - 4)/3$. So we could say that the solution set is the set of all pairs

$$\left(c, \frac{2c - 4}{3} \right)$$

where c may range over all real numbers.

EXAMPLE 4   Solve the system: $\begin{cases} x + 2y + z = 3 \\ 2x - y - z = 0 \\ 3x + 4y + 2z = 8 \end{cases}$

Solution   We will again use the process of elimination. The object is to eliminate the *same* unknown from two pairs of equations, thus obtaining two equations in two unknowns. Then one further elimination enables us to find one of the unknowns, and the others are found by substitution.

It appears that z would be particularly easy to eliminate in this case. We add corresponding sides of the first two equations:

$$
\begin{array}{l}
x + 2y + z = 3 \\
\underline{2x - y - z = 0} \\
3x + y \phantom{{}+ z} = 3
\end{array}
\qquad (1)
$$

Now returning to the original system, we must eliminate z from another pair of equations. It is essential to stay with z so that a second equation containing only x and y can be obtained. We multiply both sides of the second equation by 2 and add it to the third:

$$
\begin{array}{l}
4x - 2y - 2z = 0 \\
\underline{3x + 4y + 2z = 8} \\
7x + 2y \phantom{{}+ 2z} = 8
\end{array}
\qquad (2)
$$

Now we wish to solve equations (1) and (2) simultaneously, and to this end we multiply both sides of equation (1) by -2 and add the result to equation (2):

$$
\begin{array}{l}
-6x - 2y = -6 \\
\underline{7x + 2y = 8} \\
x \phantom{{}- 2y} = 2
\end{array}
$$

Having found x, we can substitute in either equation (1) or (2) to find y. We select (1):

$$3(2) + y = 3$$
$$y = -3$$

Finally, we substitute both x and y into any one of the original equations in order to find z. From the first equation we get

$$2 + 2(-3) + z = 3$$
$$2 - 6 + z = 3$$
$$z = 7$$

The final answer for the simultaneous solution is the triple $(2, -3, 7)$, that is, $x = 2$, $y = -3$, and $z = 7$.

Keeping track of things in the case of three equations and three unknowns is difficult enough, but with more than three the difficulty increases rapidly. For example, with four equations, three pairs are selected and from each pair the *same* unknown is eliminated. The result is a system of three equations in three unknowns, which is solved as in Example 4. When the three unknowns are found, they are substituted into any one of the original equations to find the fourth unknown.

Equations in two or more unknowns often arise in real-world problems. Sometimes problems that could be treated with one unknown are easier to handle with two, as illustrated in the following examples.

EXAMPLE 5   A saline solution containing 5% salt is to be mixed with one containing 8% salt to obtain 5 gallons of a 6% salt solution. How many gallons of each solution should be used?

Solution   Let

$$x = \text{Number of gallons of 5\% solution required}$$
$$y = \text{Number of gallons of 8\% solution required}$$

Then

$$0.05x = \text{Amount of pure salt in first solution}$$
$$0.08y = \text{Amount of pure salt in second solution}$$
$$(0.06)5 = \text{Amount of pure salt in final solution}$$

We need two equations relating x and y. Since there are to be 5 gallons in all, we must have

$$x + y = 5$$

Further, the amount of pure salt from the two components must equal the amount of pure salt in the final mixture:

$$0.05x + 0.08y = (0.06)5$$

or, after multiplying by 100 and simplifying,

$$5x + 8y = 30$$

So the system we wish to solve is

$$\begin{cases} x + y = 5 \\ 5x + 8y = 30 \end{cases}$$

By multiplying both sides of the top equation by 5 and subtracting from the bottom, we get

$$\begin{array}{r} 5x + 8y = 30 \\ 5x + 5y = 25 \\ \hline 3y = 5 \end{array}$$

$$y = \tfrac{5}{3}$$

Finally,

$$x = 5 - y = 5 - \frac{5}{3} = \frac{15 - 5}{3} = \frac{10}{3}$$

So $3\frac{1}{3}$ gallons of the 5% solution are needed, and $1\frac{2}{3}$ gallons of the 8% solution are needed.

EXAMPLE 6   A boy rows 6 miles upstream in 1 hour and 20 minutes, and the return trip takes 1 hour. Find how fast he can row in still water and find the rate of the current.

Solution   Let

$$x = \text{Rate in miles per hour that the boy can row in still water}$$

$$y = \text{Rate in miles per hour of the current}$$

Then

$$x - y = \text{Actual rate of progress upstream}$$

$$x + y = \text{Actual rate of progress downstream}$$

We apply the fundamental relation $d = rt$ two times, once for going upstream and once for going downstream:

$$\begin{cases} 6 = (x - y) \cdot \frac{4}{3} \\ 6 = (x + y) \cdot 1 \end{cases} \quad (\tfrac{4}{3} \text{ hours } = 1 \text{ hour and 20 minutes})$$

These can be simplified to the form

$$\begin{cases} 2x - 2y = 9 \\ x + y = 6 \end{cases}$$

The solution is now easily found:

$$\begin{array}{rl} 2x - 2y = & 9 \\ 2x + 2y = & 12 \\ \hline 4x \quad\quad = & 21 \\ x = & \tfrac{21}{4} \end{array}$$

So

$$y = 6 - \frac{21}{4} = \frac{24 - 21}{4} = \frac{3}{4}$$

The boy can row $5\frac{1}{4}$ miles per hour in still water, and the rate of the current is $\frac{3}{4}$ mile per hour.

EXAMPLE 7   Joan has a tractor mower and Charles has a hand-operated power mower. They regularly mow a certain large lawn in 4 hours by working together. On one occasion, after working together for 3 hours on the lawn, Joan's mower breaks down, and it takes Charles another 3 hours to finish the job alone. How long would it take each of them to do the entire job alone?

Solution   Let

$$x = \text{Number of hours for Joan alone}$$

$$y = \text{Number of hours for Charles alone}$$

Then

$$\frac{1}{x} = \text{Fractional part of the job Joan does per hour}$$

$$\frac{1}{y} = \text{Fractional part of the job Charles does per hour}$$

If each works 4 hours, the entire job is done:

$$\frac{4}{x} + \frac{4}{y} = 1$$

When Joan works only 3 hours, Charles has to work an additional 3 hours, or 6 hours in all, to accomplish the entire job:

$$\frac{3}{x} + \frac{6}{y} = 1$$

Now we have the system

$$\begin{cases} \dfrac{4}{x} + \dfrac{4}{y} = 1 \\[2mm] \dfrac{3}{x} + \dfrac{6}{y} = 1 \end{cases}$$

These are not linear equations, but we can use the same techniques we used above to solve them. It is probably easiest in this situation not to clear the fractions at the outset. We multiply the top equation by 3 and the bottom one by 4:

$$\frac{12}{x} + \frac{12}{y} = 3$$

$$\frac{12}{x} + \frac{24}{y} = 4$$

Now we subtract the top equation from the bottom one to obtain

$$\frac{12}{y} = 1$$

Clearing the fractions now yields $y = 12$. We may substitute into either of the original equations. We choose the first:

$$\frac{4}{x} + \frac{4}{12} = 1$$

$$\frac{4}{x} = \frac{2}{3}$$

$$x = 6$$

So Joan could do the job alone in 6 hours, and Charles could do it alone in 12 hours.

EXAMPLE 8   The units digit of a two-digit number is 1 less than twice the tens digit. If the digits are reversed, the new number is 27 greater than the old number. Find the original number.

Solution   Let

$$x = \text{Tens digit}$$
$$y = \text{Units digit}$$

Then

$$10x + y = \text{Original number}$$
$$10y + x = \text{Number with digits reversed}$$

Thus, from the given information,

$$y = 2x - 1$$
$$10y + x = 10x + y + 27$$

After simplifying the second equation, the system to be solved can be written in the form

$$\begin{cases} y = 2x - 1 \\ x - y = -3 \end{cases}$$

We eliminate y by substitution from the first equation into the second:

$$x - (2x - 1) = -3$$
$$-x + 1 = -3$$
$$x = 4$$

So $y = 2(4) - 1 = 7$. Thus, the original number is 47.

Remark.   In Example 8 we wrote the number as $10x + y$, since in our positional system the tens digit is understood to be multiplied by 10. Thus, 47 means $(4 \times 10) + 7$. Similarly, if x were the hundreds digit, y the tens digit, and z the units digit of a number, we would indicate the number as $100x + 10y + z$.

EXERCISE SET 1

A   In Problems 1–22 solve the systems.

1. $\begin{cases} x - y = 2 \\ x + y = 6 \end{cases}$

2. $\begin{cases} x + 2y = -4 \\ x - 3y = 6 \end{cases}$

3. $\begin{cases} 3x - y = 0 \\ 5x - 2y = 1 \end{cases}$

4. $\begin{cases} 2x - 3y = 18 \\ 3x + 4y = -7 \end{cases}$

5. $\begin{cases} 3x - 2y = 6 \\ x + 2y = 10 \end{cases}$

6. $\begin{cases} 2x - y = -2 \\ 3x + 2y = 1 \end{cases}$

7. $\begin{cases} 4x - y = 6 \\ 3x + 2y = 5 \end{cases}$

8. $\begin{cases} 5s - 2t = 4 \\ 3s - 2t = 0 \end{cases}$

9. $\begin{cases} 2x + 3y = 2 \\ 3x - 4y = 20 \end{cases}$

10. $\begin{cases} 5x + 2y = 8 \\ 7x - 3y = -12 \end{cases}$

11. $\begin{cases} 5s + 4t = -1 \\ 2s - 3t = -28 \end{cases}$

12. $\begin{cases} 5m - 2n = -2 \\ 4m + 5n = 16 \end{cases}$

13. $\begin{cases} x + 3y - 2z = -1 \\ 2x - y + 3z = 5 \\ -x + 5y - 4z = 1 \end{cases}$

14. $\begin{cases} 2x - 3y - 4z = -4 \\ 3x + 4y + 2z = 1 \\ 5x - 2y - 3z = 0 \end{cases}$

15. $\begin{cases} 2x + y + z = 7 \\ x - y + 2z = 11 \\ 5x + y - 2z = 1 \end{cases}$

16. $\begin{cases} 3x + 4y - z = 4 \\ 2x - 5y + 3z = -10 \\ 4x + 3y + z = 10 \end{cases}$

17. $\begin{cases} 2x + 3y + 3z = 9 \\ 5x - 2y + 8z = 6 \\ 4x - y + 5z = -1 \end{cases}$

18. $\begin{cases} 2x + 4y + 3z = -2 \\ x - 3y + 4z = 9 \\ 5x + 7y - z = -9 \end{cases}$

19. $\begin{cases} \dfrac{2}{x} - \dfrac{3}{y} = 4 \\ \dfrac{1}{x} + \dfrac{4}{y} = -10 \end{cases}$

20. $\begin{cases} \dfrac{2}{x} + \dfrac{3}{y} = -2 \\ \dfrac{5}{x} + \dfrac{4}{y} = \dfrac{5}{6} \end{cases}$

21. $\begin{cases} ax + by = c \\ bx - ay = d \end{cases}$

22. $\begin{cases} x + y = -1 \\ y + 3z = 2 \\ 4z - w = 0 \\ 2x - w = 0 \end{cases}$

In Problems 23 and 24 show that the system is dependent, and give the solution set. Draw the graph.

23. $\begin{cases} 3x - 4y = 5 \\ 8y - 6x + 10 = 0 \end{cases}$

24. $\begin{cases} 4x - 10y = 14 \\ -6x + 15y = -21 \end{cases}$

Show that the systems in Problems 25 and 26 are inconsistent, and draw their graphs.

25. $\begin{cases} 8x - 6y = 3 \\ 3y - 4x = 2 \end{cases}$

26. $\begin{cases} 8x - 10y = 7 \\ -12x + 15y = 11 \end{cases}$

27. Show that the following system is consistent. Draw the graph.

$$\begin{cases} 2x + 5y = 4 \\ x - 2y = -7 \\ 5x + 8y = 1 \end{cases}$$

28. Show that the lines whose equations are $2x - 3y = 0$, $4x + 5y = 22$, and $3x - 2y = 5$ intersect at a common point. Draw their graphs.

29. Find the two numbers whose sum is 127 and whose difference is 63.

30. The difference between two numbers is 5. Three times the larger number minus four times the smaller number is 1. Find the numbers.

31. The perimeter of a certain rectangle is 24. If the width were doubled and the length tripled, the perimeter would be 64. Find the length and width.

32. The length of a rectangle is 10 more than twice the width, and the perimeter is 50. Find the length and width.

33. Flying with the wind, an airplane makes a 1,500 mile trip in 2 hours and 30 minutes. The return trip takes 3 hours and 20 minutes. Find the average airspeed of the airplane and the average wind velocity.

34. Rowing with the current, a 10 mile trip on a river takes a boy 1 hour and 20 minutes. The return trip takes 4 hours. Find how fast the boy can row in still water and the rate of the current.

35. A girl covered a distance of $3\frac{1}{4}$ miles by alternating between jogging and walking. She walked a total of 15 minutes and jogged a total of 30 minutes. The next day she walked a total of 30 minutes and jogged 45 minutes, covering a total distance of $5\frac{1}{4}$ miles. Assuming her rates of walking and of jogging remained the same from one day to the next, find these rates.

36. The sum of the digits of a certain two-digit number is 10. By interchanging the units and tens digits, the value of the number is increased by 36. Find the number.

37. The sum of the digits of a two-digit number is 14. If the digits are reversed, the value of the number is decreased by 18. Find the number.

38. By working together for 4 hours Bill and his younger brother Tom weed one-half of a large garden. The next day they work together for 2 hours, and then Tom gets tired and quits. It takes Bill an additional 3 hours to complete the job. How long would it have taken Bill to weed the entire garden alone? How long would it have taken Tom?

39. A roofer estimates that he and his assistant working together can roof a house in 4 hours. They work together for $2\frac{1}{2}$ hours, when the assistant becomes ill and has to quit. It takes the roofer another $2\frac{1}{2}$ hours to finish the job. How long would it have taken the roofer alone to do the entire job?

40. A man has a total of $5,000 invested in two ways. Part is in a savings account earning 5% interest, and the rest is in a certificate of deposit earning 6% interest. If the earnings for the first year total $286, find how much is invested in each account.

41. A woman has two investments, one yielding 5% interest and one 7% interest annually. The amount at 7% lacks $1,000 of being double the amount at 5%. The total interest after 1 year is $690. Find how much she has invested in each account.

42. A customer in a coffee house wants a blend of two kinds of coffee: Colombian, selling at $4.00 per pound, and Brazilian, selling for $3.60 per pound. He gets 4 pounds of the blend, and the cost is $15.40. How many pounds of each kind did he buy?

43. A chemist wishes to mix a solution containing 10% sulfuric acid with a solution containing 25% sulfuric acid. She wants 30 cubic centimeters in all, and she wants the final mixture to be 18% sulfuric acid. How much of each concentration should she use?

44. An alloy that is 35% copper is to be combined with an alloy that is 60% copper to obtain 180 pounds of an alloy that is 50% copper. How many pounds of each kind of alloy should be used?

45. Twenty pounds of English breakfast tea is to be made by blending Indian and Ceylon teas. The Indian tea is priced at $6.25 per pound, and the Ceylon tea at $7.00 per pound. The price of the English breakfast tea is to be $6.60 per pound. How much Indian tea and how much Ceylon tea should be used?

46. The prices of tickets at a movie theater are $2.75 for adults and $1.50 for children. For a certain show there were 400 tickets sold for a total of $875. How many adults and how many children attended?

47. Four pounds of a certain brand of coffee and three dozen large eggs cost $19.70. Two pounds of the same brand of coffee and four dozen large eggs cost $12.10. Find the cost of 1 pound of the coffee and one dozen large eggs.

B In Problems 48–52 solve the systems.

48. $\begin{cases} 3x - 2y + 4z = 5 \\ 2x + 3y - 2z = 6 \\ 4x - 5y + 3z = -5 \end{cases}$

49. $\begin{cases} 3x + 5y - 7z = 6 \\ 4x - 2y + 8z = -5 \\ 6x + 7y - 3z = 4 \end{cases}$

50. $\begin{cases} 5x - 2y + 3z = 6 \\ 6x - 3y + 4z = 10 \\ -4x + 4y - 9z = 4 \end{cases}$

51. $\begin{cases} x + 2y - 3z - 2w = 1 \\ 3x - 4y + z - w = 3 \\ 2x + 4z + 3w = 4 \\ 3y - 2z + w = 10 \end{cases}$

52. $\begin{cases} 2x + 4y - z - 3w = 6 \\ x + 3y + 2z + 4w = -4 \\ 3x - 5y + 3z + 2w = 13 \\ 4x - 6y + z - 3w = 22 \end{cases}$

53. Show that the following system is inconsistent:

$$\begin{cases} 3x - 2y + 4z = 5 \\ 4x - y + 2z = 3 \\ x + y - 2z = 7 \end{cases}$$

In Problems 54 and 55 show that the system is dependent and give the solution set.

54. $\begin{cases} x - 3y + 4z = -1 \\ 2x + y - 5z = 3 \\ 7x - 7y + 2z = 3 \end{cases}$

55. $\begin{cases} 2x - 3y + 4z - w = 3 \\ x + y - 3z + 2w = -1 \\ 3x - y - 3z + 8w = 5 \end{cases}$

56. Find a, b, and c so that the parabola $y = ax^2 + bx + c$ passes through the points $(1, -5)$, $(-2, 10)$, and $(3, 5)$.

57. Let $f(x) = ax^2 + bx + c$. If $f(2) = 5$, $f(-1) = 2$, and $f(4) = -3$, find a, b, and c.

58. A car leaves town A going toward town B, which is 50 miles away on a straight road. At the same time, two cars leave B, one going toward A and one going away from A, both traveling at the same rate of speed. The car that left from town A meets the car going from B toward A after 30 minutes, and overtakes the one going from B away from A after $2\frac{1}{2}$ hours. Find the speed of each car.

59. A commuter finds that if she leaves the office 5 minutes earlier than usual, she can average 36 miles per hour and arrive home 10 minutes earlier than usual. However, if she leaves the office 5 minutes later than usual, she can only average 25 miles per hour and arrives home 11 minutes later than usual. Find the time it normally takes her to get home and the distance from the office to her home.

60. Brine containing 40% salt is diluted by adding pure water, resulting in a 25% salt solution. Then 20 more gallons of pure water are added, diluting the mixture to a 20% salt solution. Find how much brine there was originally and how many gallons of water were added the first time.

61. Pumps A, B, and C working simultaneously can fill a tank in 10 hours. Pumps B and C working together can fill it in 15 hours. If all three pumps work together for 4 hours and then pump C is shut off, it takes 8 more hours to fill the tank. How long would it take each pump alone to fill the tank?

2   Reduction to Triangular Form and the Use of Matrices

We will illustrate by means of the following example a technique for solving linear systems that makes the elimination of unknowns more systematic. This procedure is especially useful for systems of more than three equations and three unknowns.

EXAMPLE 9   Solve the system:
$$\begin{cases} x + 2y + 3z + 2w = 1 \\ x + 3y - z - 2w = -1 \\ 2x + 3y - 2z - w = 7 \\ 3x + 4y + z - w = -4 \end{cases}$$

Solution   The first step is to eliminate x from the bottom three equations. This is accomplished by adding to each of these, in turn, certain multiples of the first equation. Thus, we add to the second equation -1 times the first, to the third -2 times the first, and to the fourth -3 times the first. The result is the new system

$$x + 2y + 3z + 2w = 1$$
$$y - 4z - 4w = -2$$
$$-y - 8z - 5w = 5$$
$$-2y - 8z - 7w = -7$$

Next we eliminate y from the bottom two equations by adding the second equation to the third and then adding to the fourth equation 2 times the second:

$$x + 2y + 3z + 2w = 1$$
$$y - 4z - 4w = -2$$
$$-12z - 9w = 3$$
$$-16z - 15w = -11$$

The third equation can be simplified by dividing by -3:

$$x + 2y + 3z + 2w = 1$$
$$y - 4z - 4w = -2$$
$$4z + 3w = -1$$
$$-16z - 15w = -11$$

Now we eliminate z from the bottom equation by adding to it 4 times the third:

$$x + 2y + 3z + 2w = 1$$
$$y - 4z - 4w = -2$$
$$4z + 3w = -1$$
$$-3w = -15$$

The system is now in what is called **triangular form** and can be solved by working from the bottom up. We obtain from the last equation $w = 5$. Then, substituting this in the third equation, we get

$$4z + 15 = -1$$
$$4z = -16$$
$$z = -4$$

Substituting both w and z into the second equation yields

$$y + 16 - 20 = -2$$
$$y = 2$$

Finally, we get from the first equation

$$x + 4 - 12 + 10 = 1$$
$$x = -1$$

The desired solution, then, is given by the array $(-1, 2, -4, 5)$, where it is understood that the values shown are for x, y, z, and w, respectively.

Students should be warned that not all systems reduce to triangular form as easily as the one in Example 9. The arithmetic necessary to eliminate the desired unknowns often is more complicated. But this method always works, and it is systematic.

We will show an abbreviated version of Example 9 below, but before doing so, it will be useful to list the admissible operations for solving a system of equations. Any of the following change the system to an equivalent one, that is, to one having the same solution set:

1. Both sides of any equation may be multiplied by a nonzero number.
2. Two equations may be interchanged.
3. A multiple of any equation may be added to any other equation.

Now let us return to the original system in Example 9:

$$\begin{cases} x + 2y + 3z + 2w = 1 \\ x + 3y - z - 2w = -1 \\ 2x + 3y - 2z - w = 7 \\ 3x + 4y + z - w = -4 \end{cases}$$

We are going to represent this by means of the coefficients and the constant terms only, with the understanding that whenever it is convenient to do so, we can reinsert the letters representing the unknowns. We write this as follows:

$$\begin{bmatrix} 1 & 2 & 3 & 2 & | & 1 \\ 1 & 3 & -1 & -2 & | & -1 \\ 2 & 3 & -2 & -1 & | & 7 \\ 3 & 4 & 1 & -1 & | & -4 \end{bmatrix}$$

The vertical line is used to indicate the separation of the left-hand sides from the right. Such an array of numbers is called a **matrix,** and when used to represent a system of equations, as in the present case, it is called an **augmented matrix.** This description is used to distinguish the full matrix shown from the **coefficient matrix** only, that is,

$$\begin{bmatrix} 1 & 2 & 3 & 2 \\ 1 & 3 & -1 & -2 \\ 2 & 3 & -2 & -1 \\ 3 & 4 & 1 & -1 \end{bmatrix}$$

Now we are going to retrace the steps shown in Example 9 using the augmented matrix. We show 0 where a coefficient is eliminated. A brief indication of each step is given as a reminder of what was done to the preceding matrix to get the current one. First, we exhibit again the augmented matrix.

$$\left[\begin{array}{cccc|c} 1 & 2 & 3 & 2 & 1 \\ 1 & 3 & -1 & -2 & -1 \\ 2 & 3 & -2 & -1 & 7 \\ 3 & 4 & 1 & -1 & -4 \end{array}\right]$$

$$\left[\begin{array}{cccc|c} 1 & 2 & 3 & 2 & 1 \\ 0 & 1 & -4 & -4 & -2 \\ 0 & -1 & -8 & -5 & 5 \\ 0 & -2 & -8 & -7 & -7 \end{array}\right] \begin{array}{l} \\ \text{Row } 2 + [(-1) \cdot \text{Row } 1] \\ \text{Row } 3 + [(-2) \cdot \text{Row } 1] \\ \text{Row } 4 + [(-3) \cdot \text{Row } 1] \end{array}$$

$$\left[\begin{array}{cccc|c} 1 & 2 & 3 & 2 & 1 \\ 0 & 1 & -4 & -4 & -2 \\ 0 & 0 & -12 & -9 & 3 \\ 0 & 0 & -16 & -15 & -11 \end{array}\right] \begin{array}{l} \\ \\ \text{Row } 3 + (1 \cdot \text{Row } 2) \\ \text{Row } 4 + (2 \cdot \text{Row } 2) \end{array}$$

$$\left[\begin{array}{cccc|c} 1 & 2 & 3 & 2 & 1 \\ 0 & 1 & -4 & -4 & -2 \\ 0 & 0 & 4 & 3 & -1 \\ 0 & 0 & -16 & -15 & -11 \end{array}\right] \begin{array}{l} \\ \\ \text{Row } 3 \cdot (-\tfrac{1}{3}) \\ \\ \end{array}$$

$$\left[\begin{array}{cccc|c} 1 & 2 & 3 & 2 & 1 \\ 0 & 1 & -4 & -4 & -2 \\ 0 & 0 & 4 & 3 & -1 \\ 0 & 0 & 0 & -3 & -15 \end{array}\right] \begin{array}{l} \\ \\ \\ \text{Row } 4 + (4 \cdot \text{Row } 3) \end{array}$$

The matrix is now in triangular form, so we are ready to find the solution to the system. The last row of the matrix represents the equation

$$-3w = -15$$

so that $w = 5$. In turn we write the other equations, from the bottom up, and substitute known values as before.

EXAMPLE 10   Solve the system: $\begin{cases} 3x + 2y - 4z = -7 \\ 7x - 8y - 5z = 5 \\ -8x + 5y + 6z = -1 \end{cases}$

Solution   The augmented matrix is

$$\left[\begin{array}{ccc|c} 3 & 2 & -4 & -7 \\ 7 & -8 & -5 & 5 \\ -8 & 5 & 6 & -1 \end{array}\right]$$

It is an advantage to have 1 or -1 as the first element in Row 1, since it is then easy to choose appropriate multiples to eliminate the elements below it. We can always make the first element 1 by multiplying the first row by its reciprocal; in this case we would multiply by $\frac{1}{3}$. But this would introduce fractions, and it is generally simpler to avoid doing this. In this case we can add Row 2 to Row 3 and then interchange Rows 1 and 3:

$$\left[\begin{array}{ccc|c} 3 & 2 & -4 & -7 \\ 7 & -8 & -5 & 5 \\ -1 & -3 & 1 & 4 \end{array}\right] \quad \text{Row 3 + Row 2}$$

$$\left[\begin{array}{ccc|c} -1 & -3 & 1 & 4 \\ 7 & -8 & -5 & 5 \\ 3 & 2 & -4 & -7 \end{array}\right] \quad \begin{array}{l}\text{Row 3 interchanged with Row 1} \\ \\ \text{Row 1 interchanged with Row 3}\end{array}$$

Now we proceed to introduce zeros under the -1:

$$\left[\begin{array}{ccc|c} -1 & -3 & 1 & 4 \\ 0 & -29 & 2 & 33 \\ 0 & -7 & -1 & 5 \end{array}\right] \quad \begin{array}{l}\text{Row 2 + (7 · Row 1)} \\ \text{Row 3 + (3 · Row 1)}\end{array}$$

We are faced with messy arithmetic now unless we make some further simplifying moves. It appears that if we add -4 times Row 3 to Row 2, things will improve significantly:

$$\left[\begin{array}{ccc|c} -1 & -3 & 1 & 4 \\ 0 & -1 & 6 & 13 \\ 0 & -7 & -1 & 5 \end{array}\right] \quad \text{Row 2 + [(-4) · Row 3]}$$

Finally, we add -7 times Row 2 to Row 3:

$$\left[\begin{array}{ccc|c} -1 & -3 & 1 & 4 \\ 0 & -1 & 6 & 13 \\ 0 & 0 & -43 & -86 \end{array}\right] \quad \text{Row 3 + [(-7) · Row 2]}$$

The third equation now reads

$$-43z = -86$$

so

$$z = 2$$

The second equation is
$$-y + 6z = 13$$
which on substituting $z = 2$ yields
$$y = -1$$
The first equation can now be solved for x:
$$-x - 3y + z = 4$$
$$-x + 3 + 2 = 4$$
$$-x = -1$$
$$x = 1$$

The desired solution is $(1, -1, 2)$.

The techniques to simplify the arithmetic illustrated in this example should be kept in mind as you do problems by this method.

EXAMPLE 11   Solve the system below by reducing the augmented matrix to triangular form.

$$\begin{cases} 4x_1 + 11x_2 + 6x_3 + 22x_4 = 16 \\ x_1 + 2x_2 \qquad\quad + x_4 = 7 \\ 2x_1 + 5x_2 + x_3 + 10x_4 = 7 \end{cases}$$

Solution   This example will show that the method we are employing will work even when the number of unknowns is different from the number of equations. To save one step we will mentally interchange the first two equations before writing the matrix:

$$\begin{bmatrix} 1 & 2 & 0 & 1 & | & 7 \\ 4 & 11 & 6 & 22 & | & 16 \\ 2 & 5 & 1 & 10 & | & 7 \end{bmatrix}$$

$$\begin{bmatrix} 1 & 2 & 0 & 1 & | & 7 \\ 0 & 3 & 6 & 18 & | & -12 \\ 0 & 1 & 1 & 8 & | & -7 \end{bmatrix} \quad \begin{matrix} \text{Row } 2 + [(-4) \cdot \text{Row 1}] \\ \text{Row } 3 + [(-2) \cdot \text{Row 1}] \end{matrix}$$

We could now interchange Rows 2 and 3, but we choose to multiply Row 2 by $\frac{1}{3}$ instead:

$$\begin{bmatrix} 1 & 2 & 0 & 1 & | & 7 \\ 0 & 1 & 2 & 6 & | & -4 \\ 0 & 1 & 1 & 8 & | & -7 \end{bmatrix}$$

Now we subtract Row 2 from Row 3:

$$\begin{bmatrix} 1 & 2 & 0 & 1 & | & 7 \\ 0 & 1 & 2 & 6 & | & -4 \\ 0 & 0 & -1 & 2 & | & -3 \end{bmatrix}$$

This is as far as we can go in the triangular form. The third equation now reads

$$-x_3 + 2x_4 = -3$$

While this cannot be solved for x_3 as a specific number, it can be solved in terms of x_4:

$$x_3 = 3 + 2x_4$$

From the second equation we get

$$x_2 + 2x_3 + 6x_4 = -4$$
$$x_2 = -4 - 2(3 + 2x_4) - 6x_4$$
$$= -4 - 6 - 4x_4 - 6x_4$$
$$= -10 - 10x_4$$

and from the first,

$$x_1 + 2x_2 + x_4 = 7$$
$$x_1 = 7 - 2(-10 - 10x_4) - x_4$$
$$= 7 + 20 + 20x_4 - x_4$$
$$= 27 + 19x_4$$

Summarizing, we have found

$$\begin{cases} x_1 = 27 + 19x_4 \\ x_2 = -10 - 10x_4 \\ x_3 = 3 + 2x_4 \\ x_4 = x_4 \end{cases}$$

We may select any real value for x_4. Then x_1, x_2, and x_3 are determined. Let us designate x_4 by c. Then the solution set can be written as

$$\{(27 + 19c, -10 - 10c, 3 + 2c, c): \quad c \in R\}$$

Thus, there are infinitely many points that satisfy all three equations.

EXAMPLE 12   Determine whether the following system is consistent by reducing the augmented matrix to triangular form.

$$\begin{cases} x - y + 2z = 3 \\ 2x + y \quad\quad = 4 \\ x + 3y - z = 2 \\ -x + 4y + 2z = 3 \end{cases}$$

Solution   In this case there are more equations than unknowns. While we might expect three equations in three unknowns to have a common solution, it is a special case when this solution also works for a fourth equation. So it should come as no surprise if these equations turn out to be inconsistent.

$$\begin{bmatrix} 1 & -1 & 2 & 3 \\ 2 & 1 & 0 & 4 \\ 1 & 3 & -1 & 2 \\ -1 & 4 & 2 & 3 \end{bmatrix}$$

$$\begin{bmatrix} 1 & -1 & 2 & 3 \\ 0 & 3 & -4 & -2 \\ 0 & 4 & -3 & -1 \\ 0 & 3 & 4 & 6 \end{bmatrix}$$ Row 2 + [(−2) · Row 1]
Row 3 + [(−1) · Row 1]
Row 4 + (1 · Row 1)

$$\begin{bmatrix} 1 & -1 & 2 & 3 \\ 0 & -1 & -1 & -1 \\ 0 & 4 & -3 & -1 \\ 0 & 3 & 4 & 6 \end{bmatrix}$$ Row 2 + [(−1) · Row 3]

$$\begin{bmatrix} 1 & -1 & 2 & 3 \\ 0 & -1 & -1 & -1 \\ 0 & 0 & -7 & -5 \\ 0 & 0 & 1 & 3 \end{bmatrix}$$ Row 3 + (4 · Row 2)
Row 4 + (3 · Row 2)

$$\begin{bmatrix} 1 & -1 & 2 & 3 \\ 0 & -1 & -1 & -1 \\ 0 & 0 & 1 & 3 \\ 0 & 0 & -7 & -5 \end{bmatrix}$$ Row 4 interchanged with Row 3
Row 3 interchanged with Row 4

$$\begin{bmatrix} 1 & -1 & 2 & 3 \\ 0 & -1 & -1 & -1 \\ 0 & 0 & 1 & 3 \\ 0 & 0 & 0 & 16 \end{bmatrix}$$ Row 4 + (7 · Row 3)

The last equation now reads

$$0 = 16$$

which is not possible. Therefore, the system is inconsistent.

EXERCISE SET 2

By reducing the augmented matrix to triangular form, solve the following systems or show that no solution exists. If the system is dependent, show the solution set as in Example 11.

A 1. $\begin{cases} x + 3y - 2z = 3 \\ \quad\ y + z = 4 \\ x + 2y + 4z = 6 \end{cases}$

2. $\begin{cases} x - y + z = 1 \\ 2x + y + z = 4 \\ x + 5y - 2z = 1 \end{cases}$

3. $\begin{cases} x + 2y + z = 3 \\ 2x - y - z = 0 \\ 3x + 4y + 2z = 8 \end{cases}$

4. $\begin{cases} x + 2y + z = 1 \\ 2x + 5y + 4z = 2 \\ -x - 3y + 5z = 7 \end{cases}$

5. $\begin{cases} x + y - z = 3 \\ 3x + 4y + 2z = 3 \\ 4x - 2y - 8z = -4 \end{cases}$

6. $\begin{cases} x + 3y - 2z = -1 \\ x - 5y + 4z = -1 \\ 2x - y + 3z = 5 \end{cases}$

7. $\begin{cases} x - y = 1 \\ y + z = 1 \\ x + z = 2 \end{cases}$

8. $\begin{cases} x + 4y - 2z = -3 \\ 3x - 2y + z = 12 \\ 2x - 3y + z = 11 \end{cases}$

9. $\begin{cases} x - y + 2z = 3 \\ 2x - y + 8z = 5 \\ -3x + 5y - 3z = -11 \end{cases}$

10. $\begin{cases} 2x - 3y + z = 0 \\ 3x + 2y + 4z = 4 \\ -4x + y - 5z = -2 \end{cases}$

11. $\begin{cases} 2x - 3y + 2z = -2 \\ 3x - 8y - z = 2 \\ -5x + 10y - 3z = 4 \end{cases}$

12. $\begin{cases} 2x - 3y + 4z = 5 \\ x + y - 2z = 1 \\ 7x - 3y + 2z = 13 \end{cases}$

13. $\begin{cases} 2x + y + z = 7 \\ x - y + 2z = 11 \\ 5x + y - 2z = 1 \end{cases}$

14. $\begin{cases} 2x + 4y + 3z = -2 \\ x - 3y + 4z = 9 \\ 5x + 7y - z = -9 \end{cases}$

15. $\begin{cases} x - y - 3z = -3 \\ 2x - 3y - 2z = 0 \\ 4x - 3y + z = -1 \end{cases}$

16. $\begin{cases} x + 3y + 3z = -2 \\ 3x - 6y + 2z = 6 \\ 2x + 3y - z = 4 \end{cases}$

17. $\begin{cases} x + 2y + 3z = 9 \\ x - y + 2z = 5 \\ x + 5y + 4z = 13 \end{cases}$

18. $\begin{cases} 2x + y - 3z = 1 \\ x + 2y + z = 3 \\ 4x - y - 11z = 5 \end{cases}$

19. $\begin{cases} 2x - 3y + 4z = 0 \\ x - y - 2z = 0 \\ x - 3y + 14z = 0 \end{cases}$

20. $\begin{cases} x + y - z + w = 2 \\ 2x - 3y + 4z + 5w = -4 \\ 4x + 5y - 2z - 3w = 8 \end{cases}$

21. $\begin{cases} 4x - 2y + 3z = 5 \\ 3x + y + 2z = 3 \\ 2x - 3y + 5z = 4 \end{cases}$

22. $\begin{cases} x + 4y + z = 7 \\ 3x - 2y + 2z = 12 \\ x + 2y - z = 2 \end{cases}$

23. $\begin{cases} 3x - 2y - z = 2 \\ 5x + 3y - 4z = 6 \\ 4x - 9y + z = 3 \end{cases}$

24. $\begin{cases} 2x + 3y + 3z = 9 \\ 5x - 2y + 8z = 6 \\ 4x - y + 5z = -1 \end{cases}$

25. $\begin{cases} 5x - 2y + 3z = 6 \\ 6x - 3y + 4z = 10 \\ -4x + 4y - 9z = 4 \end{cases}$

26. $\begin{cases} 3x + 4y - z = 4 \\ 2x - 5y + 3z = -10 \\ 4x + 3y + z = 10 \end{cases}$

27. $\begin{cases} x + 2y - z + w = 1 \\ 2x - y + z - 2w = 0 \\ -x - 3y + 2z - 3w = -5 \end{cases}$

28. $\begin{cases} 2x - 3y - 2z = 1 \\ 3x - 4y - 3z = 2 \\ 4x + y + z = 10 \end{cases}$

29. $\begin{cases} 2x - 3y + 4z = 0 \\ x - y + 3z = 0 \\ 3x - 5y + 5z = 0 \end{cases}$

30. $\begin{cases} 5x - 6y + z = 1 \\ 9x - 8y + 7z = 1 \\ -3x + 4y - z = 5 \end{cases}$

31. $\begin{cases} 8x - 5y + 6z = 0 \\ 3x - 2y - 2z = -10 \\ -11x + 9y - 2z = 16 \end{cases}$

32. $\begin{cases} 2x + y + z = 1 \\ 5x + 3y - 2z = -4 \\ 3x + y + 6z = 8 \end{cases}$

33. $\begin{cases} x - y - z = -1 \\ y - z + w = 0 \\ x + 2z - 2w = 1 \\ 2x + y - w = -1 \end{cases}$

34. $\begin{cases} x + y + z + w = 1 \\ 3y - w = 0 \\ x - z = 1 \\ y - 2z - w = -2 \end{cases}$

35. $\begin{cases} x_1 + x_2 - x_3 + 2x_4 = 1 \\ 2x_1 - x_2 - 3x_3 + 2x_4 = -6 \\ -x_1 + 2x_2 + x_3 + 3x_4 = 2 \\ x_1 - x_2 - 2x_3 + x_4 = 0 \end{cases}$

36. $\begin{cases} 3x + 5y - 7z = 6 \\ 4x - 2y + 8z = -5 \\ 6x + 7y - 3z = 4 \end{cases}$

37. $\begin{cases} 3x - 2y - 2z = -1 \\ 7x + y - 2z = 0 \\ 5x - 3y + 4z = 20 \end{cases}$

B **38.** $\begin{cases} x + y + 3z - 2w = -2 \\ 2x + 4y - 2z + 4w = 1 \\ -3x - 5y - 4z - w = 4 \\ 4x + 6y - 2z + 5w = 1 \end{cases}$

39. $\begin{cases} 2x_1 - x_2 + 3x_3 + x_4 = 0 \\ 4x_1 - 3x_2 + x_3 - 2x_4 = 1 \\ -3x_1 + 2x_3 - 4x_4 = -2 \\ 6x_1 - 4x_2 - 5x_3 + 3x_4 = -3 \end{cases}$

40. $\begin{cases} x - y - 2z + w = -2 \\ 3x + 2y + 3z + 2w = 1 \\ 2x - 7y - 4z - 3w = 1 \\ -x + 6y + 5z + 2w = 1 \end{cases}$

41. $\begin{cases} 2x - 3y + 5z + 4w = -3 \\ 3x - 2y - 2z - w = 1 \\ 5x + 2y + 6z - 8w = 1 \\ -7x + 4y - 3z + 2w = 2 \end{cases}$

42. $\begin{cases} 3x - 2y + z - w = -1 \\ x + y - 3z + 3w = 3 \\ 2x - 3y + 5z - 4w = 1 \\ 4x + 2y + 4z - w = -4 \end{cases}$

43. $\begin{cases} 2x - 3y + 4z + 2w = 5 \\ -3x + 2y - z + w = 2 \\ 4x - 5y + 2z - 3w = 6 \\ 5x - 4y - z - 6w = 1 \end{cases}$

44. $\begin{cases} x_1 - 2x_2 + x_3 - x_4 = 2 \\ x_1 - x_2 - x_4 = 6 \\ 2x_1 - 2x_2 + x_3 = 9 \\ -x_1 + 5x_2 - 2x_3 + 5x_4 = 6 \end{cases}$

45. $\begin{cases} 2s - t + 3u + v = 0 \\ 3s + 2t + 4u - v = 0 \\ 5s - 2t - 2u - v = 0 \\ -2s - 3t + 7u + 5v = 0 \end{cases}$

46. $\begin{cases} x + y + 2z - w = 1 \\ -x + 2y + 3z - 2w = -3 \\ 2x - y - z + w = 4 \\ x + 2y + z - w = 1 \end{cases}$

47. $\begin{cases} x_1 - 2x_2 + 3x_4 + x_5 = 2 \\ -x_1 + 3x_2 + 2x_3 - 4x_4 + x_5 = -1 \\ x_1 - 2x_2 + x_3 + 3x_4 = 0 \\ x_2 + 2x_3 - 2x_4 + 2x_5 = 7 \\ 2x_1 - 4x_2 + 2x_3 + 6x_4 + x_5 = 5 \end{cases}$

3 Determinants

In Section 2 our approach made use of the augmented matrix of a system of equations. In Section 4 we will present an alternative procedure that uses the coefficient matrix and some other matrices associated with it. This procedure works only when the number of unknowns and the number of equations are the same, in which case the coefficient matrix is said to be **square,** since it has the same number of rows as columns. However, it will first be necessary to introduce a new concept—the **determinant** of a matrix. Various symbols are used to designate the determinant of a matrix; we will use the one that is probably the most common, namely, vertical bars on each side of the matrix. For example, the determinant of the matrix

$$\begin{bmatrix} 2 & 3 & 1 \\ 4 & -2 & 5 \\ 1 & 2 & 7 \end{bmatrix}$$

is designated by

$$\begin{vmatrix} 2 & 3 & 1 \\ 4 & -2 & 5 \\ 1 & 2 & 7 \end{vmatrix}$$

The number of rows (or columns) in a square matrix is called the **order** of the matrix. The determinant of a matrix of order 2 is defined as follows: Let

$$\begin{bmatrix} a & b \\ c & d \end{bmatrix}$$

designate an arbitrary matrix of order 2. Then

$$\begin{vmatrix} a & b \\ c & d \end{vmatrix} = ad - bc$$

The right-hand side is suggested by the arrows on the left. The diagonal elements from upper left to lower right, called the **principal diagonal elements,** are multiplied, and from their product is subtracted the product of the other diagonal elements. For example, we have

$$\begin{vmatrix} 2 & 3 \\ 4 & 5 \end{vmatrix} = 10 - 12 = -2$$

and

$$\begin{vmatrix} 3 & -1 \\ 5 & -2 \end{vmatrix} = -6 - (-5) = -6 + 5 = -1$$

The procedure for finding the determinant of a square matrix of order higher than 2 is more complicated and will require several preliminary steps. We will use a general matrix of order 3 to illustrate,

$$\begin{bmatrix} a_1 & b_1 & c_1 \\ a_2 & b_2 & c_2 \\ a_3 & b_3 & c_3 \end{bmatrix}$$

(subscripts are employed to avoid the use of too many different letters), but everything we do can be extended in a straightforward way to higher orders. We refer to the entries in the matrix as **elements** of the matrix. It is important to distinguish between rows (which go across) and columns (which go up and down). For example, consider the matrix

$$\begin{bmatrix} 2 & -1 & 3 \\ 4 & 2 & -7 \\ -1 & 0 & 5 \end{bmatrix}$$

The element in Row 2 and Column 3 is -7, and the element in Row 3 and Column 2 is 0.

If we delete the row and column of a given element in a square matrix, we are left with a submatrix of order 1 less than that of the original matrix. The determinant of this submatrix is called the **minor** of the element in question. For example, let us again consider the matrix

$$\begin{bmatrix} 2 & -1 & 3 \\ 4 & 2 & -7 \\ -1 & 0 & 5 \end{bmatrix}$$

and determine the minor of the element -7 in Row 2 and Column 3. We show the row

$$\begin{bmatrix} 2 & -1 & 3 \\ 4 & 2 & 7 \\ -1 & 0 & 5 \end{bmatrix}$$

and column containing this element as being deleted by dashed lines. The minor of -7 is therefore the determinant

$$\begin{vmatrix} 2 & -1 \\ -1 & 0 \end{vmatrix} = 2(0) - (-1)(-1) = 0 - 1 = -1$$

In a similar way we could find the minor of each of the elements.

If we multiply the minor of the element in the ith row and jth column by $(-1)^{i+j}$, we obtain what is called the **cofactor** of the element. So the cofactor of an element differs from the minor at most in sign. If $i + j$ is even, $(-1)^{i+j} = 1$, so the cofactor equals the minor. If $i + j$ is odd, $(-1)^{i+j} = -1$, so the cofactor is the negative of the minor. From this definition we can devise the following scheme for determining the sign that should be affixed to the minor to give the cofactor:

$$\begin{bmatrix} + & - & + \\ - & + & - \\ + & - & + \end{bmatrix}$$

Starting with a plus sign in the upper left, the signs alternate as we move either horizontally or vertically.

EXAMPLE 13   Find the cofactor of each element in the first row of the matrix.

$$\begin{bmatrix} 2 & 1 & -3 \\ -1 & 4 & 2 \\ 5 & 3 & 6 \end{bmatrix}$$

Solution   To get the minor of the first element, 2, we cross out (mentally or otherwise) the first row and first column:

$$\begin{bmatrix} 2 & 1 & 3 \\ -1 & 4 & 2 \\ 5 & 3 & 6 \end{bmatrix}$$

The minor is

$$\begin{vmatrix} 4 & 2 \\ 3 & 6 \end{vmatrix} = 24 - 6 = 18$$

Since by the scheme of signs given above, we multiply by $+1$ to get the cofactor, it follows that the cofactor of this element also is 18.

Similarly, we determine the minor of the second element in Row 1:

$$\begin{bmatrix} -2 & 1 & -3 \\ -1 & 4 & 2 \\ 5 & 3 & 6 \end{bmatrix}$$

$$\begin{vmatrix} -1 & 2 \\ 5 & 6 \end{vmatrix} = -6 - 10 = -16$$

To get the cofactor we must multiply by -1, according to the scheme of signs for determining cofactors. So the cofactor of this element is 16.

For the third element in Row 1 we proceed in a similar way:

$$\begin{bmatrix} -2 & 1 & -3 \\ -1 & 4 & 2 \\ 5 & 3 & 6 \end{bmatrix}$$

$$\begin{vmatrix} -1 & 4 \\ 5 & 3 \end{vmatrix} = -3 - 20 = -23$$

The cofactor is obtained by multiplying this minor by $+1$, so the cofactor is -23.

With a little practice you will probably be able to calculate cofactors of elements in a matrix of order 3 mentally. It is a good idea to write the correct factor $+1$ or -1 first, according to the position of the element, and then calculate the minor. To illustrate, consider again the matrix in Example 13. The cofactor of the element -1 in the second row and first column is $-(6 + 9) = -15$; the cofactor of the element 3 in the third row and second column is $-(4 - 3) = -1$; and so on. Note that the cofactor of an element depends on the position of the element but not on the value of the element itself.

We are now ready to state a **procedure for finding the value of the determinant of a square matrix of order 3 or greater.**

1. **Select any row or column of the matrix.**
2. **Multiply each element of that row or column by its cofactor.**
3. **Add the results.**

We will illustrate with the matrix used in Example 13.

EXAMPLE 14   Find the determinant of the matrix.

$$\begin{bmatrix} 2 & 1 & -3 \\ -1 & 4 & 2 \\ 5 & 3 & 6 \end{bmatrix}$$

Solution Let us select Row 1, since we have already found the cofactors of the elements in that row in Example 13, namely, 18, 16, and -23, respectively. Next, we multiply each element in the row by its own cofactor and then add the results. Thus,

$$\begin{vmatrix} 2 & 1 & -3 \\ -1 & 4 & 2 \\ 5 & 3 & 6 \end{vmatrix} = 2(18) + 1(16) + (-3)(-23) = 36 + 16 + 69 = 121$$

Note. In Step 1 above we specified that *any* row or column may be selected at the outset. It can be proved (See Problem 26, Exercise Set 3) that the same result will occur regardless of this initial choice, so long as Steps 2 and 3 are carried out for the row or column selected. To illustrate let us select Column 2 in the example above (we say "expand by Column 2"). The cofactors can be calculated mentally as we move down the column.

$$\begin{vmatrix} 2 & 1 & -3 \\ -1 & 4 & 2 \\ 5 & 3 & 6 \end{vmatrix} = 1[-(-6-10)] + 4[+(12+15)] + 3[-(4-3)]$$

$$= 16 + 4(27) + 3(-1) = 16 + 108 - 3 = 121$$

EXAMPLE 15 Find the determinant of the matrix

$$\begin{bmatrix} 2 & 3 & 0 \\ -1 & 2 & -4 \\ -3 & -1 & 5 \end{bmatrix}$$

Solution It is convenient to select either Row 1 or Column 3. (Why?) Let us expand by Column 3. We can ignore the cofactor of the first element 0, since whatever it is, the result will be 0 after multiplication. So we have

$$\begin{vmatrix} 2 & 3 & 0 \\ -1 & 2 & -4 \\ -3 & -1 & 5 \end{vmatrix} = 0 + (-4)[-(-2+9)] + 5[4+3]$$

$$= (-4)(-7) + 5(7) = 28 + 35 = 63$$

While the procedure outlined for finding determinants is applicable to matrices of any order greater than 2, the work involved rapidly becomes excessive as the orders increase. For example, finding the determinant of a matrix of order 4 involves multiplying each of four elements (any row or column) by their respective cofactor, each of which involves evaluating the determinant of an order 3 matrix. It is not hard to visualize how difficult it would be to find the determinant of a matrix of order 5. In these more complicated situations there are admissible preliminary steps that can be taken which introduce zeros into a given row or column without changing the value of the determinant. The more zeros present, the easier the evaluation of the determinant. We will not pursue these methods here, since in the applications of determinants to systems of equations the method to be described is practical only for relatively small systems.

EXERCISE SET 3

A   In Problems 1–3 evaluate the determinants.

1.   **a.** $\begin{vmatrix} 2 & 3 \\ -1 & 4 \end{vmatrix}$ **b.** $\begin{vmatrix} -3 & 5 \\ 12 & 6 \end{vmatrix}$ **c.** $\begin{vmatrix} 4 & -8 \\ -6 & -3 \end{vmatrix}$

2.   **a.** $\begin{vmatrix} 3 & -1 \\ 4 & -2 \end{vmatrix}$ **b.** $\begin{vmatrix} 5 & 6 \\ -8 & 3 \end{vmatrix}$ **c.** $\begin{vmatrix} -10 & 3 \\ 14 & -5 \end{vmatrix}$

3.   **a.** $\begin{vmatrix} 9 & 6 \\ 3 & 2 \end{vmatrix}$ **b.** $\begin{vmatrix} 0 & 5 \\ -1 & 4 \end{vmatrix}$ **c.** $\begin{vmatrix} 8 & 4 \\ -4 & -2 \end{vmatrix}$

Problems 4–8 refer to the matrix

$$\begin{bmatrix} 3 & -2 & 1 \\ -4 & 5 & -3 \\ -6 & -1 & 7 \end{bmatrix}$$

In each case for the position specified find:

a.   The element in that position
b.   The minor of that element
c.   The cofactor of that element

4.   Row 2, Column 3   5.   Row 3, Column 2
6.   Row 1, Column 3   7.   Row 3, Column 1
8.   Row 3, Column 3

9.   Evaluate the determinant of the matrix used in Problems 4–8 by expanding on:
     a.   The third column   **b.**   The third row

In Problems 10–18 evaluate the determinants.

10. $\begin{vmatrix} 2 & -1 & -3 \\ 4 & 2 & -5 \\ -1 & 3 & 6 \end{vmatrix}$ 11. $\begin{vmatrix} 3 & 0 & -1 \\ 2 & 5 & 0 \\ 1 & 3 & 1 \end{vmatrix}$ 12. $\begin{vmatrix} 1 & 5 & 3 \\ -2 & -4 & 10 \\ 6 & 8 & -3 \end{vmatrix}$

13. $\begin{vmatrix} 3 & -1 & -5 \\ 0 & 2 & -4 \\ 5 & 1 & 6 \end{vmatrix}$ 14. $\begin{vmatrix} 2 & 3 & -1 \\ -4 & 2 & 3 \\ 5 & -2 & 6 \end{vmatrix}$ 15. $\begin{vmatrix} 1 & -1 & 4 \\ 3 & 2 & 0 \\ 4 & -1 & 6 \end{vmatrix}$

16. $\begin{vmatrix} 5 & 1 & 2 \\ -2 & 4 & 3 \\ 7 & 2 & -1 \end{vmatrix}$ 17. $\begin{vmatrix} -2 & 1 & -5 \\ 0 & 3 & -4 \\ 6 & -1 & 2 \end{vmatrix}$ 18. $\begin{vmatrix} 4 & -2 & -3 \\ -5 & 3 & -4 \\ 2 & -1 & -2 \end{vmatrix}$

19. Find all values of x satisfying

$$\begin{vmatrix} 1 & x & 0 \\ 2 & -4 & x-1 \\ 1 & -x & 2 \end{vmatrix} = 0$$

20. Find all values of t satisfying

$$\begin{vmatrix} -t & 2 & -1 \\ 0 & t & 4 \\ t+2 & 1 & -2 \end{vmatrix} = 0$$

21. Prove that for any number k

$$\begin{vmatrix} a & b \\ c & d \end{vmatrix} = \begin{vmatrix} a+kb & b \\ c+kd & d \end{vmatrix}$$

and

$$\begin{vmatrix} a & b \\ c & d \end{vmatrix} = \begin{vmatrix} a+kc & b+kd \\ c & d \end{vmatrix}$$

22. It can be proved that the area of a triangle with vertices (x_1, y_1), (x_2, y_2), and (x_3, y_3) is given by

$$A = \frac{1}{2} \begin{vmatrix} x_1 & y_1 & 1 \\ x_2 & y_2 & 1 \\ x_3 & y_3 & 1 \end{vmatrix}$$

Use this to find the area of the triangle with vertices $(2, -3)$, $(-4, 1)$, and $(-3, -5)$.

B Evaluate the determinants in Problems 23–25.

23. $\begin{vmatrix} 3 & 2 & -1 & 5 \\ 1 & 0 & 2 & 4 \\ -1 & 1 & 3 & 0 \\ 2 & 0 & -1 & 1 \end{vmatrix}$

24. $\begin{vmatrix} 1 & 0 & -2 & 3 \\ 0 & -2 & 5 & 0 \\ 2 & 3 & -1 & 1 \\ -1 & -2 & 3 & 4 \end{vmatrix}$

25. $\begin{vmatrix} 2 & -1 & 3 & 1 \\ 4 & 2 & 1 & 3 \\ -1 & 1 & 2 & 4 \\ 3 & 5 & -2 & 6 \end{vmatrix}$

26. Verify the fact that the same final result occurs regardless of which row or column is used in the calculation of the determinant of the order 3 matrix

$$\begin{bmatrix} a_1 & b_1 & c_1 \\ a_2 & b_2 & c_2 \\ a_3 & b_3 & c_3 \end{bmatrix}$$

27. Prove that if each element in any row or column of a square matrix A is multiplied by a constant k, the determinant of the resulting matrix is $k|A|$.

28. It can be proved that if any two rows or any two columns of a square matrix A are interchanged, the determinant of the resulting matrix is $-|A|$. Use this result to show that if any two rows or columns of a square matrix A are identical, then $|A| = 0$.

29. Prove that

$$\begin{vmatrix} x & y & 1 \\ x_1 & y_1 & 1 \\ x_2 & y_2 & 1 \end{vmatrix} = 0$$

is the equation of a line passing through the points (x_1, y_1) and (x_2, y_2).

30. Prove that if all elements above or below the principal diagonal of a square matrix are 0, then the determinant of the matrix equals the product of the elements along the principal diagonal.

31. Solve for x:

$$\begin{vmatrix} x & 0 & 0 & 0 \\ 15 & x-1 & 0 & 0 \\ -5 & 24 & x+2 & 0 \\ 24 & -31 & 57 & 2x+3 \end{vmatrix} = 0$$

4 Cramer's Rule

To see how determinants can play a role in solving systems of equations, let us consider the general case of two equations in two unknowns:

$$\begin{cases} a_1 x + b_1 y = c_1 \\ a_2 x + b_2 y = c_2 \end{cases} \tag{3}$$

We eliminate y by multiplying the top equation by b_2 and the bottom by b_1 and then subtracting:

$$\begin{array}{l} a_1 b_2 x + b_1 b_2 y = c_1 b_2 \\ a_2 b_1 x + b_1 b_2 y = c_2 b_1 \\ \hline (a_1 b_2 - a_2 b_1)x = c_1 b_2 - c_2 b_1 \end{array} \tag{4}$$

Assume for the moment that the coefficient of x is not 0. Then we can solve for x to get

$$x = \frac{c_1 b_2 - c_2 b_1}{a_1 b_2 - a_2 b_1}$$

This can be written as a quotient of determinants:

$$x = \frac{\begin{vmatrix} c_1 & b_1 \\ c_2 & b_2 \end{vmatrix}}{\begin{vmatrix} a_1 & b_1 \\ a_2 & b_2 \end{vmatrix}} \tag{5}$$

Returning to the original system of equations (3), we multiply the top equation by a_2 and the bottom by a_1, and then subtract, in order to eliminate x:

$$\begin{array}{l} a_1 a_2 x + a_2 b_1 y = a_2 c_1 \\ a_1 a_2 x + a_1 b_2 y = a_1 c_2 \\ \hline (a_1 b_2 - a_2 b_1)y = a_1 c_2 - a_2 c_1 \end{array} \tag{6}$$

(We subtracted the top equation from the bottom one.) Again, assuming the coefficient of y is nonzero, we get

$$y = \frac{a_1 c_2 - a_2 c_1}{a_1 b_2 - a_2 b_1}$$

which can be written as

$$y = \frac{\begin{vmatrix} a_1 & c_1 \\ a_2 & c_2 \end{vmatrix}}{\begin{vmatrix} a_1 & b_1 \\ a_2 & b_2 \end{vmatrix}} \tag{7}$$

Let us now examine the situation when the coefficient of x in equation (4) or of y in (6) is 0:

$$a_1 b_2 - a_2 b_1 = 0 \tag{8}$$

We will suppose that neither b_1 nor b_2 is 0, because if either was 0, both would have to be (why?); thus, the lines would both be vertical, and so either parallel or coincident. We can rewrite equation (8) in the form

$$\frac{a_1}{b_1} = \frac{a_2}{b_2}$$

The slopes of the lines are $-a_1/b_1$ and $-a_2/b_2$, respectively; we conclude that the lines are parallel or coincident. If they are parallel, no solution exists; the system is inconsistent. If they are coincident, the system is dependent. We conclude that a *unique* solution to the system exists if and only if $a_1 b_2 - b_2 a_1 \neq 0$, that is,

$$\begin{vmatrix} a_1 & b_1 \\ a_2 & b_2 \end{vmatrix} \neq 0$$

EXAMPLE 16   Solve the system below using equations (5) and (7).

$$\begin{cases} 2x - 3y = 4 \\ 7x + 5y = 6 \end{cases}$$

Solution   We have immediately

$$x = \frac{\begin{vmatrix} 4 & -3 \\ 6 & 5 \end{vmatrix}}{\begin{vmatrix} 2 & -3 \\ 7 & 5 \end{vmatrix}} = \frac{20 + 18}{10 + 21} = \frac{38}{31}$$

$$y = \frac{\begin{vmatrix} 2 & 4 \\ 7 & 6 \end{vmatrix}}{\begin{vmatrix} 2 & -3 \\ 7 & 5 \end{vmatrix}} = \frac{12 - 28}{31} = -\frac{16}{31}$$

So the solution is $(\frac{38}{31}, -\frac{16}{31})$.

It is useful to analyze the results (5) and (7):

$$x = \frac{\begin{vmatrix} c_2 & b_1 \\ c_2 & b_2 \end{vmatrix}}{\begin{vmatrix} a_1 & b_1 \\ a_2 & b_2 \end{vmatrix}} \qquad y = \frac{\begin{vmatrix} a_1 & c_1 \\ a_2 & c_2 \end{vmatrix}}{\begin{vmatrix} a_1 & b_1 \\ a_2 & b_2 \end{vmatrix}}$$

The denominators are the same, and this common denominator is seen to be the determinant of the coefficient matrix of the system. In each case the numerator differs from the denominator only in that the column of coefficients of the unknown in question is replaced by the column of constants. It is convenient to introduce notation for the matrices involved. Let D denote the coefficient matrix of the system, and let D_x and D_y be matrices that are obtained from D by replacing the column of coefficients of x and of y, respectively, by the column of constants. Then we have

$$x = \frac{|D_x|}{|D|} \qquad y = \frac{|D_y|}{|D|}$$

This same pattern holds true for higher-order systems. For example, the solution to the system of three equations in three unknowns,

$$\begin{cases} a_1 x + b_1 y + c_1 z = d_1 \\ a_2 x + b_2 y + c_2 z = d_2 \\ a_3 x + b_3 y + c_3 z = d_3 \end{cases}$$

is given by

$$x = \frac{|D_x|}{|D|} \qquad y = \frac{|D_y|}{|D|} \qquad z = \frac{|D_z|}{|D|} \qquad (D \neq 0)$$

where D_x, D_y, and D_z are obtained from the coefficient matrix D by replacing the column of coefficients of x, of y, and of z, respectively, by the column of constants appearing on the right-hand side.

As an illustration of the meanings of the matrices D, D_x, D_y, and D_z, consider the system

$$\begin{cases} x + y + z = 4 \\ 2x + y - z = 0 \\ 3x - 4y - 3z = 1 \end{cases}$$

Here,

$$D = \begin{bmatrix} 1 & 1 & 1 \\ 2 & 1 & -1 \\ 3 & -4 & -3 \end{bmatrix}$$

and so

$$D_x = \begin{bmatrix} 4 & 1 & 1 \\ 0 & 1 & -1 \\ 1 & -4 & -3 \end{bmatrix} \qquad D_y = \begin{bmatrix} 1 & 4 & 1 \\ 2 & 0 & -1 \\ 3 & 1 & -3 \end{bmatrix} \qquad D_z = \begin{bmatrix} 1 & 1 & 4 \\ 2 & 1 & 0 \\ 3 & -4 & 1 \end{bmatrix}$$

Column of constants

The general result is known as **Cramer's rule,** which we state as follows:

Cramer's Rule

Let D denote the coefficient matrix in a system of n linear equations in the n unknowns $x_1, x_2, \ldots, x_n$, where each equation is written in the form

$$a_1 x_1 + a_2 x_2 + \cdots + a_n x_n = b$$

Let D_{x_i} be the matrix obtained from D by replacing the column of coefficients of x_i by the column of constants appearing on the right-hand side of the system. Then if $D \neq 0$,

$$x_1 = \frac{|D_{x_1}|}{|D|}, \quad x_2 = \frac{|D_{x_2}|}{|D|}, \quad \ldots \quad , \quad x_n = \frac{|D_{x_n}|}{|D|}$$

Note.   In applying Cramer's rule it is important that the terms involving the unknowns in each equation be written in the same order, and that the constant terms all appear on the right. For example, before applying Cramer's rule to the system

$$\begin{cases} 2y - 3x + 4z - 7 = 0 \\ x + 2z - y = 8 \\ 3z - 4y + 2x + 5 = 0 \end{cases}$$

we would rewrite it as

$$\begin{cases} -3x + 2y + 4z = 7 \\ x - y + 2z = 8 \\ 2x - 4y + 3z = -5 \end{cases}$$

EXAMPLE 17   Solve the system below by Cramer's rule.

$$\begin{cases} x + y + z = 4 \\ 2x + y - z = 0 \\ 3x - 4y - 3z = 1 \end{cases}$$

Solution

$$x = \frac{|D_x|}{|D|} = \frac{\begin{vmatrix} 4 & 1 & 1 \\ 0 & 1 & -1 \\ 1 & -4 & -3 \end{vmatrix}}{\begin{vmatrix} 1 & 1 & 1 \\ 2 & 1 & -1 \\ 3 & -4 & -3 \end{vmatrix}} = \frac{4(-7) + 1(-2)}{1(-7) - 2(1) + 3(-2)} = \frac{-30}{-15} = 2$$

We expanded by the first column in both cases.

$$y = \frac{|D_y|}{|D|} = \frac{\begin{vmatrix} 1 & 4 & 1 \\ 2 & 0 & -1 \\ 3 & 1 & -3 \end{vmatrix}}{-15} = \frac{-4(-3) - (-3)}{-15} = \frac{15}{-15} = -1$$

We expanded the numerator by the second column.

$$z = \frac{|D_z|}{|D|} = \frac{\begin{vmatrix} 1 & 1 & 4 \\ 2 & 1 & 0 \\ 3 & -4 & 1 \end{vmatrix}}{-15} = \frac{4(-11) + 1(-1)}{-15} = \frac{-45}{-15} = 3$$

Expansion was by the third column. So the solution is $(2, -1, 3)$.

Remark. Cramer's rule works only when the number of equations is the same as the number of unknowns. Furthermore, it is usually impractical for systems with more than three equations. It works well for two equations and two unknowns, and is often the easiest method in this case. Depending on how adept one becomes in expanding order 3 determinants, it can be an efficient way to solve a system of three equations in three unknowns.

EXERCISE SET 4

A Solve each of the following by Cramer's rule.

1. $\begin{cases} 2x - 3y = 7 \\ x + 4y = 3 \end{cases}$

2. $\begin{cases} 5x + 4y = 9 \\ 7x - 3y = 6 \end{cases}$

3. $\begin{cases} 3x + y = -2 \\ 2x + 5y = 4 \end{cases}$

4. $\begin{cases} x - 2y = 4 \\ 3x - 4y = 5 \end{cases}$

5. $\begin{cases} 6x - 5y = 4 \\ 8x + 7y = -3 \end{cases}$

6. $\begin{cases} 3y - 4x + 5 = 0 \\ 6x = 2y - 3 \end{cases}$

7. $\begin{cases} 7x - 8y = 10 \\ 5x - 4y = 8 \end{cases}$

8. $\begin{cases} 5x + 6y = -3 \\ 2x - 3y = 4 \end{cases}$

9. $\begin{cases} x + 2y - 3 = 0 \\ 3x - y + 4 = 0 \end{cases}$

10. $\begin{cases} 3x + 2y = 4 \\ 3y + 5x = 2 \end{cases}$

11. $\begin{cases} 3x - 5y = 2 \\ 2x - 3y = 4 \end{cases}$

12. $\begin{cases} 9y + 7x = 10 \\ 2x + 3y - 8 = 0 \end{cases}$

13. $\begin{cases} x - y + 2z = 0 \\ 2x \quad - 3z = 1 \\ y + z = 3 \end{cases}$

14. $\begin{cases} x + 2y - 3z = 1 \\ 2x - y + z = 2 \\ x + y + 4z = 0 \end{cases}$

15. $\begin{cases} x + y = 1 \\ 2y - z = -1 \\ x \quad + z = 0 \end{cases}$

16. $\begin{cases} 3x + y + z = 3 \\ x \quad - z = -5 \\ 2y - z = 0 \end{cases}$

17. $\begin{cases} 3x + 5y & = 2 \\ x + y - z = -1 \\ x - 3y - z = 1 \end{cases}$

18. $\begin{cases} x + y - z = 4 \\ 2x - y + z = 0 \\ x - 2y - 3z = 1 \end{cases}$

19. $\begin{cases} 3x - y + z = 1 \\ x \qquad - z = 2 \\ 2y + 3z = 4 \end{cases}$

20. $\begin{cases} x + y + z = 1 \\ 3x + 2y - 3z = 0 \\ y + 3z = 0 \end{cases}$

21. $\begin{cases} 3x - 4y + 5z = 7 \\ 2x + 3y - 8z = 5 \\ 4x - 2y + 3z = 6 \end{cases}$

22. $\begin{cases} 2x - 3y - 4z = -4 \\ 3x + 4y + 2z = 1 \\ 5x - 2y - 3z = 0 \end{cases}$

B   **23.** $\begin{cases} (a - 1)x + \qquad 2ay = 3 \\ (a + 1)x + (2a - 1)y = 2 \end{cases}$
Solve for x and y.

24. $\begin{cases} x^2u - 3xv = 1 \\ 2xu + 4v = 2 \end{cases}$
Solve for u and v.

25. $\begin{cases} u - 2xv = -1 \\ xu + v = x \end{cases}$
Solve for u and v.

26. $\begin{cases} ax - (a + 1)y = 2 \\ (a - 1)x - \qquad ay = 1 \end{cases}$
Solve for x and y.

27. $\begin{cases} 2m^2s + (m + 2)t = 6 \\ 4(m - 1)s + \qquad 2t = 3 \end{cases}$
Solve for s and t.

28. $\begin{cases} 3x - 4y + z = 2 \\ x + 2y - z = 3 \\ 4x - 3y - 2z = 4 \end{cases}$

29. $\begin{cases} 3x - 2y + 4z = 5 \\ 2x + 3y - 2z = 6 \\ 4x - 5y + 3z = -5 \end{cases}$

30. $\begin{cases} x + y + z + w = 1 \\ 3y \qquad - w = 0 \\ x \qquad - z \qquad = 1 \\ y - 2z - w = 2 \end{cases}$

31. $\begin{cases} x - y \qquad + w = 1 \\ 2x \qquad + 3z - w = 0 \\ 3y - z + 2w = 2 \\ x + y - z \qquad = 3 \end{cases}$

---

5   The Algebra of Matrices

A matrix with *m* **rows** and *n* **columns** is said to be an *m* × *n* (read "*m* by *n*") matrix, and the **size** (or **dimension**) of the matrix is said to be $m \times n$. In this section we will see how the operations of addition and multiplication are defined for matrices.

Two matrices are said to be **equal** provided they are of the same size and corresponding elements are equal. For example,

$$\begin{bmatrix} a & b \\ c & d \end{bmatrix} = \begin{bmatrix} 2 & 3 \\ 5 & 7 \end{bmatrix}$$

if and only if $a = 2$, $b = 3$, $c = 5$, and $d = 7$.

To **add** two matrices of the same size, we add their corresponding elements. For example,

$$\begin{bmatrix} 2 & -1 & 3 \\ 4 & 2 & -5 \end{bmatrix} + \begin{bmatrix} 3 & 7 & -8 \\ 5 & -2 & 4 \end{bmatrix} = \begin{bmatrix} 5 & 6 & -5 \\ 9 & 0 & -1 \end{bmatrix}$$

Addition is not defined for matrices of different sizes.

A **zero matrix,** denoted by **0,** is one in which all elements are 0. There are zero matrices of all sizes. For any matrix A, if **0** is a zero matrix of the same size as A, then $A + \mathbf{0} = A$. Thus, **0** is the **additive identity.**

The **additive inverse** of a matrix A, denoted by $-A$, is the matrix whose elements are the negatives of the corresponding elements of A. For example,

$$-\begin{bmatrix} 2 & 3 \\ 4 & -2 \\ -5 & 6 \end{bmatrix} = \begin{bmatrix} -2 & -3 \\ -4 & 2 \\ 5 & -6 \end{bmatrix}$$

If A and B are matrices of the same size, then the **difference** between A and B is defined by

$$A - B = A + (-B)$$

It follows that subtraction is carried out term-by-term, as in

$$\begin{bmatrix} 4 & 2 \\ 5 & 6 \end{bmatrix} - \begin{bmatrix} 3 & 1 \\ 7 & 2 \end{bmatrix} = \begin{bmatrix} 1 & 1 \\ -2 & 4 \end{bmatrix}$$

The following properties can now be proven:

$$A + B = B + A$$
$$A + (B + C) = (A + B) + C$$
$$A + (-A) = \mathbf{0}$$

In each case the matrices are understood to be of the same size. For matrices of the same size, then, the commutative and associative properties hold true for addition, and each matrix has an additive inverse.

Multiplication of matrices is more complicated. First we define multiplication of a matrix by a real number. This is referred to as **scalar multiplication,** and the real number is called a **scalar.** If A is a matrix and c is a scalar, then cA is defined as the matrix obtained by multiplying every element of A by c. For example,

$$2\begin{bmatrix} 3 & 1 & -4 \\ 2 & 3 & 6 \end{bmatrix} = \begin{bmatrix} 6 & 2 & -8 \\ 4 & 6 & 12 \end{bmatrix}$$

Now let A and B denote two matrices. In order to define the product AB we require that the number of *columns in A* equal the number of *rows in B*. Before stating the general definition of the product AB, let us consider a specific example. Suppose

$$A = \begin{bmatrix} 2 & 1 & 4 \\ 3 & -2 & 0 \end{bmatrix} \quad \text{and} \quad B = \begin{bmatrix} 5 & 4 & 3 & -2 \\ 6 & -1 & 2 & 3 \\ 2 & 3 & -2 & 4 \end{bmatrix}$$

Then A is 2×3 and B is 3×4. So the requirement that the number of columns in A equals the number of rows in B is met. To find the elements of the product AB we describe first what is known as the **inner product** of a given row of A with a given column of B. Take, for example, the first row of A and the second column of B:

$$\boxed{2 \quad 1 \quad 4} \begin{bmatrix} 4 \\ -1 \\ 3 \end{bmatrix} = 2(4) + 1(-1) + 4(3) = 8 - 1 + 12 = 19$$

Observe that we multiplied corresponding elements and then added. This is what we mean by inner product. Now to obtain the element in the ith row and the jth column of AB, we find the inner product of the ith row of A and the jth column of B. In our example, we have found that 19 is the element in the first row and second column of AB. We now calculate the other elements:

$$\begin{bmatrix} 2 & 1 & 4 \\ 3 & -2 & 0 \end{bmatrix} \begin{bmatrix} 5 & 4 & 3 & -2 \\ 6 & -1 & 2 & 3 \\ 2 & 3 & -2 & 4 \end{bmatrix} = \begin{bmatrix} 24 & 19 & 0 & 15 \\ 3 & 14 & 5 & -12 \end{bmatrix}$$

We have shaded the first row of A, the second column of B, and their inner product 19, which therefore appears in the first row and second column of AB. The other elements are obtained in a similar way. (You should check these.) Notice that the size of the product matrix is 2×4, the number of rows being the same as the number of rows in A, and the number of columns being the same as the number of columns in B.

We now generalize these concepts.

DEFINITION 1   Let A be an $m \times n$ matrix and let B be an $n \times p$ matrix, and suppose the ith row of A and the jth column of B are as shown:

$$B$$

$$i\text{th row} \rightarrow \begin{bmatrix} \cdots\cdots\cdots \\ a_1 \quad a_2 \quad a_3 \quad \cdots \quad a_n \\ \cdots\cdots\cdots \end{bmatrix}_A \qquad \begin{bmatrix} \cdot & b_1 & \cdot \\ \cdot & b_2 & \cdot \\ \cdot & b_3 & \cdot \\ \vdots & \vdots & \vdots \\ \cdot & b_n & \cdot \end{bmatrix}$$

$$\uparrow$$
$$j\text{th column}$$

The **inner product** of this row and this column is the number

$$a_1 b_1 + a_2 b_2 + a_3 b_3 + \cdots + a_n b_n$$

DEFINITION 2   If A and B are $m \times n$ and $n \times p$ matrices, respectively, then the product AB is the $m \times p$ matrix whose element in the ith row and jth column is the inner product of the ith row of A and the jth column of B.

EXAMPLE 18

$$\begin{bmatrix} 2 & -1 \\ 3 & 4 \\ 5 & -3 \end{bmatrix} \begin{bmatrix} -2 & 3 & 6 \\ 4 & 1 & -5 \end{bmatrix} = \begin{bmatrix} -8 & 5 & 17 \\ 10 & 13 & -2 \\ -22 & 12 & 45 \end{bmatrix}$$

In general, $AB \neq BA$. In fact, BA may not be defined even if AB is. For example, if

$$A = \begin{bmatrix} 3 & 1 \\ 4 & 2 \end{bmatrix} \quad \text{and} \quad B = \begin{bmatrix} 1 & 5 & 4 \\ 4 & 3 & 6 \end{bmatrix}$$

then the product AB is defined, but BA is not. Even when both AB and BA are defined, they may not be equal. For example, let

$$A = \begin{bmatrix} 3 & 4 \\ -2 & 5 \end{bmatrix} \quad \text{and} \quad B = \begin{bmatrix} -1 & 2 \\ 3 & -4 \end{bmatrix}$$

Then

$$AB = \begin{bmatrix} 3 & 4 \\ -2 & 5 \end{bmatrix}\begin{bmatrix} -1 & 2 \\ 3 & -4 \end{bmatrix} = \begin{bmatrix} 9 & -10 \\ 17 & -24 \end{bmatrix}$$

and

$$BA = \begin{bmatrix} -1 & 2 \\ 3 & -4 \end{bmatrix}\begin{bmatrix} 3 & 4 \\ -2 & 5 \end{bmatrix} = \begin{bmatrix} -7 & 6 \\ 17 & -8 \end{bmatrix}$$

So $AB \neq BA$. Matrix multiplication is therefore *noncommutative*. The following properties, however, can be shown to hold (we assume the sizes are such that all products are defined):

$$A(BC) = (AB)C \qquad \text{Associative property}$$
$$A(B + C) = AB + AC \qquad \text{Left distributive property}$$
$$(B + C)A = BA + CA \qquad \text{Right distributive property}$$

The left and right distributive properties normally yield different results because of noncommutativity.

EXERCISE SET 5

A   In Problems 1–14 perform the indicated operations.

1. $\begin{bmatrix} 3 & 4 & -2 \\ 7 & 8 & 10 \end{bmatrix} + \begin{bmatrix} -1 & 2 & 5 \\ 3 & -4 & 6 \end{bmatrix}$

2. $\begin{bmatrix} 2 & 0 \\ 7 & 3 \\ -1 & 4 \end{bmatrix} + \begin{bmatrix} 5 & -3 \\ 2 & 6 \\ 3 & -7 \end{bmatrix}$

3. $\begin{bmatrix} 1 & 4 \\ -2 & 5 \end{bmatrix} - \begin{bmatrix} 3 & 2 \\ 4 & -6 \end{bmatrix}$

4. $\begin{bmatrix} 7 & -5 & 3 \\ -6 & 4 & 2 \\ 0 & 3 & -8 \end{bmatrix} - \begin{bmatrix} 5 & -8 & 1 \\ 2 & -6 & 3 \\ 4 & 0 & -2 \end{bmatrix}$

5. $\begin{bmatrix} 2 & 5 \\ 3 & -4 \end{bmatrix} + \begin{bmatrix} -1 & 2 \\ 5 & 6 \end{bmatrix} - \begin{bmatrix} 3 & -7 \\ 8 & 2 \end{bmatrix}$

6. $\begin{bmatrix} 3 & 1 & 5 \\ 2 & -3 & -6 \end{bmatrix} - \begin{bmatrix} 8 & -4 & 3 \\ -2 & 5 & -6 \end{bmatrix} + \begin{bmatrix} 4 & 6 & 9 \\ -5 & 7 & -2 \end{bmatrix}$

7. $2A + 3B$, where $A = \begin{bmatrix} 2 & 1 \\ -1 & 4 \\ 3 & -5 \end{bmatrix}$ and $B = \begin{bmatrix} -3 & 2 \\ 4 & -3 \\ 7 & 8 \end{bmatrix}$

8. $4A - 5B$, where A and B are as given in Problem 7

9. $\begin{bmatrix} 2 & 3 \\ -1 & 4 \end{bmatrix} \begin{bmatrix} 3 & -2 \\ 5 & 6 \end{bmatrix}$

10. $\begin{bmatrix} 1 & 2 & -4 \end{bmatrix} \begin{bmatrix} 3 & 4 \\ -2 & 6 \\ 5 & -1 \end{bmatrix}$

11. $\begin{bmatrix} 2 & 1 \\ 5 & 7 \end{bmatrix} \begin{bmatrix} 4 & -2 & 5 \\ -1 & 3 & 0 \end{bmatrix}$

12. $\begin{bmatrix} 1 & 3 & -2 \\ 0 & 4 & 5 \\ -2 & 1 & 3 \end{bmatrix} \begin{bmatrix} 4 \\ 3 \\ 2 \end{bmatrix}$

13. $\begin{bmatrix} 3 & 1 \\ 2 & -4 \\ 5 & 6 \end{bmatrix} \begin{bmatrix} 1 & 3 & -4 & 2 \\ -2 & 0 & 5 & 1 \end{bmatrix}$

14. $\begin{bmatrix} 1 & 3 & -2 \\ 4 & 0 & 1 \\ -2 & 5 & 3 \end{bmatrix} \begin{bmatrix} 1 & 2 & 0 & 5 \\ -2 & 3 & 1 & 4 \\ 3 & 0 & -5 & 6 \end{bmatrix}$

In Problems 15 and 16 solve for x and y by performing the multiplications and equating elements.

15. $\begin{bmatrix} 3 & 7 \\ 2 & -1 \end{bmatrix} \begin{bmatrix} x \\ y \end{bmatrix} = \begin{bmatrix} 1 \\ 12 \end{bmatrix}$

16. $\begin{bmatrix} 3 & 7 \\ 5 & -2 \end{bmatrix} \begin{bmatrix} x \\ y \end{bmatrix} = \begin{bmatrix} 11 \\ -9 \end{bmatrix}$

17. Perform the multiplication

$$\begin{bmatrix} 2 & -3 & 1 & -1 \\ 1 & 1 & -2 & 1 \end{bmatrix} \begin{bmatrix} 1 & 2 \\ -1 & 1 \\ 5 & 4 \\ 10 & 5 \end{bmatrix}$$

What can you conclude in general about matrices A and B if $AB = 0$?

18. Let

$$A = \begin{bmatrix} 2 & 3 \\ -1 & 4 \end{bmatrix} \qquad B = \begin{bmatrix} 1 & 0 & -2 \\ 3 & -1 & 5 \end{bmatrix} \qquad C = \begin{bmatrix} 1 & -4 \\ -2 & 3 \\ 5 & -1 \end{bmatrix}$$

Determine which of the following are defined. For those that are defined, perform the indicated multiplications.
a. ABC b. ACB c. BAC d. BCA e. CAB f. CBA

B 19. Let

$$A = \begin{bmatrix} a_1 & a_2 \\ a_3 & a_4 \end{bmatrix} \qquad B = \begin{bmatrix} b_1 & b_2 \\ b_3 & b_4 \end{bmatrix} \qquad C = \begin{bmatrix} c_1 & c_2 \\ c_3 & c_4 \end{bmatrix}$$

Prove that:
a. $A + B = B + A$ b. $A + (B + C) = (A + B) + C$
c. $A(BC) = (AB)C$ d. $A(B + C) = AB + AC$

20. When AA is defined, it is customary to denote this by A^2. Prove that A^2 exists if and only if A is a square matrix.

21. For a square matrix A, we write $A^2 = AA$, $A^3 = AAA$, and so on. If

$$A = \begin{bmatrix} 1 & 1 \\ 1 & 1 \end{bmatrix}$$

find A^2, A^3, and A^4. Generalize this to A^n.

22. Use 2×2 matrices to prove that in general $(A + B)^2 \neq A^2 + 2AB + B^2$. Find a correct formula for $(A + B)^2$.

23. Use 2×2 matrices to prove that in general $(A + B)(A - B) \neq A^2 - B^2$. Find a correct formula for $(A + B)(A - B)$.

24. Let

$$A = \begin{bmatrix} 2 & -1 & 3 \\ 4 & 2 & -2 \\ -3 & 5 & 4 \end{bmatrix} \qquad X = \begin{bmatrix} x \\ y \\ z \end{bmatrix} \qquad B = \begin{bmatrix} 13 \\ -6 \\ -1 \end{bmatrix}$$

Solve the equation $AX = B$ for X.

25. The following represents a system of m linear equations in n unknowns:

$$\begin{cases} a_{11}x_1 + a_{12}x_2 + a_{13}x_3 + \cdots + a_{1n}x_n = b_1 \\ a_{21}x_1 + a_{22}x_2 + a_{23}x_3 + \cdots + a_{2n}x_n = b_2 \\ a_{31}x_1 + a_{32}x_2 + a_{33}x_3 + \cdots + a_{3n}x_n = b_3 \\ \quad\cdots\cdots\cdots\cdots\cdots\cdots\cdots\cdots \\ a_{m1}x_1 + a_{m2}x_2 + a_{m3}x_3 + \cdots + a_{mn}x_n = b_n \end{cases}$$

Find matrices A, B, and X for which the matrix equation $AX = B$ is equivalent to this system.

26. Let

$$A = \begin{bmatrix} 0 & 1 & 1 \\ 1 & 0 & 1 \\ 1 & 1 & 0 \end{bmatrix} \qquad I = \begin{bmatrix} 1 & 0 & 0 \\ 0 & 1 & 0 \\ 0 & 0 & 1 \end{bmatrix}$$

Solve the equation $|A - \lambda I| = 0$ for λ, where λ is a scalar.

---

6 Nonlinear Systems

No general theory exists to handle the great variety of nonlinear systems that can occur. One must deal with each particular situation and decide what will work, and of course there are times when nothing will give an exact solution. Nevertheless, we can consider some fairly common situations that may be solved by elementary means. We will illustrate these by examples.

EXAMPLE 19   Solve the system: $\begin{cases} 2x^2 - y + 3 = 0 \\ \qquad x + y = 6 \end{cases}$

Solution   This is a case of a quadratic equation and a linear equation. Typically, whenever one equation is linear and the other is not, the best procedure is to substitute from the linear equation into the other. In this problem we choose to solve the second equation for y (rather than x, since the substitution is simpler):

$$y = 6 - x$$
$$2x^2 - (6 - x) + 3 = 0$$
$$2x^2 + x - 3 = 0$$
$$(2x + 3)(x - 1) = 0$$
$$x = -\tfrac{3}{2} \quad | \quad x = 1$$

When $x = -\frac{3}{2}$, $y = 6 - (-\frac{3}{2}) = \frac{15}{2}$, and when $x = 1$, $y = 6 - 1 = 5$. So the solution set is $\{(1, 5), \quad (-\frac{3}{2}, \frac{15}{2})\}$. The graphs of both equations are shown in Figure 3, and the solutions are indicated.

Figure 3

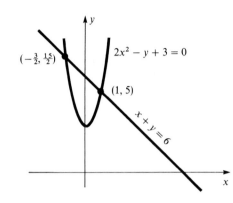

EXAMPLE 20 Solve the system: $\begin{cases} xy + 2 = 0 \\ 2x - y = 5 \end{cases}$

Solution This is handled in a similar way, although we could easily substitute from the first equation into the second:

$$y = 2x - 5$$
$$x(2x - 5) + 2 = 0$$
$$2x^2 - 5x + 2 = 0$$
$$(2x - 1)(x - 2) = 0$$
$$x = \tfrac{1}{2} \quad | \quad x = 2$$

When $x = \frac{1}{2}$, $y = 2(\frac{1}{2}) - 5 = -4$, and when $x = 2$, $y = 2(2) - 5 = -1$. So the solution set is $\{(\frac{1}{2}, -4), \quad (2, -1)\}$. The graph of the first equation is obtained by plotting a few points. It is a hyperbola (Figure 4).

Figure 4

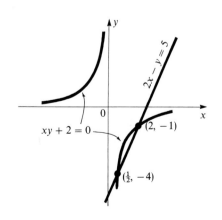

EXAMPLE 21 Solve the system: $\begin{cases} x^2 - y^2 = 5 \\ 2x^2 + y^2 = 22 \end{cases}$

Solution Even though these are quadratic equations, they are linear in form, and the techniques of solving linear systems can be used to find x^2 and y^2, from which x and y can be obtained. In particular, Cramer's rule could be used, but it is probably easier here just to eliminate y^2 by addition:

$$x^2 - y^2 = 5$$
$$\underline{2x^2 + y^2 = 22}$$
$$3x^2 = 27$$
$$x^2 = 9$$
$$x = \pm 3$$

Now substitute $x^2 = 9$ into the first equation (since it is simpler):

$$9 - y^2 = 5$$
$$y^2 = 4$$
$$y = \pm 2$$

The complete solution set is $\{(3, 2), (3, -2), (-3, 2), (-3, -2)\}$. Again, it is helpful to visualize the situation graphically (Figure 5).

Figure 5

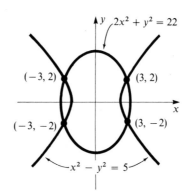

EXAMPLE 22 Solve the system: $\begin{cases} y = 3x - \dfrac{4}{x} \\ y = x^2 \end{cases}$

Solution Here, we can eliminate y to get

$$x^2 = 3x - \frac{4}{x}$$

Multiplying by x yields

$$x^3 - 3x^2 + 4 = 0*$$

We know from Chapter 6 that the only possible rational roots are $\pm 1, \pm 2, \pm 4$. We proceed by synthetic division to test these:

x				y
0	1	-3	0	4
1	1	-2	-2	2
2	1	-1	-2	0

So $x = 2$ is a root, and the depressed equation is quadratic:

$$x^2 - x - 2 = 0$$
$$(x - 2)(x + 1) = 0$$
$$x = 2 \quad | \quad x = -1$$

Thus, $x = 2$ is a double root. When $x = 2$, $y = 4$, and when $x = -1$, $y = 1$. The solution set is $\{(2, 4), (-1, 1)\}$ with $(2, 4)$ as a double point. Graphically, the significance of the double point is that the curves are tangent to each other at that point. The graph of the first equation is somewhat more complicated this time, but again we give a sketch in Figure 6 to help visualize the situation.

Figure 6

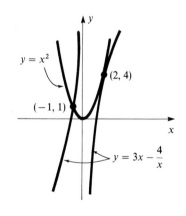

$y = x^2$

$(2, 4)$

$(-1, 1)$

$y = 3x - \dfrac{4}{x}$

EXAMPLE 23 Solve the system: $\begin{cases} y = \sqrt{x} \\ y^2 = \dfrac{2}{x} + 1 \end{cases}$

---

* This could lead to an extraneous root, since we may be unwittingly multiplying by 0. The answers should be checked to see that this is not the case.

Solution   Again, substitution can be used. We substitute for y from the first equation into the second:

$$x = \frac{2}{x} + 1$$
$$x^2 = 2 + x$$
$$x^2 - x - 2 = 0$$
$$(x - 2)(x + 1) = 0$$
$$x = 2 \quad | \quad x = -1$$

When $x = 2$, $y = \sqrt{2}$, and when $x = -1$, y is imaginary. So $(2, \sqrt{2})$ is the only real solution.

Notice that after finding $x = 2$ we substituted into the first equation to get y. Had we substituted into the second, we would have had

$$y^2 = \tfrac{2}{2} + 1 = 2$$
$$y = \pm\sqrt{2}$$

But $y = -\sqrt{2}$ is not possible, since the first equation implies that y is never negative; so we must reject $-\sqrt{2}$. This extraneous root resulted from squaring the first equation to substitute into the second. Whenever both sides of an equation are squared, there is a possibility of introducing extraneous roots, and so the answer should be checked. In general, after finding one unknown it is best to substitute this into the equation that imposes the greater restriction on the other unknown, if this can be identified. In solving for x, we also multiplied by the unknown, which again could introduce extraneous roots since we might have multiplied by 0. We can check mentally to see that this did not occur. The graphs are rather complicated, and we will not show them here.

The main device we have used in the above examples is that of elimination by substitution. This technique is especially useful whenever at least one of the unknowns appears to the first power in either of the equations. You should also be alert to the possibility of eliminating an unknown by addition or subtraction.

EXERCISE SET 6

A   Find the solution set in Problems 1–33.

1. $\begin{cases} x^2 + y^2 = 5 \\ x - y = 1 \end{cases}$

2. $\begin{cases} x^2 + y^2 = 10 \\ 2x - y = 1 \end{cases}$

3. $\begin{cases} x^2 + 2y^2 = 9 \\ x - 2y = 3 \end{cases}$

4. $\begin{cases} x^2 + y^2 = 5 \\ 3x + 4y = 5 \end{cases}$

5. $\begin{cases} y = x^2 - 4 \\ y = \dfrac{2x - 7}{3} \end{cases}$

6. $\begin{cases} y = 3 - x^2 \\ 2x - y + 4 = 0 \end{cases}$

7. $\begin{cases} x - y^2 = 4 \\ 2x - 3y = 10 \end{cases}$

8. $\begin{cases} xy = 4 \\ y - x = 3 \end{cases}$

9. $\begin{cases} 4x^2 + 3y^2 = 9 \\ y = 2x \end{cases}$

10. $\begin{cases} 3x^2 + 4y^2 = 63 \\ x - 2y + 3 = 0 \end{cases}$

11. $\begin{cases} y^2 = x \\ 2x - 3y = 5 \end{cases}$

12. $\begin{cases} y^2 = x \\ x^2 = y \end{cases}$

13. $\begin{cases} xy = 1 \\ 4x - 7y = 3 \end{cases}$

14. $\begin{cases} 2xy = -3 \\ 3x - 4y = 9 \end{cases}$

15. $\begin{cases} x^2 + y^2 = 3 \\ y^2 = 2x \end{cases}$

16. $\begin{cases} x^2 + y^2 = 6 \\ y^2 - 3x = 2 \end{cases}$

17. $\begin{cases} x^2 - y^2 = 7 \\ y = 2x - 5 \end{cases}$

18. $\begin{cases} 4x^2 - 5y^2 + 16 = 0 \\ 2x + 3y = 4 \end{cases}$

19. $\begin{cases} y^2 = 4x \\ x^2 - 2y^2 = 9 \end{cases}$

20. $\begin{cases} y = x^2 - 4 \\ 5x^2 - y^2 = 20 \end{cases}$

21. $\begin{cases} x^2 + y^2 - 2y = 24 \\ 2x - y = 4 \end{cases}$

22. $\begin{cases} 2x^2 - 3y^2 = 5 \\ x^2 - 2y^2 = 2 \end{cases}$

23. $\begin{cases} 4x^2 - y^2 = 7 \\ x^2 + 3y^2 = 31 \end{cases}$

24. $\begin{cases} xy + 6 = 0 \\ x^2 + y^2 = 13 \end{cases}$

25. $\begin{cases} 2x^2 - 3x + 4y^2 = 35 \\ x + 2y = 5 \end{cases}$

26. $\begin{cases} 3x^2 - 4y^2 = 12 \\ 7x^2 - y^2 = 8 \end{cases}$

27. $\begin{cases} 3x^2 - 7y^2 = 5 \\ 5x^2 + 3y^2 = 12 \end{cases}$

28. $\begin{cases} 3x^2 - y^2 = 1 \\ 7x^2 - 2y^2 = 5 \end{cases}$

29. $\begin{cases} y^2 = x - 3 \\ y^2 - 3y + 2x = 12 \end{cases}$

30. $\begin{cases} xy = 3 \\ 2x^2 - 3y^2 = 15 \end{cases}$

31. $\begin{cases} x^2 + y^2 + 2x - y = 8 \\ x^2 + y^2 = 5 \end{cases}$

32. $\begin{cases} x^2 + y^2 + 3x - 4y = -4 \\ x^2 + y^2 + 2x - 3y = 1 \end{cases}$

33. $\begin{cases} y = x^2 - 4 \\ x^2 + 3y^2 + 4y - 6 = 0 \end{cases}$

34. The perimeter of a rectangle is 40, and its area is 96. Find its dimensions.

35. The length of a rectangle is 4 more than twice its width, and its area is 126. Find its dimensions.

36. The difference between two numbers is 3, and their product is 270. Find the numbers.

37. The difference between two numbers is 4 and the difference between their squares is 80. Find the numbers.

38. The product of two numbers is 128, and the sum of their reciprocals is $\frac{3}{16}$. Find the numbers.

39. The perimeter of a rectangle is 34, and the length of its diagonal is 13. Find the length and width.

40. The length of a rectangle is 3 less than twice the width, and the square of the length of the diagonal is 306. Find the length and width.

41. A beam is to be sawed from a log of diameter 10 inches. If the height of the cross-section of the beam is to be twice the width, what are the maximum dimensions of the cross-section of the beam?

42. A driver travels the first 75 miles of a 95 mile trip without encountering congestion, but traffic is heavy on the remaining part, and his average speed is reduced by 10 miles per hour. The total time for the trip is 2 hours. Find his average speed on each part of the trip.

43. A man has a certain amount invested at simple interest, the annual income from which is $180. If he had invested it at another bank that pays $\frac{1}{2}\%$ more interest, he would have received an income of $198 per year. Find the amount he invested and the rate of interest.

B   Find the solution set in Problems 44–47.

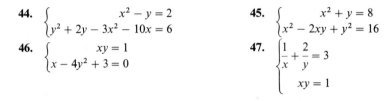

44. $\begin{cases} x^2 - y = 2 \\ y^2 + 2y - 3x^2 - 10x = 6 \end{cases}$

45. $\begin{cases} x^2 + y = 8 \\ x^2 - 2xy + y^2 = 16 \end{cases}$

46. $\begin{cases} xy = 1 \\ x - 4y^2 + 3 = 0 \end{cases}$

47. $\begin{cases} \dfrac{1}{x} + \dfrac{2}{y} = 3 \\ xy = 1 \end{cases}$

48. The height of a certain right circular cylinder is 1 inch more than twice its radius. Its volume is 9π cubic inches. Find the height and the radius.

49. A right circular cylinder of volume 72π is inscribed in a sphere of radius 5. Find the base radius and height of the cylinder.

50. The height of a right circular cone is 1 inch less than twice its base radius. If the volume is 15π cubic inches, find the base radius and the height.

51. Find the equation of the line that passes through the points of intersection of the curves whose equations are $y^2 = x$ and $x^2 + 3xy - 2y^2 = 32$.

52. Find the equation of the line passing through the points of intersection of the curves $x^2 + y^2 = 10$ and $x^2 + y^2 - 2x + 4y = 0$.

53. An isosceles triangle with base equal to its altitude is inscribed in a circle of diameter 10. Find the altitude of the triangle.

54. A boy is in a rowboat 4 miles from the point A nearest him on the shore. In order to get to the point B on the shore 10 miles from A, he rows to point C, between A and B, and jogs the rest of the way to B. If he can row 5 miles per hour and jog 7 miles per hour, and if his time rowing equals his time jogging, find how far C is from A (see sketch).

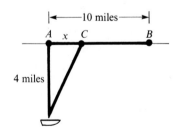

---

7   Systems of Inequalities

We will illustrate in this section how to determine graphic solutions to systems of inequalities. First we must see how to solve graphically a single inequality in two variables. This is illustrated in the next two examples.

EXAMPLE 24   Show the graphic solution to the inequality $2x - y < 3$.

Solution   The inequality can be written in the equivalent form $y > 2x - 3$. We wish to describe as a region in the plane the set of all points (x, y) satisfying this. One technique is to sketch first the line $y = 2x - 3$ and to observe that the solution to the given inequality consists of all points *above* this line (Figure 7). The line is dashed in the figure to indicate that it is not part of the solution set.

Figure 7

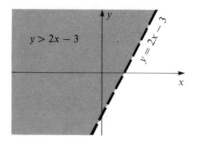

Remark.   We could equally well write the inequality in the form $x < (y + 3)/2$ and conclude that the solution set consists of all points to the *left* of the line $x = (y + 3)/2$, and this is seen to agree with what we have shown.

EXAMPLE 25   Show the graphic solution of the inequality $y \leq x^2 + 4$.

Solution   The graph of $y = x^2 + 4$ is the parabola pictured in Figure 8. All points *on the parabola or below* it satisfy the given inequality. The parabola is shown as a solid curve to indicate that it is part of the solution set.

Figure 8

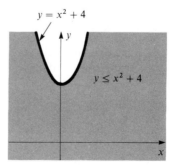

The next two examples illustrate graphic solutions to systems of linear inequalities in two variables.

EXAMPLE 26   Find the common solution set to the system: $\begin{cases} 2x + 3y > 4 \\ x + y < 3 \end{cases}$

Solution   We rewrite the inequalities in the form

$$y > \frac{4 - 2x}{3}$$

$$y < 3 - x$$

and next draw the bounding lines, as shown in Figure 9. The simultaneous solution is the shaded region in the figure, since that is where y is greater than $(4 - 2x)/3$ and at the same time less than $3 - x$.

Figure 9

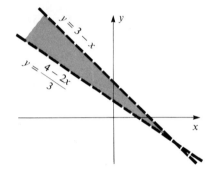

EXAMPLE 27   Exhibit graphically the simultaneous solution set to the system:

$$\begin{cases} x + 3y \le 12 \\ x + y \le 6 \\ 2x + y \le 10 \\ x \ge 0 \\ y \ge 0 \end{cases}$$

Solution   The last two inequalities restrict the solution set to the first quadrant. We rewrite the other three inequalities in the form

$$y \le \frac{12 - x}{3}$$

$$y \le 6 - x$$

$$y \le 10 - 2x$$

and draw the lines represented by the corresponding equalities (Figure 10). The simultaneous solution is the shaded region in the figure, including the boundary lines. In applications, it is usually important to know the vertices of the boundary. These are found by solving appropriate pairs of equations simultaneously and are shown in the figure.

Figure 10

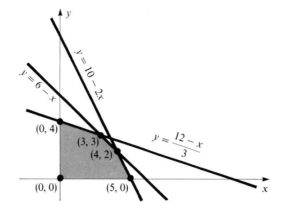

The next example involves absolute values and is equivalent to a system of simultaneous inequalities.

EXAMPLE 28   Show the graphic solution of the inequality $|x| - |y| \leq 1$.

Solution   We may rewrite the inequality in the form

$$|x| \leq |y| + 1$$

which is equivalent to

$$-|y| - 1 \leq x \leq |y| + 1$$

So if $y \geq 0$, this gives

$$-y - 1 \leq x \leq y + 1$$

whereas if $y < 0$, we obtain

$$y - 1 \leq x \leq -y + 1$$

Equivalently, we may say that when $y \geq 0$, the inequality is satisfied by all pairs (x, y) which simultaneously satisfy

$$\begin{cases} x \leq y + 1 \\ x \geq -y - 1 \end{cases}$$

and when $y < 0$, by the pairs (x, y) satisfying

$$\begin{cases} x \leq -y + 1 \\ x \geq y - 1 \end{cases}$$

So for $y \geq 0$ we draw in the bounding lines $x = y + 1$ and $x = -y - 1$ and include all points to the *left* of the first line and at the same time to the *right* of the second; we also include points on the boundaries.

Similarly, when $y < 0$, we draw the bounding lines $x = -y + 1$ and $x = y - 1$ and include all points between as well as on these lines. The solution set is shown in Figure 11.

Figure 11

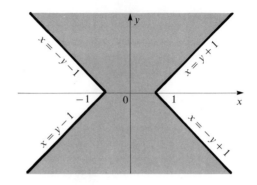

EXERCISE SET 7

Give the graphic solution to each of the following.

A **1.** $y < 3x - 5$

2. $y > 4 - x$

3. $x + y < 2$

4. $2x - y > 3$

5. $3x + 2y \le 4$

6. $5x - 4y \ge 3$

7. $3y - 4 < 2x - 5$

8. $5x + 2 > 3y - 4$

9. $4y - x^2 > 0$

10. $y - 1 < x^2$

11. $x > y^2$

12. $y - x^2 > 4$

13. $16y \le x^2$

14. $\begin{cases} x - y > 0 \\ y - 2x < 4 \end{cases}$

15. $\begin{cases} x - 2y + 4 > 0 \\ x + y > -3 \end{cases}$

16. $\begin{cases} 3x - 2y > 4 \\ x + y \ge 1 \end{cases}$

17. $\begin{cases} 3x + 4y > 12 \\ 2x - y + 3 > 0 \end{cases}$

18. $\begin{cases} x \ge 2 \\ x \le 2 + y \end{cases}$

19. $\begin{cases} y \le 2x \\ 2x - y < 3 \end{cases}$

20. $\begin{cases} 2x - 5y < 6 \\ x + 2y \ge 4 \end{cases}$

21. $\begin{cases} y \le 1 \\ x - y \le 2 \end{cases}$

22. $\begin{cases} x < 1 \\ y > 1 \\ x - y + 3 > 0 \end{cases}$

23. $\begin{cases} x \ge 0 \\ x + y \le 2 \\ 2x - 3y \le 6 \end{cases}$

24. $|y| \le x - 1$

25. $|2x - y| \le 3$

26. $|y| > x - 1$

27. $|x| < 2y + 3$

28. $|x| \ge 2 - y$

B **29.** $\begin{cases} y \le 1 + x \\ y \le 1 - x \\ y \ge x - 1 \\ y \ge -x - 1 \end{cases}$

30. $\begin{cases} 2x + y \le 8 \\ 2x + 3y \le 12 \\ x \ge 0 \\ y \ge 0 \end{cases}$

31. $\begin{cases} x^2 - 4y \le 0 \\ 2y - x < 2 \end{cases}$

32. $\begin{cases} y - \sqrt{x} < 0 \\ 4y - x > 0 \end{cases}$

33. $\begin{cases} x^2 + y < 4 \\ 2x - y < 4 \\ 2x + y + 4 > 0 \end{cases}$

34. $\begin{cases} 4y - x^2 > 0 \\ y^2 - 4x < 0 \end{cases}$

35. $\begin{cases} x^2 - 2x + y \leq 0 \\ x + y > 0 \end{cases}$

36. $|x - y| \geq 1$

37. $|x| + |y| \leq 1$

---

8 Linear Programming

An important application of systems of linear inequalities occurs in a technique developed in the 1940's for solving problems in which a quantity is to be maximized or minimized, subject to certain limitations called **constraints.** For example, a company might want to maximize the profit on some manufactured item or to minimize the cost. Applications of this technique, called **linear programming,** can be found in many areas but are especially important in business and industry.

In a typical linear programming problem the quantity to be maximized or minimized is a linear function of two or more variables. The method we will illustrate is applicable to functions of two variables only, but other procedures exist for dealing with larger numbers of variables. This function is known as the **objective function,** and has the form $F = ax + by + c$. The constraints on x and y are expressed in the form of linear inequalities.

For example, suppose we wish to find the maximum value of the objective function $F = 3x + 4y$, subject to the constraints given by

$$\begin{cases} x + y \leq 7 \\ 2x + 5y \leq 20 \\ x \geq 0 \\ y \geq 0 \end{cases}$$

The four constraints have the simultaneous solution set shown by the shaded region in Figure 12. This region is called the **set of feasible solutions** to the problem. We have shown the coordinates of each vertex of the polygon bounding the region. Finding these vertices is an essential part of the problem and involves solving two or more sets of equations simultaneously. To find the values of x and y among the feasible solutions for which the objective function F is a

Figure 12

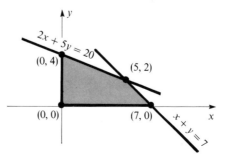

maximum, we make use of the following fundamental result, which we state without proof:

THEOREM   **The maximum and minimum values of the objective function in a linear programming problem, if they exist, occur at a vertex of the graph of the set of feasible solutions.***

Thus, to find the maximum value of F we test its value at each of the vertices $(0, 0)$, $(7, 0)$, $(5, 2)$, and $(0, 4)$. The maximum value is found to be 23, and this occurs when x is 5 and y is 2.

Vertex	F
$(0, 0)$	0
$(7, 0)$	21
$(5, 2)$	23
$(0, 4)$	16

The next two examples illustrate applications in economics.

EXAMPLE 29   A company manufactures two kinds of pocket calculators, one with and one without rechargeable batteries. The company can make at most 50 of the type with rechargeable batteries (type A) per day and 60 of the type without rechargeable batteries (type B) per day. Type A requires 3 work-hours to produce and type B requires 2 work-hours. The work force provides a total of 180 work-hours available per day. If the profit on each type A calculator is $2.50, and the profit on each type B calculator is $2.00, find how many of each type of calculator should be produced per day to give the maximum profit.

Solution   Let x be the number of type A and y the number of type B calculators to be produced. Then the profit P is given by

$$P = 2.50x + 2.00y$$

This is the objective function, which is to be maximized, subject to the constraints

$$\begin{cases} 0 \le x \le 50 \\ 0 \le y \le 60 \\ 3x + 2y \le 180 \end{cases}$$

---

* Both maximum and minimum values do exist if the feasibility set is bounded and is **convex**, which means that for any two points of the set the line segment joining them lies entirely within the set.

Figure 13

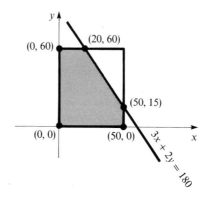

The set of feasible solutions is shown in Figure 13. Testing the value of P at each vertex, we find the maximum value to be \$170, and this occurs when $x = 20$ and $y = 60$.

Vertex	P
(0, 0)	0
(50, 0)	125
(50, 15)	155
(20, 60)	170
(0, 60)	120

EXAMPLE 30

A machine tool company produces two kinds of parts, A and B. The weekly demand for each requires that at least 20 type A parts and 10 type B parts be produced. Limitations on capacity require that at most 60 type A and 40 type B parts be produced each week. In order to keep the work force fully employed, the combination of A and B parts produced must be at least 50 each week. If it costs \$3 to produce each A part and \$2 to produce each B part, how many parts of each type should be produced per week to minimize cost?

Solution

Let x be the number of type A parts and y the number of type B parts to be produced each week. The objective function is the cost C, given by

$$C = 3x + 2y$$

and this is to be minimized, subject to the constraints

$$\begin{cases} 20 \le x \le 60 \\ 10 \le y \le 40 \\ x + y \ge 50 \end{cases}$$

Figure 14

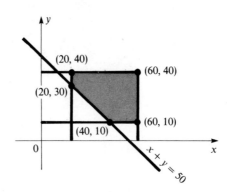

Figure 14 shows the set of feasible solutions. Calculating C at each vertex shows that the minimum value of $120 occurs when 20 type A parts and 30 type B parts are produced each week.

Vertex	C
(60, 10)	200
(60, 40)	260
(20, 40)	140
(20, 30)	120
(40, 10)	140

EXERCISE SET 8

A In Problems 1–8 find the maximum or minimum value, as specified, of the objective function, subject to the given constraints. Give the values of x and y that produce this value.

1. Maximize $F = 2x + y$, subject to $\begin{cases} 3x + 4y \le 24 \\ x \ge 2 \\ y \ge 3 \end{cases}$

2. Minimize $C = 2x - 3y + 8$, subject to $\begin{cases} x - y \ge 0 \\ x \le 6 \\ y \ge 2 \end{cases}$

3. Maximize $G = 2x + 7y - 4$, subject to $\begin{cases} x - y \le 1 \\ 2x + 3y \le 12 \\ x \ge 0 \\ y \ge 0 \end{cases}$

4. Minimize $F = 4x - 3y + 7$, subject to $\begin{cases} y - 2x \le 0 \\ x - 2y \le 0 \\ 2 \le y \le 4 \end{cases}$

5. Minimize $H = 3x + 5y$, subject to $\begin{cases} x \leq 2y + 2 \\ x \geq 6 - 2y \\ y \leq x \\ x \leq 6 \end{cases}$

6. Maximize $P = 2x + 3y$, subject to $\begin{cases} y \leq 3x - 2 \\ 3x + 4y \leq 22 \\ y \geq 1 \end{cases}$

7. Minimize $C = 4x - y + 5$, subject to $\begin{cases} x + y \leq 5 \\ x + 2y \geq 6 \\ x \geq 0 \\ y \geq 0 \end{cases}$

8. Maximize $P = 10x + 6y$, subject to $\begin{cases} x + 3y \leq 18 \\ x + y \leq 8 \\ 2x + y \leq 14 \\ x \geq 0 \\ y \geq 0 \end{cases}$

9. A grocer expects to sell between 40 and 60 cases of canned apple sauce during a 6 month period. He knows from experience that he can sell at least 20 cases of brand A and at most 30 cases of brand B apple sauce in this time period. His profit on a case of brand A is \$3.00 and on a case of brand B is \$3.25. Assuming he sells all the apple sauce, how much should he stock for maximum profit?

10. An oil refinery needs to produce at least 100,000 gallons of unleaded gasoline and 150,000 gallons of regular leaded gasoline each month to meet demand. In order to use the plant capacity efficiently at least 300,000 gallons of the two types combined must be produced each month, but no more than 400,000 gallons can be produced. The cost of producing unleaded gasoline is \$300 per 1,000 gallons and for regular leaded gasoline it is \$250 per 1,000 gallons. How many gallons of each should be produced to minimize cost?

11. A certain machine can produce two types of products, A and B. It can produce 100 type A products per hour and 150 type B products per hour. The machine can run between 6 and 12 hours per day. It is to be used at most 8 hours per day to produce product A and at least 1 but no more than 5 hours per day in the production of B. If the profit on each A item is \$0.05 and on each B item \$0.04, how many hours per day should the machine be used in the production of each item in order to maximize profit?

12. A woman has \$20,000 which she wishes to divide between two investments, one a certificate of deposit paying $8\frac{1}{2}\%$ annual interest and requiring a minimum of \$5,000, and the other a utility stock paying 10% annual interest. Because the latter involves greater risk, she wants to put no more than half as much in it as in the certificate of deposit, but she wants at least 100 shares, and it costs \$20 per share. Assuming the stock price does not fluctuate, how should she invest the money for maximum income?

13. Two hardware store owners wish to buy paint of a certain kind from a wholesaler who has a stock of 1,000 gallons of the paint. The first owner will buy between 200 and 500 gallons, and the second will buy between 300 and 600 gallons. The wholesaler makes a profit of \$1.10 on each gallon, but this is reduced by the cost of delivery. He estimates the cost at 5¢ per gallon for delivery to the first hardware store and 10¢ per gallon to the second. How many gallons should he sell to each owner to maximize his profits?

14. A dairy farmer plans to purchase two types of feed for his 100 cows. Type A contains 10% protein and 30% carbohydrates (by weight), and type B contains 30% protein

and 40% carbohydrates. He wants to provide each cow with a minimum of 1 pound of protein and 2 pounds of carbohydrates. If type A costs 30¢ per pound and type B costs 50¢ per pound, how many pounds of each type should he buy to minimize the cost? What is the minimum cost?

15. A machine tool company produces two types of parts that require time on a drill press and on a lathe. Part A requires 5 minutes on the drill press and 10 minutes on the lathe. Part B requires 6 minutes on the drill press and 4 minutes on the lathe. The time per day available on the drill press for producing these parts is 1 hour and 40 minutes, and the time available on the lathe is 2 hours. To meet demand, at least 5 parts of each type should be produced each day. If the profit on part A is $3 and on part B is $2, how many parts per day of each type should be made for maximum profit?

16. A baker has orders for 15 angel food cakes and 20 chocolate cakes. Each angel food cake uses 12 eggs and 1 cup of sugar, whereas each chocolate cake uses 3 eggs and 2 cups of sugar. He has available for these two items 25 dozen eggs and 36 pounds of sugar (each pound equals $2\frac{1}{4}$ cups). His profit on each angel food cake is $0.50 and on each chocolate cake is $0.75. How many cakes of each type should he make in order to meet the demand and maximize profit (assuming he can sell all the cakes he makes)?

17. A manufacturer of washing machines has two plants where three models are produced: standard, superior, and luxury. There are orders for 600 standard, 340 superior, and 200 luxury models. Plant A can produce 10 standard, 8 superior, and 3 luxury models per day, whereas plant B can produce 20 standard, 9 superior, and 8 luxury models per day. It costs $500 per day to operate plant A and $800 per day to operate plant B. How many days should each plant be used in the production of these machines to meet demand and minimize cost?

B   18. Consider the objective function $P = 5x + 2y$ and the feasibility set defined by

$$\begin{cases} x + 3y \geq 10 \\ 3x + 2y \leq 34 \\ 2x - 5y + 28 \geq 0 \\ x \geq 1 \\ y \geq 2 \end{cases}$$

Graph the feasibility set and show several members of the family obtained by setting the objective function equal to a constant, $5x + 2y = C$, for increasing values of C (if P represents profit, these are called **isoprofit** lines). Explain from what you observe why the objective function must assume both its maximum and minimum values at a vertex of the graph of the feasibility set. At which vertex do each of the extreme values of P occur and what are the maximum and minimum values of P?

19. A dairy farmer has determined that each of his cows should get three types of food supplements, A, B, and C, each day. The minimum requirements are 10 units of A, 14 units of B, and 16 units of C. There are two kinds of feed that contain all three supplements. Feed number 1 contains 2 units each of A, B, and C per pound. Feed number 2 contains 2 units of A, 3 units of B, and 4 units of C per pound. Feed number 1 costs 10¢ per pound and feed number 2 costs 12¢ per pound. How many pounds of each type of feed should the farmer purchase to meet the daily need per cow in order to minimize cost?

20. At a furniture factory, desks and tables each go through four stations where different aspects of fabrication take place. The number of work-hours required at each station, as well as the number of work-hours available each day, are shown in the table. The profit on each item also is shown. How many desks and how many tables should be produced each day to maximize profit?

	Work-hours needed				
	Station 1	Station 2	Station 3	Station 4	Profit
Desk	1	1	2	3	$80
Table	3	1	1	2	$70
Available work-hours	9	7	10	18	

21. A woman owns two orchard supply stores. She has orders from two customers for a certain type of insecticide. Customer A wants 350 gallons and customer B wants 400 gallons. The store owner has 500 gallons of this insecticide on hand at store number 1 and 600 gallons at store number 2. The cost of delivery is estimated as shown in the table. How many gallons should be sent to each customer from store 1 and from store 2 to minimize the shipping cost?

From store	To customer	Delivery cost per gallon
1	A	$0.10
2	A	$0.12
1	B	$0.11
2	B	$0.15

Review Exercise Set

A In Problems 1–7 solve the systems using the method of elimination by addition or subtraction.

1. $\begin{cases} 3x - 2y = 5 \\ x + 4y = 11 \end{cases}$

2. $\begin{cases} 5x + 3y = 6 \\ 2x - 4y = 5 \end{cases}$

3. $\begin{cases} 4x + 7y = 3 \\ 5x + 8y = 0 \end{cases}$

4. $\begin{cases} x - 2y + 4z = 1 \\ 3x + y + z = 2 \\ 2x + y - z = -1 \end{cases}$

5. $\begin{cases} x - 2y + z = 2 \\ 2x - 3y - z = 0 \\ x + y - 4z = 2 \end{cases}$

6. $\begin{cases} 2x - 3y - 5z = -1 \\ x - y - 2z = 1 \\ -4x + 5y - 6z = 14 \end{cases}$

7. $\begin{cases} x - y = 1 \\ y + z = 1 \\ x - 2w = 0 \\ x + y + 3z + w = 2 \end{cases}$

Solve Problems 8–19 by reducing the augmented matrix to triangular form.

8. $\begin{cases} x - 2y + 4z = -3 \\ 3x + y - 2z = 12 \\ 2x + y - 3z = 7 \end{cases}$

9. $\begin{cases} 2x + y - z = 1 \\ x + y + z = -3 \\ -5x - 2y + 3z = -1 \end{cases}$

10.
$$\begin{cases} x + 2y + 3z = 1 \\ 2x - y - z = 5 \\ -x + y - 2z = 10 \end{cases}$$

11.
$$\begin{cases} 2x + 4y - 3z = 8 \\ 3x + 2y - z = 8 \\ 5x - 2y + 4z = 0 \end{cases}$$

12.
$$\begin{cases} 4x + 2y - z = 0 \\ 3x - 5y - 7z = 5 \\ 7x + 3y - 5z = -1 \end{cases}$$

13.
$$\begin{cases} 2x - y + z = 3 \\ x + 3y - 2z = 4 \\ 4x - 9y + 7z = 1 \end{cases}$$

14.
$$\begin{cases} 3x - 2y + 4z = 6 \\ 5x - 7y - 2z = 3 \\ 7x - 23y - 34z = -20 \end{cases}$$

15.
$$\begin{cases} x - y + z = 0 \\ y - z + 3w = 1 \\ x + z + 2w = 0 \\ x + y + 3z + w = 1 \end{cases}$$

16.
$$\begin{cases} x + y - 2z + w = 2 \\ 3x + 2y - 4z - 2w = -19 \\ 2x - y + z + 3w = 7 \\ -x - 3y + 4z + w = 4 \end{cases}$$

17.
$$\begin{cases} x + y - z + 3w = 1 \\ 2x - y + 3z - 4w = -13 \\ 3x + 2y + z + w = 5 \end{cases}$$

18.
$$\begin{cases} x - 2y - z - w = -4 \\ 3x + y + 2z + w = 3 \\ 2x - 3y - z + 2w = 7 \\ x - 3y - 2z - 5w = -19 \end{cases}$$

19.
$$\begin{cases} 2x + y + 3z = 1 \\ x - 2y - z = 3 \\ 4x - 5y + 6z = 11 \\ x - y + z = 2 \end{cases}$$

Evaluate the determinants in Problems 20–22.

20. a. $\begin{vmatrix} 2 & -1 & 1 \\ 1 & 3 & 4 \\ -2 & 1 & -5 \end{vmatrix}$ **b.** $\begin{vmatrix} 4 & 1 & -2 \\ 0 & 2 & -3 \\ 5 & -3 & 6 \end{vmatrix}$

21. a. $\begin{vmatrix} 3 & -5 & 4 \\ 2 & -3 & 6 \\ -1 & 2 & 3 \end{vmatrix}$ **b.** $\begin{vmatrix} -2 & 7 & 3 \\ 3 & -8 & 4 \\ 6 & -10 & -5 \end{vmatrix}$

22. a. $\begin{vmatrix} 4 & 3 & -5 \\ 7 & 8 & 2 \\ -3 & 2 & -6 \end{vmatrix}$ **b.** $\begin{vmatrix} 2 & -1 & 0 & 3 \\ 0 & 5 & -1 & 2 \\ 1 & 6 & 3 & -2 \\ -3 & 4 & 0 & -1 \end{vmatrix}$

23. Solve for x:

$$\begin{vmatrix} 1 & 0 & x \\ 2 & -x & 1 \\ 3 & x+1 & 2 \end{vmatrix} = 3$$

Find the solution to the systems in Problems 24–28 by Cramer's rule.

24. a. $\begin{cases} 5x - 11y = 6 \\ 7x - 8y = 3 \end{cases}$ **b.** $\begin{cases} 3x + 7y + 8 = 0 \\ 2x - 3y - 4 = 0 \end{cases}$

25. a. $\begin{cases} 9x - 5y = 11 \\ -6x + 3y = 7 \end{cases}$ **b.** $\begin{cases} 4x + 3y - 5 = 0 \\ 8x - 5y + 2 = 0 \end{cases}$

26.
$$\begin{cases} 2x + 4y + 3z = 1 \\ 3x - 5y - 2z = 5 \\ 4x - 6y - 8z = -5 \end{cases}$$

27.
$$\begin{cases} 3x - 5y - z = 4 \\ 2x + 4y + 3z = -1 \\ 2x + y + z = 0 \end{cases}$$

28.
$$\begin{cases} x + 3y + z = -2 \\ 4x - 2y - 5z = -1 \\ 5x + 7y + 3z = -4 \end{cases}$$

29. Two drivers are initially 120 miles apart, and they drive toward each other. One averages 50 miles per hour and the other 40 miles per hour. Find how far each has driven when they meet.

30. A man in a motorboat makes a trip of 12 miles upstream in 1 hour. The return trip with the current takes 40 minutes. If he held the throttle wide open all the time, find how fast he would have gone in still water and the rate of the current.

31. Thirty cubic centimeters of a 25% sulfuric acid solution are obtained by mixing a 40% sulfuric acid solution with a 15% solution. How much of each was needed?

32. The total income from a community concert was $2,325. Admission for adults was $3.00 and for children $1.50. If 900 persons attended, how many were adults and how many were children?

33. The length of a certain rectangle is 1 more than twice the width, and the perimeter is 32. Find the length and the width.

34. Find how much nickel and how much zinc should be added to 200 kilograms of an alloy which is 15% nickel and 25% zinc in order to obtain an alloy which is 25% nickel and 30% zinc?

35. Fifty gallons of a 25% salt solution are obtained by mixing pure water with a 40% salt solution. Find how much pure water and how much of the 40% salt solution were used.

36. The sum of the digits of a two-digit number is 11. If the digits were reversed, the number would be decreased by 45. Find the number.

37. The income from an amount invested at simple interest for 1 year is $108. If the same amount were invested in an account yielding $1\frac{1}{2}\%$ more interest, the income for the year would be $135. Find the interest rate and the amount invested.

38. Three men working together can do a certain job in 2 hours. Mr. Jones and Mr. Smith can do the job in 3 hours, and Mr. Smith and Mr. Robinson together can do it in 4 hours. How long would it take each man working alone?

39. Two pumps working simultaneously can fill a tank in $2\frac{1}{2}$ hours. Both pumps are started at the same time, but the larger one breaks down after $1\frac{1}{2}$ hours, and it takes the smaller one an additional $2\frac{1}{2}$ hours to fill the tank. Find how long it would take each pump alone to fill the tank.

Solve the systems of equations in Problems 40–48.

40. $\begin{cases} y = x^2 - 2x \\ x + 2y = 5 \end{cases}$

41. $\begin{cases} xy + 2 = 0 \\ 2x + 3y = 4 \end{cases}$

42. $\begin{cases} x^2 - 2y^2 = 4 \\ 3x^2 + 4y^2 = 12 \end{cases}$

43. $\begin{cases} y^2 - 7x^2 = 36 \\ 3x + y = 2 \end{cases}$

44. $\begin{cases} 3y^2 - x^2 = 26 \\ x + y^2 = 12 \end{cases}$

45. $\begin{cases} x^2 - y^2 - 2x = 0 \\ 2x - 3y = 4 \end{cases}$

46. $\begin{cases} 8xy - 6x - 5y = 3 \\ 10x + y = 1 \end{cases}$

47. $\begin{cases} \dfrac{1}{x} + \dfrac{2}{y} = 4 \\ 3x + 5y = 6 \end{cases}$

48. $\begin{cases} x^2 + 3xy - 2y^2 - 8 = 0 \\ 2x - y = 3 \end{cases}$

49. Find two numbers whose difference is 8 and the sum of whose squares is 274.

50. Find two numbers whose difference is 13 and the difference of whose squares is 65.

51. The diagonal of a rectangle is 25 inches and its perimeter is 62 inches. Find the length and width.

52. The length of a certain rectangle is 4 less than 3 times the width, and the area is 160. Find the length and the width.

Show the graphic solutions of the inequalities in Problems 53–56.

53. a. $3x - 5y < 8$ **b.** $y \geq x^2 - 3$

54. a. $\begin{cases} x + 2y \leq 8 \\ 2y - x > 3 \end{cases}$ **b.** $\begin{cases} y + x^2 \leq 4 \\ 2x + y \geq 0 \end{cases}$

55. a. $\begin{cases} 2y - x < 8 \\ 2x - y < 5 \\ x + y > 1 \end{cases}$ **b.** $\begin{cases} y - 2x \leq 3 \\ x + 2y \geq 2 \end{cases}$

56. a. $\begin{cases} -6 \leq x - 3y \leq 6 \\ |y - 2| \leq 1 \end{cases}$ **b.** $|x + y| > 2$

In Problems 57–60 find the maximum or minimum value, as specified, subject to the given constraints.

57. Maximize $F = 10x + 5y$, subject to $\begin{cases} 2x + 3y \leq 12 \\ x + y \geq 2 \\ x \geq 1 \\ y \geq 0 \end{cases}$

58. Minimize $C = 2x + 3y + 5$, subject to $\begin{cases} 2x + y \geq 4 \\ x \leq 8 \\ 1 \leq y \leq 4 \end{cases}$

59. Minimize $G = 3x - y + 7$, subject to $\begin{cases} 3x + 2y \geq 12 \\ x + 2y \geq 5 \\ x + y \leq 8 \\ x \leq 6 \\ y \leq 4 \end{cases}$

60. Maximize $P = 60x + 100y$, subject to $\begin{cases} x + 2y \leq 14 \\ 6 \leq x + y \leq 10 \\ y \geq 2 \\ x \geq 0 \end{cases}$

61. A farmer has 300 acres which he will divide between corn and oats. In order to meet commitments he has made he must plant at least 150 acres of corn and 50 acres of oats, and the demand for each suggests that he should plant at least twice as many acres in corn as in oats. If the profit from corn is $40 per acre and from oats is $30 per acre, how many acres should he plant of each for maximum profit?

62. A company manufactures two types of rotary lawn mowers, a standard model and a deluxe model. At most 100 lawn mowers can be produced each week, and no more than 40 of these can be the deluxe model. To meet demand, at least 40 standard and between 20 and 40 deluxe models should be produced each week. If the profit on the standard model is $50 and on the deluxe model $70, how many of each type should be produced each week for maximum profit?

63. The owner of an old-fashioned excursion train offers a short trip at $1.50 per person and a long trip at $5.00 per person. Each long trip he makes per day reduces by 3 the number of short trips he can make. If he makes no long trips, he can make 12 short trips, and if he makes no short trips, he can make 4 long trips. He agrees to offer at least 3 short trips and 1 long trip per day. The train holds 200 passengers, and it is 80% full on the average for short trips and 60% full for long trips. How many trips of each type should he make per day to maximize his income?

B   Solve the systems in Problems 64–69 by reducing the augmented matrix to triangular form.

64.
$$\begin{cases} 3x + 4y - 2z + w = 25 \\ x + 2y - z - w = -3 \\ 2x - y - 3z + 4w = 7 \\ -x - 3y + 5z - 2w = -7 \end{cases}$$

65.
$$\begin{cases} x - y + 2z - w = 4 \\ x + 2y - z = 5 \\ 2x - 3y + z - 4w = 7 \\ x + y - 4z - 2w = 4 \end{cases}$$

66.
$$\begin{cases} 2x - y + 5z - 4w = 0 \\ 3x - 2y - 3z + 2w = 2 \\ -5x + 3y - 4z + 6w = 10 \\ 4x + y + 2z + 3w = -1 \end{cases}$$

67.
$$\begin{cases} 2x - y + 4z - w = 0 \\ -3x + 2y + 3z - 2w = 0 \\ 4x - 3y - 10z + 5w = 0 \\ 5x + 2y + 7z + 8w = 0 \end{cases}$$

68.
$$\begin{cases} 4x - 3y + 2z - w = 0 \\ 2x - y + z - 4w = 0 \\ -5x + 6y - 8z + w = 0 \\ 7x + 2y - 5z - 5w = 0 \end{cases}$$

69.
$$\begin{cases} x + y - u = 1 \\ 2x - y + 2v - w = 3 \\ u + v + w = -1 \\ 3y - u + v - 2w = 2 \\ x + 2u + 5w = 5 \end{cases}$$

70.   Solve only for y by Cramer's rule:
$$\begin{cases} 2x - 3y = 3 \\ 4x + z = 5 \\ 2y - z + w = 0 \\ x + w = 4 \end{cases}$$

71.   Solve for u and v by Cramer's rule:
$$\begin{cases} (a^2 - 1)u - 3av = 3 \\ au - 4v = 2 \end{cases}$$

72.   Solve for x, y, and z by Cramer's rule:
$$\begin{cases} nx - my + mnz = 0 \\ x + mny - 2mz = 3n \\ 2mnx - y + 3nz = 4m \end{cases}$$

73.   The points $(-1, 0)$, $(2, -3)$, and $(3, 4)$ lie on a curve whose equation is of the form $y = ax^2 + bx + c$. Find a, b, and c.
    Hint.   Each point must satisfy the equation.

74.   The points $(5, 1)$, $(-2, 0)$, and $(2, 2)$ lie on a circle whose equation is of the form $x^2 + y^2 + ax + by + c = 0$. Find a, b, and c.

In Problems 75 and 76 find the solution to the system.

75.
$$\begin{cases} x^2 + y^2 - 2x - 5y + 1 = 0 \\ x^2 + y^2 + 3x - 8y - 2 = 0 \end{cases}$$

76.
$$\begin{cases} x^2 - 16y^2 = 5x - 20y \\ xy = 1 \end{cases}$$

77.   A 40% salt solution is mixed with a 25% solution, resulting in a 30% solution. Then 4 gallons of fresh water are added to this mixture, and the final solution is 20% salt. Find how many gallons of each original solution there were.

78.   A man rows 8 miles upstream and back in 3 hours. The next day the current is twice as strong, and he makes the same trip upstream and back in 4 hours and 48 minutes. Assuming no change in his rowing effort from one day to the next, find how fast he can row in still water and the rate of the current the first day.

79.   Find all points of intersection of the curves whose equations are $5x^2 - 2xy + y^2 - 12x - 16 = 0$ and $x^2 - y = 2$.

80. An isosceles triangle is inscribed in a circle of radius a. The two equal sides each have length $3a/2$. Find the length of the base and the altitude of the triangle in terms of a.

81. A girl rides a bicycle a distance of 5 miles and then has a flat tire. She has to walk 2 miles to a service station. If the total elapsed time for riding and walking was 1 hour and if she rode at an average speed of 12 miles per hour faster than she walked, find her rate of riding and her rate of walking.

82. Two open-top boxes each have square bases. The base of the larger box is 2 inches larger on each side than that of the smaller, and the height of the larger box is 1 inch greater than that of the smaller box. If the volume of the larger box exceeds that of the smaller by 320 cubic inches, and the total surface area of the larger box is 124 square inches greater than that of the smaller, find the dimensions of the smaller box. (There are two solutions.)

83. A company produces two kinds of fertilizer. The first kind, for lawns, contains 20% nitrogen, 5% phosphorous, and 5% potash. The second kind, for gardens, contains 8% nitrogen, 10% phosphorous, and 8% potash. Each kind is packaged in 100 pound bags. The company has 1960 pounds of nitrogen, 850 pounds of phosphorous, and 730 pounds of potash. The owner wants to have at least 30 bags of each type of fertilizer available. His profit on the lawn fertilizer is $6 per bag and on the garden fertilizer $5 per bag. How many bags of each type should he produce to maximize his profit?

11 Further Aids to Graphing

The ability to analyze and sketch the graphs of a great variety of equations is expected of students in calculus. Techniques studied in this chapter form the basis for such an analysis; additional aids to graphing will be studied in calculus itself. The following example employs both precalculus and calculus concepts, but we have omitted the part that requires calculus:*

$$y^2(x^2 - x) = x^2 + 1$$

We solve for y, to obtain

$$y = \pm \sqrt{\frac{x^2 + 1}{x(x - 1)}}.$$

The expression under the radical must not be negative, so no portion of the curves lies between the lines $x = 0$ and $x = 1$. But all values of $x > 1$ and all negative values of x, that is, $x < 0$, are permissible . . . .

We now have quite a bit of information about the curve . . . . It is immediately evident that the curve is symmetric about the x-axis, because y may be replaced by $-y$ without changing the equation. There is no curve between $x = 0$ and $x = 1$. The lines $x = 0$, $x = 1$, $y = 1$, and $y = -1$ are asymptotes of the curve. It crosses $y = 1$ and $y = -1$ at $x = -1$. There is no x-intercept . . . .

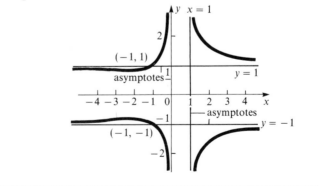

You will learn in this chapter how to test for symmetry, intercepts, asymptotes, and excluded regions, all of which are involved in this example.

---

* George B. Thomas, Jr., and Ross L. Finney, *Calculus and Analytic Geometry*, 5th ed. (Reading, Mass.: Addison-Wesley, 1979), pp. 390–392. Reprinted by permission.

417

1 Introduction

We have seen how to graph straight lines, circles, certain parabolas, ellipses, and hyperbolas; exponential and logarithmic functions; and the trigonometric functions and their inverses. The ability to recognize and sketch these basic curves rapidly is important in calculus. Beyond this, however, you will need to be able to graph a host of other curves about which you have no prior knowledge. To rely solely on plotting points is not only inefficient, it often is simply not enough to enable you to draw the curve correctly. This chapter is devoted to a study of ways of analyzing equations, which when coupled with the plotting of a small number of points, will enable you to graph a great variety of equations.

2 Intercepts and Symmetry

The **x and y intercepts** of a curve can be quite helpful in drawing the graph, and they are often relatively easy to find. The x intercepts are found by setting $y = 0$ and solving for x, and the y intercepts are found by setting $x = 0$ and solving for y. It is not always feasible to do this, but when it can be done, it should be.

EXAMPLE 1   Find the x and y intercepts of the graph of the equation: $y^2 = \dfrac{2x^2 - 3x - 9}{(x - 1)^3}$

Solution   We set $y = 0$ and solve for x to find the x intercepts:

$$0 = \frac{2x^2 - 3x - 9}{(x - 1)^3}$$

$$2x^2 - 3x - 9 = 0$$

$$(2x + 3)(x - 3) = 0$$

$$x = -\tfrac{3}{2} \quad \Big| \quad x = 3$$

The x intercepts are 3 and $-\tfrac{3}{2}$.
If we put $x = 0$, we get $y^2 = 9$. So the y intercepts are ± 3.

EXAMPLE 2   Discuss the intercepts of: $x^2 - 2y^5 + 3xy - 4y^3 + y - 4 = 0$

Solution   The x intercepts are ± 2, found by setting $y = 0$ and solving the resulting equation, $x^2 - 4 = 0$.
If x is put equal to 0, the resulting equation is

$$-2y^5 - 4y^3 + y - 4 = 0$$

and the solutions are not readily available. We could try the technique of Chapter 6, but we would quickly discover that there are no rational roots. So in this problem finding the y intercepts evidently involves more work than it is worth for the purpose of graphing the equation.

Determining whether a curve has certain types of symmetry can be another valuable aid in drawing its graph. We will consider two types of symmetry: **symmetry with respect to a line, and symmetry with respect to a point.**

DEFINITION 1   A curve is said to be **symmetric with respect to a line** l if for each point P on the curve there is another point P_1 also on the curve so that the line l is the perpendicular bisector of the line segment PP_1 (Figure 1).

Figure 1

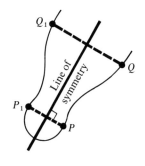

When the line of symmetry is the x axis, the definition implies that for each point (x, y) on the curve, the point $(x, -y)$ is also on the curve. Similarly, a curve is symmetric with respect to the y axis if whenever (x, y) is on the curve, $(-x, y)$ is also on the curve. This is illustrated in Figure 2. The x axis and the y axis are the primary lines of symmetry we will consider.

Figure 2

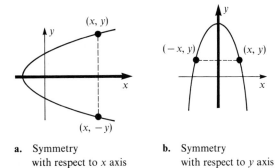

a. Symmetry
with respect to x axis

b. Symmetry
with respect to y axis

DEFINITION 2   A curve is said to be **symmetric with respect to a point** Q if for each point P on the curve there is another point P_1 also on the curve such that Q is the midpoint of the line segment PP_1.

We will be concerned with **symmetry with respect to the origin**, and in this case the definition requires that whenever (x, y) is on the curve, $(-x, -y)$ is also on the curve (Figure 3).

Figure 3

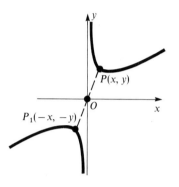

In terms of the equation of the curve the tests for symmetry with respect to the two axes and to the origin can be stated as follows:

Tests for Symmetry

1. A curve is symmetric with respect to the x axis if replacing y by $-y$ in its equation leaves the equation unchanged.
2. A curve is symmetric with respect to the y axis if replacing x by $-x$ in its equation leaves the equation unchanged.
3. A curve is symmetric with respect to the origin if replacing x and y by $-x$ and $-y$, respectively, leaves the equation unchanged.

For example, the graph of $y^2 = 2x$ is symmetric to the x axis, since replacing y by $-y$ produces no change. Thus, whenever a point (x, y) satisfies the equation, so does the point $(x, -y)$. In general, if y occurs only to even powers, then replacing y by $-y$ causes no change, so the curve is symmetric to the x axis. Similarly, if x appears to even powers only, the curve is symmetric to the y axis. These are sufficient conditions only, however; the graph of $y = \cos x$, for example, is symmetric with respect to the y axis, since $\cos(-x) = \cos x$.

Some examples of curves symmetric with respect to the origin are those whose equations are

$$y = x^3 - 2x$$

$$xy = 4$$

$$y = \sin x$$

$$2x^2y - 3y^3 - x^5 = 0$$

$$y = \frac{x^2 - 1}{2x}$$

since in each case replacement of both x and y by their negatives results in an equivalent equation, so that whenever (x, y) is on the graph, so is $(-x, -y)$.

EXAMPLE 3   Test for symmetry with respect to the x axis, y axis, and origin:

a. $y^2 = \dfrac{x - 1}{x + 2}$ **b.** $xy^2 - 4x^3 + 3x^2y^5 = 0$ **c.** $y = \dfrac{\sqrt{4 - x^2}}{x^4}$

d. $9x^2 - 4y^2 = 36$ **e.** $y = \dfrac{2x^2 + 1}{x - 2}$

Solution **a.** If y is replaced by $-y$, the equation is unchanged. So the curve is symmetric to the x axis. If x is replaced by $-x$, the resulting equation is not equivalent to the original. It follows that the curve is not symmetric to the y axis or the origin.
  b. If *either x or y* is replaced by its negative, the equation is altered, but if *both x and y* are replaced by their negatives, the result is

$$(-x)(-y)^2 - 4(-x)^3 + 3(-x)^2(-y)^5 = 0$$
$$-xy^2 + 4x^3 - 3x^2y^5 = 0$$

which, on multiplying both sides by -1, becomes

$$xy^2 - 4x^3 + 3x^2y^5 = 0$$

This is exactly the same as the original. Therefore, the curve is symmetric only with respect to the origin. A quicker way to see this is to observe that each term of the polynomial has *odd* degree (remember the degree of a term is the sum of the exponents of the variables in that term), so if the signs of x and y are both changed, *every* term will have its sign changed, and since the right-hand side is 0, this yields an equivalent equation.
  c. Since x appears to even powers only, replacing x by $-x$ leaves the equation unchanged. So the curve is symmetric to the y axis. The equation is altered if y is replaced by $-y$, so there is no other symmetry.
  d. There is no change if either x or y is replaced by its negative, so the curve is symmetric to the x axis, y axis, and origin.
  e. If either x or y is replaced by its negative, the equation is changed. If both x and y are replaced by their negatives, the equation again is changed. So there is no symmetry to either axis or to the origin.

Remark.   If a curve has any two of the three types of symmetry (x axis, y axis, origin), it will also have the third. (Why?) So if you find that a curve has one type of symmetry, but not a second, then it will not have the third type either.

When attempting to find the graph of an equation, it is helpful to discover in advance if it possesses some sort of symmetry. If it is known that a curve is symmetric, say, with respect to the x axis, it is necessary only to determine its shape above the x axis; then below that axis the curve is the mirror image of the part above.

EXERCISE SET 2

A   In Problems 1–12 find the x intercepts and the y intercepts.

1. $y = \dfrac{x - 2}{x + 1}$

2. $y = \dfrac{4 - x^2}{x + 3}$

3. $y = 2x^2 - 3x - 5$

4. $y = 15 - x - 6x^2$

5. $y = \dfrac{x^2 - 2x + 1}{x^2 - 4}$

6. $y = \dfrac{3x^2 - 4x + 1}{2x^2 + 5x - 3}$

7. $y = \sqrt{\dfrac{x - 2}{x - 8}}$

8. $y = \dfrac{\sqrt{4 - x^2}}{x + 1}$

9. $y^2 = \dfrac{8x - 3}{2x + 5}$

10. $y^2 = \dfrac{x^2 - 9}{x^2 - 1}$

11. $x^2 - 2xy + 3y^2 - x = 12$

12. $x^2y - 2y^2 + 3x - 4y + 6 = 0$

In Problems 13–29 test for symmetry with respect to both axes and the origin.

13. **a.** $y = x^2 - 1$ **b.** $y = x^3 - 2x$

14. **a.** $y^2 = x + 2$ **b.** $2x^2 + 3y^2 = 4$

15. **a.** $y = \sqrt{x^2 + 4}$ **b.** $x^2 = y^3$

16. **a.** $y^2 - x^2 = 4$ **b.** $x = \sqrt{1 + y^2}$

17. **a.** $xy = 3$ **b.** $y = x + \dfrac{1}{x}$

18. **a.** $y = \dfrac{x^2}{x^4 - 16}$ **b.** $y = \dfrac{2x}{x^2 + 4}$

19. **a.** $3x^2 - 4y^2 = 7$ **b.** $y = \dfrac{x^2 - 1}{x}$

20. **a.** $y = |x|$ **b.** $y = x(x^2 - 1)$

21. **a.** $x^2 + 2xy + 3y^2 = 4$ **b.** $y = x^5 - 2x^3$

22. **a.** $y = \sin x \cos x$ **b.** $y = \dfrac{\cos x - 1}{x}$

23. **a.** $y^2 = 1 - x$ **b.** $y = \pm\sqrt{4 - x^2}$

24. **a.** $y = \sin x$ **b.** $y = \cos x$

25. **a.** $y = \tan x$ **b.** $y = e^{x^2}$

26. **a.** $y = \log_e |x|$ **b.** $|x| + |y| = 1$

27. **a.** $y = \dfrac{\sin x}{x}$ **b.** $y = \dfrac{x(x - 1)}{x + 2}$

28. **a.** $y = 1 - e^{-|x|}$ **b.** $y = \tan^{-1} x$

29. **a.** $y = \sin^{-1} x$ **b.** $y = \dfrac{1 - \cos x}{x^2}$

B In Problems 30–33 find the x intercepts and the y intercepts.

30. $y = \dfrac{x^3 + 4x^2 + x - 6}{3x^2 - 10x + 8}$

31. $y = \dfrac{3x^3 - 7x^2 + 4}{x^4 - 2x^3 + 4x - 5}$

32. $y = \sqrt{\dfrac{x^4 - 6x^2 - 8x - 3}{x^3 - 1}}$

33. $x^2 - 3xy^2 + y^3 - 4x - 2y^2 + 3 = 0$

In Problems 34 and 35 test for symmetry with respect to both axes and the origin.

34. **a.** $x^{2/3} + y^{2/3} = 1$ **b.** $y = \ln \dfrac{1 - x}{1 + x}$ $(|x| < 1)$

35. **a.** $y = \dfrac{e^x - e^{-x}}{2}$ **b.** $y = x^2 \sin \dfrac{1}{x}$ $(x \neq 0)$

36. Show that a curve is symmetric with respect to the line $x = y$ if when x and y are interchanged in its equation, the equation is not altered.

37. Use the result of Problem 36 to test each of the following for symmetry with respect to the line $x = y$:
  a. $x^3 + y^3 - 2xy = 4$ **b.** $3xy = -4$ **c.** $xy - x = y$

  d. $y = \dfrac{x}{x - 1}$ **e.** $y = \dfrac{7 - 4x}{3x + 4}$

3 Asymptotes

In Chapter 6 we discussed vertical and horizontal asymptotes of rational functions, and we also have encountered asymptotes in connection with the hyperbola, certain trigonometric functions, exponential functions and logarithmic functions. All these share the common property that **the asymptote is a line such that as the curve recedes indefinitely, the distance between the curve and the line approaches 0.** In a sense, the curve and the line approach coincidence far out in the plane. We are not able to give a more precise definition without going more deeply into the theory of curves, but this description is sufficient for our purposes.

We are going to give a more detailed analysis now of vertical and horizontal asymptotes of rational functions and then extend the ideas to certain irrational functions and other curves. We begin with the rational function

$$y = \frac{P(x)}{Q(x)} \tag{1}$$

where $P(x)$ and $Q(x)$ are polynomials having no nonconstant factor in common and where $Q(x) \neq 0$. Suppose $x = a$ is a zero of the denominator. Then, we can factor $Q(x)$ as $(x - a)Q_1(x)$, so that

$$y = \frac{P(x)}{(x - a)Q_1(x)}$$

Now let x approach a. The numerator approaches the nonzero value $P(a)$ [**Why is $P(a) \neq 0$?**], and the denominator approaches 0. Therefore, y becomes arbitrarily large in absolute value. Whether y is positive or negative depends upon the signs of the factors $P(x)$ and $Q_1(x)$ in the neighborhood of $x = a$, and upon whether x approaches a from the right or from the left. It follows that the vertical line $x = a$ is an asymptote to the curve, since when the distance $|x - a|$ between a point (x, y) on the curve and the line $x = a$ becomes arbitrarily small, the ordinate y becomes arbitrarily large in absolute value, so that in the remote upper or lower (or both) parts of the plane the curve approaches coincidence with the line (Figures 4 and 5). The curve may be asymptotic to the line from above on one side and below on the other side, as in Figure 4, or it may be asymptotic in the same direction from both sides, as Figure 5. The first situation occurs when the factor $(x - a)$ appears to an odd power in the denominator, and the second when it appears to an even power.

Figure 4

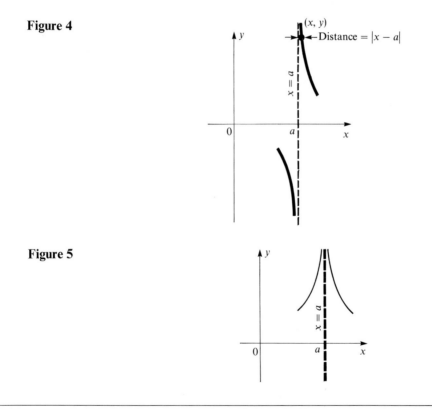

Figure 5

EXAMPLE 4 Find all vertical asymptotes to the curves defined by:

a. $y = \dfrac{2x + 3}{x - 2}$ b. $y = \dfrac{x^2 + 1}{x^2 - 4x + 4}$ c. $y = \dfrac{x^2 - 4}{x^2 - 9}$

Solution a. $x = 2$ is the only vertical asymptote. We note that when x approaches 2 from the right (such as 2.1, 2.01, 2.001, . . .), both the numerator and the denominator are positive, so y is positive. When x approaches 2 from the left (such as 1.9, 1.99, 1.999, . . .), the numerator is again positive, but the denominator is negative, so that y is negative. The situation near the asymptote therefore is similar to that shown in Figure 4.

b. Factoring the denominator gives

$$y = \frac{x^2 + 1}{(x - 2)^2}$$

So again $x = 2$ is the only vertical asymptote. This time whether x approaches 2 from the right or the left, both the numerator and the denominator are positive. So the situation resembles that in Figure 5.

c. We factor the denominator to get

$$y = \frac{x^2 - 4}{(x + 3)(x - 3)}$$

So $x = 3$ and $x = -3$ are vertical asymptotes. In the immediate vicinity of $x = 3$, y is negative when $x < 3$ and positive when $x > 3$, so the curve is asymptotic to $x = 3$ in a manner similar to that in Figure 4. In the immediate vicinity of $x = -3$, y is positive when $x < -3$ and negative when $x > -3$ (check this), so the situation is the reverse of that in Figure 4.

To determine whether there is a horizontal asymptote to the curve defined by equation (1), we permit x to become large (either positively or negatively) and observe what happens to y.

There are only three possibilities:

1. y approaches 0, in which case $y = 0$ is a horizontal asymptote
2. y approaches some number $b \neq 0$, in which case $y = b$ is a horizontal asymptote
3. y becomes arbitrarily large numerically, in which case there is no horizontal asymptote

That $y = 0$ and $y = b$ are horizontal asymptotes in cases 1 and 2 follows immediately from the definition. The three cases are illustrated by the following examples:

CASE 1

$$y = \frac{x}{x^2 - 4}$$

To determine what happens to y as x becomes arbitrarily large (we say "x approaches infinity"), it is a standard technique to divide the numerator and the denominator of the right-hand side by the highest power of x. In this example we get

$$y = \frac{\dfrac{1}{x}}{1 - \dfrac{4}{x^2}}$$

Now as x becomes arbitrarily large in absolute value, the terms $1/x$ and $4/x^2$ both approach 0. So y approaches $\frac{0}{1} = 0$. Thus, $y = 0$ (the x axis) is an asymptote. We might observe that since y approaches 0 as x gets large either positively or negatively, the curve is asymptotic to both the positive and negative ends of the x axis. There are times when curves approach their asymptotes from one end only. We will see this in later examples.

CASE 2

$$y = \frac{x^2}{x^2 - 4}$$

As before, we divide the numerator and the denominator by x^2:

$$y = \frac{1}{1 - \dfrac{4}{x^2}}$$

So as $|x|$ approaches infinity, y approaches 1. Thus, $y = 1$ is a horizontal asymptote.

CASE 3

$$y = \frac{x^3}{x^2 - 4}$$

Dividing the numerator and the denominator on the right by x^3 yields

$$y = \frac{1}{\dfrac{1}{x} - \dfrac{4}{x^3}}$$

Now as $|x|$ increases without limit, the denominator approaches 0; so y gets large numerically (the sign of y depends on the sign of the denominator, which in turn depends on whether x becomes large positively or negatively). So y does not approach a constant, and there is no horizontal asymptote.

You may already have observed a quick way to detect which of the three cases applies, and even how to see what the asymptote is if it exists. Everything depends on the highest powers in the numerator and the denominator. The degree of a polynomial is the degree of the highest power of the variable appearing in it. So we can say that we have case 1 if deg $P(x) <$ deg $Q(x)$ [read "the degree of $P(x)$ is less than the degree of $Q(x)$"], case 2 if deg $P(x) =$ deg $Q(x)$, and case 3 if deg $P(x) >$ deg $Q(x)$. Furthermore, when deg $P(x) =$ deg $Q(x)$, the horizontal asymptote is the line $y = b$, where b is the ratio of the highest-degree terms of $P(x)$ and $Q(x)$.

To verify this last statement in general, let us suppose that $P(x)$ and $Q(x)$ are both of degree n and that

$$P(x) = A_n x^n + A_{n-1} x^{n-1} + \cdots + A_1 x + A_0$$

$$Q(x) = B_n x^n + B_{n-1} x^{n-1} + \cdots + B_1 x + B_0$$

Then

$$y = \frac{P(x)}{Q(x)} = \frac{A_n x^n + A_{n-1} x^{n-1} + \cdots + A_1 x + A_0}{B_n x^n + B_{n-1} x^{n-1} + \cdots + B_1 x + B_0}$$

$$= \frac{A_n + \dfrac{A_{n-1}}{x} + \cdots + \dfrac{A_1}{x^{n-1}} + \dfrac{A_0}{x^n}}{B_n + \dfrac{B_{n-1}}{x} + \cdots + \dfrac{B_1}{x^{n-1}} + \dfrac{B_0}{x^n}}$$

where in the last expression A_n and B_n are the only terms without some power of x in the denominator. As in the examples, then, as x becomes infinite, y approaches A_n/B_n. Thus, $y = A_n/B_n$ is a horizontal asymptote.

So, in our example

$$y = \frac{x^2}{x^2 - 4}$$

we could have concluded immediately that since the numerator and the denominator are of the same degree, the line $y = 1$ is the horizontal asymptote.

It is instructive also to analyze case 3 for rational functions somewhat more carefully. Consider the example

$$y = \frac{x^3}{x^2 - 4}$$

We can rewrite this after long division:

$$
\begin{array}{r}
x \\
x^2 - 4 \overline{)x^3 } \\
\underline{x^3 - 4x} \\
4x
\end{array}
$$

The result is

$$y = x + \frac{4x}{x^2 - 4}$$

Now as x becomes arbitrarily large, the term $4x/(x^2 - 4)$ approaches 0, and the curve approaches the line $y = x$. This line is therefore an **inclined asymptote.** This situation occurs when the degree of the numerator is *exactly 1 more* than the degree of the denominator.

Here is another example:

$$y = \frac{x^2 - 2x + 1}{x - 4}$$

This time we can use synthetic division:

$$
\begin{array}{r|rrr}
4 & 1 & -2 & 1 \\
  &   & 4 & 8 \\
\hline
  & 1 & 2 & 9
\end{array}
$$

The result is

$$y = x + 2 + \frac{9}{x - 4}$$

This shows that the vertical distance between the curve and the line $y = x + 2$ is $9/(x - 4)$, and since this approaches 0 as x approaches infinity, we conclude that $y = x + 2$ is an asymptote.

The methods employed for finding vertical and horizontal asymptotes for a rational function extend to some irrational expressions and also expressions involving nonalgebraic functions. We illustrate with some examples.

EXAMPLE 5   Find all vertical and horizontal asymptotes: $y = \dfrac{x}{\sqrt{x^2 - 4}}$

Solution   The reasoning is exactly the same as before for vertical asymptotes. They are the lines $x = \pm 2$. For horizontal asymptotes, we proceed as follows:

$$\frac{x}{\sqrt{x^2 - 4}} = \frac{x}{\sqrt{x^2\left(1 - \dfrac{4}{x^2}\right)}} = \frac{x}{\sqrt{x^2}\sqrt{1 - \dfrac{4}{x^2}}} \qquad (x \neq 0)$$

$$= \frac{x}{|x|\sqrt{1 - \dfrac{4}{x^2}}}$$

If $x > 0$, then $|x| = x$, so that the fraction becomes

$$\frac{1}{\sqrt{1 - \dfrac{4}{x^2}}}$$

and this approaches 1 as x increases indefinitely. So $y = 1$ is a horizontal asymptote on the right. If $x < 0$, then $|x| = -x$, so that

$$y = \frac{-1}{\sqrt{1 - \dfrac{4}{x^2}}}$$

and as x increases indefinitely through negative values, y approaches -1. So $y = -1$ is a horizontal asymptote on the left.

Had the problem above been

$$y = \frac{x}{\sqrt{4 - x^2}}$$

there would be no horizontal asymptotes, since we cannot allow x to become arbitrarily large. In fact, the domain of y is restricted to those values of x satisfying $|x| < 2$.

EXAMPLE 6   Find all vertical and horizontal asymptotes: $y = \dfrac{x}{\sqrt{x^3 - 8}}$

Solution   The vertical asymptote is $x = 2$, since

$$x^3 - 8 = (x - 2)(x^2 + 2x + 4)$$

The factor $x^2 + 2x + 4$ has no real zeros, since $B^2 - 4AC = 4 - 16 < 0$. In seeking horizontal asymptotes, we observe first that the domain of y is limited to $x > 2$. Thus,

$$y = \frac{x}{\sqrt{x^3 - 8}} = \frac{x}{\sqrt{x^3\left(1 - \dfrac{8}{x^3}\right)}} = \frac{x}{\sqrt{x^3}\sqrt{1 - \dfrac{8}{x^3}}}$$

$$= \frac{x}{x^{3/2}\sqrt{1 - \dfrac{8}{x^3}}} = \frac{1}{x^{1/2}\sqrt{1 - \dfrac{8}{x^3}}}$$

which approaches 0 as x approaches infinity. So $y = 0$ is a horizontal asymptote.

EXAMPLE 7   Find all vertical and horizontal asymptotes: $y = \dfrac{\sin x}{x}$

Solution   We saw in Section 11 of Chapter 9 that as x approaches 0, $\sin x/x$ approaches 1. So $x = 0$ is not an asymptote. If we allow x to become infinite, the numerator oscillates between $+1$ and -1, while the denominator becomes arbitrarily large. So y approaches 0, and the curve has $y = 0$ as a horizontal asymptote. The graph is shown in Figure 6. This is called a **damped sine curve.** Note that the curve crosses its asymptote infinitely many times.

Figure 6

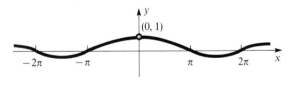

Asymptotes, when they exist, are among the most valuable aids in sketching curves.

EXERCISE SET 3

Find all vertical, horizontal, and inclined asymptotes.

A   **1.** $y = \dfrac{1}{x - 2}$ **2.** $y = \dfrac{x}{x - 2}$

3. $y = \dfrac{3}{4 - x}$

4. $y = \dfrac{x - 1}{x + 3}$

5. $y = \dfrac{2x + 3}{x - 4}$

6. $y = \dfrac{3x}{x^2 - 9}$

7. $y = \dfrac{x^2}{x^2 - 4}$

8. $y = \dfrac{x^2 - 4}{x^2 - 1}$

9. $y = \dfrac{x^4 + 1}{x^2 - 4}$

10. $y = \dfrac{x - 1}{x^2 - 2x - 3}$

11. $y = \dfrac{2x^2 - 3x + 1}{x^2 - 9}$

12. $y = \dfrac{2x^2 - 3x + 1}{3x^2 + 5x - 2}$

13. $y = \dfrac{4 - x}{2x^3 + x^2 - 3x}$

14. $y = \dfrac{x^2 + 3x - 10}{4x^2 - 5x - 6}$

15. $y = \dfrac{x^3 + 1}{x^3 - 8}$

16. $y = \dfrac{x^2}{x - 1}$

17. $y = 3x - 4 + \dfrac{2}{x}$

18. $y = \dfrac{2x^2 + 3x + 1}{x - 3}$

19. $y = \dfrac{x^3 - 1}{x^2 - 4}$

20. $y = \dfrac{x^3 + 2x^2 - 3x - 4}{x^2 + x - 2}$

21. $y = \dfrac{1}{\sqrt{x^2 - 1}}$

22. $y = \sqrt{\dfrac{x}{x + 4}}$

23. $y = \dfrac{1}{\sqrt{1 - x}}$

24. $y = \dfrac{2x}{\sqrt{3x - 2}}$

25. $y = \dfrac{x}{\sqrt{x^2 - 16}}$

26. $y = \dfrac{x - 1}{\sqrt{9 - x^2}}$

27. $y = \dfrac{2x}{\sqrt[3]{x^4 + 1}}$

28. $y = \dfrac{\sqrt{x^2 + 1}}{x}$

29. $y = \dfrac{x^2 - 1}{\sqrt{x^4 - 13x^2 + 36}}$

30. $y^2 = \dfrac{x}{4 - x^2}$

31. $y^2 = \dfrac{x^2 - 4}{x^2 - 9}$

B 32. $y = \dfrac{x^3 + 2x - 1}{2x^3 - x^2 - 8x + 4}$

33. $y = \dfrac{x^2 - 5x + 3}{x^3 - 12x - 16}$

34. $y = \dfrac{x^4 - 15x^2 + 10x + 24}{2x^3 - 5x^2 + 7x - 8}$

35. $y = \dfrac{2x - 3}{\sqrt[3]{x^3 - 2x - 1}}$

36. $y = \dfrac{\sqrt{x^2 - 4}}{\sqrt[3]{9x - x^3}}$

37. $y = \dfrac{\cos x}{x}$

38. $y = 1 - e^{-x}$

39. $y = \dfrac{\tan x}{x}$ (Be careful!)

40. $y = \dfrac{\log_{10} x}{x}$ (Find y for $x = 10$, 100, and $1{,}000$. Then conjecture what happens when x gets arbitrarily large.)

4   Excluded Regions

Often, we can determine certain regions in the plane in which a given curve cannot lie. We refer to these as **excluded regions**. They are most often found by observing where either x or y is imaginary. Some examples will illustrate the technique.

EXAMPLE 8   Find the excluded regions for the graph of: $y = \dfrac{x}{\sqrt{1 - x^2}}$

Solution   We see that y will be imaginary if $x^2 > 1$; that is, if $x > 1$ or $x < -1$. So these regions are excluded. The lines $x = \pm 1$ are vertical asymptotes and form the boundaries of the excluded regions. They are a part of the regions, since y is not defined when $x = \pm 1$. The entire curve lies in the strip between $x = -1$ and $x = +1$ shown in Figure 7.

Figure 7

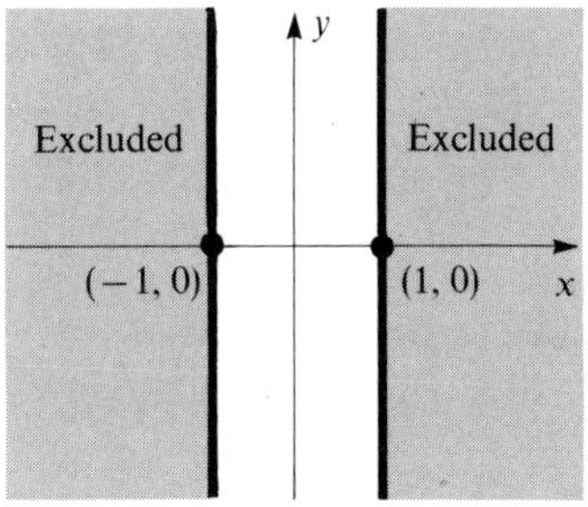

EXAMPLE 9   Find the excluded regions for the graph of: $y^2 = \dfrac{x^2 - x}{x^2 - 9}$

Solution   In order for y to be real it is necessary for the right-hand side to be nonnegative. So the regions for which

$$\frac{x^2 - x}{x^2 - 9} < 0$$

are excluded. We use the techniques of Chapter 3 to solve this inequality:

$$\frac{x^2 - x}{x^2 - 9} = \frac{x(x - 1)}{(x + 3)(x - 3)} < 0$$

The regions excluded are those for which

$$-3 \leq x < 0 \qquad \text{and} \qquad 1 < x \leq 3$$

as shown in Figure 8. The end points -3 and $+3$ are part of the excluded regions, since y is undefined there. So the graph lies wholly in the regions that are not excluded (Figure 9).

Figure 8

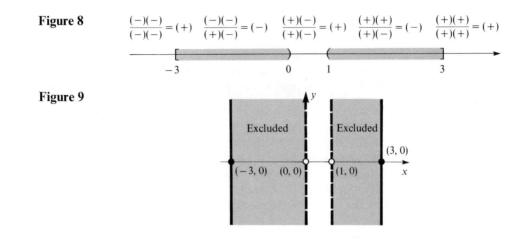

Figure 9

If the equation can be conveniently solved for x in terms of y, the same considerations enable us to find horizontal strips that are excluded. This is illustrated in the next example.

EXAMPLE 10 Find the excluded regions for the graph of: $y^2 = \dfrac{x^2 - 4}{x^2 + 9}$

Solution By inspection we see that the vertical strip for which $-2 < x < 2$ is excluded. On solving for x^2, we get

$$x^2 = \frac{4 + 9y^2}{1 - y^2}$$

from which we conclude that the horizontal strips for which $y \geq 1$ and $y \leq -1$ are excluded (Figure 10).

Figure 10

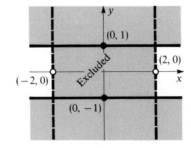

It should be noted that by this method we are finding only complete vertical or horizontal strips that are excluded. In fact, there are obviously many regions in space not occupied by the curve, but the types of regions we have found are usually found easily and are helpful in drawing graphs.

EXERCISE SET 4

Find all excluded regions.

A 1. $y = \dfrac{1}{\sqrt{x^2 - 4}}$

2. $y^2 = \dfrac{x}{x + 1}$

3. $y = \sqrt{\dfrac{x + 1}{x - 2}}$

4. $y^2 = \dfrac{x + 2}{x - 4}$

5. $y^2 = \dfrac{x^2}{x^2 - 1}$

6. $y^2 = \dfrac{x}{x^2 - 4}$

7. $y^2 = \dfrac{x - 1}{x + 1}$

8. $y^2 = \dfrac{2}{1 - x^2}$

9. $y^2 = \dfrac{x^2 + 4}{x}$

10. $y = \dfrac{1}{\sqrt{x^2 - 2x - 8}}$

11. $y^2 = \dfrac{x}{x^2 - 9}$

12. $y^2 = \dfrac{x^2 - 9}{x^2 + 4}$

13. $y = \dfrac{\sqrt{2x - 1}}{x}$

14. $x^2 = \dfrac{2y}{y - 1}$

15. $x^2 = \dfrac{y^2 - 4}{y^2 + 1}$

16. $y = \dfrac{1}{\sqrt{x^3 - x}}$

B 17. $y^2 = \dfrac{x - 1}{x^2 - 2x}$

18. $y = \sqrt{\dfrac{x^2 - 2x - 3}{x^2 + 3x}}$

19. $x^2 - 2xy + y^2 + 2x - y = 0$

20. $y = \sqrt{\ln \dfrac{1 + x}{1 - x}}$

---

5 Curve Sketching

We are ready now to put together all the graphing aids discussed in the preceding sections. We wish to emphasize that these are supposed to be *aids* in drawing the graph. If in some instances the work involved in finding intercepts, asymptotes, or excluded regions becomes too great, common sense should be used to determine when to abandon the effort, which will probably mean finding a more extensive table of values to compensate for the information not obtained.

EXAMPLE 11   Discuss and sketch the graph of: $y = \dfrac{x^2 - 4}{x^2 - 1}$

Solution   **Intercepts.**   When $x = 0$, then $y = 4$, so 4 is the y intercept. When $x = \pm 2$, then $y = 0$, so 2 and -2 are x intercepts.

Symmetry.   Since x appears to even powers only, no change occurs when x is replaced by $-x$. Thus, the curve is symmetric with respect to the y axis. There is no other symmetry.

Asymptotes. $x = \pm 1$ are vertical asymptotes, and $y = 1$ is a horizontal asymptote.

Excluded Region. If we solve for x^2, we get

$$x^2 = \frac{y - 4}{y - 1}$$

so the region for which $1 \leq y < 4$ is excluded (Figure 11). If this information is shown on a set of axes, it is possible to sketch the curve with some degree of confidence (Figure 12). If more accuracy is desired, a few points on the curve can be found by substitution.

Figure 11

$$\frac{(-)}{(-)} = (+) \qquad \frac{(-)}{(+)} = (-) \qquad \frac{(+)}{(+)} = (+)$$

Figure 12

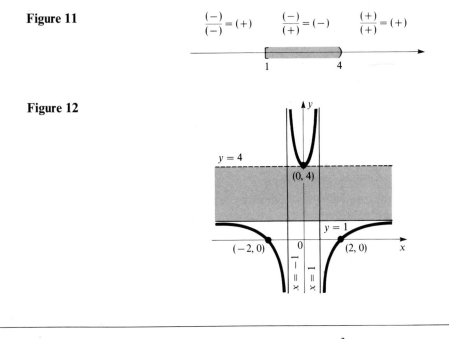

EXAMPLE 12   Discuss and sketch the graph of: $y^2 = \dfrac{4x^2}{x^2 - 4}$

Solution   **Intercept.**   When $x = 0$, then $y = 0$, and conversely. So the x intercept is 0 and the y intercept is 0.

Symmetry.   Replacement of x or y by their negatives leaves the equation unchanged. So the graph is symmetric to the x axis, y axis, and origin.

Asymptotes. $x = \pm 2$, and $y = \pm 2$

Excluded Regions.   Except when $x = 0$, we must have $x^2 > 4$, so that $x > 2$ or $x < -2$. Solving for x^2, we find also that y must satisfy $y > 2$ or $y < -2$, except that y may be 0. Thus, the regions for which $-2 \leq x \leq 2$ and for which $-2 \leq y \leq 2$ are excluded, with the single exception of the point $(0, 0)$. The origin is an **isolated point** of the graph (Figure 13).

Figure 13

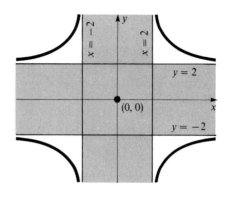

EXAMPLE 13 Discuss and sketch the graph of: $y = \dfrac{x}{x^2 - 1}$

Solution **Intercept.** When $x = 0$, then $y = 0$, and conversely. So the origin is the only intercept on the coordinate axes.

Symmetry. The curve is not symmetric to either the x or the y axis, since replacing either x or y by its negative alters the equation. However, if x and y simultaneously are replaced by their negatives, we get

$$-y = \frac{-x}{(-x)^2 - 1}$$

which, on multiplication by -1, becomes

$$y = \frac{x}{x^2 - 1}$$

Thus, $(-x, -y)$ lies on the curve whenever (x, y) does, so the curve is symmetric with respect to the origin.

Asymptotes. The vertical asymptotes are $x = \pm 1$, and the horizontal asymptote is $y = 0$.

Excluded Regions. There are no excluded vertical strips, since no value of x results in an imaginary value for y. To determine if there are horizontal strips that are excluded, we solve for x by the quadratic formula:

$$x^2y - y = x$$
$$x^2y - x - y = 0$$
$$x = \frac{1 \pm \sqrt{1 + 4y^2}}{2y}$$

Since $1 + 4y^2$ is never negative, x is never imaginary. So there are no excluded regions in the horizontal direction.

Since there is insufficient information to make a reasonable sketch, we obtain a few points. The region between $x = -1$ and $x = 1$ is especially in doubt, so we substitute $x = \frac{1}{2}$ and get $y = -\frac{2}{3}$. So $(\frac{1}{2}, -\frac{2}{3})$ is on the graph, and by symmetry we know that $(-\frac{1}{2}, \frac{2}{3})$ is also. When $x = 2$, then $y = \frac{2}{3}$, so both $(2, \frac{2}{3})$ and $(-2, -\frac{2}{3})$ are on the graph. Now a sketch can be completed with reasonable confidence (Figure 14).

Figure 14

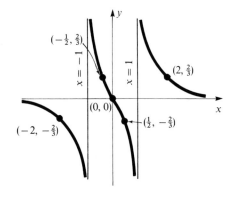

EXAMPLE 14   Discuss and sketch: $y = \dfrac{x^2 - 3x}{x^2 + 3}$

Solution   **Intercepts.**   When $x = 0$, then $y = 0$, so the y intercept is 0. The x intercepts are 0 and 3.

Symmetry.   The curve is not symmetric to the x axis, y axis, or origin. The first two of these are easy to see, since it is obvious that changing the sign of either x or y alters the equation. To see that there is also no symmetry with respect to the origin, when the equation is cleared of fractions we get

$$x^2 y + 3y = x^2 - 3x$$

Now in order that there be no effective change when both x and y are replaced by their negatives, it is necessary that all terms be of even degree (so that no sign changes result) or that all terms be of odd degree (so that all terms change in sign, resulting in an equivalent equation). In this case, three terms are of odd degree, but one is even. In general, this check can be performed mentally, without actually carrying out the multiplications.

Asymptotes.   There are no vertical asymptotes, since no value of x causes y to become infinite (the denominator is never 0). Allowing x to become infinite shows that $y = 1$ is a horizontal asymptote.

Excluded Regions.   No value of x produces an imaginary y, so there are no vertical strips that are excluded. Again we solve for x by means of the quadratic formula:

$$x^2y + 3y = x^2 - 3x$$

$$x^2(y - 1) + 3x + 3y = 0$$

$$x = \frac{-3 \pm \sqrt{9 - 12y(y - 1)}}{2(y - 1)}$$

$$= \frac{-3 \pm \sqrt{9 - 12y^2 + 12y}}{2(y - 1)}$$

Now x will be imaginary if the expression under the radical is negative. So the excluded regions are those for which

$$9 + 12y - 12y^2 < 0$$

$$3(3 + 4y - 4y^2) < 0$$

$$3(3 - 2y)(1 + 2y) < 0$$

and the regions for which $y < -\frac{1}{2}$ and for which $y > \frac{3}{2}$ are excluded, as shown in Figure 15.

Figure 15

$(+)(-) = (-)$ $(+)(+) = (+)$ $(-)(+) = (-)$

$-\frac{1}{2}$ $\frac{3}{2}$

If we attempt our sketch, we again see that our information is a bit skimpy. So we substitute several points. These are shown in the following table:

x	1	2	4	5	-1	-2	-3	-4
y	$-\frac{1}{2}$	$-\frac{2}{7}$	$\frac{4}{19}$	$\frac{5}{14}$	1	$\frac{10}{7}$	$\frac{3}{2}$	$\frac{28}{19}$

Since y can never be greater than $\frac{3}{2}$ and $y = \frac{3}{2}$ when $x = -3$, we conclude that the rounded high point on the left occurs when $x = -3$. Similarly, the low point is at $(1, -\frac{1}{2})$, since y can never be less than $-\frac{1}{2}$ (Figure 16). Finding such high and low points on curves is an important endeavor in calculus, and techniques are developed there that make this possible in many situations where elementary techniques such as we used here would not work. In fact, we were lucky that these points happened to occur at integral values of x, and we encountered them by accident when we were substituting points.

Figure 16

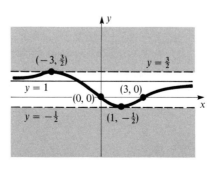

EXAMPLE 15   Discuss and sketch: $y = \dfrac{x^2 - 2x}{x - 1}$

Solution   **Intercepts.** $x = 0$, $x = 2$, $y = 0$

Symmetry.   Not symmetric to either axis or the origin.

Asymptotes. $x = 1$ is a vertical asymptote. There is no horizontal asymptote, but since the degree of the numerator is exactly 1 greater than the degree of the denominator, there is an inclined asymptote, which we find by division:

$$
\begin{array}{r|rrr}
1 & 1 & -2 & 0 \\
  &   & 1 & -1 \\
\hline
  & 1 & -1 & -1
\end{array}
$$

So

$$y = x - 1 - \frac{1}{x - 1}$$

and thus, $y = x - 1$ is an asymptote.

Excluded Regions.   Solving for x yields

$$x = \frac{y + 2 \pm \sqrt{y^2 + 4}}{2}$$

and since $y^2 + 4$ is never negative, there are no excluded horizontal regions. Nor are there any vertical ones.

Sometimes, it is useful to analyze the sign of y in various regions. In this problem we can do this by factoring the numerator:

$$y = \frac{x(x - 2)}{x - 1}$$

So, from Figure 17, $y < 0$ when $x < 0$ or $1 < x < 2$, and $y > 0$ when $0 < x < 1$ or $x > 2$. The sketch is shown in Figure 18.

Figure 17

$$\frac{(-)(-)}{(-)} = (-) \qquad \frac{(+)(-)}{(-)} = (+) \qquad \frac{(+)(-)}{(+)} = (-) \qquad \frac{(+)(+)}{(+)} = (+)$$

0	1	2

Figure 18

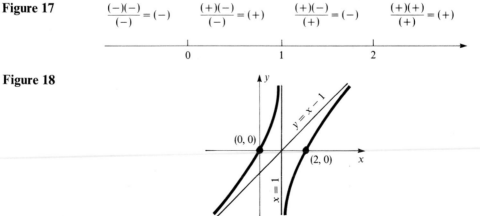

EXAMPLE 16   Discuss and sketch: $xy^2 - x - 4y^2 + 1 = 0$

Solution   We first solve for y^2:

$$y^2 = \frac{x-1}{x-4}$$

Intercepts. $x = 1, y = \pm\frac{1}{2}$

Symmetry.   The graph is symmetric with respect to the x axis.

Asymptotes. $x = 4, y = \pm 1$ (as $|x|$ approaches infinity, y^2 approaches 1, so y approaches ± 1).

Excluded Regions.   The region for which $(x - 1)/(x - 4) < 0$ is excluded, which is the region for which $1 < x \le 4$ (Figure 19). The graph is shown in Figure 20.

Figure 19

$$\frac{(-)}{(-)} = (+) \qquad \frac{(+)}{(-)} = (-) \qquad \frac{(+)}{(+)} = (+)$$

Figure 20

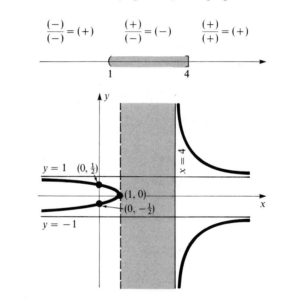

EXERCISE SET 5

Discuss and sketch the graph of each of the following:

A   **1.** $y = \dfrac{x+2}{x}$ **2.** $y = \dfrac{x+1}{x-1}$ **3.** $y = \dfrac{2-x}{x+3}$

       4. $y = \dfrac{x}{x-1}$ **5.** $y = \dfrac{x}{x^2-1}$ **6.** $y = \dfrac{x}{1+x^2}$

       7. $y = \dfrac{x^2-4}{x^2-9}$ **8.** $y = \dfrac{x-1}{x^2-4}$ **9.** $y = \dfrac{x^2}{x-1}$

      10. $y = \dfrac{x^2}{x^2-4}$ **11.** $y = \dfrac{x^2-1}{x^2}$ **12.** $y = \dfrac{x^2-4}{x}$

13. $y = \dfrac{x^2 - 1}{x + 2}$

14. $y = \dfrac{x}{\sqrt{x - 2}}$

15. $y = \sqrt{\dfrac{x - 1}{x + 1}}$

16. $y = \dfrac{x}{\sqrt{x^2 - 1}}$

17. $y^2 = \dfrac{x^2 - 4}{x^2 + 1}$

18. $y^2 = \dfrac{1}{x^2 - 4}$

B 19. $y = x - 3 + \dfrac{4}{x^2}$

20. $y^2 = \dfrac{x^2 - 9}{x^2 - x - 20}$

21. $y = \dfrac{x - 2}{\sqrt{x^2 - 3x - 4}}$

22. $y^2 = \dfrac{x^2(x - 1)}{x - 4}$

23. $y^2 = \dfrac{x - 1}{x^3}$

24. $y = \dfrac{\ln x}{x}$

25. $y = \dfrac{\sin^2 x}{x^2}$

Review Exercise Set

A   In Problems 1–5 test for symmetry with respect to the x axis, y axis, and origin.

1. **a.** $y = \dfrac{x}{1 + x^2}$ **b.** $y^2 = x^3$

2. **a.** $y^2 = \dfrac{1}{x^2 - 1}$ **b.** $y = \dfrac{x^2 + 4}{2x}$

3. **a.** $y = \dfrac{x^2 - 1}{x^2 - 4}$ **b.** $y^2 = \dfrac{x^4}{9 - x^2}$

4. **a.** $y = \sqrt{\dfrac{1 - x^2}{1 + x^2}}$ **b.** $y = \dfrac{2 \sin x}{1 + x^2}$

5. **a.** $x^2y - y^3 = x + y$ **b.** $y = \dfrac{e^x + e^{-x}}{2}$

In Problems 6–15 find all vertical, horizontal, and inclined asymptotes.

6. **a.** $y = \dfrac{1}{2x - 3}$ **b.** $y = \dfrac{2x - 1}{x + 4}$

7. **a.** $y = \dfrac{x}{x^2 - 1}$ **b.** $y = \dfrac{2x^2 + 1}{x^2 - 4}$

8. **a.** $y = \dfrac{x^2 - 3x}{x + 2}$ **b.** $y = \dfrac{x - 5}{1 - x^2}$

9. **a.** $y = \dfrac{x^2 - 2x - 3}{2x^2 + x - 3}$ **b.** $y = \dfrac{1}{x^3 + 8}$

10. **a.** $y^2 = \dfrac{x^2}{x^2 - 4}$ **b.** $y = 1 + \dfrac{1}{x - 2}$

11. **a.** $y = \dfrac{x^2}{x - 1}$ **b.** $y = \dfrac{1}{\sqrt{x^2 - 4}}$

12. **a.** $y^2 = \dfrac{4x + 3}{x - 2}$ **b.** $y = \sqrt{\dfrac{2x - 1}{2x + 1}}$

13. **a.** $y = \dfrac{x^2 - 9}{x^2 + x - 2}$ **b.** $y = \dfrac{x^2 - 2x + 5}{x - 3}$

14. **a.** $y = \dfrac{x^3 + 2}{x^2 - 1}$ **b.** $y = 2x + 3 - \dfrac{1}{x^2}$

15. **a.** $y = \dfrac{2x}{\sqrt{x^2 - 1}}$ **b.** $y = \dfrac{x^3 - 8}{x + 2}$

In Problems 16–29 discuss symmetry, intercepts, asymptotes, and excluded regions, and sketch the graph.

16. $y = \dfrac{x - 1}{x + 2}$

17. $y = \dfrac{x}{x + 4}$

18. $y = \dfrac{3 - 2x}{x - 1}$

19. $y = \dfrac{x^2 - 1}{x^2 - 4}$

20. $y = \dfrac{4 - x^2}{4 + x^2}$

21. $y^2 = \dfrac{x^2}{1 - x^2}$

22. $y^2 = \dfrac{4(x + 2)}{x - 4}$

23. $y = \sqrt{\dfrac{x - 2}{x}}$

24. $y = \dfrac{x + 2}{x^2 - 2x}$

B 25. $y^2 = \dfrac{x^2 - 4}{x^3}$

26. $y = \dfrac{x^2 - 2x - 8}{x^3 - 3x^2 + 4}$

27. $x = \dfrac{1}{y^2 - x^2}$

28. $y = \dfrac{e^x - e^{-x}}{e^x + e^{-x}}$

29. $y = \dfrac{\cos x}{x}$

12 The Conic Sections

The conic sections are important in astronomy, optics, acoustics, architecture, engineering, and art. They occur in calculus in diverse ways,—often as an incidental part of some larger problem. The following excerpt from a calculus textbook illustrates an interesting property of one of these curves:*

Like the parabola, the ellipse has an interesting reflecting property. To derive it, we consider the ellipse

$$\frac{x^2}{a^2} + \frac{y^2}{b^2} = 1$$

. . . We can now show the following:

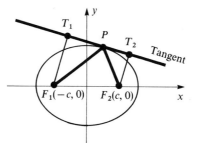

At each point P of the ellipse, the focal radii $\overline{F_1 P}$ and $\overline{F_2 P}$ make equal angles with the tangent.

. . . The result we just proved has the following physical consequence:

An elliptical mirror takes light or sound originating at one focus and converges it at the other focus.

The details of the calculus, as usual, have been omitted. You will recall from Chapter 5 that the given equation is that of an ellipse with center at the origin. In this chapter we will study the ellipse as well as the other conic sections in greater detail. We will learn the meaning of the **focus** of an ellipse. The physical property cited explains a phenomenon that has been observed in some rooms. If the room is elliptical and the walls are made of some acoustically reflecting material, a person who stands at one focus can whisper and be heard clearly by someone at the other focus, but not at points in between.

* Saturnino L. Salas and Einar Hille, *Calculus of One and Several Variables*, 3d ed. (Lexington, Mass.: Xerox College Publishing, 1978), pp. 381–383. Reprinted by permission.

1 Introduction

In Chapter 5 we introduced the curves known as the conic sections and gave the standard forms of the equations when the curves were in certain positions. With the exception of the circle, whose equation we derived, our basic approach was to determine the graph, given the equation. In this chapter we will go more deeply into the conic sections. Our approach will be to begin with the curves themselves and to derive the equations of them. Thus, we begin with the *geometric* description of the curve, and we obtain an *algebraic* description (its equation). The idea introduced in the next section will be helpful in this process.

2 The Locus of a Point

In deriving the equations of some curves, it is convenient to consider the curve as having been traced out by a moving point. A graph is really a static thing, just a collection of points, but there is no harm in supposing it to be the path of a moving point. In the final analysis, it is the equation of the path that we determine, and the notion of the moving point will disappear. To illustrate this, let us consider a circle of radius 2 with its center at the origin. One way of describing the circle is to say it is the collection of all points that are 2 units from the origin. This is the static description. It is equally valid, and useful in the derivation, to say the circle is the path traced out by a point moving so as always to be 2 units from the origin. We might call this the **kinematic description.** The path traced out by a moving point is called the **locus of the point.**

In what follows we will frequently need to use the formula for the distance between two points, which we derived in Chapter 4. It states that if the points have coordinates (x_1, y_1) and (x_2, y_2), the distance d between them is given by

$$d = \sqrt{(x_1 - x_2)^2 + (y_1 - y_2)^2}$$

EXAMPLE 1   Find the equation of the locus of a point that moves so as to be always equidistant from $A(3, 4)$ and $B(-2, 1)$.

Solution   Let P denote the moving point, and denote its coordinates by (x, y). Our problem is to determine what restrictions must be imposed on x and y (this will be in the form of an equation) so that P will satisfy the given condition. The condition requires that $\overline{PA} = \overline{PB}$, which, translated in terms of coordinates, is equivalent to

$$\sqrt{(x - 3)^2 + (y - 4)^2} = \sqrt{(x + 2)^2 + (y - 1)^2}$$

On squaring both sides,* expanding, and then simplifying, we obtain the

---

* Squaring introduces nothing extraneous in this case, because for positive numbers A and B, $\sqrt{A} = \sqrt{B}$ if and only if $A = B$, since $\sqrt{A} = -\sqrt{B}$ is impossible.

equivalent equations

$$x^2 - 6x + 9 + y^2 - 8y + 16 = x^2 + 4x + 4 + y^2 - 2y + 1$$
$$10x + 6y - 20 = 0$$
$$5x + 3y - 10 = 0$$

This is the equation of the locus.

Remark. We recognize the answer as the equation of a straight line. It is, in fact, the equation of the perpendicular bisector of AB.

EXAMPLE 2 Find the equation of the locus of a point that moves so as to be always 2 units from the origin.

Solution This is the circle referred to above. Let $P(x, y)$ be the moving point. Then we must have

$$\overline{OP} = 2$$

or equivalently,

$$\overline{OP}^2 = 4$$

so that

$$x^2 + y^2 = 4$$

is the required equation.

EXERCISE SET 2

In Problems 1–17 find the equation of the locus of a point that moves so as to satisfy the given condition.

A 1. Its distance from (1, 2) is always 3 units; describe the locus.
2. It is always equidistant from $A(2, -1)$ and $B(-3, -4)$.
3. Its distance from $A(-1, 3)$ is always twice its distance from $B(2, -1)$.
4. It is always half as far from $A(2, 4)$ as from $B(-1, -3)$.
5. Its distance from $A(3, -4)$ is half its distance from $B(0, 2)$.
6. The ratio of its distance from $A(-2, -4)$ to its distance from $B(3, -1)$ is $\frac{3}{2}$.
7. Its distance from the point (3, 0) is always equal to its distance from the line $x + 3 = 0$.
8. Its distance from $A(1, 2)$ is one-third its distance from $B(-3, 4)$.
9. Its distance from the line $y - 2 = 0$ is equal to its distance from the point $(1, -2)$.
10. Its distance from the line $x - 2 = 0$ is equal to its distance from the point $(-2, 3)$.
11. The sum of its distances from $P_1(2, 0)$ and $P_2(-2, 0)$ is always 6 units.
12. The difference of its distances from $P_1(0, 4)$ and $P_2(0, -4)$ is always 2 units (that is, $|\overline{P_1 P} - \overline{P_2 P}| = 2$).
13. Its distance from (0, 1) equals its distance from the line $y = -1$.
14. Its distance from the line $x = 6$ is half its distance from the point (0, 4).
15. Its distance from the point (0, 2) is half its distance from the line $y - 3 = 0$.

B 16. The sum of its distances from (1, 4) and (1, -2) is 10 units.
17. The difference of its distances from (2, 1) and (-3, 1) is always 4 units.

18. Derive the following formula for the distance d from the line $Ax + By + C = 0$ to the point (x_1, y_1):

$$d = \frac{|Ax_1 + By_1 + C|}{\sqrt{A^2 + B^2}}$$

Hint.   Find the equation of the line through (x_1, y_1) and perpendicular to the given line; then get the point of intersection.

19. Find the locus of a point that moves so that its distance from the line $3x + 4y + 7 = 0$ equals its distance from the point $(2, 1)$. (Use the result of Problem 18.)

20. Find the equation of the locus of the point whose distance from $x - 2y = 3$ is always twice its distance from the point $(-1, -2)$.

3 The Circle

Let C be a fixed point and let r be a fixed positive number. Then **the circle with center C and radius r can be defined as the locus of a point that moves in a plane so that its distance from C is always r.** Suppose C has coordinates (h, k). Then if we let $P(x, y)$ be the moving point, we must have $\overline{CP} = r$, or equivalently, $\overline{CP^2} = r^2$, which translates to

$$(x - h)^2 + (y - k)^2 = r^2 \qquad (1)$$

This is the standard form that we found in Chapter 5. If C happens to be the origin, then equation (1) assumes the particularly simple form

$$x^2 + y^2 = r^2$$

The equation of the circle of radius 3 with its center at $(2, -4)$ is

$$(x - 2)^2 + (y + 4)^2 = 9$$

So in the case of the circle, going from the geometric description to the algebraic description is easy. Let us consider the reverse procedure. Given an equation, how can we tell if it is the equation of a circle? And when it is a circle, how can we determine which circle? That is, how can we determine its center and radius? We take our cue from equation (1), which when expanded becomes

$$x^2 + y^2 - 2hx - 2ky + h^2 + k^2 - r^2 = 0$$

We know that every circle has an equation of this form. Let us consider a general equation of the form

$$x^2 + y^2 + Ax + By + C = 0$$

If this can be put into form (1), with $r > 0$, then we will know it is a circle and can read off its center and radius. The procedure is to complete the square in x and y. Rather than analyze this further, we will illustrate the procedure with some examples.

EXAMPLE 3   Determine the nature of the graph of the equation: $x^2 + y^2 + 2x - 4y + 1 = 0$

Solution   We first complete the squares in x and y, taking care to add to both sides the necessary quantities (the circled terms):

$$x^2 + 2x \fbox{$+1$} + y^2 - 4y \fbox{$+4$} = -1 \fbox{$+1$} \fbox{$+4$}$$

This can be written as

$$(x + 1)^2 + (y - 2)^2 = 4$$

which is in the form of equation (1) with $(h, k) = (-1, 2)$ and $r = 2$. So this describes a circle of radius 2 with its center at $(-1, 2)$.

EXAMPLE 4   Find the graph of: $x^2 + y^2 - 6x + 2y + 14 = 0$

Solution   Proceeding as in Example 3, we have

$$x^2 - 6x + 9 + y^2 + 2y + 1 = -14 + 9 + 1$$
$$(x - 3)^2 + (y + 1)^2 = -4$$

This is not satisfied by any pair (x, y) since the left-hand side of the equation is always nonnegative and the right-hand side is negative. So the equation has no graph.

EXAMPLE 5   Find the graph of: $x^2 + y^2 + 10x + 8y + 41 = 0$

Solution   Completing the square, we obtain

$$x^2 + 10x + 25 + y^2 + 8y + 16 = -41 + 25 + 16$$

or $(x + 5)^2 + (y + 4)^2 = 0$. This is satisfied by $(-5, -4)$ and no other point, since otherwise the left-hand side is positive. We could say this is a circle having zero radius; it is often referred to as a **degenerate circle.**

EXAMPLE 6   Find the graph of: $3x^2 + 3y^2 - 2x + 5y - 4 = 0$

Solution   The coefficient 3 of x^2 and y^2 is an annoyance, but we remove it by dividing both sides of the equation by 3:

$$x^2 - \tfrac{2}{3}x + \tfrac{1}{9} + y^2 + \tfrac{5}{3}y + \tfrac{25}{36} = \tfrac{4}{3} + \tfrac{1}{9} + \tfrac{25}{36}$$
$$(x - \tfrac{1}{3})^2 + (y + \tfrac{5}{6})^2 = \tfrac{77}{36}$$

This is the equation of a circle with center $(\tfrac{1}{3}, -\tfrac{5}{6})$ and radius $\sqrt{77}/6$. The only thing different about this problem was the messy arithmetic.

These examples serve to illustrate all possibilities for the equation

$$x^2 + y^2 + Ax + By + C = 0$$

Either it has no graph (as in Example 4), or its graph is a circle or a point (a degenerate circle). When you see the combination $x^2 + y^2$ or $Ax^2 + Ay^2$ in an equation that otherwise involves only linear terms, you should immediately realize that it is a circle (or degenerate) if it has a graph at all.

EXERCISE SET 3

A Find the equation of the circle having the given center and radius in Problems 1 and 2.

1. **a.** Center $(1, -3)$, radius 2 **b.** Center $(-2, -4)$, radius 4
2. **a.** Center $(-2, 3)$, radius 5 **b.** Center $(4, 0)$, radius 4

In Problems 3–9 write the equation in standard form, and determine whether the graph is a circle or a point, or if there is no graph. When there is a graph, draw it.

3. $x^2 + y^2 - 4x + 6y + 4 = 0$ 4. $x^2 + y^2 + 4x - 12 = 0$
5. $x^2 + y^2 + 2x - 8y + 1 = 0$ 6. $x^2 + y^2 - 6x + 2y + 1 = 0$
7. $x^2 + y^2 + 8x - 4y + 20 = 0$ 8. $x^2 + y^2 + 4x - 6y + 14 = 0$
9. $2x^2 + 2y^2 + 3x - 5y - 3 = 0$

10. Find the equation of the circle of radius 5 whose center is in the fourth quadrant and which is tangent to both coordinate axes.

11. The line segment joining $(-2, 1)$ and $(4, 7)$ is a diameter of a certain circle. Find the equation of the circle.

12. The portion of the line $2x + 3y = 12$ that is cut off by the x and y axes is a diameter of a certain circle. Find the equation of the circle.

13. Find the equation of the tangent line to the circle $(x - 2)^2 + (y + 3)^2 = 25$ at the point $(-1, 1)$.
    Hint. The tangent line to a circle is perpendicular to the radius drawn to the point of tangency.

14. Find the equation of the tangent line to the circle $x^2 + y^2 + 4x - 10y + 19 = 0$ at the point $(-3, 2)$. (See hint to Problem 13.)

15. The center of a circle of radius 6 lies on the line $x - 2y = 4$, and it is tangent to the y axis. Find its equation. (There are two solutions.)

16. Find the equations of the two circles, each having radius 8, with center on the line $y = 2x + 4$, and tangent to the x axis.

B 17. Show that the circles $x^2 + y^2 - 6x + 8y = 0$ and $x^2 + y^2 + 4x - 16y + 4 = 0$ are tangent to one another. Find the equation of the line joining their centers and the equation of the common tangent line.
    Hint. Find the distance between the centers and compare with the radii.

18. Show that the locus of a point that moves so that its distance from $(3, 4)$ is twice its distance from $(-1, 2)$ is a circle. Find its center and radius.

19. Find the equation of the circle passing through the three points $(2, 2)$, $(-5, 1)$, and $(4, -2)$.
    Hint. One approach is to observe that each point must satisfy $x^2 + y^2 + Ax + By + C = 0$, so A, B, and C can be determined.

20. Find the equation of the circle that circumscribes the triangle whose sides are given by $4x - 3y = 21$, $x + 3y = 9$, and $2x + y = 3$.

21. Find the points of intersection of the circles $x^2 + y^2 + 2x + 3y = 4$ and $x^2 + y^2 + x + 2y = 3$. Find the equation of the common chord and the equation of its perpendicular bisector.

22. Find the smaller area between the line $x + 2y = 5$ and the circle $x^2 + y^2 - 4x + 2y - 5 = 0$.

4 The Parabola

A parabola is defined as the locus of a point that moves in a plane so as to be always equidistant from a fixed line and a fixed point in that plane. The fixed line is called the **directrix** of the parabola and the fixed point is called the **focus.** The line through the focus and perpendicular to the directrix is called the **axis** of the parabola, and the point where the parabola crosses its axis is called its **vertex.** The vertex is halfway between the focus and the directrix.

If we know the equation of the directrix of a parabola and the coordinates of its focus, its equation can be obtained by requiring of the point $P(x, y)$ that $\overline{PF} = \overline{PP'}$, as shown in Figure 1. In general, this will result in a rather messy equation. If the directrix happens to be either vertical or horizontal, however, the situation is considerably simplified, and if the vertex is also at the origin, the equation is very easily obtained. We begin with this simplest position. First, suppose the directrix is vertical and is p units to the left of the y axis. Then the focus F must be p units to the right of the y axis, since the vertex is at the origin. Also, the axis of the parabola is the x axis. If we consider an arbitrary point $P(x, y)$ on the parabola, then we must have $\overline{PF} = \overline{PP'}$ where P' as shown in Figure 2 is the intersection of the directrix and a horizontal line through P. In terms of coordinates, this requirement becomes

$$\sqrt{(x - p)^2 + y^2} = x + p$$

Figure 1

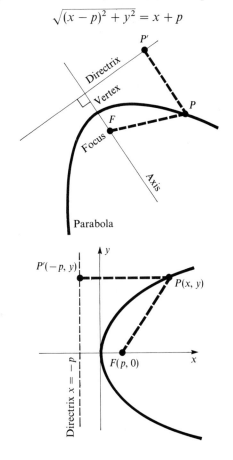

Figure 2

These two nonnegative numbers are equal if and only if their squares are equal:

$$x^2 - 2px + p^2 + y^2 = x^2 + 2px + p^2$$
$$y^2 = 4px$$

This, then, is the equation of the parabola described. The number p represents the distance from the vertex to the focus, or equivalently, the distance from the vertex to the directrix. This position is sometimes referred to as the **first standard position.** The second, third, and fourth standard positions are those obtained by three successive counterclockwise rotations of $90°$ each (Figure 3). It is a relatively easy matter to show that the equations for these other three positions are $x^2 = 4py$ (II), $y^2 = -4px$ (III), and $x^2 = -4py$ (IV). Verification of these is left for the exercises. It is important to observe that when the axis is horizontal (positions I and III), y appears to the second power and x to the first; when the axis is vertical (positions II and IV), this situation is reversed. Also, the presence of a minus sign indicates a left-opening parabola when the axis is horizontal and a downward-opening one when the axis is vertical.

Figure 3

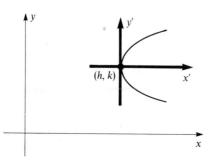

a. Position I   **b.** Position II   **c.** Position III   **d.** Position IV

We now wish to find the equations of parabolas in the four standard positions but with vertices removed from the origin. Suppose, for example, we have a parabola in position I with vertex at the point (h, k), as in Figure 4. We introduce a new set of axes, say, x' and y', parallel to the x and y axes, respectively, and such that the new origin is at (h, k). Now with respect to the new axes we know the equation of the parabola—it is $y'^2 = 4px'$. So all we need to do to get its equation with respect to the original axes is to replace x' and y' by their equivalents in terms of x and y. Suppose P is a given point having coordinates (x', y') with respect to the new axes. But P also has coordinates

Figure 4

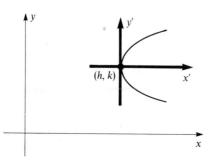

(x, y) with respect to the original axes. The relationship between the two sets of coordinates is evident from Figure 5:

$$\begin{cases} x = x' + h \\ y = y' + k \end{cases} \tag{2}$$

or equivalently,

$$\begin{cases} x' = x - h \\ y' = y - k \end{cases} \tag{3}$$

So, we replace x' by $x - h$ and y' by $y - k$ to obtain the desired equation:

Position I: $(y - k)^2 = 4p(x - h)$

The analogous results for the other positions are:

Position II: $(x - h)^2 = 4p(y - k)$

Position III: $(y - k)^2 = -4p(x - h)$

Position IV: $(x - h)^2 = -4p(y - k)$

Figure 5

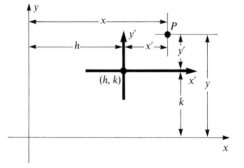

Remark. Equations (2) and (3), relating the new coordinates to the old under what is called a **translation of axes,** are general results not limited to the parabola. We can say that if the equation of any curve is known when that curve is in a given orientation with respect to the x and y axes, then the equation of the congruent curve that has the same orientation with respect to new axes with origin at (h, k) is obtained from the original equation by replacing x by $x - h$ and y by $y - k$. For example, a circle of radius r with center at the origin has equation $x^2 + y^2 = r^2$. A circle of radius r with center at the point (h, k) therefore has equation $(x - h)^2 + (y - k)^2 = r^2$. We will use this result in considering other curves later.

Parabolas having horizontal axes have equations of positions I or III, and those with vertical axes have equations of positions II or IV. If the left-hand sides of these equations are expanded and terms are collected, we find that two basic forms emerge:

Vertical axis: $x^2 + Cx + Dy + E = 0$

Horizontal axis: $y^2 + Cx + Dy + E = 0$

We wish now to reverse the procedure; that is, if we are given an equation of one of the two types above, can we conclude that it represents a parabola? The answer is yes, and the proof consists of showing, by completing the square, that every such equation can be put in one of the forms of positions I–IV.* We illustrate this with examples.

EXAMPLE 7

Discuss the graph of: $x^2 - 2x + 4y - 7 = 0$

Solution

First complete the square in x:

$$x^2 - 2x \;\boxed{+1} = -4y + 7 \;\boxed{+1}$$
$$(x - 1)^2 = -4y + 8$$
$$(x - 1)^2 = -4(y - 2)$$

This is the form of position IV, so we know it represents a parabola having a vertical axis opening downward with its vertex at the point (1, 2).

For applications, it is often sufficient to obtain a sketch of a curve, as opposed to an accurate drawing; for this purpose, locating the vertex of a parabola and recognizing its position are usually enough. If more accuracy is desired, we can get a few points on the curve by substitution. The location of the focus is seldom necessary, but it is easy to do when needed, since the distance from the vertex to the focus is p, which can be read off when the equation is in standard form. In our example, we have $p = 1$. The focus is therefore at the point (1, 1) and the graph is shown in Figure 6.

Figure 6

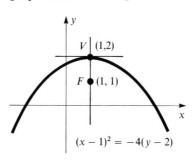

$$(x - 1)^2 = -4(y - 2)$$

One way the focus can be useful in making a rapid sketch is by observing the length of the line segment through the focus, perpendicular to the axes, and terminating on the parabola. This segment is called the **latus rectum.** The length of the latus rectum is found easily, since by definition the distance of one of its end points from the directrix must equal its distance from the focus. But the distance from the directrix is $2p$. So the length of the latus rectum is $4p$ (Figure 7). This enables us to see how "fat" or "skinny" the parabola is.

---

* This assumes that in the equation $x^2 + Cx + Dy + E = 0$, $D \neq 0$ and that in $y^2 + Cx + Dy + E = 0$, $C \neq 0$. Otherwise, the graph would be two parallel or coincident lines (a degenerate parabola), or would not exist.

Figure 7

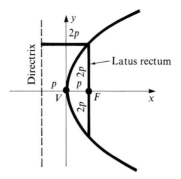

To make a rapid sketch we proceed as follows. From the standard form we locate the vertex and observe the position. The number p tells us the distance of the focus from the vertex. We proceed $2p$ units in either direction from the focus and perpendicular to the axis to get two more points on the curve, namely, the end points of the latus rectum. In Example 7, after having determined that $p = 1$ so that the focus is at $(1, 1)$, we know that the end points of the latus rectum are at $(-1, 1)$ and $(3, 1)$. With these points and the vertex $(1, 2)$ we are able to sketch the curve.

EXAMPLE 8   Discuss and sketch the graph of: $y^2 - 6x + 4y + 16 = 0$

Solution

$$y^2 + 4y \; \boxed{+\,4} = 6x - 16 \; \boxed{+\,4}$$
$$(y + 2)^2 = 6x - 12$$
$$(y + 2)^2 = 6(x - 2)$$

This is the equation of a parabola in position I, with vertex $(2, -2)$ and $4p = 6$, so that $p = \frac{3}{2}$. The latus rectum extends 3 units to either side of the focus (Figure 8).

Figure 8

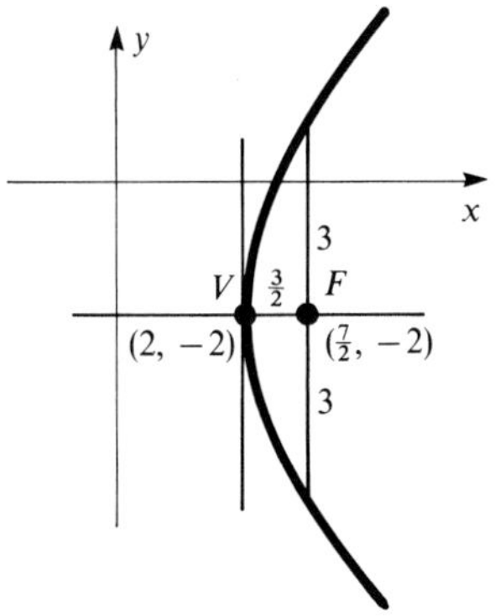

The location of the focus and the length of the latus rectum are helpful, but primary emphasis should be placed on recognition of the general nature of the curve. For example, in the equation $y^2 - 6x + 4y + 16 = 0$, we should see immediately that this represents a parabola having a horizontal axis (since y^2, rather than x^2, appears). Next in importance is the location of the vertex, which is accomplished by completing the square in the squared variable and writing the equation in one of the standard forms so that (h, k) can be determined. Remember that regardless of which variable is squared, h is found with the x term and k with the y term. For example, in

$$(y - 2)^2 = -10(x + 3)$$

$h = -3$ and $k = 2$, whereas in

$$(x - 2)^2 = -10(y + 3)$$

$h = 2$ and $k = -3$.

We give one more example to show that the arithmetic does not always work out so nicely.

EXAMPLE 9   Discuss and sketch: $3x^2 - 5x + 2y - 4 = 0$

Solution   We see immediately that this is a parabola with vertical axis. To get it in standard form, first divide by 3 and then complete the square:

$$x^2 - \tfrac{5}{3}x + \tfrac{2}{3}y - \tfrac{4}{3} = 0$$
$$x^2 - \tfrac{5}{3}x \;\boxed{+ \tfrac{25}{36}} = -\tfrac{2}{3}y + \tfrac{4}{3} \;\boxed{+ \tfrac{25}{36}}$$
$$(x - \tfrac{5}{6})^2 = -\tfrac{2}{3}y + \tfrac{73}{36}$$
$$(x - \tfrac{5}{6})^2 = -\tfrac{2}{3}(y - \tfrac{73}{24})$$

So the vertex is at the point $(\tfrac{5}{6}, \tfrac{73}{24})$, and it opens downward (position IV). Since $4p = \tfrac{2}{3}$, we have that $p = \tfrac{1}{6}$. The focus is therefore $\tfrac{1}{6}$ unit down from the vertex, and the latus rectum extends $\tfrac{1}{3}$ unit to either side. This time, since the focus is so close to the vertex, the latus rectum is not much help in making the sketch. The point $(0, 2)$ is seen to lie on the curve, however, by setting $x = 0$ in the original equation. Since the curve is symmetric to its axis, this effectively gives another point at the same distance on the other side of the axis, namely, $(\tfrac{5}{3}, 2)$ (Figure 9).

Figure 9

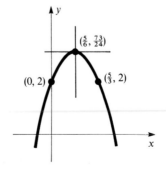

The parabola is useful in structural design and in art and architecture. The best-known example of the latter is probably the graceful Jefferson Memorial Arch in St. Louis, Missouri, designed by Eero Saarinen. Parabolas also possess an interesting and useful focusing property. If a parabolic arc is rotated about its axis, a surface called a **paraboloid** is obtained. When sound waves or light rays strike such a surface, they are all reflected to the focus of the parabola (Figure 10). Reflecting telescopes make use of this principle, for example, as do radiotelescopes. Reversing this, if light is emitted from the focus, then all rays are reflected off the surface in rays parallel to the axis. For this reason headlights on a car are in the shape of paraboloids. Since it is impossible to have a point source of light precisely at the focus, the rays do not reflect exactly in the ideal way, but they do approximate this situation.

Figure 10

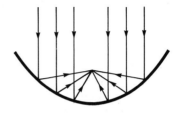

EXERCISE SET 4

A   In Problems 1–11 find the equation of the parabola described.

1. **a.** Focus $(2, 0)$; directrix $x = -2$ **b.**   Focus $(0, -3)$; directrix $y = 3$
2. **a.** Vertical axis; vertex at the origin; passing through $(-3, 2)$
    b. Horizontal axis; vertex at the origin; passing through $(-4, -3)$
3. **a.** Vertex $(1, -2)$; directrix $y = -4$ **b.**   Vertex $(-3, 1)$; focus $(-3, -1)$
4. **a.** Focus $(2, 4)$; directrix $x = 4$ **b.**   Focus $(-4, -2)$; vertex $(-5, -2)$
5. **a.** Vertex $(2, -3)$; end points of latus rectum $(-4, 0)$ and $(8, 0)$
    b. Directrix $x + 2 = 0$; end points of latus rectum $(1, 2)$ and $(1, -4)$
6. Vertex $(2, 3)$; focus $(2, -1)$
7. Focus $(0, 4)$; directrix $x = 6$
8. Vertex $(2, -4)$; axis vertical; passing through $(-2, 0)$
9. Vertex $(3, -2)$; axis horizontal; passing through $(-1, 2)$
10. Axis horizontal; passing through $(0, 3)$, $(-2, 1)$, and $(6, -3)$
    Hint.   Substitute the given points in the equation $y^2 + Ax + By + C = 0$ to obtain three equations in the unknown constants A, B, and C.
11. Axis vertical; passing through $(1, 1)$, $(-1, 3)$, and $(3, 5)$

In Problems 12–18 write the equation in one of the standard forms I–IV and sketch the graph. Give the coordinates of the focus.

12. **a.** $y^2 = 8x$ **b.** $y^2 + 4x = 0$ **c.** $x^2 = -6y$ **d.** $x^2 - 8y = 0$
13. $x^2 - 4x - 12y - 8 = 0$ 14. $y^2 + 8x - 6y + 41 = 0$
15. $y = x^2 - 2x$ 16. $2y^2 - 5x + 3y - 4 = 0$
17. $4y^2 + 4y + 24x - 35 = 0$ 18. $x = y^2 - 3y + 4$

19. If f is a quadratic function for which $f(0) = 4$, $f(2) = 0$, and $f(-2) = 0$, find $f(x)$ and draw its graph.

20. If g is a quadratic function for which $g(1) = -1$, $g(2) = 0$, and $g(-1) = 3$, find $g(x)$ and draw its graph.

In Problems 21–24 draw the graphs and shade the area described. Find the points of intersection of the curves.

21. Area between $y = 3x - x^2$ and $y = x - 3$

22. Area between the parabola $y = 4 - x^2$ and inside the circle $x^2 + y^2 - 8y + 14 = 0$

23. Area between $y = \sqrt{x}$ and $x - 2y = 0$

24. Area between $x^2 - 6x + 4y - 11 = 0$ and $x^2 - 6x - 8y + 1 = 0$

B   25. Find the equation of the circle that has its center at the focus of the parabola $x^2 - 2x + 8y - 23 = 0$ and that is tangent to this parabola at its vertex.

26. Verify that the equations given for parabolas with vertices at the origin in positions II, III, and IV are correct.

In Problems 27 and 28 draw the graphs and shade the area described. Find the points of intersection of the curves.

27. Area between $y^2 = 8x$ and $8x^2 + 5y = 12$

28. Area inside the circle $x^2 + y^2 - 2x + 4y + 1 = 0$ and above the parabola $x^2 - 2x - 3y = 5$

29. Use the definition of a parabola and the result of Problem 18 in Exercise Set 2 to find the equation of the parabola having directrix $x - 2y - 4 = 0$ and focus $(-2, 0)$. Sketch the parabola.

---

5 The Ellipse

An **ellipse is the locus of a point that moves in a plane so that the sum of its distances from two fixed points in the plane is constant.** The fixed points are called the **foci** of the ellipse, and the line through them is sometimes called the **focal axis.** The point on the focal axis midway between the foci is the **center,** and the points where the ellipse crosses its focal axis are called the **vertices.** The line segment joining the two vertices is called the **major axis** (so the major axis is a part of the focal axis); the line segment through the center, perpendicular to the major axis, and terminating at the ellipse is called the **minor axis.** These definitions are illustrated in Figure 11 and are intrinsic to the ellipse itself and independent of any particular orientation with respect to a coordinate system.

Figure 11

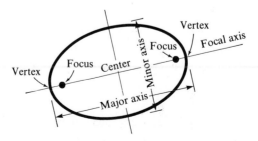

According to the definition, the sum of the distances of the point that traces out the ellipse from the two foci is constant. The size of this constant and the distance between the foci determine the shape of the ellipse. It is convenient in deriving equations of ellipses in standard positions with respect to the x and y axes to designate the constant referred to in the definition by $2a$ (this avoids fractions).

The simplest positions are those in which the focal axis is either horizontal or vertical and the center is at the origin. We will derive the equation for the horizontal case and leave the vertical case as an exercise. Let the coordinates of the foci be $(\pm c, 0)$, as shown in Figure 12. The point P that traces out the ellipse must satisfy

$$\overline{PF_1} + \overline{PF_2} = 2a \tag{4}$$

Figure 12

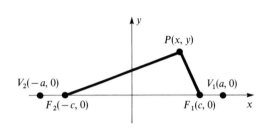

We must have $2a > 2c$, or $a > c$, since the sum of the lengths of two sides of a triangle $(\overline{PF_1} + \overline{PF_2} = 2a)$ must be greater than the length of the third side $(\overline{F_1F_2} = 2c)$. The coordinates of the end points of the major axis (that is, the vertices) and of the minor axis are obtained as follows. When P is at V_1, equation (4) gives

$$(x - c) + (x + c) = 2a$$

or $2x = 2a$, so that $x = a$. Thus, V_1 has coordinates $(a, 0)$. Similarly, V_2 is at $(-a, 0)$. Let the end points of the minor axis be $(0, b)$ and $(0, -b)$. Then when P coincides with $(0, b)$, we have that $\overline{PF_1} = \overline{PF_2}$; since $\overline{PF_1} + \overline{PF_2} = 2a$, it follows that $\overline{PF_1} = a$ and $\overline{PF_2} = a$. Then, by the Pythagorean theorem (see Figure 13),

$$b^2 = a^2 - c^2 \tag{5}$$

Now consider an arbitrary location of P. Condition (4) in terms of coordinates is

$$\sqrt{(x - c)^2 + y^2} + \sqrt{(x + c)^2 + y^2} = 2a$$

Figure 13

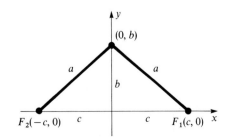

We seek an equation that is free of radicals and so square twice, after suitable rearrangement of terms:

$$\sqrt{(x-c)^2 + y^2} = 2a - \sqrt{(x+c)^2 + y^2}$$
$$x^2 - 2cx + c^2 + y^2 = 4a^2 - 4a\sqrt{(x+c)^2 + y^2} + x^2 + 2cx + c^2 + y^2$$
$$a\sqrt{(x+c)^2 + y^2} = a^2 + cx$$
$$a^2(x^2 + 2cx + c^2 + y^2) = a^4 + 2a^2cx + c^2x^2$$
$$(a^2 - c^2)x^2 + a^2y^2 = a^2(a^2 - c^2)$$

Or, in view of equation (5),

$$b^2x^2 + a^2y^2 = a^2b^2$$

Finally, we divide both sides by a^2b^2:

$$\frac{x^2}{a^2} + \frac{y^2}{b^2} = 1 \tag{6}$$

The fact that the squaring operations in this procedure led to equivalent equations can be demonstrated by showing that each time this was done, both sides of the equation represented positive numbers, and since two positive numbers are equal if and only if their squares are equal, the equivalence follows.

The corresponding equation for a vertical focal axis is

$$\frac{x^2}{b^2} + \frac{y^2}{a^2} = 1 \tag{7}$$

It should be emphasized that in equations (6) and (7) the number a is greater than b, since $b^2 = a^2 - c^2$. Whether the ellipse has a horizontal or vertical major axis is recognized from the equation in form (6) or (7) by observing which denominator is greater. For example,

$$\frac{x^2}{4} + \frac{y^2}{9} = 1$$

has a vertical major axis with $a = 3$, $b = 2$, whereas

$$\frac{x^2}{4} + \frac{y^2}{1} = 1$$

has a horizontal major axis with $a = 2$, $b = 1$.

If the center is at the point (h, k), then equations (6) and (7) become, in accordance with the discussion in the previous section,

$$\frac{(x-h)^2}{a^2} + \frac{(y-k)^2}{b^2} = 1 \tag{8}$$

and

$$\frac{(x-h)^2}{b^2} + \frac{(y-k)^2}{a^2} = 1 \tag{9}$$

The shape of an ellipse, whether "fat" or "skinny," depends on the relative sizes of c and a. To get a measure of this, we define what is known as the eccentricity e by

$$e = \frac{c}{a}$$

Since $0 < c < a$, we have $0 < e < 1$. A small eccentricity indicates that the ellipse tends toward being circular, whereas an eccentricity close to 1 indicates that the ellipse is elongated. These facts can be seen by observing that when c/a is near 0, c is small in comparison to a, so that b, which equals $\sqrt{a^2 - c^2}$, is only slightly less than a; hence, the major and minor axes are nearly equal. On the other hand, when c/a is near 1, c and a are nearly equal, and so $b = \sqrt{a^2 - c^2}$ is close to 0. Thus, the minor axis is small in relation to the major axis.

The following examples illustrate how equations of ellipses can be found when certain properties are known.

EXAMPLE 10   Find the equation of the ellipse with center at $(3, -4)$, a focus at $(0, -4)$, and a vertex at $(8, -4)$.

Solution   The information given enables us to conclude the following:

1. The major axis is horizontal, so the equation is of the form

$$\frac{(x - 3)^2}{a^2} + \frac{(y + 4)^2}{b^2} = 1$$

2. The distance from the center to a focus is 3, so $c = 3$.
3. The distance from the center to a vertex is 5, so $a = 5$.

From this we get that $b^2 = a^2 - c^2 = 25 - 9 = 16$, so $b = 4$. The equation is therefore

$$\frac{(x - 3)^2}{25} + \frac{(y + 4)^2}{16} = 1$$

EXAMPLE 11   Find the equation of the ellipse with vertices at $(1, 8)$ and $(1, -4)$ and with eccentricity $\frac{2}{3}$.

Solution   The center is at the midpoint of the major axis, which is $(1, 2)$. Also, $a = 6$. Since $e = c/a$, we have $c/6 = \frac{2}{3}$, so that $c = 4$. Also, $b^2 = a^2 - c^2 = 36 - 16 = 20$. Since the ellipse has a vertical major axis, its equation is

$$\frac{(x - 1)^2}{20} + \frac{(y - 2)^2}{36} = 1$$

If equations (8) and (9) are multiplied out and terms are rearranged, each results in an equation of the type

$$Ax^2 + By^2 + Cx + Dy + E = 0 \qquad (10)$$

where A and B are like in sign but unequal. Now we reverse the procedure and ask whether an equation in the form of (10) with these restrictions on A and B always represents an ellipse. To answer the question in any given case, we have only to complete the squares in x and y and write the equation in the form of equation (8) or (9), if possible. The following examples serve to illustrate the possibilities that may occur.

EXAMPLE 12   Discuss the graph of the equation: $4x^2 + 9y^2 + 16x - 18y - 11 = 0$

Solution   To complete the squares, we first factor out the coefficient of x^2 from terms involving x and the coefficient of y^2 from those involving y:

$$4(x^2 + 4x + \boxed{4}) + 9(y^2 - 2y + \boxed{1}) = 11 + \boxed{16} + \boxed{9}$$

Notice that we added 4 and 1 inside the parentheses to complete the squares, but this had the effect of adding 16 and 9, which had to be balanced off accordingly. Simplifying, we get

$$4(x + 2)^2 + 9(y - 1)^2 = 36$$

which on dividing by 36 becomes

$$\frac{(x + 2)^2}{9} + \frac{(y - 1)^2}{4} = 1$$

This is in standard form (8) and so is the equation of an ellipse with center at $(-2, 1)$, major axis horizontal, and $a = 3$, $b = 2$. If it is desired, we could determine the coordinates of the foci and the eccentricity, since $c^2 = a^2 - b^2 = 9 - 4 = 5$. A sketch of the ellipse is given in Figure 14.

Figure 14

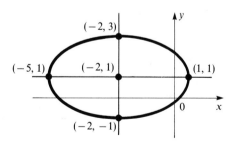

EXAMPLE 13   Discuss the graph of the equation: $9x^2 + 4y^2 - 54x - 16y + 61 = 0$

Solution   We proceed as before:

$$9(x^2 - 6x + 9) + 4(y^2 - 4y + 4) = -61 + 81 + 16$$
$$9(x - 3)^2 + 4(y - 2)^2 = 36$$
$$\frac{(x - 3)^2}{4} + \frac{(y - 2)^2}{9} = 1$$

Figure 15

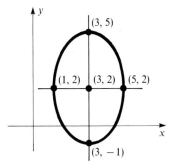

So this is an ellipse with center at $(3, 2)$, major axis vertical, and $a = 3$, $b = 2$ (Figure 15).

EXAMPLE 14   Discuss the graph of: $4x^2 + 5y^2 - 8x + 20y + 24 = 0$

Solution   Again, as above, we have

$$4(x^2 - 2x + 1) + 5(y^2 + 4y + 4) = -24 + 4 + 20$$
$$4(x - 1)^2 + 5(y + 2)^2 = 0$$

This cannot be put in either of the standard forms (8) or (9), so it does not represent an ellipse. In fact, the left-hand side is a positive number except when $x = 1$ and $y = -2$, and at that point the equation is satisfied. The entire graph therefore consists of the single point $(1, -2)$. This is an example of a **degenerate ellipse.**

EXAMPLE 15   Discuss the graph of: $4x^2 + 5y^2 - 8x + 20y + 25 = 0$

Solution   This is just like the preceding problem except for the constant. The final form is

$$4(x - 1)^2 + 5(y + 2)^2 = -1$$

and this is not satisfied by any point. So there is no graph.

EXAMPLE 16   Discuss the graph of: $3x^2 + 5y^2 - 2x + 7y - 11 = 0$

Solution   This example is given to show that the numbers need not always work out nicely.

$$3(x^2 - \tfrac{2}{3}x + \tfrac{1}{9}) + 5(y^2 + \tfrac{7}{5}y + \tfrac{49}{100}) = 11 + \tfrac{1}{3} + \tfrac{49}{20}$$
$$3(x - \tfrac{1}{3})^2 + 5(y + \tfrac{7}{10})^2 = \tfrac{827}{60}$$
$$\frac{3(x - \tfrac{1}{3})^2}{\frac{827}{60}} + \frac{5(y + \tfrac{7}{10})^2}{\frac{827}{60}} = 1$$
$$\frac{(x - \tfrac{1}{3})^2}{\frac{827}{180}} + \frac{(y + \tfrac{7}{10})^2}{\frac{827}{300}} = 1$$

Notice that we divided the numerator and the denominator of the first term by 3 and of the second term by 5 in order to get the final form. We recognize this to be the equation of an ellipse in standard form with center at $(\frac{1}{3}, -\frac{7}{10})$, major axis horizontal, and $a = \sqrt{827/180} \approx 2.14$, $b = \sqrt{827/300} \approx 1.66$. Its graph is shown in Figure 16.

Figure 16

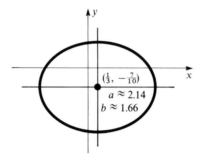

These examples serve to illustrate all possibilities for an equation of the form

$$Ax^2 + By^2 + Cx + Dy + E = 0$$

where A and B are either both positive or both negative, but unequal. Its graph is either an ellipse or a single point (degenerate ellipse), or there is no graph at all.

EXERCISE SET 5

A   In Problems 1–8 find the equation of the ellipse with the given properties. Sketch the graph.

1. **a.** x intercepts ± 4; y intercepts ± 2 **b.** x intercepts ± 5; y intercepts ± 8
2. Foci at $(-4, 3)$ and $(2, 3)$; a vertex at $(4, 3)$
3. Foci at $(-2, 0)$ and $(-2, 8)$; a vertex at $(-2, -1)$
4. Foci at $(2, 4)$ and $(2, -2)$; end point of minor axis $(4, 1)$
5. End points of minor axis $(4, -3)$ and $(4, -7)$; a focus at $(8, -5)$
6. Major axis vertical; a vertex at $(-3, 4)$; end point of minor axis $(-5, 1)$
7. A vertex at $(1, 3)$; corresponding focus at $(1, 0)$; eccentricity $\frac{1}{2}$
8. End points of minor axis $(-3, 3)$ and $(-3, -5)$; eccentricity $\frac{3}{5}$

In Problems 9–15 find the standard form of the equation. Identify and sketch the graph.

9. **a.** $25x^2 + 4y^2 = 100$ **b.** $4x^2 + 5y^2 = 20$
10. **a.** $x^2 = 4(1 - y^2)$ **b.** $y^2 = 4(4 - x^2)$
11. **a.** $x^2 + 9y^2 = 4$ **b.** $5x^2 = 3(1 - 5y^2)$
12. $9x^2 + 4y^2 + 36x - 24y + 36 = 0$ 13. $x^2 + 4y^2 - 2x + 16y + 13 = 0$
14. $4x^2 + 7y^2 - 48x - 70y + 319 = 0$ 15. $2x^2 + 4y^2 - 5x + 6y - 4 = 0$

16. Find the equation of the parabola whose vertex and focus coincide with the upper vertex and focus of the ellipse $16x^2 + 12y^2 - 64x - 24y - 116 = 0$.
17. An arch is to be made in the shape of a semiellipse. It is to be 12 feet from end to end and 4 feet high at the center. Find how high the arch is at a point halfway from the center to one end.

B **18.** The definition of the latus rectum for the ellipse is the same as for the parabola—a line segment through a focus, perpendicular to the major axis, and terminating on the curve. Show that the length of each latus rectum for an ellipse is $2b^2/a$.
Hint. Since this is independent of the orientation of the ellipse, you may take the ellipse with its center at the origin.

19. Find the equations of the inscribed and circumscribed circles to the ellipse $x^2 + 2y^2 - 4x + 12y - 3 = 0$. Sketch all three curves.

20. Show graphically the area bounded by the curves $y = \sqrt{25 - 4x^2}$, $3x - y + 3 = 0$, and $x = 0$. Find the point of intersection of the first two of these curves.

21. Find the equation of the ellipse with foci at $(-1, 2)$ and $(1, -2)$ and the length of major axis 10.
Hint. Use the definition of the ellipse.

6 The Hyperbola

The hyperbola is defined as the locus of a point that moves in a plane so that the absolute value of the difference between its distances from two fixed points in the plane is constant. Just as with the ellipse, we call the two fixed points the **foci** and the point halfway between them the **center.** We also again denote the fixed constant by $2a$. The **focal axis** is the line through the foci, and the **vertices** are the points where the hyperbola crosses the focal axis. The **transverse axis** is that portion of the focal axis between the two vertices. When the center is at the origin and the foci are on the x axis with coordinates $(\pm c, 0)$, the equation of the hyperbola is obtained in a manner similar to that of the ellipse. We require of the point P that $|\overline{PF_1} - \overline{PF_2}| = 2a$. The absolute values are used because we do not care which of the two distances is greater. In terms of coordinates this becomes

$$\left| \sqrt{(x - c)^2 + y^2} - \sqrt{(x + c)^2 + y^2} \right| = 2a$$

or equivalently,

$$\sqrt{(x - c)^2 + y^2} - \sqrt{(x + c)^2 + y^2} = \pm 2a$$

We isolate one radical, square, isolate the remaining radical, and square again. After simplification the result is

$$(c^2 - a^2)x^2 - a^2y^2 = a^2(c^2 - a^2) \tag{11}$$

From Figure 17 we see that $\overline{F_1F_2} + \overline{PF_1} > \overline{PF_2}$, or $\overline{F_1F_2} > \overline{PF_2} - \overline{PF_1}$. Similarly, $\overline{F_1F_2} + \overline{PF_2} > \overline{PF_1}$, or $\overline{F_1F_2} > \overline{PF_1} - \overline{PF_2}$. So

$$2c = \overline{F_1F_2} > |\overline{PF_1} - \overline{PF_2}| = 2a$$

Figure 17

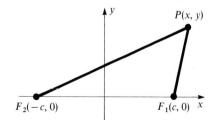

giving $c > a$. So $c^2 - a^2$ is a positive number, and we choose to designate it by b^2, that is, $b^2 = c^2 - a^2$. If we make the substitution in equation (11) and divide both sides by a^2b^2, the equation becomes

$$\frac{x^2}{a^2} - \frac{y^2}{b^2} = 1 \tag{12}$$

When $y = 0$, $x^2/a^2 = 1$, so that $x = \pm a$. The hyperbola therefore crosses the x axis at $(a, 0)$ and $(-a, 0)$. These points are the vertices. The curve does not cross the y axis, since setting $x = 0$ would lead to the impossibility $-y^2/b^2 = 1$. We name the line segment through the center, perpendicular to the tranverse axis, and b units to either side the **conjugate axis.** So, in the present case, the conjugate axis is the line segment from $(0, -b)$ to $(0, b)$. This will be helpful in drawing the graph.

If we solve equation (12) for y, we get

$$y = \pm \frac{b}{a} \sqrt{x^2 - a^2} = \pm \frac{bx}{a} \sqrt{1 - \frac{a^2}{x^2}} \tag{13}$$

Now, when x is very large in absolute value, then a^2/x^2 is small, since a^2 is fixed. The larger $|x|$ becomes, the closer to zero a^2/x^2 becomes. So $\sqrt{1 - (a^2/x^2)} \approx 1$, and

$$y \approx \pm \frac{bx}{a}$$

The lines $y = bx/a$ and $y = -bx/a$ are asymptotes to the hyperbola.

To make matters somewhat more precise, let us consider the distance between the first-quadrant portion of the hyperbola (13) and the line $y = bx/a$. Let $d(x)$ denote this distance for any value of x (for $x > a$) (Figure 18). Then

$$d(x) = \frac{bx}{a} - \frac{b}{a} \sqrt{x^2 - a^2}$$

$$= \frac{b}{a} (x - \sqrt{x^2 - a^2})$$

$$= \frac{b}{a} \left(\frac{x - \sqrt{x^2 - a^2}}{1} \right) \left(\frac{x + \sqrt{x^2 - a^2}}{x + \sqrt{x^2 - a^2}} \right)$$

$$= \frac{b}{a} \left[\frac{x^2 - (x^2 - a^2)}{x + \sqrt{x^2 - a^2}} \right]$$

$$= \frac{b}{a} \left[\frac{a^2}{x + \sqrt{x^2 - a^2}} \right]$$

$$= \frac{ab}{x + \sqrt{x^2 - a^2}}$$

Now, since the numerator is fixed, it follows that as x becomes arbitrarily large, $d(x)$ approaches 0. So the vertical distance between the curve and the line approaches 0 as the curve recedes indefinitely, and this is what is meant by the line being asymptotic to the curve. The situation is similar in the other quadrants.

Figure 18

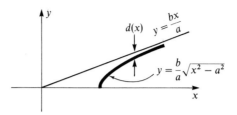

The asymptotes $y = \pm(b/a)x$ are the diagonals of the rectangle built on the transverse and conjugate axes (Figure 19). We call this the **fundamental rectangle.** This provides a guide for sketching the hyperbola. Furthermore, the description of the asymptotes as the diagonals of the fundamental rectangle is independent of the location or orientation of the hyperbola. Of course, the equations of the asymptotes will be different when the hyperbola is in a different position.

Figure 19

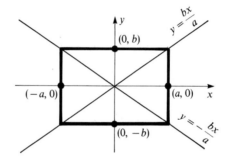

A sketch of the hyperbola described by equation (12) is given in Figure 20. It consists of two distinct branches symmetrically placed with respect to the y axis. Observe how the fundamental rectangle and its diagonals are used to aid in making the sketch.

Figure 20

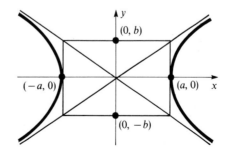

The corresponding equation for the hyperbola with its center at the origin but with transverse axis vertical is

$$\frac{y^2}{a^2} - \frac{x^2}{b^2} = 1 \tag{14}$$

and its graph is as shown in Figure 21.

Figure 21

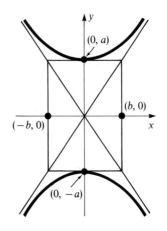

It is important to note that the position, whether horizontal or vertical, is determined by which term on the left-hand side of the equation is positive after it is in standard form (12) or (14), *not* by which of the numbers a^2 or b^2 is larger, as was true with the ellipse. For example,

$$\frac{x^2}{4} - \frac{y^2}{9} = 1$$

is a hyperbola with a horizontal transverse axis. Here, $a^2 = 4$ and $b^2 = 9$. Note that a^2 is always the denominator of the positive term. Contrast this with the ellipse

$$\frac{x^2}{4} + \frac{y^2}{9} = 1$$

which has a vertical major axis and $a^2 = 9$, $b^2 = 4$. In the case of the ellipse, it is the *size* of the denominator that determines the position, whereas in the case of the hyperbola, it is the *sign* of the term that is the determining factor. For the ellipse, a^2 is always larger than b^2, since

$$a^2 = b^2 + c^2$$

but for the hyperbola, either of the quantities a^2 or b^2 may be the larger, since

$$c^2 = a^2 + b^2$$

When the center is shifted to the point (h, k), we have:

Horizontal axis: $\quad \dfrac{(x - h)^2}{a^2} - \dfrac{(y - k)^2}{b^2} = 1$

Vertical axis: $\quad \dfrac{(y - k)^2}{a^2} - \dfrac{(x - h)^2}{b^2} = 1$

If either of these equations is cleared of fractions and terms are rearranged, an equation of the form

$$Ax^2 + By^2 + Cx + Dy + E = 0 \qquad (15)$$

is obtained, where A and B are opposite in sign. So every hyperbola with horizontal or vertical transverse axis has an equation of this form. In considering the converse, we again look at examples, since all possibilities are easily illustrated.

EXAMPLE 17   Discuss and sketch the graph of: $4x^2 - 9y^2 - 16x - 18y - 29 = 0$

Solution   We complete the squares, as in the case of the ellipse:

$$4(x^2 - 4x + 4) - 9(y^2 \boxed{+} 2y + 1) = 29 + 16 - 9$$

$$\text{Be careful here}$$

$$4(x - 2)^2 - 9(y + 1)^2 = 36$$

$$\frac{(x - 2)^2}{9} - \frac{(y + 1)^2}{4} = 1$$

So this is a hyperbola in standard position with a horizontal transverse axis, center at $(2, -1)$, and $a = 3$, $b = 2$ (Figure 22). We could find the equations of the asymptotes if desired. The slopes are $\frac{2}{3}$ and $-\frac{2}{3}$; so the equations are

$$y + 1 = \pm\tfrac{2}{3}(x - 2)$$

Figure 22

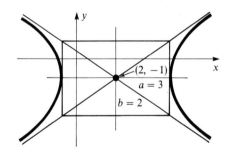

EXAMPLE 18   Discuss and sketch the graph of: $x^2 - 4y^2 + 4x + 24y - 28 = 0$

Solution

$$x^2 + 4x + 4 - 4(y^2 - 6y + 9) = 28 + 4 - 36$$

$$(x + 2)^2 - 4(y - 3)^2 = -4$$

$$\frac{(y - 3)^2}{1} - \frac{(x + 2)^2}{4} = 1$$

So this is a hyperbola with a vertical transverse axis, center at $(-2, 3)$, and $a = 1, b = 2$ (Figure 23). The slopes of the asymptotes are $\pm\frac{1}{2}$, so their equations are

$$y - 3 = \pm\tfrac{1}{2}(x + 2)$$

Figure 23

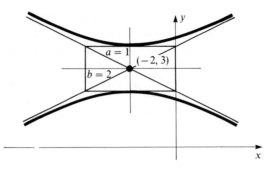

EXAMPLE 19   Discuss and sketch the graph of: $16x^2 - 25y^2 - 64x + 200y - 336 = 0$

Solution $16(x^2 - 4x + 4) - 25(y^2 - 8y + 16) = 336 + 64 - 400$
$$16(x - 2)^2 - 25(y - 4)^2 = 0$$

Since the right-hand side is 0, this is not the equation of a hyperbola. The graph can be determined, however, as follows:

$$16(x - 2)^2 = 25(y - 4)^2$$
$$4(x - 2) = \pm 5(y - 4)$$
$$y - 4 = \pm \tfrac{4}{5}(x - 2)$$

This is seen to be two lines with slopes $\tfrac{4}{5}$ and $-\tfrac{4}{5}$, and passing through the point (2, 4). In fact, these lines are the asymptotes of all hyperbolas we would get if anything other than 0 had occurred on the right-hand side of the equation $16(x - 2)^2 - 25(y - 4)^2 = 0$. We might imagine the two lines as being the limiting position of a sequence of hyperbolas of the form

$$16(x - 2)^2 - 25(y - 4)^2 = K$$

where K gets smaller and smaller in absolute value. The vertices of the two branches of the hyperbolas come closer and closer as K approaches 0. The limiting position, namely, the two lines, is called a **degenerate hyperbola.**

These three examples are typical and illustrate all possibilities for equations of the form (15) where A and B are opposite in sign. So we can say that such an equation always represents a hyperbola or a degenerate of one.

The eccentricity of a hyperbola is defined just as for the ellipse, $e = c/a$, but in this case, since $c > a$, we see that $e > 1$.

EXERCISE SET 6

A   In Problems 1–8 find the equation of the hyperbola with the given properties. Sketch the graph.

1. **a.** Center at the origin; a vertex at (2, 0); a focus at $(-3, 0)$
    b. Center at the origin; a vertex at $(0, -3)$; a focus at (0, 4)

2. Center at (2, 3); a vertex at (5, 3); a focus at (−3, 3)
3. Center at (−2, −4); a vertex at (−2, −7); a focus at (−2, −9)
4. Ends of transverse axis at (−1, 2) and (−1, −4); one end of conjugate axis at (3, −1)
5. Ends of conjugate axis at (2, −7) and (2, 5); a vertex at (0, −1)
6. Foci at (0, 2) and (6, 2); eccentricity $\frac{3}{2}$
7. Ends of conjugate axis at (3, 5) and (3, −1); eccentricity $\frac{5}{4}$
8. Center at (4, −2); axis vertical; passing through (5, −4) and (−3, 8)

In Problems 9–15 write the equation in standard form and sketch the graph. In each problem give the coordinates of the vertices, the eccentricity, and the equations of the asymptotes.

9. **a.** $4x^2 - 9y^2 = 36$ **b.** $4x^2 - 9y^2 + 36 = 0$
10. **a.** $x^2 = 4(y^2 + 1)$ **b.** $y^2 = x^2 + 9$
11. **a.** $9x^2 - 4y^2 = -36$ **b.** $25(x^2 - 9) = 9y^2$
12. **a.** $36x^2 - 4y^2 = 9$ **b.** $y^2 = 3x^2 + 16$
13. $4x^2 - y^2 - 32x + 4y + 56 = 0$ 14. $x^2 - 4y^2 - 2x - 16y - 19 = 0$
15. $9x^2 - 16y^2 + 36x + 96y - 108 = 0$

16. Find the equation of the ellipse having the same foci as the hyperbola

$$\frac{x^2}{9} - \frac{y^2}{27} = 1$$

and with eccentricity the reciprocal of that of the hyperbola.

In Problems 17–19 draw the graphs and shade the area described. Find all points of intersection of the curves involved.

17. Bounded by the curves $y = \sqrt{9 + x^2}$ and $x - 2y + 6 = 0$
18. Bounded by the curves $x^2 - y^2 = 1$ and $y = 2 - (x^2/2)$ (two areas)
19. Bounded by $y = \sqrt{1 + x^2}$ and $y = \sqrt{4 - 2x^2}$

B 20. Find the equation of the parabola whose vertex and focus coincide with the upper vertex and focus, respectively, of the hyperbola $(y^2/100) - (x^2/44) = 1$. Sketch both curves carefully and compare them. Do you think any parabola would fit this branch of the hyperbola exactly? Explain your reasoning.
21. Discuss and sketch the graph of the equation $3x^2 - 12y^2 - 3x - 18y - 10 = 0$.
22. Show the area bounded by $y = 2x - x^2$ and $11x^2 - 3y^2 + 16 = 0$, and find the points of intersection.
23. Find the equation of the locus of a point that moves so that the absolute value of the difference of its distances from (1, 1) and (5, 3) is always 4. Find the coordinates of the vertices and the end points of the conjugate axis, sketch the asymptotes, and draw the graph of the equation.

7 The General Second-Degree Equation and the Rotation of Axes

The most general equation of second degree in x and y is of the form

$$Ax^2 + Bxy + Cy^2 + Dx + Ey + F = 0 \qquad (16)$$

If $B = 0$, we now know how to identify the graph of the resulting equation, because in this case we have

$$Ax^2 + Cy^2 + Dx + Ey + F = 0$$

and the identity of the graph is dependent on the coefficients of the squared terms.

1. If $A = C$, the curve is a circle or a degenerate of a circle, or there is no graph.
2. If $AC = 0$ (so that either A or C is 0, but not both, since it would not be of second degree if both were 0), the curve is a parabola in standard position or a degenerate of a parabola. If x^2 is present, the axis is vertical, and if y^2 is present, the axis is horizontal.
3. If $AC > 0$, but $A \neq C$, then A and C are like in sign, and so the graph is an ellipse in standard position or a degenerate, or there is no graph.
4. If $AC < 0$, so that A and C are unlike in sign, the graph is a hyperbola in standard position or a degenerate (two intersecting lines).

The primary object of this section is to show that when $B \neq 0$ in equation (16), the curve is still a conic section but with a rotated axis.

Consider a cartesian coordinate system with axes x and y, and let a second cartesian coordinate system with axes x' and y' be superimposed upon the first, so that the origins coincide and the angle from the x axis to the x' axis is θ, as shown in Figure 24. It is sufficient to consider $0 < \theta < \pi/2$. Let P be a point in the plane. It has coordinates (x, y) with respect to the original axes and coordinates (x', y') with respect to the rotated axes. We wish to determine the relationships among x, y, x' and y'. Consider the construction as shown in Figure 24. It can be seen that triangle PST is similar to triangle OSR, so that the angle at P in the triangle PST is also θ. From that triangle we have

$$\sin \theta = \frac{\overline{ST}}{y'} \qquad \cos \theta = \frac{\overline{PT}}{y'}$$

so that

$$\overline{ST} = y' \sin \theta \qquad \overline{PT} = y' \cos \theta$$

Also, from triangle OSR, since $\overline{OS} = \overline{NP} = x'$,

$$\sin \theta = \frac{\overline{RS}}{\overline{OS}} = \frac{\overline{RS}}{\overline{NP}} = \frac{\overline{RS}}{x'} \qquad \cos \theta = \frac{\overline{OR}}{\overline{OS}} = \frac{\overline{OR}}{x'}$$

and thus,

$$\overline{RS} = x' \sin \theta \qquad \overline{OR} = x' \cos \theta$$

Figure 24

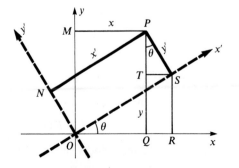

From Figure 24 we also see that

$$x = \overline{MP} = \overline{OQ} = \overline{OR} - \overline{QR} = \overline{OR} - \overline{ST} = x' \cos \theta - y' \sin \theta$$
$$y = \overline{QP} = \overline{QT} + \overline{TP} = \overline{RS} + \overline{TP} = x' \sin \theta + y' \cos \theta$$

These are the desired equations relating the old coordinates to the new:

$$\begin{cases} x = x' \cos \theta - y' \sin \theta \\ y = x' \sin \theta + y' \cos \theta \end{cases} \qquad (17)$$

Even though we showed P in the first quadrant with respect to both sets of axes, the equations in (17) hold true regardless of the location of P.

Now suppose we are given a second-degree equation in x and y,

$$Ax^2 + Bxy + Cy^2 + Dx + Ey + F = 0 \qquad (18)$$

in which $B \neq 0$. We will show that the equation of this curve when referred to a new set of axes x' and y' that are rotated through an angle θ from the old will be free of the $x'y'$ term if θ is appropriately chosen. The procedure will be as follows. We consider θ as unknown, substitute from (17) into (18), and obtain an equation of the form

$$A'x'^2 + B'x'y' + C'y'^2 + D'x' + E'y' + F' = 0 \qquad (19)$$

where the coefficients are functions of θ. Then we set $B' = 0$ and solve for θ. The details of the calculations are cumbersome. The result after substituting and rearranging terms is that equation (19) is obtained, with

$$A' = A \cos^2 \theta + B \sin \theta \cos \theta + C \sin^2 \theta$$
$$B' = B(\cos^2 \theta - \sin^2 \theta) - 2(A - C)(\sin \theta \cos \theta)$$
$$C' = A \sin^2 \theta - B \sin \theta \cos \theta + C \cos^2 \theta$$
$$D' = D \cos \theta + E \sin \theta$$
$$E' = E \cos \theta - D \sin \theta$$
$$F' = F$$

Our goal is to determine θ so that $B' = 0$. Note first that B' can be simplified by use of the identities

$$\cos^2 \theta - \sin^2 \theta = \cos 2\theta \qquad \text{and} \qquad 2 \sin \theta \cos \theta = \sin 2\theta$$

Using these, we obtain

$$B' = B \cos 2\theta - (A - C) \sin 2\theta \qquad (20)$$

We set this equal to 0:

$$B \cos 2\theta - (A - C) \sin 2\theta = 0$$
$$B \cos 2\theta = (A - C) \sin 2\theta$$
$$\frac{B}{A - C} = \frac{\sin 2\theta}{\cos 2\theta}$$

or finally,

$$\tan 2\theta = \frac{B}{A - C} \qquad (21)$$

provided $A \neq C$. If $A = C$, then we see from equation (20) that B' will be 0 if $\cos 2\theta = 0$, that is, if $2\theta = 90°$, or $\theta = 45°$. So if $A = C$, we take $\theta = 45°$; otherwise, find 2θ from equation (21) and from this determine $\sin \theta$ and $\cos \theta$. With θ chosen in this way, equation (19) becomes

$$A'x'^2 + C'y'^2 + D'x' + E'y' + F' = 0$$

which we know to be the equation of a conic section (or a degenerate) if it has a graph at all.

We will apply this technique to several examples.

EXAMPLE 20   By a suitable rotation of axes identify and sketch the curve having the equation:
$$x^2 - 2xy + y^2 - 2x - 2y = 0$$

Solution   In this case, $A = C$, so we choose $\theta = 45°$. The equations (17) of rotation then become

$$x = \frac{1}{\sqrt{2}}(x' - y') \quad \text{and} \quad y = \frac{1}{\sqrt{2}}(x' + y')$$

Substituting, we obtain

$$\tfrac{1}{2}(x'^2 - 2x'y' + y'^2) - \tfrac{2}{2}(x'^2 - y'^2) + \tfrac{1}{2}(x'^2 + 2x'y' + y'^2)$$
$$-\frac{2}{\sqrt{2}}(x' - y') - \frac{2}{\sqrt{2}}(x' + y') = 0$$

which upon simplification becomes

$$y'^2 = \sqrt{2}x'$$

We recognize this as the equation of a parabola in standard position I with respect to the rotated axes. A sketch is shown in Figure 25. It should be emphasized that the rotated axes are a means to an end. What we really wanted was to graph the original equation with respect to the x and y axes. But this has now been done. We could erase the x' and y' axes and leave the curve intact.

Figure 25

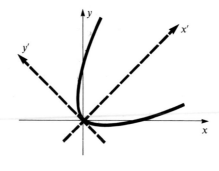

EXAMPLE 21   Identify and sketch the curve whose equation is: $x^2 + 24xy - 6y^2 = 30$

Solution   Again, we wish to rotate axes so as to eliminate the term involving $x'y'$. Using equation (21), we want to choose θ so that

$$\tan 2\theta = \frac{B}{A - C} = \frac{24}{7}$$

It is not necessary to find θ, but rather $\sin \theta$ and $\cos \theta$ in order to apply equations (17). To do this we use the trigonometric identities

$$\sin^2 \theta = \frac{1 - \cos 2\theta}{2} \qquad \text{and} \qquad \cos^2 \theta = \frac{1 + \cos 2\theta}{2}$$

Since we are restricting θ to be between $0°$ and $90°$, we can sketch the angle 2θ as shown in Figure 26. We determine the hypotenuse, 25, by the Pythagorean theorem, and then read

$$\cos 2\theta = \tfrac{7}{25}$$

So

$$\sin^2 \theta = \frac{1 - \tfrac{7}{25}}{2} = \frac{9}{25} \qquad \text{and} \qquad \cos^2 \theta = \frac{1 + \tfrac{7}{25}}{2} = \frac{16}{25}$$

from which it follows that (since θ is acute) $\sin \theta = \tfrac{3}{5}$ and $\cos \theta = \tfrac{4}{5}$. Thus, equations (17) become

$$x = \tfrac{1}{5}(4x' - 3y') \qquad \text{and} \qquad y = \tfrac{1}{5}(3x' + 4y')$$

We substitute these into the original equation to obtain

$$\tfrac{1}{25}(16x'^2 - 24x'y' + 9y'^2) + \tfrac{24}{25}(12x'^2 + 7x'y' - 12y'^2)$$
$$- \tfrac{6}{25}(9x'^2 + 24x'y' + 16y'^2) = 30$$

which upon simplification becomes

$$10x'^2 - 15y'^2 = 30$$

or in standard form,

$$\frac{x'^2}{3} - \frac{y'^2}{2} = 1$$

Figure 26

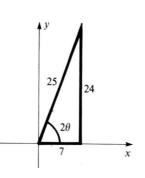

Figure 27

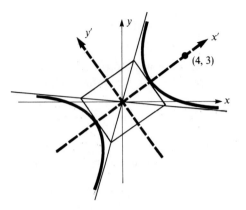

This is seen to be a hyperbola in standard position with vertices on the x' axis. To draw the rotated axes, we note that since $\sin \theta = \frac{3}{5}$ and $\cos \theta = \frac{4}{5}$, the point $(4, 3)$ lies on the x' axis (Figure 27).

EXAMPLE 22   By a suitable rotation of axes, sketch the graph of: $8x^2 - 4xy + 5y^2 = 36$

Solution   The angle θ is to be such that

$$\tan 2\theta = \frac{B}{A - C} = \frac{-4}{8 - 5} = -\frac{4}{3}$$

Since we are limiting θ to be between $0°$ and $90°$, it follows that $0° \le 2\theta \le 180°$. In this case, therefore, 2θ must lie in the second quadrant. We can now determine that $\cos 2\theta = -\frac{3}{5}$ (see Figure 28). So

$$\sin^2 \theta = \frac{1 - \cos 2\theta}{2} = \frac{1 + \frac{3}{5}}{2} = \frac{4}{5}$$

and

$$\cos^2 \theta = \frac{1 + \cos 2\theta}{2} = \frac{1 - \frac{3}{5}}{2} = \frac{1}{5}$$

and thus,

$$\sin \theta = \frac{2}{\sqrt{5}} \qquad \cos \theta = \frac{1}{\sqrt{5}}$$

So the equations of transformation are

$$x = \frac{1}{\sqrt{5}} (x' - 2y') \qquad \text{and} \qquad y = \frac{1}{\sqrt{5}} (2x' + y')$$

Figure 28

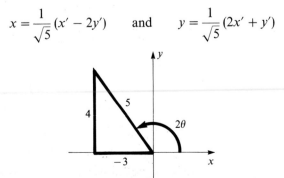

On substituting these into the original equation, we obtain

$$\tfrac{8}{5}(x'^2 - 4x'y' + 4y'^2) - \tfrac{4}{5}(2x'^2 - 3x'y' - 2y'^2) + \tfrac{5}{5}(4x'^2 + 4x'y' + y'^2) = 36$$

which becomes, after simplification,

$$4x'^2 + 9y'^2 = 36$$

or finally,

$$\frac{x'^2}{9} + \frac{y'^2}{4} = 1$$

This is seen to be the equation of an ellipse having its major axis on the x' axis. As in the previous example, we sketch the new axes in relation to the old by observing that since $\cos\theta = 1/\sqrt{5}$ and $\sin\theta = 2/\sqrt{5}$, the point $(1, 2)$ lies on the x' axis (Figure 29).

Figure 29

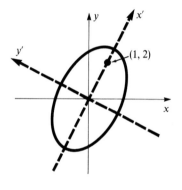

One particular type of equation with an xy term should be singled out. It is an equation of the form

$$xy = K \tag{22}$$

In this case, $A = C = 0$, so we rotate through an angle of $45°$. Thus,

$$x = \frac{1}{\sqrt{2}}(x' - y') \qquad \text{and} \qquad y = \frac{1}{\sqrt{2}}(x' + y')$$

and equation (22) becomes

$$\tfrac{1}{2}(x'^2 - y'^2) = K$$
$$x'^2 - y'^2 = 2K \tag{23}$$

This is seen to be a hyperbola with axes of equal length. It is called an **equilateral hyperbola.** If $K > 0$, its transverse axis is the x' axis, and if $K < 0$, the y' axis. If the center of the hyperbola is at the point (h, k), then equation (22) is replaced by

$$(x - h)(y - k) = K \tag{24}$$

The following examples illustrate problems of this type.

EXAMPLE 23   Sketch the curve: $xy = 2$

   Solution   After rotating $45°$, the equation becomes

$$x'^2 - y'^2 = 4$$

according to equation (23). The graph is sketched as shown in Figure 30. Note that the x and y axes are asymptotes.

Figure 30

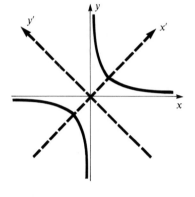

EXAMPLE 24   Discuss and sketch: $xy + 2x - y = 10$

   Solution   We could substitute $x = (1/\sqrt{2})(x' - y')$ and $y = (1/\sqrt{2})(x' + y')$. Alternately, we could try to put this in the form of equation (24) by factoring. To do this, we factor x from the first two terms and then complete the factor on y:

$$x(y + 2) - (y + 2) = 10 - 2$$

That is, we add to both sides the appropriate quantity to obtain a common binomial factor involving y, in this case $y + 2$. Now the left-hand side can be factored as

$$(x - 1)(y + 2) = 8$$

Figure 31

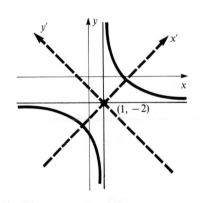

This is in the form of equation (24) and so is an equilateral hyperbola with center $(1, -2)$ and axis inclined at $45°$ (Figure 31).

EXERCISE SET 7

A   In Problems 1–8 rotate the axes so as to remove the $x'y'$ term. Identify the curve and sketch.

1. $x^2 + 2xy + y^2 + 2\sqrt{2}(x - y) = 0$ **2.** $x^2 - 3xy + 5y^2 = 22$

3. $6x^2 - 24xy - y^2 = 150$ **4.** $4x^2 + 4xy + y^2 + 5x - 10y = 0$

5. $2x^2 - 5xy + 2y^2 = 18$ **6.** $20x^2 - 8xy + 5y^2 = 84$

7. $9x^2 - 12xy + 4y^2 + 8\sqrt{13}x + 12\sqrt{13}y = 52$

8. $x^2 + 3xy - 3y^2 = 42$

In Problems 9–15 sketch the equilateral hyperbolas.

9. $xy + 2 = 0$ **10.** $y = 4/x$

11. $(x - 3)(y + 2) = 8$ **12.** $(2x + 1)(3x - 4) = 12$

13. $xy - 2x + y = 0$ **14.** $xy - 3x - 2y = 2$

15. $xy - 4x + 3y = 4$

B   Identify and sketch the following curves after a suitable rotation of axes.

16. $2x^2 - \sqrt{3}xy + y^2 + 5\sqrt{3}x - 5y = 0$

17. $9x^2 + 24xy + 16y^2 - 110x + 20y + 125 = 0$

18. $x^2 - 4xy - 2y^2 + 22x + 4y = 5$

8   Invariance of the Discriminant

It is often helpful to be able to identify the conic represented by an equation of the form (16), that is,

$$Ax^2 + Bxy + Cy^2 + Dx + Ey + F = 0$$

in which $B \neq 0$, without going through the complete rotation process described above. This is made possible by an analysis of what is called the **discriminant** of the equation, namely,

$$B^2 - 4AC$$

In the procedure described in Section 7, equation (16) can be transformed to an equation of form (19), namely,

$$A'x'^2 + B'x'y' + C'y'^2 + D'x' + E'y' + F' = 0$$

when referred to axes that have been rotated through an angle θ, where the coefficients are given by the equations in Section 7. While it was our object in that section to find θ so that $B' = 0$, the equations for the new coefficients hold true whatever the value of θ.

It is an interesting and useful fact that under any rotation whatever (that is, for all values of θ), the discriminant of equation (19) is the same as the discriminant of equation (16):

$$B'^2 - 4A'C' = B^2 - 4AC \qquad (25)$$

This result may be stated in the following way:

> The discriminant of a quadratic equation in two variables is invariant under rotation of axes.

The proof of this is straightforward but tedious (so it will be left for the exercises!). The values of A', B', and C' as given by the equations for coefficients in Section 7 are substituted into the left-hand side of equation (25), and by appropriate use of trigonometric identities and algebraic simplification, this eventually reduces to the right-hand side.

Now let us suppose that θ is chosen so as to cause $B' = 0$. Then equation (19) is

$$A'x'^2 + C'y'^2 + D'x' + E'y' + F' = 0$$

and we know this to be:

1. A parabola if either A or C is 0
2. An ellipse if A and C are like in sign
3. A hyperbola if A and C are opposite in sign

(The graph may also be a degenerate of these, or there may be no graph.) Furthermore, the discriminant becomes

$$B'^2 - 4A'C' = -4A'C'$$

since $B' = 0$. Thus, in case 1 the discriminant is 0; in case 2 it is negative; and in case 3 it is positive. Since the new discriminant equals the old, we may ascertain the following by looking at the discriminant of the original equation: If equation (16) has a graph, then it is:

1. A parabola if $B^2 - 4AC = 0$
2. An ellipse if $B^2 - 4AC < 0$
3. A hyperbola if $B^2 - 4AC > 0$

In each case, allowance must be made for a degenerate of the given conic.

EXAMPLE 25 Identify each of the following, assuming that a graph exists.

a. $2x^2 - 3xy + 5y^2 = 7$ b. $x^2 + 9xy + 3y^2 - 2x + 5y = 4$
c. $8x^2 - xy - 2y^2 + x + 3y = 0$ d. $9x^2 - 24xy + 16y^2 + 7x - 5y = 8$

Solution We calculate the discriminant in each case.

a. $B^2 - 4AC = 9 - 40 = -31 < 0$; ellipse
b. $B^2 - 4AC = 81 - 12 = 69 > 0$; hyperbola

 c. $B^2 - 4AC = 1 + 64 = 65 > 0;$ hyperbola
 d. $B^2 - 4AC = 576 - 576 = 0;$ parabola

EXERCISE SET 8

A In Problems 1 and 2 identify the graph, assuming a graph exists.

 1. **a.** $3x^2 + 2xy - y^2 + 4x - 3y = 7$ **b.** $x^2 - 4xy + 4y^2 - 3x + 2y - 5 = 0$
   c. $2x^2 - 3xy + 5y^2 = 4$ **d.** $x^2 + 3xy - 7x + 4y = 0$
   e. $5x^2 + 7xy + 3y^2 - 2x = 8$

 2. **a.** $2xy - y^2 + 4x - 5y = 7$ **b.** $xy = x^2 - 2y^2 + 4$
   c. $y^2 = 3x - 10x^2 + 4xy - 9$ **d.** $x^2 + 4y^2 = 4xy - 2x + 3y + 11$
   e. $x(3x - 2y) + y(x + 5y) = 3$

B **3.** Prove that the discriminant is invariant under rotation, that is, that $B'^2 - 4A'C' = B^2 - 4AC$.

  4. Prove that the quantity $A + C$ in the general quadratic equation is invariant under rotation, that is, that $A' + C' = A + C$.

Review Exercise Set

A In Problems 1 and 2 find the equation of the locus of a point that moves in the prescribed manner.

 1. **a.** Its distance from $A(2, 3)$ is twice its distance from $B(-1, 4)$.
  b. Its distance from the point $(2, 1)$ equals its distance from the line $x + 2 = 0$.
 2. **a.** The sum of its distances from $F_1(0, 3)$ and $F_2(0, -3)$ is 10 units.
  b. Its distance from $A(3, -1)$ is one-fourth of its distance from $B(-2, 3)$.

In Problems 3–13 write the equations in standard form and draw their graphs.

 3. **a.** $y^2 = 4 - x^2$ **b.** $y^2 = 4 + x^2$ **c.** $y^2 = 4 - x$ **d.** $y = 4 + x^2$
 4. **a.** $16x^2 + 25y^2 = 400$ **b.** $y^2 = 16 - 4x^2$
 5. **a.** $x = y^2 - 2y$ **b.** $4x^2 = y^2 + 4$
 6. $x^2 + y^2 - 2x + 4y - 20 = 0$ **7.** $x^2 + 4y^2 + 6x - 16y + 21 = 0$
 8. $y^2 - 4x - 10y + 13 = 0$ **9.** $4x^2 - 9y^2 - 16x - 18y + 43 = 0$
 10. $25x^2 + 16y^2 - 192y + 176 = 0$ **11.** $x^2 + y^2 - 10x + 8y + 41 = 0$
 12. $y^2 - x^2 + 4x + 6y + 9 = 0$ **13.** $2x^2 - 12x - 15y - 42 = 0$

In Problems 14–26 find the equations of the conics described.

 14. Circle with center $(4, -3)$ and radius 3
 15. Parabola with focus $(1, -2)$ and vertex $(3, -2)$
 16. Ellipse with vertices $(3, 2)$ and $(-5, 2)$, and foci $(1, 2)$ and $(-3, 2)$
 17. Hyperbola with vertices $(2, 5)$ and $(2, -1)$, and eccentricity $\frac{5}{3}$
 18. Parabola with vertical axis, vertex at $(-2, 5)$, and passing through the point $(2, -1)$
 19. Hyperbola with same vertices as the vertices of the ellipse $9x^2 + 4y^2 = 36$ and with asymptotes $y = \pm 3x/2$
 20. Ellipse with eccentricity $\frac{1}{2}$, and foci $(-1, 3)$ and $(-11, 3)$
 21. Parabola with directrix $y + 2 = 0$ and focus $(3, -6)$

22. Hyperbola with vertices $(4, 3)$ and $(10, 3)$, and foci $(3, 3)$ and $(11, 3)$
23. Circle having as a diameter the latus rectum of the parabola $x^2 - 6x - 16y + 9 = 0$
24. Ellipse whose foci coincide with the foci of the hyperbola $x^2 - 3y^2 - 2x - 12y + 1 = 0$ and whose eccentricity is the reciprocal of that of the hyperbola
25. Parabola with vertical axis and passing through the points $(0, 4)$, $(3, 1)$, and $(-2, -4)$
26. Circle having as diameter the transverse axis of the hyperbola $2x^2 - y^2 + 8x + 4y - 1 = 0$
27. Find the equation of the tangent line to the circle $x^2 + y^2 - 6x + 8y + 17 = 0$ at the point $(1, -2)$.
28. The underneath side of a masonry bridge is in the form of a semiellipse whose span is 80 feet and whose maximum height is 30 feet. Find the height at 10 foot intervals from one end to the other.
29. Assuming there is a graph and it is not of a degenerate conic, identify the curves whose equations are as follows;
    a. $2x^2 - 5xy - 3y^2 + 5x - 7y + 8 = 0$ b. $4x^2 - 20xy + 2y^2 - 3x + 8y - 7 = 0$
    c. $9x^2 - 24xy + 16y^2 - 3 = 0$ d. $x(x + 5y) = y(3 - 8y)$

In Problems 30–33 rotate the axes so as to eliminate the $x'y'$ term. Identify the curve and sketch.

30. $2x^2 + xy + 2y^2 - 30 = 0$ 31. $9x^2 - 24xy + 16y^2 + 4x + 3y = 0$
32. $2x^2 - 3xy + 6y^2 - 39 = 0$ 33. $x^2 - 4xy + y^2 - 3 = 0$

34. Sketch the equilateral hyperbolas.
    a. $(x - 1)(x + 3) = 4$ b. $xy - 4x + 2y - 6 = 0$

B 35. Find the equation of the locus of a point that moves so that the difference of its distances from $(2, -1)$ and $(-1, 0)$ is equal to 2. Identify the locus.
   36. Find the equation of the locus of a point that moves so that its distance from the point $(3, -1)$ is equal to its distance from the line $4x - 3y + 2 = 0$. Identify the locus. (See Problem 18, Exercise Set 2.)
   37. The end points of the conjugate axes of a hyperbola are $(4, -1)$ and $(4, 3)$, and the eccentricity is $\frac{3}{2}$. Find the equation of the hyperbola and draw its graph.
   38. A parabola that opens upward has its vertex at the center of the circle $x^2 + y^2 - 4x + 6y - 7 = 0$, and its latus rectum is a chord of the circle. Find the equation of the parabola.

13 Sequences and Series

Sequences and series play an important role in calculus and its applications. The following example from calculus involves infinite series:*

Consider the function f defined by $f(t) = 1/(1 + t)$. A power-series representation of this function is given by . . .

$$\frac{1}{1+t} = 1 - t + t^2 - t^3 + \cdots + (-1)^n t^n + \cdots \qquad \text{if } |t| < 1$$

. . . . We integrate term by term and obtain

$$\int_0^x \frac{dt}{1+t} = \sum_{n=0}^{+\infty} \int_0^x (-1)^n t^n \, dt \qquad \text{if } |x| < 1$$

and so

$$\ln(1 + x) = x - \frac{x^2}{2} + \frac{x^3}{3} - \frac{x^4}{4} + \cdots + (-1)^n \frac{x^{n+1}}{n+1} + \cdots \qquad \text{if } |x| < 1$$

or, equivalently,

$$\ln(1 + x) = \sum_{n=1}^{+\infty} (-1)^{n-1} \frac{x^n}{n} \qquad \text{if } |x| < 1$$

Note that because $|x| < 1$, $|1 + x| = (1 + x)$. Thus, the absolute-value bars are not needed when writing $\ln(1 + x)$.

We will learn in this chapter why the first equation is true. The right-hand side is an example of an *infinite geometric series*, which we will study. We will also learn the meaning of the *summation symbol*, $\sum$, used in the example.

---

* Louis Leithold, *The Calculus with Analytic Geometry*, 3d ed. (New York: Harper & Row, 1976), p. 724. Copyright © 1968, 1972, 1976 by Louis Leithold. Reprinted by permission of Harper & Row, Publishers, Inc.

1 Sequences

A sequence is defined as follows:

DEFINITION 1
A **sequence** is a function whose domain is the set N of natural numbers. If f is such a function, then for each natural number n, $f(n)$ is called the **nth term** of the sequence.

We will be concerned primarily with sequences whose terms are real numbers. It is customary to use subscripted variables to designate the terms of a sequence, rather than the usual functional notation. For example, we frequently use a_n instead of $f(n)$ for the nth term. For n arbitrary, the nth term of a sequence is also called the **general term.**

As an example, consider a sequence whose general term is given by

$$a_n = \frac{n}{2n + 1}$$

The sequence is now completely defined, and we can find as many of the terms as we wish. For $n = 1$, 2, and 3 we have

$$a_1 = \frac{1}{2 \cdot 1 + 1} = \frac{1}{3}$$

$$a_2 = \frac{2}{2 \cdot 2 + 1} = \frac{2}{5}$$

$$a_3 = \frac{3}{2 \cdot 3 + 1} = \frac{3}{7}$$

It is customary to exhibit the terms in the following way:

$$\frac{1}{3}, \frac{2}{5}, \frac{3}{7}, \cdots \cdots \frac{n}{2n + 1}, \cdots \cdots$$

The terms are separated by commas, three dots are shown following the last specific term, and the nth term is shown, followed by three more dots to indicate indefinite continuation. Often, we refer to such a listing of the terms of a sequence as the sequence itself, although strictly speaking the sequence is the function, and the terms constitute the range of the function.

When no ambiguity is likely to result, it is satisfactory to omit the nth term in exhibiting a sequence. For instance, we might write

$$2, 4, 6, 8, \ldots$$

and it would be understood that the sequence continues in the manner which is suggested by the specific terms shown. It should be emphasized that there may be many possibilities for the continuation of this sequence but only one *reasonable* way to continue, and we will understand this to be what is intended. Thus, in this case we would infer that the general term is given by $a_n = 2n$. Contrast this with

$$1, 13, 18, 36, \ldots$$

Here there is no clear pattern established, and this would not be considered a valid representation of a sequence.

To summarize, we can define a sequence either by giving the general term or by listing the first few terms of the sequence. When listing the terms, all uncertainty is removed if the nth term is shown in the list; however, it is permissible to omit it if the pattern is clearly established by the specific terms listed.

There is another way of defining a sequence which makes use of what is called a **recursion formula.** Consider, for example, the sequence for which

$$a_1 = 2$$

$$a_n = \frac{a_{n-1}}{3} \qquad n \geq 2$$

The second equation, in which a_n is given as a function of a_{n-1}, is the recursion formula. We are given a_1. Then, by the recursion formula with $n = 2$, we find

$$a_2 = \frac{a_1}{3} = \frac{2}{3}$$

Now using the recursion formula with $n = 3$, we get

$$a_3 = \frac{a_2}{3} = \frac{\frac{2}{3}}{3} = \frac{2}{9}$$

We can continue in this way to find as many terms as are desired.

It is sometimes useful to employ a **finite sequence,** by which we mean a sequence that terminates at some stage. More precisely, we have the following definition:

DEFINITION 2   A **finite sequence** is a function whose domain is a subset of N of the form $\{1, 2, 3, \ldots, n\}$ for some fixed natural number n.

In contrast to this, a sequence as defined in Definition 1 is sometimes called an **infinite sequence,** but when we say *sequence* we will always mean *infinite sequence* unless otherwise specified.

EXAMPLE 1   Write the first five terms of the sequence whose nth term is given.

a. $a_n = 2n - 1$ **b.** $a_n = \dfrac{(-1)^{n-1}}{n}$

c. $a_n = \dfrac{2^n - 1}{2^n}$ **d.** $a_n = \dfrac{(-1)^{n-1}n}{n+1}$

Solution   In each case we substitute $n = 1, 2, 3, 4,$ and 5:

a. $1, 3, 5, 7, 9, \ldots$ **b.** $1, -\frac{1}{2}, \frac{1}{3}, -\frac{1}{4}, \frac{1}{5}, \ldots$

c. $\frac{1}{2}, \frac{3}{4}, \frac{7}{8}, \frac{15}{16}, \frac{31}{32}, \ldots$ **d.** $\frac{1}{2}, -\frac{2}{3}, \frac{3}{4}, -\frac{4}{5}, \frac{5}{6}, \ldots$

Remark. Part a above shows that the nth odd natural number is $2n - 1$. Similarly, the nth even natural number is $2n$. Parts b and d illustrate a standard way of handling alternating signs. The factor $(-1)^{n-1}$ equals $+1$ for n odd and -1 for n even. These are useful facts in constructing the nth term of certain sequences.

EXAMPLE 2   Determine the nth term of each of the following sequences:

    a. $1, \frac{1}{4}, \frac{1}{9}, \frac{1}{16}, \ldots$ **b.** $3, 6, 9, 12, \ldots$

    c. $1, -\frac{1}{3}, \frac{1}{5}, -\frac{1}{7}, \ldots$ **d.** $5, 9, 13, 17, \ldots$

    e. $1, -\frac{1}{2}, \frac{1}{4}, -\frac{1}{8}, \ldots$

Solution   **a.** The denominators are perfect squares. We could rewrite the sequence as

$$\frac{1}{1^2}, \frac{1}{2^2}, \frac{1}{3^2}, \frac{1}{4^2}, \ldots$$

and conclude that the general term is

$$a_n = \frac{1}{n^2}$$

b. Each term is a multiple of 3; in fact, the terms are

$$3 \cdot 1, 3 \cdot 2, 3 \cdot 3, 3 \cdot 4, \ldots$$

and the general term is

$$a_n = 3n$$

c. The denominators are the odd natural numbers in order, and the signs alternate. By the remark following Example 1, we can therefore write

$$a_n = \frac{(-1)^{n-1}}{2n - 1}$$

d. Observe that each term after the first is 4 greater than the preceding term. This suggests that the terms are related to multiples of 4. In fact, each term is exactly 1 more than a multiple of 4. We can rewrite the terms as

$$4(1) + 1, 4(2) + 1, 4(3) + 1, 4(4) + 1, \ldots$$

and infer that the nth term is

$$a_n = 4n + 1$$

e. The denominators appear to be powers of 2, starting with 2^0:

$$\frac{1}{2^0}, -\frac{1}{2^1}, \frac{1}{2^2}, \frac{-1}{2^3}, \ldots$$

So we have

$$a_n = \frac{(-1)^{n-1}}{2^{n-1}}$$

EXAMPLE 3   Write the first six terms of the sequence defined recursively by:

a. $a_1 = -2$ **b.** $a_1 = 1$

$a_n = 1 - 2a_{n-1}, \quad n \geq 2$ $a_2 = 3$

$a_n = \dfrac{a_{n-2} + a_{n-1}}{2}, \quad n \geq 3$

Solution **a.** $a_2 = 1 - 2a_1 = 1 - 2(-2) = 5$

$a_3 = 1 - 2a_2 = 1 - 2(5) = -9$

$a_4 = 1 - 2a_3 = 1 - 2(-9) = 19$

$a_5 = 1 - 2a_4 = 1 - 2(19) = -37$

$a_6 = 1 - 2a_5 = 1 - 2(-37) = 75$

b. $a_3 = \dfrac{a_1 + a_2}{2} = \dfrac{1 + 3}{2} = 2$

$a_4 = \dfrac{a_2 + a_3}{2} = \dfrac{3 + 2}{2} = \dfrac{5}{2}$

$a_5 = \dfrac{a_3 + a_4}{2} = \dfrac{2 + \frac{5}{2}}{2} = \dfrac{4 + 5}{4} = \dfrac{9}{4}$

$a_6 = \dfrac{a_4 + a_5}{2} = \dfrac{\frac{5}{2} + \frac{9}{4}}{2} = \dfrac{10 + 9}{8} = \dfrac{19}{8}$

EXERCISE SET 1

A   In Problems 1–10, write the first five terms of the sequence whose nth term is given by the indicated expression.

1. $a_n = \dfrac{2n}{3n + 1}$

2. $a_n = \dfrac{1}{n^2 + 1}$

3. $a_n = \dfrac{(-1)^{n-1}}{n(n + 1)}$

4. $a_n = \dfrac{(-1)^{n-1}n}{2n + 1}$

5. $a_n = \left(\dfrac{2}{3}\right)^{n-1}$

6. $a_n = \dfrac{(-1)^{n-1}n^2}{1 + n}$

7. $a_n = \dfrac{2^n}{3^n + 1}$

8. $a_n = \dfrac{(-1)^{n-1}}{2n^2 - 3n + 5}$

9. $a_n = \dfrac{2}{3n + 4}$

10. $a_n = \dfrac{(-1)^{n-1}}{e^n}$

In Problems 11–20 write the next three terms of the sequence.

11. $1, \frac{1}{3}, \frac{1}{9}, \frac{1}{27}, \frac{1}{81}, \ldots$

12. $2, 4, 6, 8, 10, \ldots$

13. $1, \frac{1}{3}, \frac{1}{5}, \frac{1}{7}, \frac{1}{9}, \ldots$

14. $1, -1, 1, -1, 1, \ldots$

15. $1, \frac{1}{2}, \frac{1}{4}, \frac{1}{8}, \frac{1}{16}, \ldots$

16. $1, -\frac{1}{4}, \frac{1}{7}, -\frac{1}{10}, \frac{1}{13}, \ldots$

17. $1, \frac{4}{5}, \frac{6}{10}, \frac{8}{17}, \frac{10}{26}, \ldots$

18. $\frac{1}{2}, \frac{2}{3}, \frac{3}{4}, \frac{4}{5}, \ldots$

19. $\frac{1}{5}, -\frac{2}{7}, \frac{3}{9}, -\frac{4}{11}, \frac{5}{13}, \ldots$

20. $\frac{1}{2}, \frac{3}{4}, \frac{9}{8}, \frac{27}{16}, \frac{81}{32}, \ldots$

In Problems 21–30 determine an expression for the nth term of the sequence.

21. $1, \frac{1}{3}, \frac{1}{5}, \frac{1}{7}, \ldots$

22. $1, -\frac{1}{2}, \frac{1}{4}, -\frac{1}{8}, \ldots$

23. $\frac{1}{3}, \frac{1}{6}, \frac{1}{9}, \frac{1}{12}, \ldots$

24. $\frac{1}{2}, -\frac{2}{3}, \frac{3}{4}, -\frac{4}{5}, \ldots$

25. $\frac{1}{2}, \frac{3}{4}, \frac{7}{8}, \frac{15}{16}, \ldots$

26. $\frac{1}{2}, \frac{2}{5}, \frac{3}{10}, \frac{4}{17}, \frac{5}{26}, \ldots$

27. $1, \frac{1}{4}, \frac{1}{7}, \frac{1}{10}, \ldots$

28. $\frac{1}{7}, -\frac{1}{11}, \frac{1}{15}, -\frac{1}{19}, \frac{1}{23}, \ldots$

29. $0.5, 0.05, 0.005, 0.0005, \ldots$

30. $-\frac{2}{5}, \frac{4}{7}, -\frac{6}{9}, \frac{8}{11}, \ldots$

In Problems 31–35 write the first five terms of the sequence.

31. $a_1 = 2$

$a_n = \frac{2}{3}a_{n-1}, \quad n \geq 2$

32. $a_1 = 1$

$a_n = 2a_{n-1} - 3, \quad n \geq 2$

33. $a_1 = 4$

$a_n = -\frac{a_{n-1}}{3}, \quad n \geq 2$

34. $a_1 = 1$

$a_n = \frac{a_{n-1}}{n}, \quad n \geq 2$

35. $a_1 = 2$

$a_n = -na_{n-1}, \quad n \geq 2$

B   In Problems 36–40 determine an expression for the nth term of the sequence.

36. $\frac{1}{7}, \frac{3}{10}, \frac{5}{13}, \frac{7}{16}, \ldots$

37. $-1, \frac{2}{5}, \frac{3}{15}, \frac{4}{29}, \frac{5}{47}, \ldots$

38. $1, -\frac{1}{2}, \frac{1}{6}, -\frac{1}{24}, \frac{1}{120}, -\frac{1}{720}, \ldots$

39. $\frac{1}{4}, -\frac{2}{10}, \frac{4}{28}, -\frac{8}{82}, \frac{16}{244}, \ldots$

40. $\frac{1}{2}, -\frac{3}{6}, \frac{5}{12}, -\frac{7}{20}, \frac{9}{30}, \ldots$

In Problems 41–43 write the first five terms of the sequence.

41. $a_1 = 1$

$a_2 = 1$

$a_n = a_{n-1} + a_{n-2}, \quad n \geq 3$

This is called the **Fibonacci sequence.**

42. $a_1 = 2$

$a_n = \sqrt{2 + a_{n-1}}, \quad n \geq 2$

43. $a_1 = 1$

$a_2 = 2$

$a_n = \frac{a_{n-1}}{a_{n-2}}, \quad n \geq 3$

44. Let $f(n) = (2i)^n$, where $n \in N$. Write the values of f for $n = 1, 2, 3, 4, 5$.

45. Let

$$x_n = \frac{n+4}{n} \quad \text{and} \quad y_n = \frac{2n}{n^2+1}$$

Plot the sequence of points (x_n, y_n) for $n = 1, 2, 3, 4, 5$. By taking $n = 10, 100, 1,000$, conjecture what point the sequence is approaching as a limit.

2   Arithmetic and Geometric Sequences

When each term after the first term of a sequence is formed by adding a fixed amount to the preceding term, the sequence is called **arithmetic.** A more precise definition is given below.

DEFINITION 3 The sequence $a_1, a_2, a_3, \ldots, a_n, \ldots$ is said to be an **arithmetic sequence** if there exists a constant d such that for all $n \geq 1$,

$$a_{n+1} - a_n = d$$

The number d is called the **common difference.**

Note. An arithmetic sequence is also sometimes called an **arithmetic progression.**

EXAMPLE 4 Show that the sequence $2, 5, 8, 11, \ldots, 3n - 1, \ldots$ is arithmetic, and find the common difference.

Solution From the first four terms we observe that the difference between successive terms is the constant 3. To test to see if this pattern continues, we use Definition 3. We know that $a_n = 3n - 1$. To find a_{n+1} we replace n by $n + 1$. This gives $a_{n+1} = 3(n + 1) - 1 = 3n + 2$. So

$$a_{n+1} - a_n = (3n + 2) - (3n - 1) = 3$$

and since this is a constant, independent of n, it follows that the given sequence is arithmetic, with common difference 3.

If $a_1, a_2, a_3, \ldots, a_n, \ldots$ is an arithmetic sequence with common difference d, then by Definition 3 we have

$$
\begin{array}{lll}
a_2 - a_1 = d & \text{or} & a_2 = a_1 + d \\
a_3 - a_2 = d & \text{or} & a_3 = a_2 + d = a_1 + 2d \\
a_4 - a_3 = d & \text{or} & a_4 = a_3 + d = a_1 + 3d \\
\quad \vdots & & \qquad \vdots \\
a_n - a_{n-1} = d & \text{or} & a_n = a_{n-1} + d = a_1 + (n - 1)d
\end{array}
$$

The last formula enables us to find the nth term when we know a_1 and d. We restate the result for emphasis.

In an arithmetic sequence with first term a_1 and common difference d, the nth term is given by

$$a_n = a_1 + (n - 1)d \tag{1}$$

Note. While this result seems evident from the pattern exhibited above, we have not *proved* it. A proof can be based on mathematical induction, and you will be asked to carry out the details in Problem 20, Exercise Set 5.

EXAMPLE 5   Find the twenty-first term of the arithmetic sequence 2, 5, 8, 11, ..., $3n - 1$, ...

Solution   As we saw in Example 4, the common difference is 3. So by equation (1),

$$a_{21} = 2 + (21 - 1) \cdot 3 = 2 + (20) \cdot 3 = 62$$

EXAMPLE 6   The fifth term of an arithmetic progression is 22 and the fourteenth term is 67. Find the common difference and the first term.

Solution   By equation (1), we have $a_5 = a_1 + 4d$ and $a_{14} = a_1 + 13d$. From the given values, then,

$$a_1 + 4d = 22$$
$$a_1 + 13d = 67$$

We eliminate a_1 by subtracting the members of the top equation from those of the bottom:

$$9d = 45$$
$$d = 5$$

Substituting in the equation $a_1 + 4d = 22$, we get

$$a_1 = 22 - 4(5) = 2$$

EXAMPLE 7   Insert three arithmetic means between -4 and 20.

Solution   The instructions mean to find a_2, a_3, and a_4, so that

$$a_1, a_2, a_3, a_4, a_5, \ldots$$

where $a_1 = -4$ and $a_5 = 20$, will be an arithmetic sequence. By equation (1), $a_5 = a_1 + 4d$, so that

$$20 = -4 + 4d$$
$$4d = 24$$
$$d = 6$$

So

$$a_2 = a_1 + d = -4 + 6 = 2$$
$$a_3 = a_2 + d = \quad 2 + 6 = 8$$
$$a_4 = a_3 + d = \quad 8 + 6 = 14$$

The five terms in arithmetic progression are -4, 2, 8, 14, 20.

A **geometric sequence** (or **geometric progression**) is a sequence in which each term after the first is obtained by multiplying the preceding term by a nonzero constant. We can state this in the following equivalent form:

DEFINITION 4   The sequence $a_1, a_2, a_3, \ldots, a_n, \ldots$ is said to be a **geometric sequence** if there exists a constant $r \neq 0$ such that for all $n \geq 1$,

$$\frac{a_{n+1}}{a_n} = r$$

The number r is called the **common ratio.**

EXAMPLE 8   Show that the sequence

$$1, \frac{1}{3}, \frac{1}{9}, \frac{1}{27}, \ldots, \frac{1}{3^{n-1}}, \ldots$$

is geometric, and find the common ratio.

Solution   It appears from the first four terms that a given term after the first is found by multiplying the preceding term by $\frac{1}{3}$, that is, that the common ratio is $\frac{1}{3}$. To show this is always true, we use Definition 4. We are given that

$$a_n = \frac{1}{3^{n-1}}$$

So

$$a_{n+1} = \frac{1}{3^{(n+1)-1}} = \frac{1}{3^n}$$

Thus,

$$\frac{a_{n+1}}{a_n} = \frac{1/3^n}{1/3^{n-1}} = \frac{3^{n-1}}{3^n} = \frac{1}{3}$$

and since this is a nonzero constant, independent of n, it follows that the given sequence is geometric, with $r = \frac{1}{3}$.

If $a_1, a_2, a_3, \ldots, a_n, \ldots$ is a geometric sequence with common ratio r, then

$$\frac{a_2}{a_1} = r \qquad \text{or} \qquad a_2 = a_1 r$$

$$\frac{a_3}{a_2} = r \qquad \text{or} \qquad a_3 = a_2 r = a_1 r^2$$

$$\frac{a_4}{a_3} = r \qquad \text{or} \qquad a_4 = a_3 r = a_1 r^3$$

$$\vdots \qquad\qquad\qquad \vdots$$

$$\frac{a_n}{a_{n-1}} = r \qquad \text{or} \qquad a_n = a_{n-1} r = a_1 r^{n-1}$$

We restate the last general result:

In a geometric sequence with first term a_1 and common ratio r, the nth term is given by

$$a_n = a_1 r^{n-1} \tag{2}$$

As with formula (1), this result can be proved formally by techniques we will study in Section 5.

EXAMPLE 9   Find the tenth term of the geometric sequence $1, -\frac{1}{2}, \frac{1}{4}, -\frac{1}{8}, \ldots$.

Solution   We see by inspection that the common ratio r is $-\frac{1}{2}$. So

$$a_{10} = a_1 r^9 = 1\left(-\frac{1}{2}\right)^9 = -\frac{1}{2^9} = -\frac{1}{512}$$

EXERCISE SET 2

A   In Problems 1–5 show that the given sequence is arithmetic. Find the common difference, and write the next five terms.

1. $3, 6, 9, 12, \ldots, 3n, \ldots$

2. $1, \dfrac{3}{2}, 2, \dfrac{5}{2}, \ldots, \dfrac{n+1}{2}, \ldots$

3. $5, 2, -1, -4, \ldots, 8 - 3n, \ldots$

4. $-3, 1, 5, 9, \ldots, 4n - 7, \ldots$

5. $5, \dfrac{7}{2}, 2, \dfrac{1}{2}, \ldots, \dfrac{13 - 3n}{2}, \ldots$

In Problems 6–10 show that the given sequence is geometric. Find the common ratio, and write the next four terms.

6. $\dfrac{2}{3}, \dfrac{4}{9}, \dfrac{8}{27}, \ldots, \left(\dfrac{2}{3}\right)^n, \ldots$

7. $1, -\dfrac{1}{2}, \dfrac{1}{4}, \ldots, \left(-\dfrac{1}{2}\right)^{n-1}, \ldots$

8. $0.2, 0.02, 0.002, \ldots, 2(0.1)^n, \ldots$

9. $\dfrac{1}{9}, \dfrac{1}{3}, 1, \ldots, \dfrac{3^{n-1}}{9}, \ldots$

10. $-2, 3, -\dfrac{9}{2}, \ldots, (-1)^n\left(\dfrac{3^{n-1}}{2^{n-2}}\right), \ldots$

Each sequence in Problems 11–16 is either arithmetic or geometric. Determine the nature of each, and find the specified term.

11. $-1, 2, 5, 8, \ldots;$ a_{16}

12. $\frac{1}{4}, \frac{1}{2}, 1, 2, \ldots,$ a_{12}

13. $-3, 1, -\frac{1}{3}, \frac{1}{9}, \ldots;$ a_{10}

14. $1, \frac{1}{2}, 0, -\frac{1}{2}, \ldots;$ a_{30}

15. $-10, -6, -2, 2, \ldots;$ a_{21}

16. $8, 6, \frac{9}{2}, \frac{27}{8}, \ldots;$ a_7

In Problems 17–30 identify the sequence as being arithmetic, geometric, or neither. If arithmetic, give the common difference, and if geometric, give the common ratio.

17. $-12, -6, -3, -\frac{3}{2}, \ldots$ **18.** $1, \frac{1}{2}, \frac{1}{3}, \frac{1}{4}, \ldots$

19. $3, 7, 11, 15, \ldots$ **20.** $1, 2, 3, 4, 5, \ldots$

21. $1, \frac{1}{4}, \frac{1}{9}, \frac{1}{16}, \ldots$ **22.** $1, a^2, a^4, a^6, \ldots$

23. $\frac{1}{2}, \frac{3}{4}, \frac{7}{8}, \frac{15}{16}, \ldots$ **24.** $0.1, 0.01, 0.001, \ldots$

25. $0.1, 0.11, 0.111, 0.1111, \ldots$ **26.** $2, \frac{1}{2}, -1, -\frac{5}{2}, \ldots$

27. $36, 24, 16, \frac{32}{3}, \ldots$ **28.** $-3, -\frac{19}{4}, -\frac{13}{2}, -\frac{33}{4}, \ldots$

29. $\frac{1}{2}, \frac{2}{3}, \frac{3}{4}, \frac{4}{5}, \ldots$ **30.** $x, x + 2, x + 4, x + 6, \ldots$

31. In a certain arithmetic sequence, $a_{20} = 32$ and $d = 3$. Find a_1.

32. In a certain geometric sequence, $a_8 = \frac{729}{512}$, and $r = \frac{3}{2}$. Find a_1.

33. For a certain arithmetic sequence, $a_{32} = 48$ and $a_{17} = 18$. Find a_1 and d.

34. If $x_1, x_2, x_3, \ldots$ is an arithmetic progression with $x_{15} = 19$ and $x_{28} = -\frac{1}{2}$, find the first five terms of the progression.

35. In a certain geometric sequence, $a_3 = -\frac{4}{9}$ and $a_6 = \frac{32}{243}$. Find a_1 and r.

36. If $t_1, t_2, t_3, \ldots$ is a geometric progression with $t_6 = 0.32$ and $t_{11} = 0.0001024$, find the first five terms of the progression.

37. Find x in the sequence $5, x, 20$ so that the sequence will be:
    a. Arithmetic   **b.** Geometric (two solutions)

38. Find x and y in the sequence $1, x, y, -64$ so that the sequence will be:
    a. Arithmetic   **b.** Geometric

39. Insert six arithmetic means between 11 and 32.

40. Insert four geometric means between 27 and $-\frac{32}{9}$.

B   **41.** If $a_1, a_2, a_3, \ldots, a_n, \ldots$ is an arithmetic sequence, and for each $n \in N$, $x_n = a_n + 5$, prove that $x_1, x_2, x_3, \ldots, x_n, \ldots$ is also an arithmetic sequence.

42. If $a_1, a_2, a_3, \ldots, a_n, \ldots$ is a geometric sequence, and for each $n \in N$, $y_n = 8a_n$, prove that $y_1, y_2, y_3, \ldots, y_n, \ldots$ is also a geometric sequence.

43. Determine the nature of the sequence $\ln 2, \ln 4, \ln 8, \ln 16, \ldots$.

44. Determine the nature of the sequence for which

$$a_n = \log_{10} \frac{2}{3^n}$$

45. Prove that if $a_1, a_2, a_3, \ldots$ is a geometric sequence, then $\log a_1, \log a_2, \log a_3, \ldots$ is an arithmetic sequence. How are the common ratio of the first sequence and the common difference of the second related?

46. A ball is dropped to the ground from a height of 60 feet, and each time it bounces it goes two-thirds as high as it was previously. How high does it go on the eighth bounce?

47. Every person has 2 parents, 4 grandparents, 8 great-grandparents, and so on. How many ancestors does a person have in the tenth preceding generation?

48. A woman wishes to plan a program of jogging in which she will jog 10 minutes the first day and a fixed number of additional minutes each day until the twenty-first day, when she wishes to be jogging for 1 hour. How many minutes of jogging should she add each day?

49. Show that if $\$P$ are invested at $r\%$ compounded annually, the amount present after 1 year, 2 years, 3 years, $\ldots$, forms a geometric sequence. What is the common ratio? What is the nth term?

50. A woman's starting salary at a new job 5 years ago was $10,000 annually, and she has received raises of 6% each year since then. What is her salary now?

51. A piece of machinery that costs $30,000 depreciates by 20% each year. What will be its value in 6 years?

3 Series

A **series** is an indicated sum of a sequence. Some examples of series are:

a. $2 + 4 + 6 + 8$
b. $1 + \frac{1}{2} + \frac{1}{4} + \frac{1}{8} + \cdots$
c. $1 + 2 + 3 + 4 + 5 + \cdots$

The first is a **finite series,** and the other two are **infinite,** as indicated by the three dots. It is clear in series **a** how to find the sum, but in series **b** and **c** the question of whether a meaningful interpretation can be given to the sum is less clear. By adding up more and more terms in series **b** you might guess that the infinite sum is 2 (and you would be right), but this requires further explanation. It is probably also not too difficult to conclude that series **c** does not add up to any finite number. So we see that for some infinite series a number can be reasonably assigned as a sum, whereas for others this is not possible. Infinite series are studied in detail in certain advanced mathematics courses. Here, we will limit ourselves primarily to **infinite geometric series,** which are series formed by adding the terms of geometric sequences. First, however, we consider *finite* arithmetic and geometric series.

It is convenient to introduce a shorthand symbol for a series, namely the **summation symbol** $\sum$. When we write, for example,

$$\sum_{k=1}^{n} a_k$$

this means the sum of the numbers a_k, where k takes on the values from 1 to n. That is,

$$\sum_{k=1}^{n} a_k = a_1 + a_2 + a_3 + \cdots + a_n$$

The letter k as used here is called the **index of summation.** It is a "dummy variable," so-called because it does not appear in the final result, and this result would be unchanged if some other letter were used. For example,

$$\sum_{i=1}^{n} a_i = a_1 + a_2 + a_3 + \cdots + a_n$$

So

$$\sum_{k=1}^{n} a_k = \sum_{i=1}^{n} a_i$$

and, in fact, any other letter could be used as the index of summation.

EXAMPLE 10   Give the expanded form of the sums indicated.

a. $\displaystyle\sum_{k=1}^{4} \frac{1}{2k}$ b. $\displaystyle\sum_{i=1}^{6} 2^{i-1}$ c. $\displaystyle\sum_{n=1}^{5} (3n-1)$

Solution a. $\displaystyle\sum_{k=1}^{4} \frac{1}{2k} = \frac{1}{2 \cdot 1} + \frac{1}{2 \cdot 2} + \frac{1}{2 \cdot 3} + \frac{1}{2 \cdot 4} = \frac{1}{2} + \frac{1}{4} + \frac{1}{6} + \frac{1}{8}$

b. $\displaystyle\sum_{i=1}^{6} 2^{i-1} = 2^0 + 2^1 + 2^2 + 2^3 + 2^4 + 2^5 = 1 + 2 + 4 + 8 + 16 + 32$

c. $\displaystyle\sum_{n=1}^{5} (3n-1) = (3 \cdot 1 - 1) + (3 \cdot 2 - 1) + (3 \cdot 3 - 1) + (3 \cdot 4 - 1) + (3 \cdot 5 - 1)$

$= (3 - 1) + (6 - 1) + (9 - 1) + (12 - 1) + (15 - 1)$

$= 2 + 5 + 8 + 11 + 14$

Consider now a finite arithmetic sequence,

$$a_1, a_2, a_3, \ldots, a_n$$

The corresponding arithmetic series is

$$\sum_{k=1}^{n} a_k = a_1 + a_2 + a_3 + \cdots + a_n \tag{3}$$

Let us represent the sum by S_n. We wish to develop a formula for finding S_n. If d is the common difference, we can write series (3) in the form

$$S_n = a_1 + (a_1 + d) + (a_1 + 2d) + (a_1 + 3d) + \cdots + [a_1 + (n-1)d] \tag{4}$$

Now we are going to rewrite series (3) once again—this time in reverse order, observing that if we begin with a_n and go in reverse, we must subtract d each time to get the next term. So we obtain

$$S_n = a_n + (a_n - d) + (a_n - 2d) + \cdots + [a_n - (n-1)d] \tag{5}$$

Finally, we add (4) and (5), noting the terms that add to 0:

$$S_n + S_n = (a_1 + a_n) + (a_1 + a_n) + (a_1 + a_n) + \cdots + (a_1 + a_n)$$

Since there are n terms altogether, this can be written

$$2S_n = n(a_1 + a_n)$$

Therefore, we have the following:

The sum S_n of the terms of the finite arithmetic sequence $a_1, a_2, a_3, \ldots, a_n$ is

$$S_n = n\left(\frac{a_1 + a_n}{2}\right) \tag{6}$$

An easy way to remember this is to think of multiplying the number of terms, n, by the average of the first and last term, $(a_1 + a_n)/2$.

Remark. Although formula (6) was developed for the sum of a *finite* arithmetic sequence, we may also interpret it as the *sum of the first n terms* of an infinite arithmetic sequence.

An alternate formula for the sum S_n can be found by substituting for a_n in (6) its value from equation (1). This gives

$$S_n = n\left[\frac{a_1 + a_1 + (n-1)d}{2}\right]$$

So we have:

$$S_n = \frac{n}{2}[2a_1 + (n-1)d] \tag{7}$$

Whether to use (6) or (7) depends on the given information.

EXAMPLE 11 Find the sum of the first ten terms of the arithmetic sequence $2, 5, 8, \ldots$.

Solution Since $a_1 = 2$ and $d = 3$, we can use equation (7) to get

$$S_{10} = \tfrac{10}{2}[2(2) + 9(3)] = 5(4 + 27) = 5(31) = 155$$

EXAMPLE 12 The first term of an arithmetic sequence is 3 and the fifteenth term is 45. Find the sum of the first fifteen terms.

Solution This time we use equation (6):

$$S_{15} = 15\left(\frac{3 + 45}{2}\right) = 15(24) = 360$$

EXAMPLE 13 The first term of an arithmetic progression is -2, and the sum of the first ten terms is 20. Find the common difference.

Solution By equation (7), we have

$$20 = \tfrac{10}{2}[2(-2) + 9d]$$
$$20 = 5(-4 + 9d)$$
$$4 = -4 + 9d$$
$$9d = 8$$
$$d = \tfrac{8}{9}$$

For a finite geometric sequence

$$a_1, a_2, a_3, \ldots, a_n$$

we wish also to develop a formula for S_n, where S_n represents the sum

$$S_n = \sum_{k=1}^{n} a_k = a_1 + a_2 + a_3 + \cdots + a_n$$

If r is the common ratio, we know that each term after the first is formed by multiplying the preceding one by r. So

$$S_n = a_1 + a_1 r + a_1 r^2 + a_1 r^3 + \cdots + a_1 r^{n-1} \tag{8}$$

Now if we multiply both sides of equation (8) by r,

$$r S_n = a_1 r + a_1 r^2 + a_1 r^3 + \cdots + a_1 r^{n-1} + a_1 r^n$$

and then subtract this from equation (8), we get

$$S_n - r S_n = a_1 - a_1 r^n$$

or

$$S_n(1 - r) = a_1(1 - r^n)$$

If $r \neq 1$, we can solve this for S_n:

$$S_n = \frac{a_1(1 - r^n)}{1 - r} \qquad r \neq 1$$

If $r = 1$, we can find S_n from equation (8), because then it reads

$$S_n = a_1 + a_1 + a_1 + \cdots + a_1$$

and since there are n terms, this gives

$$S_n = na_1 \qquad r = 1$$

We summarize these results in the box.

If $a_1, a_2, a_3, \ldots, a_n$ is a finite geometric sequence with common ratio r, then

$$S_n = \frac{a_1(1 - r^n)}{1 - r} \quad \text{if } r \neq 1 \tag{9}$$

$$= na_1 \qquad \text{if } r = 1$$

where $S_n = a_1 + a_2 + \cdots + a_n$.

Again we may interpret equation (9) as being the sum of the first n terms of an infinite geometric sequence.

EXAMPLE 14   Find the sum of the first eight terms of the geometric sequence $1, -2, 4, -8, \ldots$.

Solution   We see that $r = -2$. So by equation (9),

$$S_8 = \frac{1[1 - (-2)^8]}{1 - (-2)} = \frac{1 - 256}{3} = -\frac{255}{3} = -85$$

EXAMPLE 15   Find the sum

$$\sum_{k=1}^{10} \frac{1}{2^k}$$

Solution   The first thing to note is that this is a geometric series. The expanded form of the sum is

$$S_{10} = \frac{1}{2} + \frac{1}{2^2} + \frac{1}{2^3} + \cdots + \frac{1}{2^{10}}$$

The common ratio is $\frac{1}{2}$, so by equation (9) we have

$$S_{10} = \frac{\frac{1}{2}[1 - (\frac{1}{2})^{10}]}{1 - \frac{1}{2}} = 1 - \frac{1}{2^{10}} = 1 - \frac{1}{1{,}024} = \frac{1{,}023}{1{,}024}$$

We conclude this section with a summary of the formulas we have developed for arithmetic and geometric sequences and series.

	Arithmetic Sequence	**Geometric Sequence**
For all $n \geq 1$:	$d = a_{n+1} - a_n$	$r = \dfrac{a_{n+1}}{a_n}$
nth term:	$a_n = a_1 + (n-1)d$	$a_n = a_1 r^{n-1}$
Sum of first n terms,	$S_n = n\left(\dfrac{a_1 + a_2}{2}\right)$	$S_n = \dfrac{a_1(1 - r^n)}{1 - r}$ if $r \neq 1$
$S_n = \sum_{k=1}^{n} a_k$:	$= \dfrac{n}{2}[2a_1 + (n-1)d]$	$= na_1$ if $r = 1$

EXERCISE SET 3

A   Write the expanded forms of the series in Problems 1–10.

1. $\displaystyle\sum_{k=1}^{4} \frac{3k+1}{2k-1}$

2. $\displaystyle\sum_{i=1}^{5} \frac{i}{i^2+1}$

3. $\displaystyle\sum_{n=1}^{6} \frac{(-1)^{n-1}}{n(n+1)}$

4. $\displaystyle\sum_{j=1}^{5} \frac{2^{j-1}}{3^j}$

5. $\displaystyle\sum_{k=1}^{8} \frac{(-1)^{k-1}}{k^2+1}$

6. $\displaystyle\sum_{m=1}^{6} \frac{2^m}{m!}$

7. $\displaystyle\sum_{k=1}^{10} (5k-3)$

8. $\displaystyle\sum_{n=1}^{5} \frac{\ln n}{n}$

9. $\displaystyle\sum_{n=1}^{6} ne^{-n}$

10. $\displaystyle\sum_{k=1}^{5} \frac{k^k}{k!}$

In Problems 11–18 find the sum of the first n terms of the given sequence for the specified value of n.

11. $3, 8, 13, 18, \ldots ; \quad n = 20$

12. $-2, 4, -8, 16, \ldots ; \quad n = 10$

13. $-6, -4, -2, 0, \ldots ; \quad n = 100$

14. $1, \frac{1}{3}, \frac{1}{9}, \ldots ; \quad n = 8$

15. $0.2, 0.02, 0.002, \ldots ; \quad n = 12$

16. $5, \frac{7}{2}, 2, \frac{1}{2}, \ldots ; \quad n = 30$

17. $0.2, 0.22, 0.24, 0.26, \ldots ; \quad n = 50$

18. $-27, 18, -12, 8, \ldots ; \quad n = 8$

In Problems 19–26 evaluate the indicated sums.

19. $\displaystyle\sum_{k=1}^{200} (2+3k)$

20. $\displaystyle\sum_{k=1}^{10} (-3)^{k-1}$

21. $\displaystyle\sum_{n=1}^{20} (2n-1)$

22. $\displaystyle\sum_{k=1}^{12} \frac{1}{2^{k+1}}$

23. $\displaystyle\sum_{i=1}^{8} \frac{2^i}{3^{i-1}}$

24. $\displaystyle\sum_{m=1}^{50} \left(\frac{m+3}{2}\right)$

25. $\displaystyle\sum_{j=1}^{6} \frac{(-1)^{j-1}3^j}{4^{j-1}}$

26. $\displaystyle\sum_{n=1}^{15} \left(\frac{n}{2}-3\right)$

27. Find the sum of the first 100 natural numbers.

B Find the sums in Problems 28 and 29.

28. $\displaystyle\sum_{k=1}^{6} (2^k + 3k - 1)$

29. $\displaystyle\sum_{n=0}^{5} [3^n - 2(n+1)]$

30. Find a formula for the sum of the first n odd positive integers.

31. Some pipes are stacked so that the bottom layer has 15 pipes, the next layer 14 pipes, the next 13, and so on, until there is only one pipe at the top. How many pipes are there in all?

32. A ball is dropped from a height of 40 feet, and each time it bounces it rises to a height three-fourths as high as previously. Find the total distance the ball has covered when it hits the ground for the fifth time.

33. A man is offered two jobs, one starting at $12,000 with constant annual raises of $700, and the other starting at $10,000 with annual raises of 8%. What would be his salary at each job during the sixth year? What would be his accumulated earnings at each job through the end of the sixth year? How many years would it be before the salary of the second job exceeded the first? How many years would it be before the accumulated earnings of the second job exceeded the first?

34. A pendulum swings a distance of 20 inches initially from one side to the other, and on each subsequent swing it goes 0.8 of the distance on the previous swing. Find the total distance covered by the pendulum after ten swings.

35. If a person were given a choice of working 30 days for 1¢ the first day, 2¢ the next, 4¢ the next, and so on, each day doubling the amount earned on the previous day, or of being paid $1,000 the first day, $2,000 the next, $3,000 the next, and so on, each day adding $1,000 to the amount on the previous day, which offer should he take? How much would he make under each arrangement?

4   Infinite Geometric Series

Consider an infinite sequence

$$a_1, a_2, a_3, \ldots, a_n, \ldots$$

We designate the corresponding **infinite series** by

$$\sum_{k=1}^{\infty} a_k = a_1 + a_2 + a_3 + \cdots + a_n + \cdots$$

The symbol ∞ is read "infinity." To arrive at a reasonable definition of what is meant by such an infinite sum we consider the finite sums

$$S_1 = a_1$$
$$S_2 = a_1 + a_2$$
$$S_3 = a_1 + a_2 + a_3$$
$$\cdots\cdots\cdots\cdots\cdots$$
$$S_n = a_1 + a_2 + \cdots + a_n$$

We call S_n the **nth partial sum** of the series $\sum_{n=1}^{\infty} a_n$, and we call the sequence $S_1, S_2, S_3, \ldots, S_n, \ldots$ **the sequence of partial sums.** If it happens that for larger and larger values of n, the partial sums approach as a limit some number S, then we define S to be the sum of the infinite series. This is written symbolically as

$$S = \sum_{n=1}^{\infty} a_n = \lim_{n \to \infty} S_n$$

The symbol "$\lim_{n \to \infty} S_n$" is read "the limit of S_n as n goes to infinity." Limits of sequences are studied in detail in calculus, and it is not appropriate in this course to give a precise definition. It is sufficient to say that

$$\lim_{n \to \infty} S_n = S$$

means that the partial sums S_n are arbitrarily close to S for all sufficiently large values of n.

To make these concepts more concrete let us consider the infinite geometric series

$$1 + \frac{1}{2} + \frac{1}{4} + \frac{1}{8} + \cdots + \frac{1}{2^{n-1}} + \cdots \tag{10}$$

The partial sums are

$$S_1 = 1$$
$$S_2 = 1 + \tfrac{1}{2} = \tfrac{3}{2}$$
$$S_3 = 1 + \tfrac{1}{2} + \tfrac{1}{4} = \tfrac{7}{4}$$
$$S_4 = 1 + \tfrac{1}{2} + \tfrac{1}{4} + \tfrac{1}{8} = \tfrac{15}{8}$$

. . . .

By equation (9) we can write the nth partial sum as

$$S_n = 1 + \frac{1}{2} + \frac{1}{4} + \frac{1}{8} + \cdots + \frac{1}{2^{n-1}} = \frac{1[1 - (\tfrac{1}{2})^n]}{1 - \tfrac{1}{2}} = 2\left[1 - \left(\frac{1}{2}\right)^n\right]$$

The sequence of partial sums,

$$1, \frac{3}{2}, \frac{7}{4}, \frac{15}{8}, \ldots, 2\left[1 - \left(\frac{1}{2}\right)^n\right], \ldots$$

appears to be approaching 2 as a limit. In fact,

$$\lim_{n \to \infty} S_n = \lim_{n \to \infty} 2\left[1 - \left(\frac{1}{2}\right)^n\right]$$

and since $(\tfrac{1}{2})^n$ can be made arbitrarily close to 0 by choosing n sufficiently large, it follows that $\lim\limits_{n \to \infty} S_n = 2$.

It should be emphasized that the partial sums of an infinite series do not always approach a finite limit. If a limit is approached, we say the series **converges**; otherwise, it **diverges.** The series

$$\sum_{n=1}^{\infty} 2n = 2 + 4 + 6 + 8 + \cdots$$

clearly diverges. In fact, this is the sum of an arithmetic sequence, and by equation (6)

$$S_n = 2 + 4 + 6 + 8 + \cdots + 2n = n\left[\frac{2 + (2n)}{2}\right] = n(1 + n)$$

As n becomes arbitrarily large, S_n is unbounded.

Now we want to determine under what conditions the infinite geometric series

$$\sum_{n=1}^{\infty} a_1 r^{n-1} = a_1 + a_1 r + a_1 r^2 + a_1 r^3 + \cdots \tag{11}$$

converges. By equation (9) we know that if $r \neq 1$, the nth partial sum is given by

$$S_n = \frac{a_1(1 - r^n)}{1 - r}$$

We want to examine what happens as n gets arbitrarily large. The key lies in the term r^n. When r is less than 1 in absolute value (for example, when $r = \tfrac{1}{2}$), it is shown by calculus that r^n approaches 0. So we conclude that

$$\lim_{n \to \infty} S_n = \frac{a_1}{1 - r} \quad \text{if} \quad |r| < 1$$

On the other hand, if $|r| \geq 1$, it can be proved that no finite limit exists. For example, if $r = 2$, we see that 2^n gets arbitrarily large, so that S_n cannot approach a finite limit. Similarly, if $r = -2$, then $(-2)^n$ gets larger and larger numerically, but with alternating signs. So again, S_n does not approach a finite limit.

The following general result can be proved:

> The infinite geometric series (11) converges if and only if $|r| < 1$. Furthermore, when $|r| < 1$, the sum S is
>
> $$S = \frac{a_1}{1 - r} \tag{12}$$

EXAMPLE 16   Show that the series below converges, and find its sum.

$$\sum_{n=1}^{\infty} \left(-\frac{2}{3}\right)^{n-1}$$

Solution   This is a geometric series with ratio $-\frac{2}{3}$, and since this is less than 1 in absolute value, the series converges. In expanded form the series is

$$1 - \frac{2}{3} + \frac{4}{9} - \frac{8}{27} + \cdots$$

By equation (12) the sum is

$$S = \frac{a_1}{1 - r} = \frac{1}{1 - (-\frac{2}{3})} = \frac{1}{1 + \frac{2}{3}} = \frac{3}{5}$$

It should be emphasized that we can never reach $\frac{3}{5}$ exactly by adding up any finite number of terms, but by definition, $\frac{3}{5}$ is the sum of the infinite series.

EXAMPLE 17   Show that each of the following series is divergent:

a. $1 + \frac{3}{2} + \frac{9}{4} + \frac{27}{8} + \cdots$ **b.** $\sum_{k=1}^{\infty} (-1)^{k-1}$

Solution   **a.**   The common ratio is $\frac{3}{2}$, which is greater than 1. So the series diverges.
b.   The expanded series is $1 - 1 + 1 - 1 + \cdots$, which is a geometric series with ratio -1. Since $|-1| = 1$, the series diverges. Notice the behavior of the partial sums in this example:

$$S_1 = 1, \quad S_2 = 0, \quad S_3 = 1, \quad S_4 = 0, \ldots$$

Although these do not get arbitrarily large, no specific limit is approached.

The formula (12) for the sum of an infinite geometric series can be used to find the rational number represented by a repeating decimal. This is an alternative to the procedure given in Chapter 1. The next two examples illustrate this technique.

EXAMPLE 18   Find the rational number represented by the repeating decimal 0.242424. . . .

Solution   We treat this as the sum of the infinite series

$$0.24 + 0.0024 + 0.000024 + \cdots$$

which is geometric, with $a_1 = 0.24$ and $r = 0.01$. Thus, the sum is

$$S_n = \frac{a_1}{1 - r} = \frac{0.24}{1 - 0.01} = \frac{0.24}{0.99} = \frac{24}{99} = \frac{8}{33}$$

EXAMPLE 19   Express 2.135135135. . . as a rational number.

Solution   We write this as

$$2 + [0.135 + 0.000135 + 0.000000135 + \cdots]$$

The portion in brackets is a geometric series with $a_1 = 0.135$ and $r = 0.001$. So its sum is

$$\frac{a_1}{1 - r} = \frac{0.135}{1 - 0.001} = \frac{0.135}{0.999} = \frac{135}{999} = \frac{5}{37}$$

and the answer to the problem is

$$2 + \tfrac{5}{37} = \tfrac{79}{37}$$

We conclude this section with a glimpse of the larger problem of convergence of infinite series in general (not just geometric series). At this stage we cannot go into this subject in depth, since to do so would require a knowledge of calculus. It is a fascinating part of mathematics and presents some intriguing questions. There are, in fact, some famous unsolved problems in this field.

To get a hint of one aspect of the problem, consider the following infinite series:

a. $1 + \dfrac{1}{2} + \dfrac{1}{3} + \dfrac{1}{4} + \cdots$

b. $1 - \dfrac{1}{2} + \dfrac{1}{3} - \dfrac{1}{4} + \cdots$

c. $1 + \dfrac{1}{2^2} + \dfrac{1}{3^2} + \dfrac{1}{4^2} + \cdots$

d. $1 + \dfrac{1}{2^3} + \dfrac{1}{3^3} + \dfrac{1}{4^3} + \cdots$

Without any attempt at a proof, we state the following facts:

a. The series diverges. By choosing n large enough we can make S_n arbitrarily large, even though the sum of the first billion terms is only about 21. So the divergence is quite slow.

b. This series converges, and its sum is $\ln 2$.

c. This series converges, and its sum is $\pi^2/6$. How this result was first arrived at by the famous seventeenth century Swiss mathematician Leonhard Euler is a fascinating chapter in the history of mathematics.

d. This series converges, but no one knows exactly what the sum is. This is one of the unsolved problems.

EXERCISE SET 4

A Determine which of the infinite geometric series in Problems 1–19 converge and which diverge. For those that converge, find the sum.

1. $1 + \frac{1}{3} + \frac{1}{9} + \frac{1}{27} + \cdots$

2. $1 - \frac{3}{4} + \frac{9}{16} - \frac{27}{64} + \cdots$

3. $1 + \frac{4}{3} + \frac{16}{9} + \frac{64}{27} + \cdots$

4. $\displaystyle\sum_{k=1}^{\infty} (-1.1)^{k-1}$

5. $\displaystyle\sum_{n=1}^{\infty} (-0.9)^{n-1}$

6. $\displaystyle\sum_{i=1}^{\infty} \frac{2^i}{3^{i+1}}$

7. $\displaystyle\sum_{n=1}^{\infty} \frac{(-1)^{n-1} 5^n}{6^{n-1}}$

8. $\displaystyle\sum_{n=1}^{\infty} 2^{-n}$

9. $\displaystyle\sum_{n=0}^{\infty} \left(\frac{3}{2}\right)^n$

10. $\displaystyle\sum_{n=0}^{\infty} \left(\frac{3}{2}\right)^{-n}$

11. $\displaystyle\sum_{n=1}^{\infty} 2\left(\frac{3}{5}\right)^{n-1}$

12. $\displaystyle\sum_{n=1}^{\infty} 3\left(-\frac{2}{3}\right)^{n-1}$

13. $36 - 12 + 4 + \cdots$

14. $0.01 + 0.02 + 0.04 + \cdots$

15. $\displaystyle\sum_{n=1}^{\infty} (\sqrt{2})^n$

16. $\displaystyle\sum_{k=1}^{\infty} (\sqrt{3})^{-k+1}$

17. $\displaystyle\sum_{n=1}^{\infty} 2^{-n/2}$

18. $1 + 0.1 + 0.01 + 0.001 + \cdots$

19. $1 - 0.1 + 0.01 - 0.001 + \cdots$

In Problems 20–25 find the rational number corresponding to the given decimal number using infinite series.

20. $0.333\ldots$

21. $0.151515\ldots$

22. $0.272727\ldots$

23. $1.545454\ldots$

24. $0.243243243\ldots$

25. $3.162162162\ldots$

26. Find the sum of the series $\displaystyle\sum_{n=1}^{\infty} x^{n-1}$ as a function of x. For what values of x is this valid?

27. Find the sum of the series $\displaystyle\sum_{n=0}^{\infty} (x^n/2^n)$. For what values of x is this valid?

28. If S_n is the nth partial sum of the series $\sum\limits_{n=1}^{\infty} a_n$, show that $a_1 = S_1$, and for all $n \geq 2$, show that $a_n = S_n - S_{n-1}$.

29. Use the result of Problem 28 to construct an infinite series for which $S_n = (n + 1)/n$. Show that the series converges and that its sum is 1.
Hint. Write S_n in the form $1 + (1/n)$ to show convergence.

B **30.** Show that

$$\frac{1}{n(n + 1)} = \frac{1}{n} - \frac{1}{n + 1}$$

Use this result to find the sum of the series

$$\sum_{n=1}^{\infty} \frac{1}{n(n + 1)}$$

Hint. Use the first result, and write the expanded form of S_n. Observe the terms that add to 0.

31. Find the sum of the series

$$\sum_{n=1}^{\infty} \frac{(-1)^{n-1} x^{2n-2}}{3^n}$$

Show that this is valid only when $|x| < \sqrt{3}$.

32. A ball is dropped from a height of 50 feet, and on each bounce it goes three-fourths as high as before. Approximate the total distance traveled by the ball in coming to rest.

33. The theory of infinite geometric series continues to hold true for complex values of r, and formula (12) is valid as long as $|r| < 1$. Using this fact, find the sum of the series

$$1 + \frac{i}{2} - \frac{1}{4} - \frac{i}{8} + \frac{1}{16} + \frac{i}{32} - \cdots$$

34. Find the sum of the series $\sum\limits_{n=0}^{\infty} (2x - 1)^n$. Find the range of values of x for which this result is valid.

In Problems 35–38 make use of formula (12) to find the infinite geometric series having the specified sum, and state the domain of validity.

35. $\dfrac{1}{1 - 2x}$ **36.** $\dfrac{1}{1 + (x/3)}$

37. $\dfrac{1}{2 - x}$ **38.** $\dfrac{3}{4 - x^2}$
Hint. Divide the numerator and denominator by 2.

39. For a certain series $\sum\limits_{n=1}^{\infty} a_n$, the nth partial sum is $4 - (1/3^n)$. What is the sum of the series? Show that the series is geometric.
Hint. Use Problem 28 and Definition 4.

40. In calculus it is shown that

$$e^x = 1 + x + \frac{x^2}{2!} + \frac{x^3}{3!} + \cdots + \frac{x^n}{n!} + \cdots$$

With the aid of a calculator use the first ten terms of this series to estimate the value of e by taking $x = 1$. The error in this estimate can be shown to be no greater than $3/10!$. What can you conclude is the accuracy of your estimate?

---

5 Mathematical Induction

In this section we present a method of proving certain statements, or formulas, about natural numbers. To illustrate how we might arrive at one such formula, let us consider the sum of the first n odd natural numbers. We observe the values for $n = 1, 2, 3,$ and 4:

$$n = 1: \quad 1 = 1$$
$$n = 2: \quad 1 + 3 = 4$$
$$n = 3: \quad 1 + 3 + 5 = 9$$
$$n = 4: \quad 1 + 3 + 5 + 7 = 16$$

Since the sum in each case is the square of the number of terms added, this suggests the following general statement:

The sum of the first n odd natural numbers is n^2.

Equivalently, we could write this as the formula

$$1 + 3 + 5 + \cdots + (2n - 1) = n^2$$

But have we *proved* this formula? The answer is no. We have shown it to be true for $n = 1, 2, 3,$ and 4, but that is all. Even if we continued for many more values of n and found the result true in every case, we still would not know for sure whether it was true for *all* natural numbers. It is just such situations as this where the method of proof called **mathematical induction** comes to our rescue. (We might note that the example we have used happens to be an arithmetic series, and we know a formula for the sum; however, the method of mathematical induction applies to a much wider class of problems.)

Principle of Mathematical Induction

If a statement involving natural numbers is true for $n = 1$, and if its truth for an arbitrary natural number k implies its truth for $k + 1$, then the statement is true for all natural numbers.

We will take this as an axiom, although it is possible to prove it based on other fundamental properties of the natural numbers. We will, however, show its plausibility. If we have shown that the statement in question is true for

$n = 1$, and if we have shown that its truth for $n = k$ implies its truth for $n = k + 1$, then we can let $k = 1$ and conclude that it is true also for $k + 1 = 2$. Then we can let $k = 2$ and conclude that it is true for $k + 1 = 3$. Next, letting $k = 3$, we see that it is true for $k + 1 = 4$, and so on and on. It can be seen, then, to be true for any given natural number, and hence for every natural number.

In order to state the principle of mathematical induction using fewer words, we introduce the symbol $P(n)$ for the statement to be proved (this can be thought of as a function with the natural numbers as domain and a set of statements as the range). The principle of mathematical induction can then be restated as follows:

> If (i) $P(1)$ is true, and (ii) $P(k)$ implies $P(k + 1)$ for every natural number k, then $P(n)$ is true for all natural numbers.

Both conditions (i) and (ii) are essential. To prove (i) we simply substitute $n = 1$ in the statement and see if it is true. For (ii), we *assume* the statement is true for an unspecified natural number k and show that it follows from this that the statement is true for $k + 1$. In general, it is more difficult to prove (ii) than (i).

Mathematical induction is often compared to lining up dominoes in such a way that if the first one is knocked down, all succeeding ones will fall (Figure 1). This is a pretty good analogy (provided we imagine infinitely many dominoes!). To say that the first one is knocked down is analogous to saying $P(1)$ is true. Having the dominoes spaced so that whenever any one of them (say the kth one) falls, the next one (the $k + 1$st one) will also fall, is analogous to saying that $P(k)$ implies $P(k + 1)$.

Figure 1

We illustrate the technique with several examples, the first of which is the one introduced at the beginning of this section.

EXAMPLE 20   Use mathematical induction to prove the formula

$$1 + 3 + 5 + \cdots + (2n - 1) = n^2 \tag{13}$$

Solution   Let $P(n)$ denote the given statement, that is, formula (13). To test (i) we look at $P(1)$. The left-hand side of the formula says "add all odd natural numbers up through the nth one" (which equals $2n - 1$). But for $n = 1$ the nth term is the first term, so the left-hand side consists of the single number 1. The right-hand side is 1^2, and since $1 = 1^2$, it follows that $P(1)$ is true.

Now we assume that $P(k)$ is true and try to show that $P(k + 1)$ follows, where k represents an arbitrary natural number. So we assume that

$$1 + 3 + 5 + \cdots + (2k - 1) = k^2 \qquad (14)$$

We wish to work from this and by means of valid mathematical operations obtain $P(k + 1)$. Now $P(k + 1)$ is formula (13) with n replaced by $k + 1$. The left-hand side is the sum of the first $k + 1$ odd numbers and so contains all the terms on the left-hand side of $P(k)$ plus one more term, namely $2k + 1$. So we are going to add $2k + 1$ to *both* sides of formula (14)—adding it to the left to make it what we want and adding it to the right to balance this off. This yields

$$1 + 3 + 5 + \cdots + (2k - 1) + (2k + 1) = k^2 + 2k + 1$$

or on factoring the right-hand side,

$$1 + 3 + 5 + \cdots + (2k + 1) = (k + 1)^2$$

This is exactly what formula (13) gives for $n = k + 1$; in other words, it is $P(k + 1)$. So we have shown that the truth of $P(k)$ implies that of $P(k + 1)$, and both parts (i) and (ii) of the principle of mathematical induction have been shown to be true. Therefore, formula (13) is true for all natural numbers n.

EXAMPLE 21   Prove that for all natural numbers n, $2^n \geq 2n$.

Solution   Let $P(n)$ denote the given statement. We test $P(1)$, which says

$$2^1 \geq 2 \cdot 1$$

This is true, since both sides equal 2, and since $2 \geq 2$, we see that $P(1)$ is true.

Now assume $P(k)$ is true for an arbitrary natural number k:

$$2^k \geq 2k$$

To work toward $P(k + 1)$ we can multiply both sides by 2 in order to make the left-hand side read 2^{k+1}. This gives

$$2 \cdot 2^k \geq 2 \cdot 2k$$

or

$$2^{k+1} \geq 2k + 2k$$

Since $k \geq 1$, it follows that $2k \geq 2$. Thus,

$$2^{k+1} \geq 2k + 2 = 2(k + 1)$$

Hence,

$$2^{k+1} \geq 2(k + 1)$$

which is $P(k + 1)$. Therefore the proof by mathematical induction is complete, and we conclude that $2^n \geq 2n$ for all natural numbers n.

EXAMPLE 22   Prove by mathematical induction that for all natural numbers n, the quantity $n^2 + n$ is an even number.

Solution   Let $P(n)$ represent the statement "$n^2 + n$ is an even number." Then $P(1)$ says that $1^2 + 1$ is an even number, which is true, since $1^2 + 1 = 2$. Now assume that $P(k)$ is true, that is, $k^2 + k$ is an even number. Since by definition an even number is a multiple of 2, we can say that

$$k^2 + k = 2m$$

for some integer m. For $P(k + 1)$ we want to examine $(k + 1)^2 + (k + 1)$. This can be written

$$k^2 + 2k + 1 + k + 1 = k^2 + k + 2k + 2$$
$$= (k^2 + k) + 2(k + 1)$$

By our assumption $k^2 + k = 2m$, so

$$(k + 1)^2 + (k + 1) = 2m + 2(k + 1) = 2(m + k + 1)$$

The right-hand side is a multiple of 2 and so is even. Thus, we have shown that whenever $k^2 + k$ is even, $(k + 1)^2 + (k + 1)$ is also even. So both parts of the proof by mathematical induction are complete, and we conclude that $n^2 + n$ is even for all natural numbers n.

EXAMPLE 23   Prove that

$$1^2 + 2^2 + 3^2 + \cdots + n^2 = \frac{n(n + 1)(2n + 1)}{6}$$

for all natural numbers n.

Solution   Again let $P(n)$ denote the formula to be proved. For $n = 1$ this reads

$$1^2 = \frac{1(1 + 1)(2 + 1)}{6}$$

which is seen to be true.

Now assume that $P(k)$ is true:

$$1^2 + 2^2 + 3^2 + \cdots + k^2 = \frac{k(k + 1)(2k + 1)}{6}$$

Add $(k + 1)^2$ to both sides:

$$1^2 + 2^2 + 3^2 + \cdots + k^2 + (k + 1)^2 = \frac{k(k + 1)(2k + 1)}{6} + (k + 1)^2$$

This brings the left-hand side to the proper form for $P(k + 1)$. We want to see if the right-hand side also is in the proper form.

$$\frac{k(k + 1)(2k + 1)}{6} + (k + 1)^2 = \frac{k(k + 1)(2k + 1) + 6(k + 1)^2}{6}$$

$$= \frac{(k + 1)[k(2k + 1) + 6(k + 1)]}{6}$$

$$= \frac{(k + 1)(2k^2 + 7k + 6)}{6}$$

$$= \frac{(k + 1)(k + 2)(2k + 3)}{6}$$

$$= \frac{(k + 1)[(k + 1) + 1][2(k + 1) + 1]}{6}$$

A reexamination of the right-hand side of the original formula $P(n)$ shows that the last expression arrived at is exactly what is obtained if n is replaced by $k + 1$. So by assuming $P(k)$ to be true, we have shown that $P(k + 1)$ also is true. Therefore, the proof by induction is complete.

EXERCISE SET 5

Using mathematical induction, prove that each of the following is true for all natural numbers n.

A   1. $2 + 4 + 6 + \cdots + (2n) = n(n + 1)$

2. $\dfrac{1}{1 \cdot 2} + \dfrac{1}{2 \cdot 3} + \dfrac{1}{3 \cdot 4} + \cdots + \dfrac{1}{n(n + 1)} = \dfrac{n}{n + 1}$

3. $1 + 5 + 9 + \cdots + (4n - 3) = n(2n - 1)$

4. $n^2 + 1 \geq 2n$

5. $3 + 7 + 11 + \cdots + (4n - 1) = n(2n + 1)$

6. $1 + 2 + 3 + \cdots + n = \dfrac{n(n + 1)}{2}$

7. $1 + 2^1 + 2^2 + 2^3 + \cdots + 2^{n-1} = 2^n - 1$

8.   3 is a factor of $n^3 + 2n$

9.   If $a > 1$, then $a^n > 1$.

10.   If $0 < a < 1$, then $0 < a^n < 1$.

11. $\dfrac{1}{1 \cdot 3} + \dfrac{1}{3 \cdot 5} + \dfrac{1}{5 \cdot 7} + \cdots + \dfrac{1}{(2n - 1)(2n + 1)} = \dfrac{n}{2n + 1}$

12.   6 is a factor of $n(n + 1)(n + 2)$

13. $1 \cdot 2 + 2 \cdot 3 + 3 \cdot 4 + \cdots + n(n + 1) = \dfrac{n(n + 1)(n + 2)}{3}$

14. $\displaystyle\sum_{i=1}^{n} (3i - 2) = \dfrac{n(3n - 1)}{2}$ 15. $(ab)^n = a^n b^n$

16.   2 is a factor of $n^2 - n + 2$ 17. $\displaystyle\sum_{i=1}^{n} (6i - 5) = 3n^2 - 2n$

18. 3 is a factor of $4^n - 1$

19. $\sum_{k=1}^{n} (-1)^{k-1} = \dfrac{1 - (-1)^n}{2}$

20. The nth term of an arithmetic sequence having first term a_1 and common difference d is $a_1 + (n-1)d$.

21. The nth term of a geometric sequence having first term a_1 and common ratio r is $a_1 r^{n-1}$.

B 22. $1^2 + 3^2 + 5^2 + \cdots + (2n-1)^2 = \dfrac{n(4n^2 - 1)}{3}$

23. If $x > -1$, then $(1 + x)^n \geq 1 + nx$.

24. $1^3 + 3^3 + 5^3 + \cdots + (2n-1)^3 = n^2(2n^2 - 1)$

25. $1^3 + 2^3 + 3^3 + \cdots + n^3 = (1 + 2 + 3 + \cdots + n)^2$
    Hint. See Problem 6.

26. $a - b$ is a factor of $a^n - b^n$
    Hint. For the second part of the proof write $a^{k+1} - b^{k+1} = a(a^k - b^k) + b^k(a - b)$.

27. $a + b$ is a factor of $a^{2n-1} + b^{2n-1}$
    Hint. For the second part of the proof write $a^{2k+1} + b^{2k+1} = a^2(a^{2k-1} + b^{2k-1}) - b^{2k-1}(a^2 - b^2)$.

---

6 Proof of the Binomial Theorem

In Chapter 2 we stated the binomial theorem without proving it. Mathematical induction provides one means of proving this important result. The formula to be proved can be written as follows:

$$(a + b)^n = a^n + na^{n-1}b + \frac{n(n-1)}{1 \cdot 2} a^{n-2}b^2$$

$$+ \frac{n(n-1)(n-2)}{1 \cdot 2 \cdot 3} a^{n-3}b^3 + \cdots + b^n \tag{15}$$

We wish to show that this is true for all natural numbers n.

First we introduce a convenient symbol called the **binomial coefficient symbol**:

For $k = 1, 2, 3, \ldots, n$ we define

$$\binom{n}{k} = \frac{n(n-1)(n-2) \cdots (n-k+1)}{1 \cdot 2 \cdot 3 \cdots k} \tag{16}$$

and for $k = 0$ we define $\binom{n}{0} = 1$.

The binomial formula can now be written as

$$(a + b)^n = \binom{n}{0} a^n + \binom{n}{1} a^{n-1}b + \binom{n}{2} a^{n-2}b^2 + \cdots + \binom{n}{n} b^n$$

$$= \sum_{r=0}^{n} \binom{n}{r} a^{n-r}b^r \tag{17}$$

Note that the ith term of the expansion is

$$\binom{n}{i-1} a^{n-(i-1)} b^{i-1}$$

Factorial notation enables us to give an alternate form of the binomial coefficient which is sometimes convenient. Recall that $r!$ means $1 \cdot 2 \cdot 3 \cdots r$, for any natural number r, and that we define $0! = 1$. We may therefore write

$$\binom{n}{k} = \frac{n(n-1)(n-2)\cdots(n-k+1)}{1 \cdot 2 \cdot 3 \cdots k}$$

$$= \frac{n(n-1)(n-2)\cdots(n-k+1)}{k!} \cdot \frac{(n-k)(n-k-1)\cdots 2 \cdot 1}{(n-k)(n-k-1)\cdots 2 \cdot 1}$$

and thus:

$$\binom{n}{k} = \frac{n!}{k!(n-k)!} \qquad k = 0, 1, 2, \ldots, n \qquad (18)$$

EXAMPLE 24   Evaluate each of the following in two ways, first by equation (16) and second by (18).

a. $\binom{4}{2}$ b. $\binom{5}{1}$ c. $\binom{6}{4}$ d. $\binom{8}{8}$ e. $\binom{11}{0}$

Solution   a. $\binom{4}{2} = \frac{4 \cdot 3}{1 \cdot 2} = 6$ $\qquad \binom{4}{2} = \frac{4!}{2!2!} = \frac{4 \cdot 3 \cdot 2 \cdot 1}{2 \cdot 1 \cdot 2 \cdot 1} = 6$

b. $\binom{5}{1} = \frac{5}{1} = 5$ $\qquad \binom{5}{1} = \frac{5!}{1!4!} = \frac{5 \cdot 4 \cdot 3 \cdot 2 \cdot 1}{1 \cdot 4 \cdot 3 \cdot 2 \cdot 1} = 5$

c. $\binom{6}{4} = \frac{6 \cdot 5 \cdot 4 \cdot 3}{1 \cdot 2 \cdot 3 \cdot 4} = 15$ $\qquad \binom{6}{4} = \frac{6!}{4!2!} = \frac{6 \cdot 5 \cdot 4 \cdot 3 \cdot 2 \cdot 1}{4 \cdot 3 \cdot 2 \cdot 1 \cdot 2 \cdot 1} = 15$

d. $\binom{8}{8} = \frac{8 \cdot 7 \cdot 6 \cdot 5 \cdot 4 \cdot 3 \cdot 2 \cdot 1}{1 \cdot 2 \cdot 3 \cdot 4 \cdot 5 \cdot 6 \cdot 7 \cdot 8} = 1$ $\qquad \binom{8}{8} = \frac{8!}{8!0!} = \frac{8!}{(8!)1} = 1$

e. $\binom{11}{0}$ cannot be evaluated by equation (16), but it is defined to be 1; by (18) we have

$$\binom{11}{0} = \frac{11!}{0!11!} = 1$$

In the proof of the binomial theorem we will need the following result, which you will be asked to prove in Problem 24, Exercise Set 6:

$$\binom{k}{r} + \binom{k}{r-1} = \binom{k+1}{r} \qquad (19)$$

Proof of the Binomial Theorem

We wish to show that

$$(a + b)^n = \binom{n}{0}a^n + \binom{n}{1}a^{n-1}b + \binom{n}{2}a^{n-2}b^2 + \cdots + \binom{n}{n-1}ab^{n-1} + \binom{n}{n}b^n$$

is true for all natural numbers n. Let this formula be designated by $P(n)$. Then $P(1)$ reads

$$(a + b)^1 = \binom{1}{0}a + \binom{1}{1}b$$

and since $\binom{1}{0} = 1$ and $\binom{1}{1} = 1$, we see that this is true.

Now assume $P(k)$ is true:

$$(a + b)^k = \binom{k}{0}a^k + \binom{k}{1}a^{k-1}b + \binom{k}{2}a^{k-2}b^2 + \cdots + \binom{k}{k-1}ab^{k-1} + \binom{k}{k}b^k$$

In order to make the left-hand side what we want for $P(k + 1)$, we multiply both sides by $a + b$. This gives

$$(a + b)^{k+1} = \binom{k}{0}a^{k+1} + \binom{k}{1}a^k b + \binom{k}{2}a^{k-1}b^2 + \cdots + \binom{k}{k}ab^k + \binom{k}{0}a^k b$$

$$+ \binom{k}{1}a^{k-1}b^2 + \cdots + \binom{k}{k-1}ab^k + \binom{k}{k}b^{k+1}$$

$$= \binom{k}{0}a^{k+1} + \left[\binom{k}{1} + \binom{k}{0}\right]a^k b + \left[\binom{k}{2} + \binom{k}{1}\right]a^{k-1}b^2$$

$$+ \cdots + \left[\binom{k}{k} + \binom{k}{k-1}\right]ab^k + \binom{k}{k}b^{k+1}$$

We can replace $\binom{k}{0}$ by $\binom{k+1}{0}$, since the value is 1 in either case. Also, $\binom{k}{k} = \binom{k+1}{k+1}$, since again each symbol has the value 1. Using these facts, together with equation (19), we get

$$(a + b)^{k+1} = \binom{k+1}{0}a^{k+1} + \binom{k+1}{1}a^k b + \binom{k+2}{2}a^{k-1}b^2$$

$$+ \cdots + \binom{k+1}{k}ab^k + \binom{k+1}{k+1}b^{k+1}$$

which is exactly $P(k + 1)$. Thus, $P(k)$ implies $P(k + 1)$. The proof by induction is therefore complete.

EXERCISE SET 6

A **1.** Find the value of the given binomial coefficients using equation (16).

a. $\binom{3}{2}$ **b.** $\binom{6}{3}$ **c.** $\binom{8}{2}$ **d.** $\binom{10}{3}$ **e.** $\binom{9}{9}$

f. $\dbinom{10}{5}$ g. $\dbinom{30}{2}$ h. $\dbinom{16}{5}$ i. $\dbinom{7}{6}$ j. $\dbinom{8}{6}$

2. Use equation (18) to evaluate each of the binomial coefficients in Problem 1.

Simplify the expressions in Problems 3 and 4.

3. a. $(k+1)k!$ b. $\dfrac{(k+1)!}{k!}$ c. $\dfrac{n!}{(n-2)!}$

4. a. $\dfrac{1}{k!}+\dfrac{1}{(k-1)!}$ b. $\dbinom{k}{2}+\dbinom{k}{1}$

5. Show that $\dbinom{r}{r}=1$ for any natural number r.

6. Show that $\dbinom{k}{k}=\dbinom{k+1}{k+1}$

7. Show that $\dbinom{n}{k}=\dbinom{n}{n-k}$

In Problems 8–15 expand, using equation (17), and simplify.

8. $(x+y)^6$

9. $(x-y)^5$
   Hint. Write $x-y=[x+(-y)]$.

10. $(2a+b)^7$

11. $(a-3b)^4$

12. $(3x+4y)^5$

13. $(s-2t)^6$

14. $(3r+s^2)^4$

15. $(x^2-2y)^6$

In Problems 16–20 find the specified term of the expansion.

16. $(x+y)^{10}$; eighth term

17. $(x-y)^{12}$; fifth term

18. $(2x+3y)^{12}$; ninth term

19. $(3a-b)^7$; fourth term

20. $(a-2b)^{13}$; third term

21. Evaluate $(1.02)^6$ to three decimal places of accuracy using the binomial theorem. **Hint.** Write $(1.02)^6$ as $(1+0.02)^6$.

22. Evaluate $(0.99)^8$ to the nearest thousandth using the binomial theorem.

23. Show that $\dbinom{n}{0}+\dbinom{n}{1}+\dbinom{n}{2}+\cdots+\dbinom{n}{n}=2^n$.
    Hint. Consider $(1+1)^n$.

B 24. Prove the identity (19).

25. Prove the identity

$$\frac{n-r+1}{r}\cdot\binom{n}{r-1}=\binom{n}{r}$$

26. It can be shown that if $|x|<1$ and α is not a positive integer, then

$$(1+x)^\alpha=\sum_{r=0}^{\infty}\binom{\alpha}{r}x^r$$

where $\binom{\alpha}{r}$ is given by equation (16). Use this to find the first five terms of $(1+x)^{-2}$, where $|x|<1$.

27. Use Problem 26 to write the first five terms of the expansion of $\sqrt{1 + x}$, assuming $|x| < 1$.

In Problems 28 and 29 use the expansion of Problem 27 to obtain the answer to five decimal places of accuracy.

28. $\sqrt{1.02}$ $\qquad\qquad\qquad\qquad\qquad\qquad$ **29.** $\sqrt{0.99}$

30. Use Problem 26 to evaluate $1/(1.01)^2$ to six decimal places of accuracy.

Review Exercise Set

A   Write the first four terms of the sequence whose nth term is given in Problems 1–3.

1. **a.** $a_n = \dfrac{n}{n+1}$ $\qquad\qquad\qquad\qquad$ **b.** $a_n = 4n - 3$

2. **a.** $a_n = \dfrac{(-1)^{n-1}}{\sqrt{2n-1}}$ $\qquad\qquad\quad$ **b.** $a_n = \dfrac{(-1)^{n-1}n}{n^2 + 1}$

3. **a.** $a_n = \dfrac{(-3)^{n-1}}{4^n}$ $\qquad\qquad\qquad$ **b.** $a_n = \dfrac{(-1)^{n-1}}{n \log(n+1)}$

In Problems 4–6 find an expression for the nth term of each sequence.

4. **a.** $1, -\frac{1}{3}, \frac{1}{5}, -\frac{1}{7}, \ldots$ $\qquad\qquad$ **b.** $1, \frac{3}{4}, \frac{9}{16}, \frac{27}{64}, \cdots$

5. **a.** $\dfrac{1}{1 \cdot 2}, \dfrac{1}{2 \cdot 3}, \dfrac{1}{3 \cdot 4}, \dfrac{1}{4 \cdot 5}, \cdots$ $\quad$ **b.** $\frac{1}{2}, \frac{2}{3}, \frac{3}{4}, \frac{4}{5}, \cdots$

6. **a.** $1, -\frac{1}{4}, \frac{1}{16}, -\frac{1}{64}, \cdots$ $\qquad$ **b.** $1, 4, 7, 10, \cdots$

In Problems 7–10 identify the given sequence as being arithmetic or geometric, and find the term specified.

7. $1, \frac{3}{2}, 2, \frac{5}{2}, \cdots;\quad a_{20}$ $\qquad\qquad$ **8.** $1, \frac{3}{2}, \frac{9}{4}, \frac{27}{8}, \cdots;\quad a_8$

9. $-2, \frac{1}{2}, -\frac{1}{8}, \frac{1}{32}, \cdots;\quad a_7$ $\qquad$ **10.** $-1, 3, 7, 11, \cdots;\quad a_{15}$

11. Write the first five terms of the sequence defined by:
 a. $a_1 = 2;\quad a_n = 2a_{n-1} + 1,\quad n \geq 2$ $\quad$ **b.** $a_1 = 1;\quad a_n = na_{n-1},\ n \geq 2$
12. In a certain arithmetic sequence, $a_{12} = 7$ and $a_{18} = -5$. Find a_{30}.
13. Find y so that the sequence 3, y, 15 will be:
 a. Arithmetic $\qquad$ **b.** Geometric
14. Insert five arithmetic means between 3 and 11.
15. On each swing after the first a certain pendulum swings three-fourths as far as on the preceding swing. If it swings 16 inches initially, find:
 a. How far it swings on the sixth swing
 b. The total distance swung during the first six swings
 c. The approximate distance covered before coming to rest
16. Write the expanded form of each of the following sums:

 a. $\displaystyle\sum_{k=1}^{5} \frac{k^2}{2k+3}$ $\quad$ **b.** $\displaystyle\sum_{n=1}^{6} \frac{(-1)^{n-1}n}{n^2+1}$ $\quad$ **c.** $\displaystyle\sum_{m=0}^{5} \frac{(-3)^m}{(1+m)^2}$

 d. $\displaystyle\sum_{i=2}^{6} \frac{2i-3}{3i+4}$ $\quad$ **e.** $\displaystyle\sum_{n=1}^{10} e^{-n}\ln(n+1)$

Find the sums in Problems 17–26.

17. The sum of the first thirty terms of the sequence $-5, -1, 3, 7, \ldots$.

18. The sum of the first ten terms of the sequence $-4, 2, -1, \frac{1}{2}, \ldots$.

19. $\displaystyle\sum_{n=1}^{50} (3 - 2n)$

20. $\displaystyle\sum_{k=1}^{6} \frac{(-1)^{k-1}}{3^k}$

21. $\displaystyle\sum_{j=1}^{60} \left(\frac{1 - 2j}{3}\right)$

22. $\displaystyle\sum_{n=1}^{8} (-2)^{n-3}$

23. $\displaystyle\sum_{k=1}^{12} \left(3 - \frac{k}{4}\right)$

24. $\displaystyle\sum_{n=1}^{\infty} \left(\frac{5}{6}\right)^{n-1}$

25. $\displaystyle\sum_{k=1}^{\infty} (-1)^{k-1} \left(\frac{3}{5}\right)^k$

26. $\displaystyle\sum_{n=1}^{\infty} (0.1)^{n-1}$

27. Use a geometric series to find the rational number represented by each of the following:
   a. $0.545454\ldots$ **b.** $0.148148148\ldots$ **c.** $2.181818\ldots$
   d. $27.135135135\ldots$ **e.** $1.020202\ldots$

28. A manufacturer estimates that a piece of equipment costing $60,000 depreciates by 25% each year. Find its value after 5 years.

29. Logs are stacked so that each layer after the bottom one has one log fewer than the layer below it. If there are 20 logs on the bottom layer and 8 logs on the top, how many logs are in the pile?

30. A culture of bacteria doubles in size every 2 hours. If initially it contains 200 bacteria, how many will it contain 10 hours later?

In Problems 31–38 use mathematical induction to prove that the given statement is true for all natural numbers.

31. $4 + 8 + 12 + \cdots + 4n = 2n(n + 1)$

32. $3 + 5 + 7 + \cdots + (2n + 1) = n^2 + 2n$

33. $\displaystyle\sum_{r=1}^{n} \frac{3}{2r(r + 1)} = \frac{3n}{2n + 2}$

34. $n^2 \geq 4(n - 1)$

35. $n < a^n$ if $a \geq 2$

36. 2 is a factor of $n^2 + 3n$

37. 4 is a factor of $2n^2 + 6n$

38. $n^2 + 1 < (n + 1)^2$

39. Evaluate each of the following binomial coefficients:

   a. $\dbinom{25}{3}$ **b.** $\dbinom{9}{6}$ **c.** $\dbinom{4}{3}$ **d.** $\dbinom{3}{0}$ **e.** $\dbinom{12}{10}$

In Problems 40 and 41 use the binomial theorem to expand, and simplify the result.

40. **a.** $(x + 2y)^8$ **b.** $(3a - 4b)^5$

41. **a.** $\left(x - \frac{y}{2}\right)^7$ **b.** $(1 + x^2)^{10}$

42. Find the value of each of the following correct to five decimal places, using the binomial theorem:
   a. $(1.01)^{10}$ **b.** $(0.98)^6$

43. Find the specified term of the given expansion:
   a. $(2x - y)^{12}$;   eighth term   **b.** $(x - 2)^{20}$;   tenth term

B   Find the sums in Problems 44 and 45.

44. $\displaystyle\sum_{k=1}^{8} (100k - 2^k)$

45. $\displaystyle\sum_{n=1}^{6} \frac{2 - n \cdot 3^n}{3^n}$

46. Find a formula for the sum of the first n positive integers. Use mathematical induction to prove your result.

47. In a certain city, housing costs have risen at an annual rate of 10% for the past 4 years. If, 4 years ago, a particular house cost \$50,000, what would be its fair market value now?

48. A man started to work 10 years ago at a salary of \$10,000 per year and has received annual raises of 6%. What is his salary now, and what is the total amount of money he has received during the past 10 years?

49. A ladder with 13 rungs is to be constructed so that each rung is shorter than the preceding one by a constant amount. If the initial rung is to be 24 inches long and the final one 15 inches long, what are the lengths of the intermediate rungs?

50. After college a boy agreed to pay off his debt of \$2,550 to his father by paying \$100 the first month, \$110 the second month, \$120 the third month, and so on. How many months were required to pay off the debt?

51. Prove the identity

$$\binom{n}{k} - \binom{n-1}{k} = \binom{n-1}{k-1}$$

52. Find the sum of the given infinite geometric series, and state the domain of validity:

a. $1 - \dfrac{x}{2} + \dfrac{x^2}{4} - \dfrac{x^3}{8} + \cdots$ **b.** $\displaystyle\sum_{n=1}^{\infty} \dfrac{x^{2n-1}}{3^{n-1}}$

53. Find the infinite geometric series having the given sum, and state the domain of validity:

a. $\dfrac{1}{1+x}$ **b.** $\dfrac{3}{2-x^2}$

In Problems 54–56 prove the given statement by mathematical induction.

54. $1 + 2\cdot 2 + 3\cdot 2^2 + 4\cdot 2^3 + \cdots + n\cdot 2^{n-1} = 1 + (n-1)\cdot 2^n$

55. $\dfrac{1}{1\cdot 2\cdot 3} + \dfrac{1}{2\cdot 3\cdot 4} + \dfrac{1}{3\cdot 4\cdot 5} + \cdots + \dfrac{1}{n(n+1)(n+2)} = \dfrac{n(n+3)}{4(n+1)(n+2)}$

56. $|a_1 + a_2 + a_3 + \cdots + a_n| \le |a_1| + |a_2| + |a_3| + \cdots + |a_n|$

Cumulative Review Exercise Set IV (Chapters 10–13)

1. Solve the following system in three ways:
   a. By elimination using addition or subtraction and substitution
   b. By Cramer's rule
   c. By reducing the augmented matrix to triangular form

$$\begin{cases} 2x - 6y - 3z = 8 \\ x + 2y - 4z = -1 \\ 3x + 4y + z = 5 \end{cases}$$

2. a. Multiply:

$$\begin{bmatrix} 2 & -1 & 4 & 1 \\ 3 & 2 & -5 & -2 \\ -6 & 0 & 2 & 3 \end{bmatrix} \begin{bmatrix} 1 & 4 & -3 \\ 2 & 0 & 5 \\ -1 & 3 & 2 \\ -4 & 6 & -3 \end{bmatrix}$$

   b. Find X such that $AX = B$, where

$$A = \begin{bmatrix} 2 & 3 & 1 \\ 1 & -2 & -2 \\ 4 & 9 & 3 \end{bmatrix} \qquad X = \begin{bmatrix} x \\ y \\ z \end{bmatrix} \qquad B = \begin{bmatrix} 4 \\ -1 \\ 6 \end{bmatrix}$$

3. Discuss intercepts, symmetry, excluded regions, asymptotes, and sketch the graph.

$$y^2 = \frac{4(x - 4)}{x^2 - 9}$$

4. Find the equation of the parabola having horizontal axis and passing through the vertices of the triangle whose sides have equations $2x - y = 3$, $2x + y + 7 = 0$, and $x + 2y = 4$. Show that this is a right triangle.

5. a. If the average annual inflation rate is 12% for the 5 year period beginning in 1980, what will be the purchasing power of $1 in 1985 as compared to its value in 1980?
   b. The area bounded by the curve $y = 2^x$, the x axis, and the lines $x = 0$ and $x = 2$ is to be approximated by inscribing rectangles of width 0.1 (see sketch). Find the sum of the areas of the rectangles as a finite series and give an expression for the sum of the series. Use a calculator to approximate the result.

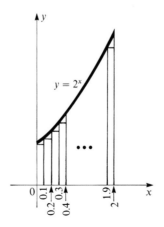

6. Find the equation of the circle whose center is the upper focus of the hyperbola $9x^2 - 16y^2 - 54x + 64y + 161 = 0$, and which passes through the center of the hyperbola. Draw both curves.

517

7. The proprietor of a coffee house prepared 100 pounds of a special blend of coffee using three different varieties: Colombian, Sumatran, and Brazilian. Colombian sells for $5.30 per pound, Sumatran for $5.00 per pound, and Brazilian for $4.80 per pound. He calculates that the price of the blend should be $5.04 per pound. If he used 50% more Colombian than Brazilian coffee, how many pounds of each did he use?

8. Discuss intercepts, symmetry, excluded regions, and asymptotes, and sketch the graph.

$$y = \frac{\sqrt{x^2 - 16}}{x}$$

9. The owner of a sporting goods store sells two types of roller skates: the standard skate on which she makes a profit of $5.00 per pair, and the deluxe model on which the profit is $8.00 per pair. From experience she expects to sell at least 50 but not more than 100 standard pairs and between 30 and 60 deluxe pairs each month. Because of limitations on storage capacity she can handle at most 140 pairs of skates at a time. Assuming she sells all she has in stock, how many should she stock per month to maximize profit?

10. Solve the following system by matrix methods:

$$\begin{cases} 2x + y - 3z + 2w = 4 \\ 5x - 8y - 6z + 3w = 7 \\ 3x - 2y - 4z + w = 5 \end{cases}$$

11. Solve graphically.

a. $\begin{cases} y \le 2x + 5 \\ y \le 6 - x \\ y \ge x^2 - 2x \end{cases}$

   Find the points of intersection of the boundary curves.

b. $|x| + |y - 1| < 2$

12. An amusement park charges $1.20 admission for children under 18, $3.00 for adults between 18 and 65, and $1.50 for senior citizens over 65. On a given Saturday there were 3,950 tickets sold, and the receipts totaled $7,110. If the number of children exceeded the total number of adults and senior citizens by 550, how many tickets of each type were sold?

13. Find the equation of the parabola whose vertex and focus, respectively, coincide with the upper vertex and focus of the ellipse

$$4x^2 + 3y^2 + 16x + 6y - 29 = 0$$

   Draw both curves.

14. Discuss intercepts, symmetry, excluded regions, asymptotes, and sketch the graph.

$$y = \frac{x^2 - 3x - 4}{4(1 - x)}$$

15. Let

$$A = \begin{bmatrix} 3 & 2 & 2 \\ 1 & 4 & 1 \\ -2 & -4 & -1 \end{bmatrix} \qquad I = \begin{bmatrix} 1 & 0 & 0 \\ 0 & 1 & 0 \\ 0 & 0 & 1 \end{bmatrix}$$

   Solve the equation $|A - \lambda I| = 0$ for λ.

16. Identify and draw the graph of the equation $x^2 - 3xy - 3y^2 - 21 = 0$ after a suitable rotation of axes.

17. Find the indicated sums.

a. $\displaystyle\sum_{k=1}^{20} \left(\frac{3k - 5}{6}\right)$ b. $\displaystyle\sum_{n=0}^{\infty} e^{-2n}$

18. In a small machine shop two types of parts are produced, each of which requires time on the lathe. The first type requires 10 minutes and the second type 6 minutes. Material for the first type costs $0.50 per part and material for the second type costs $0.75 per part. Labor and equipment charges for the lathe amount to $15.00 per hour. The lathe is to be used between 6 and 8 hours per day in the production of these parts. If it is necessary to produce at least 21 parts of the first type and 15 parts of the second type each day, how many parts of each type should be produced to minimize cost?

19. Find an infinite geometric series whose sum is $1/(4 + x^2)$. For what values of x is the result valid?

20. Solve the following system and draw the graph:

$$\begin{cases} x^2 + 2y^2 = 6 \\ 2x - y = 3 \end{cases}$$

21. Find the equation of the hyperbola with eccentricity $3/\sqrt{5}$ and with transverse axis that is the line segment joining $(-4, 3)$ and $(-4, -1)$. Find the equations of the asymptotes and draw the curve.

22. Prove the formula below in two ways.

$$\sum_{k=1}^{n} \log 2^k = \left(\frac{n^2 + n}{2} \right) \log 2$$

a. By mathematical induction
b. By showing that the sum is an arithmetic progression

23. Rotate the axes so as to eliminate the $x'y'$ term and draw the graph of the equation $16x^2 - 24xy + 9y^2 - 25x + 50y = 75$.

24. A sequence is defined by

$$\begin{cases} a_1 = \sqrt{2} \\ a_n = \sqrt{2 + a_{n-1}} \end{cases} \quad (n \geq 2)$$

Write the first five terms of the sequence. Prove by mathematical induction that $a_n \leq 2$ for all n.

25. Discuss intercepts, symmetry, excluded regions, and asymptotes, and sketch the graph.

$$y^2 = x^2 - \frac{3}{x + 2}$$

26. Nineteen pieces of pipe are to be cut so that the first piece is 2 feet long and the last piece is 14 feet long. Each piece after the first is longer than the preceding one by a fixed amount. Find this amount. Find the sum of the lengths of all the pieces of pipe.

27. a. Write the sixth term of the expansion of $(x^2 - 2y)^9$ and simplify your result.
b. Expand $(1 + x)^{-1}$ to five terms using the binomial formula. If $|x| < 1$, it can be shown that the infinite series generated by the continuation of this expansion converges to the given function. Find the nth term of the expansion. What kind of series is this? Find the sum of the series and show that this is in agreement with the original function.

28. The ends of a barn are in the shape of a square surmounted by an isosceles trapezoid (see sketch). The top of the trapezoid is half as long as the bottom, and the overall

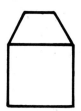

height of the barn is 32 feet. If the total area of the square and trapezoid is 580 square feet, find the dimensions of the side of the square and the height of the trapezoid. What is the overall perimeter?

29.   Prove by mathematical induction:

$$\sin x + \sin 2x + \cdots + \sin nx = \frac{\cos \frac{1}{2}x - \cos(n + \frac{1}{2})x}{2 \sin \frac{1}{2}x} \qquad (x \neq 2k\pi, k = 0, \pm 1, \pm 2, \ldots)$$

Hint.   Use the sum and product formulas for the sine and cosine.

Tables

Table I

Squares, Square Roots, and Prime Factors

No.	Sq.	Sq. Rt.	Factors	No.	Sq.	Sq. Rt.	Factors
1	1	1.000		51	2,601	7.141	$3 \cdot 17$
2	4	1.414	2	52	2,704	7.211	$2^2 \cdot 13$
3	9	1.732	3	53	2,809	7.280	53
4	16	2.000	2^2	54	2,916	7.348	$2 \cdot 3^3$
5	25	2.236	5	55	3,025	7.416	$5 \cdot 11$
6	36	2.449	$2 \cdot 3$	56	3,136	7.483	$2^3 \cdot 7$
7	49	2.646	7	57	3,249	7.550	$3 \cdot 19$
8	64	2.828	2^3	58	3,364	7.616	$2 \cdot 29$
9	81	3.000	3^2	59	3,481	7.681	59
10	100	3.162	$2 \cdot 5$	60	3,600	7.746	$2^2 \cdot 3 \cdot 5$
11	121	3.317	11	61	3,721	7.810	61
12	144	3.464	$2^2 \cdot 3$	62	3,844	7.874	$2 \cdot 31$
13	169	3.606	13	63	3,969	7.937	$3^2 \cdot 7$
14	196	3.742	$2 \cdot 7$	64	4,096	8.000	2^6
15	225	3.873	$3 \cdot 5$	65	4,225	8.062	$5 \cdot 13$
16	256	4.000	2^4	66	4,356	8.124	$2 \cdot 3 \cdot 11$
17	289	4.123	17	67	4,489	8.185	67
18	324	4.243	$2 \cdot 3^2$	68	4,624	8.246	$2^2 \cdot 17$
19	361	4.359	19	69	4,761	8.307	$3 \cdot 23$
20	400	4.472	$2^2 \cdot 5$	70	4,900	8.367	$2 \cdot 5 \cdot 7$
21	441	4.583	$3 \cdot 7$	71	5,041	8.426	71
22	484	4.690	$2 \cdot 11$	72	5,184	8.485	$2^3 \cdot 3^2$
23	529	4.796	23	73	5,329	8.544	73
24	576	4.899	$2^3 \cdot 3$	74	5,476	8.602	$2 \cdot 37$
25	625	5.000	5^2	75	5,625	8.660	$3 \cdot 5^2$
26	676	5.099	$2 \cdot 13$	76	5,776	8.718	$2^2 \cdot 19$
27	729	5.196	3^3	77	5,929	8.775	$7 \cdot 11$
28	784	5.292	$2^2 \cdot 7$	78	6,084	8.832	$2 \cdot 3 \cdot 13$
29	841	5.385	29	79	6,241	8.888	79
30	900	5.477	$2 \cdot 3 \cdot 5$	80	6,400	8.944	$2^4 \cdot 5$
31	961	5.568	31	81	6,561	9.000	3^4
32	1,024	5.657	2^5	82	6,724	9.055	$2 \cdot 41$
33	1,089	5.745	$3 \cdot 11$	83	6,889	9.110	83
34	1,156	5.831	$2 \cdot 17$	84	7,056	9.165	$2^2 \cdot 3 \cdot 7$
35	1,225	5.916	$5 \cdot 7$	85	7,225	9.220	$5 \cdot 17$
36	1,296	6.000	$2^2 \cdot 3^2$	86	7,396	9.274	$2 \cdot 43$
37	1,369	6.083	37	87	7,569	9.327	$3 \cdot 29$
38	1,444	6.164	$2 \cdot 19$	88	7,744	9.381	$2^3 \cdot 11$
39	1,521	6.245	$3 \cdot 13$	89	7,921	9.434	89
40	1,600	6.325	$2^3 \cdot 5$	90	8,100	9.487	$2 \cdot 3^2 \cdot 5$
41	1,681	6.403	41	91	8,281	9.539	$7 \cdot 13$
42	1,764	6.481	$2 \cdot 3 \cdot 7$	92	8,464	9.592	$2^2 \cdot 23$
43	1,849	6.557	43	93	8,649	9.644	$3 \cdot 31$
44	1,936	6.633	$2^2 \cdot 11$	94	8,836	9.695	$2 \cdot 47$
45	2,025	6.708	$3^2 \cdot 5$	95	9,025	9.747	$5 \cdot 19$
46	2,116	6.782	$2 \cdot 23$	96	9,216	9.798	$2^5 \cdot 3$
47	2,209	6.856	47	97	9,409	9.849	97
48	2,304	6.928	$2^4 \cdot 3$	98	9,604	9.899	$2 \cdot 7^2$
49	2,401	7.000	7^2	99	9,801	9.950	$3^2 \cdot 11$
50	2,500	7.071	$2 \cdot 5^2$	100	10,000	10.000	$2^2 \cdot 5^2$

$x^p = a \log_a x^p$

Table II

Common Logarithms

x	0	1	2	3	4	5	6	7	8	9
1.0	.0000	.0043	.0086	.0128	.0170	.0212	.0253	.0294	.0334	.0374
1.1	.0414	.0453	.0492	.0531	.0569	.0607	.0645	.0682	.0719	.0755
1.2	.0792	.0828	.0864	.0899	.0934	.0969	.1004	.1038	.1072	.1106
1.3	.1139	.1173	.1206	.1239	.1271	.1303	.1335	.1367	.1399	.1430
1.4	.1461	.1492	.1523	.1553	.1584	.1614	.1644	.1673	.1703	.1732
1.5	.1761	.1790	.1818	.1847	.1875	.1903	.1931	.1959	.1987	.2014
1.6	.2041	.2068	.2095	.2122	.2148	.2175	.2201	.2227	.2253	.2279
1.7	.2304	.2330	.2355	.2380	.2405	.2430	.2455	.2480	.2504	.2529
1.8	.2553	.2577	.2601	.2625	.2648	.2672	.2695	.2718	.2742	.2765
1.9	.2788	.2810	.2833	.2856	.2878	.2900	.2923	.2945	.2967	.2989
2.0	.3010	.3032	.3054	.3075	.3096	.3118	.3139	.3160	.3181	.3201
2.1	.3222	.3243	.3263	.3284	.3304	.3324	.3345	.3365	.3385	.3404
2.2	.3424	.3444	.3464	.3483	.3502	.3522	.3541	.3560	.3579	.3598
2.3	.3617	.3636	.3655	.3674	.3692	.3711	.3729	.3747	.3766	.3784
2.4	.3802	.3820	.3838	.3856	.3874	.3892	.3909	.3927	.3945	.3962
2.5	.3979	.3997	.4014	.4031	.4048	.4065	.4082	.4099	.4116	.4133
2.6	.4150	.4166	.4183	.4200	.4216	.4232	.4249	.4265	.4281	.4298
2.7	.4314	.4330	.4346	.4362	.4378	.4393	.4409	.4425	.4440	.4456
2.8	.4472	.4487	.4502	.4518	.4533	.4548	.4564	.4579	.4594	.4609
2.9	.4624	.4639	.4654	.4669	.4683	.4698	.4713	.4728	.4742	.4757
3.0	.4771	.4786	.4800	.4814	.4829	.4843	.4857	.4871	.4886	.4900
3.1	.4914	.4928	.4942	.4955	.4969	.4983	.4997	.5011	.5024	.5038
3.2	.5051	.5065	.5079	.5092	.5105	.5119	.5132	.5145	.5159	.5172
3.3	.5185	.5198	.5211	.5224	.5237	.5250	.5263	.5276	.5289	.5302
3.4	.5315	.5328	.5340	.5353	.5366	.5378	.5391	.5403	.5416	.5428
3.5	.5441	.5453	.5465	.5478	.5490	.5502	.5514	.5527	.5539	.5551
3.6	.5563	.5575	.5587	.5599	.5611	.5623	.5635	.5647	.5658	.5670
3.7	.5682	.5694	.5705	.5717	.5729	.5740	.5752	.5763	.5775	.5786
3.8	.5798	.5809	.5821	.5832	.5843	.5855	.5866	.5877	.5888	.5899
3.9	.5911	.5922	.5933	.5944	.5955	.5966	.5977	.5988	.5999	.6010
4.0	.6021	.6031	.6042	.6053	.6064	.6075	.6085	.6096	.6107	.6117
4.1	.6128	.6138	.6149	.6160	.6170	.6180	.6191	.6201	.6212	.6222
4.2	.6232	.6243	.6253	.6263	.6274	.6284	.6294	.6304	.6314	.6325
4.3	.6335	.6345	.6355	.6365	.6375	.6385	.6395	.6405	.6415	.6425
4.4	.6435	.6444	.6454	.6464	.6474	.6484	.6493	.6503	.6513	.6522
4.5	.6532	.6542	.6551	.6561	.6571	.6580	.6590	.6599	.6609	.6618
4.6	.6628	.6637	.6646	.6656	.6665	.6675	.6684	.6693	.6702	.6712
4.7	.6721	.6730	.6739	.6749	.6758	.6767	.6776	.6785	.6794	.6803
4.8	.6812	.6821	.6830	.6839	.6848	.6857	.6866	.6875	.6884	.6893
4.9	.6902	.6911	.6920	.6928	.6937	.6946	.6955	.6964	.6972	.6981
5.0	.6990	.6998	.7007	.7016	.7024	.7033	.7042	.7050	.7059	.7067
5.1	.7076	.7084	.7093	.7101	.7110	.7118	.7126	.7135	.7143	.7152
5.2	.7160	.7168	.7177	.7185	.7193	.7202	.7210	.7218	.7226	.2735
5.3	.7243	.7251	.7259	.7267	.7275	.7284	.7292	.7300	.7308	.7316
5.4	.7324	.7332	.7340	.7348	.7356	.7364	.7372	.7380	.7388	.7396
x	0	1	2	3	4	5	6	7	8	9

$\log_a x^p = p \frac{1}{x} \frac{\partial x}{\partial x}$

$\log_a x^p = p \log_a x$

Table II   Common Logarithms (continued)

x	0	1	2	3	4	5	6	7	8	9
5.5	.7404	.7412	.7419	.7427	.7435	.7443	.7451	.7459	.7466	.7474
5.6	.7482	.7490	.7497	.7505	.7513	.7520	.7528	.7536	.7543	.7551
5.7	.7559	.7566	.7574	.7582	.7589	.7597	.7604	.7612	.7619	.7627
5.8	.7634	.7642	.7649	.7657	.7664	.7672	.7679	.7686	.7694	.7701
5.9	.7709	.7716	.7723	.7731	.7738	.7745	.7752	.7760	.7767	.7774
6.0	.7782	.7789	.7796	.7803	.7810	.7818	.7825	.7832	.7839	.7846
6.1	.7853	.7860	.7868	.7875	.7882	.7889	.7896	.7903	.7910	.7917
6.2	.7924	.7931	.7938	.7945	.7952	.7959	.7966	.7973	.7980	.7987
6.3	.7993	.8000	.8007	.8014	.8021	.8028	.8035	.8041	.8048	.8055
6.4	.8062	.8069	.8075	.8082	.8089	.8096	.8102	.8109	.8116	.8122
6.5	.8129	.8136	.8142	.8149	.8156	.8162	.8169	.8176	.8182	.8189
6.6	.8195	.8202	.8209	.8215	.8222	.8228	.8235	.8241	.8248	.8254
6.7	.8261	.8267	.8274	.8280	.8287	.8293	.8299	.8306	.8312	.8319
6.8	.8325	.8331	.8338	.8344	.8351	.8357	.8363	.8370	.8376	.8382
6.9	.8388	.8395	.8401	.8407	.8414	.8420	.8426	.8432	.8439	.8445
7.0	.8451	.8457	.8463	.8470	.8476	.8482	.8488	.8494	.8500	.8506
7.1	.8513	.8519	.8525	.8531	.8537	.8543	.8549	.8555	.8561	.8567
7.2	.8573	.8579	.8585	.8591	.8597	.8603	.8609	.8615	.8621	.8627
7.3	.8633	.8639	.8645	.8651	.8657	.8663	.8669	.8675	.8681	.8686
7.4	.8692	.8698	.8704	.8710	.8716	.8722	.8727	.8733	.8739	.8745
7.5	.8751	.8756	.8762	.8768	.8774	.8779	.8785	.8791	.8797	.8802
7.6	.8808	.8814	.8820	.8825	.8831	.8837	.8842	.8848	.8854	.8859
7.7	.8865	.8871	.8876	.8882	.8887	.8893	.8899	.8904	.8910	.8915
7.8	.8921	.8927	.8932	.8938	.8943	.8949	.8954	.8960	.8965	.8971
7.9	.8976	.8982	.8987	.8993	.8998	.9004	.9009	.9015	.9020	.9025
8.0	.9031	.9036	.9042	.9047	.9053	.9058	.9063	.9069	.9074	.9079
8.1	.9085	.9090	.9096	.9101	.9106	.9112	.9117	.9122	.9128	.9133
8.2	.9138	.9143	.9149	.9154	.9159	.9165	.9170	.9175	.9180	.9186
8.3	.9191	.9196	.9201	.9206	.9212	.9217	.9222	.9227	.9232	.9238
8.4	.9243	.9248	.9253	.9258	.9263	.9269	.9274	.9279	.9284	.9289
8.5	.9294	.9299	.9304	.9309	.9315	.9320	.9325	.9330	.9335	.9340
8.6	.9345	.9350	.9355	.9360	.9365	.9370	.9375	.9380	.9385	.9390
8.7	.9395	.9400	.9405	.9410	.9415	.9420	.9425	.9430	.9435	.9440
8.8	.9445	.9450	.9455	.9460	.9465	.9469	.9474	.9479	.9484	.9489
8.9	.9494	.9499	.9504	.9509	.9513	.9518	.9523	.9528	.9533	.9538
9.0	.9542	.9547	.9552	.9557	.9562	.9566	.9571	.9576	.9581	.9586
9.1	.9590	.9595	.9600	.9605	.9609	.9614	.9619	.9624	.9628	.9633
9.2	.9638	.9643	.9647	.9652	.9657	.9661	.9666	.9671	.9675	.9680
9.3	.9685	.9689	.9694	.9699	.9703	.9708	.9713	.9717	.9722	.9727
9.4	.9731	.9736	.9741	.9745	.9750	.9754	.9759	.9763	.9768	.9773
9.5	.9777	.9782	.9786	.9791	.9795	.9800	.9805	.9809	.9814	.9818
9.6	.9823	.9827	.9832	.9836	.9841	.9845	.9850	.9854	.9859	.9863
9.7	.9868	.9872	.9877	.9881	.9886	.9890	.9894	.9899	.9903	.9908
9.8	.9912	.9917	.9921	.9926	.9930	.9934	.9939	.9943	.9948	.9952
9.9	.9956	.9961	.9965	.9969	.9974	.9978	.9983	.9987	.9991	.9996
x	0	1	2	3	4	5	6	7	8	9

Table III

Exponential Functions

x	e^x	e^{-x}	x	e^x	e^{-x}
0.00	1.0000	1.0000	1.5	4.4817	0.2231
0.01	1.0101	0.9901	1.6	4.9530	0.2019
0.02	1.0202	0.9802	1.7	5.4739	0.1827
0.03	1.0305	0.9705	1.8	6.0496	0.1653
0.04	1.0408	0.9608	1.9	6.6859	0.1496
0.05	1.0513	0.9512	2.0	7.3891	0.1353
0.06	1.0618	0.9418	2.1	8.1662	0.1225
0.07	1.0725	0.9324	2.2	9.0250	0.1108
0.08	1.0833	0.9331	2.3	9.9742	0.1003
0.09	1.0942	0.9139	2.4	11.023	0.0907
0.10	1.1052	0.9048	2.5	12.182	0.0821
0.11	1.1163	0.8958	2.6	13.464	0.0743
0.12	1.1275	0.8869	2.7	14.880	0.0672
0.13	1.1388	0.8781	2.8	16.445	0.0608
0.14	1.1503	0.8694	2.9	18.174	0.0550
0.15	1.1618	0.8607	3.0	20.086	0.0498
0.16	1.1735	0.8521	3.1	22.198	0.0450
0.17	1.1853	0.8437	3.2	24.533	0.0408
0.18	1.1972	0.8353	3.3	27.113	0.0369
0.19	1.2092	0.8270	3.4	29.964	0.0334
0.20	1.2214	0.8187	3.5	33.115	0.0302
0.21	1.2337	0.8106	3.6	36.598	0.0273
0.22	1.2461	0.8025	3.7	40.447	0.0247
0.23	1.2586	0.7945	3.8	44.701	0.0224
0.24	1.2712	0.7866	3.9	49.402	0.0202
0.25	1.2840	0.7788	4.0	54.598	0.0183
0.30	1.3499	0.7408	4.1	60.340	0.0166
0.35	1.4191	0.7047	4.2	66.686	0.0150
0.40	1.4918	0.6703	4.3	73.700	0.0136
0.45	1.5683	0.6376	4.4	81.451	0.0123
0.50	1.6487	0.6065	4.5	90.017	0.0111
0.55	1.7333	0.5769	4.6	99.484	0.0101
0.60	1.8221	0.5488	4.7	109.95	0.0091
0.65	1.9155	0.5220	4.8	121.51	0.0082
0.70	2.0138	0.4966	4.9	134.29	0.0074
0.75	2.1170	0.4724	5.0	148.41	0.0067
0.80	2.2255	0.4493	5.5	244.69	0.0041
0.85	2.3396	0.4274	6.0	403.43	0.0025
0.90	2.4596	0.4066	6.5	665.14	0.0015
0.95	2.5857	0.3867	7.0	1096.6	0.0009
1.0	2.7183	0.3679	7.5	1808.0	0.0006
1.1	3.0042	0.3329	8.0	2981.0	0.0003
1.2	3.3201	0.3012	8.5	4914.8	0.0002
1.3	3.6693	0.2725	9.0	8103.1	0.0001
1.4	4.0552	0.2466	10.0	22026	0.00005

Table IV

Natural Logarithms of Numbers

n	$\log_e n$	n	$\log_e n$	n	$\log_e n$
	*	4.5	1.5041	9.0	2.1972
0.1	7.6974	4.6	1.5261	9.1	2.2083
0.2	8.3906	4.7	1.5476	9.2	2.2192
0.3	8.7960	4.8	1.5686	9.3	2.2300
0.4	9.0837	4.9	1.5892	9.4	2.2407
0.5	9.3069	5.0	1.6094	9.5	2.2513
0.6	9.4892	5.1	1.6292	9.6	2.2618
0.7	9.6433	5.2	1.6487	9.7	2.2721
0.8	9.7769	5.3	1.6677	9.8	2.2824
0.9	9.8946	5.4	1.6864	9.9	2.2925
1.0	0.0000	5.5	1.7047	10	2.3026
1.1	0.0953	5.6	1.7228	11	2.3979
1.2	0.1823	5.7	1.7405	12	2.4849
1.3	0.2624	5.8	1.7579	13	2.5649
1.4	0.3365	5.9	1.7750	14	2.6391
1.5	0.4055	6.0	1.7918	15	2.7081
1.6	0.4700	6.1	1.8083	16	2.7726
1.7	0.5306	6.2	1.8245	17	2.8332
1.8	0.5878	6.3	1.8405	18	2.8904
1.9	0.6419	6.4	1.8563	19	2.9444
2.0	0.6931	6.5	1.8718	20	2.9957
2.1	0.7419	6.6	1.8871	25	3.2189
2.2	0.7885	6.7	1.9021	30	2.4012
2.3	0.8329	6.8	1.9169	35	3.5553
2.4	0.8755	6.9	1.9315	40	3.6889
2.5	0.9163	7.0	1.9459	45	3.8067
2.6	0.9555	7.1	1.9601	50	3.9120
2.7	0.9933	7.2	1.9741	55	4.0073
2.8	1.0296	7.3	1.9879	60	4.0943
2.9	1.0647	7.4	2.0015	65	4.1744
3.0	1.0986	7.5	2.0149	70	4.2485
3.1	1.1314	7.6	2.0281	75	4.3175
3.2	1.1632	7.7	2.0412	80	4.3820
3.3	1.1939	7.8	2.0541	85	4.4427
3.4	1.2238	7.9	2.0669	90	4.4998
3.5	1.2528	8.0	2.0794	100	4.6052
3.6	1.2809	8.1	2.0919	110	4.7005
3.7	1.3083	8.2	2.1041	120	4.7875
3.8	1.3350	8.3	2.1163	130	4.8676
3.9	1.3610	8.4	2.1282	140	4.9416
4.0	1.3863	8.5	2.1401	150	5.0106
4.1	1.4110	8.6	2.1518	160	5.0752
4.2	1.4351	8.7	2.1633	170	5.1358
4.3	1.4586	8.8	2.1748	180	5.1930
4.4	1.4816	8.9	2.1861	190	5.2470

* Subtract 10 for $n < 1$. Thus, $\log_e 0.1 = 7.6974 - 10 = -2.3026$.

Table V Values of Trigonometric Functions

Angle θ									
Degrees	Radians	sin θ	csc θ	tan θ	cot θ	sec θ	cos θ		
0° 00′	.0000	.0000	No value	.0000	No value	1.000	1.0000	1.5708	90° 00′
10	029	029	343.8	029	343.8	000	000	679	50
20	058	058	171.9	058	171.9	000	000	650	40
30	087	087	114.6	087	114.6	000	1.0000	621	30
40	116	116	85.95	116	85.94	000	.9999	592	20
50	145	145	68.76	145	68.75	000	999	563	10
1° 00′	.0175	.0175	57.30	.0175	57.29	1.000	.9998	1.5533	89° 00′
10	204	204	49.11	204	49.10	000	998	504	50
20	233	233	42.98	233	42.96	000	997	475	40
30	262	262	38.20	262	38.19	000	997	446	30
40	291	291	34.38	291	34.37	000	996	417	20
50	320	320	31.26	320	31.24	001	995	388	10
2° 00′	.0349	.0349	28.65	.0349	28.64	1.001	.9994	1.5359	88° 00′
10	378	378	26.45	378	26.43	001	993	330	50
20	407	407	24.56	407	24.54	001	992	301	40
30	436	436	22.93	437	22.90	001	990	272	30
40	465	465	21.49	466	21.47	001	989	243	20
50	495	494	20.23	495	20.21	001	988	213	10
3° 00′	.0524	.0523	19.11	.0524	19.08	1.001	.9986	1.5184	87° 00′
10	553	552	18.10	553	18.07	002	985	155	50
20	582	581	17.20	582	17.17	002	983	126	40
30	611	610	16.38	612	16.35	002	981	097	30
40	640	640	15.64	641	15.60	002	980	068	20
50	669	669	14.96	670	14.92	002	978	039	10
4° 00′	.0698	.0698	14.34	.0699	14.30	1.002	.9976	1.5010	86° 00′
10	727	727	13.76	729	13.73	003	974	981	50
20	756	756	13.23	758	13.20	003	971	952	40
30	785	785	12.75	787	12.71	003	969	923	30
40	814	814	12.29	816	12.25	003	967	893	20
50	844	843	11.87	846	11.83	004	964	864	10
5° 00′	.0873	.0872	11.47	.0875	11.43	1.004	.9962	1.4835	85° 00′
10	902	901	11.10	904	11.06	004	959	806	50
20	931	929	10.76	934	10.71	004	957	777	40
30	960	958	10.43	963	10.39	005	954	748	30
40	.0989	.0987	10.13	.0992	10.08	005	951	719	20
50	.1018	.1016	9.839	.1022	9.788	005	948	690	10
6° 00′	.1047	.1045	9.567	.1051	9.514	1.006	.9945	1.4661	84° 00′
10	076	074	9.309	080	9.255	006	942	632	50
20	105	103	9.065	110	9.010	006	939	603	40
30	134	132	8.834	139	8.777	006	936	573	30
40	164	161	8.614	169	8.556	007	932	544	20
50	193	190	8.405	198	8.345	007	929	515	10
7° 00′	.1222	.1219	8.206	.1228	8.144	1.008	.9925	1.4486	83° 00′
10	251	248	8.016	257	7.953	008	922	457	50
20	280	276	7.834	287	7.770	008	918	428	40
30	309	305	7.661	317	7.596	009	914	399	30
40	338	334	7.496	346	7.429	009	911	370	20
50	367	363	7.337	376	7.269	009	907	341	10
8° 00′	.1396	.1392	7.185	.1405	7.115	1.010	.9903	1.4312	82° 00′
10	425	421	7.040	435	6.968	010	899	283	50
20	454	449	6.900	465	827	011	894	254	40
30	484	478	765	495	691	011	890	224	30
40	513	507	636	524	561	012	886	195	20
50	542	536	512	554	435	012	881	166	10
9° 00′	.1571	.1564	6.392	.1584	6.314	1.012	.9877	1.4137	81° 00′
		cos θ	sec θ	cot θ	tan θ	csc θ	sin θ	Radians	Degrees
								Angle θ	

Table V Values of Trigonometric Functions (continued)

Degrees	Radians	sin θ	csc θ	tan θ	cot θ	sec θ	cos θ		
9° 00′	.1571	.1564	6.392	.1584	6.314	1.012	.9877	1.4137	81° 00′
10	600	593	277	614	197	013	872	108	50
20	629	622	166	644	6.084	013	868	079	40
30	658	650	6.059	673	5.976	014	863	050	30
40	687	679	5.955	703	871	014	858	1.4021	20
50	716	708	855	733	769	015	853	1.3992	10
10° 00′	.1745	.1736	5.759	.1763	5.671	1.015	.9848	1.3963	80° 00′
10	774	765	665	793	576	016	843	934	50
20	804	794	575	823	485	016	838	904	40
30	833	822	487	853	396	017	833	875	30
40	862	851	403	883	309	018	827	846	20
50	891	880	320	914	226	018	822	817	10
11° 00′	.1920	.1908	5.241	.1944	5.145	1.019	.9816	1.3788	79° 00′
10	949	937	164	.1974	5.066	019	811	759	50
20	.1978	965	089	.2004	4.989	020	805	730	40
30	.2007	.1994	5.016	035	915	020	799	701	30
40	036	.2022	4.945	065	843	021	793	672	20
50	065	051	876	095	773	022	787	643	10
12° 00′	.2094	.2079	4.810	.2126	4.705	1.022	.9781	1.3614	78° 00′
10	123	108	745	156	638	023	775	584	50
20	153	136	682	186	574	024	769	555	40
30	182	164	620	217	511	024	763	526	30
40	211	193	560	247	449	025	757	497	20
50	240	221	502	278	390	026	750	468	10
13° 00′	.2269	.2250	4.445	.2309	4.331	1.026	.9744	1.3439	77° 00′
10	298	278	390	339	275	027	737	410	50
20	327	306	336	370	219	028	730	381	40
30	356	334	284	401	165	028	724	352	30
40	385	363	232	432	113	029	717	323	20
50	414	391	182	462	061	030	710	294	10
14° 00′	.2443	.2419	4.134	.2493	4.011	1.031	.9703	1.3265	76° 00′
10	473	447	086	524	3.962	031	696	235	50
20	502	476	4.039	555	914	032	689	206	40
30	531	504	3.994	586	867	033	681	177	30
40	560	532	950	617	821	034	674	148	20
50	589	560	906	648	776	034	667	119	10
15° 00′	.2618	.2588	3.864	.2679	3.732	1.035	.9659	1.3090	75° 00′
10	647	616	822	711	689	036	652	061	50
20	676	644	782	742	647	037	644	032	40
30	705	672	742	773	606	038	636	1.3003	30
40	734	700	703	805	566	039	628	1.2974	20
50	763	728	665	836	526	039	621	945	10
16° 00′	.2793	.2756	3.628	.2867	3.487	1.040	.9613	1.2915	74° 00′
10	822	784	592	899	450	041	605	886	50
20	851	812	556	931	412	042	596	857	40
30	880	840	521	962	376	043	588	828	30
40	909	868	487	.2944	340	044	580	799	20
50	938	896	453	.3026	305	045	572	770	10
17° 00′	.2967	.2924	3.420	.3057	3.271	1.046	.9563	1.2741	73° 00′
10	.2996	952	388	089	237	047	555	712	50
20	.3025	.2979	357	121	204	048	546	683	40
30	054	.3007	326	153	172	048	537	654	30
40	083	035	295	185	140	049	528	625	20
50	113	062	265	217	108	050	520	595	10
18° 00′	.3142	.3090	3.236	.3249	3.078	1.051	.9511	1.2566	72° 00′
		cos θ	sec θ	cot θ	tan θ	csc θ	sin θ	Radians	Degrees
									Angle θ

Table V Values of Trigonometric Functions (continued)

Angle θ

Degrees	Radians	sin θ	csc θ	tan θ	cot θ	sec θ	cos θ		
18° 00′	.3142	.3090	3.236	.3249	3.078	1.051	.9511	1.2566	72° 00′
10	171	118	207	281	047	052	502	537	50
20	200	145	179	314	3.018	053	492	508	40
30	229	173	152	346	2.989	054	483	479	30
40	258	201	124	378	960	056	474	450	20
50	287	228	098	411	932	057	465	421	10
19° 00′	.3316	.3256	3.072	.3443	2.904	1.058	.9455	1.2392	71° 00′
10	345	283	046	476	877	059	446	363	50
20	374	311	3.021	508	850	060	436	334	40
30	403	338	2.996	541	824	061	426	305	30
40	432	365	971	574	798	062	417	275	20
50	462	393	947	607	773	063	407	246	10
20° 00′	.3491	.3420	2.924	.3640	2.747	1.064	.9397	1.2217	70° 00′
10	520	448	901	673	723	065	387	188	50
20	549	475	878	706	699	066	377	159	40
30	578	502	855	739	675	068	367	130	30
40	607	529	833	772	651	069	356	101	20
50	636	557	812	805	628	070	346	072	10
21° 00′	.3665	.3584	2.790	.3839	2.605	1.071	.9336	1.2043	69° 00′
10	694	611	769	872	583	072	325	1.2014	50
20	723	638	749	906	560	074	315	1.1985	40
30	752	665	729	939	539	075	304	956	30
40	782	692	709	.3973	517	076	293	926	20
50	811	719	689	.4006	496	077	283	897	10
22° 00′	.3840	.3746	2.669	.4040	2.475	1.079	.9272	1.1868	68° 00′
10	869	773	650	074	455	080	261	839	50
20	898	800	632	108	434	081	250	810	40
30	927	827	613	142	414	082	239	781	30
40	956	854	595	176	394	084	228	752	20
50	985	881	577	210	375	085	216	723	10
23° 00′	.4014	.3907	2.559	.4245	2.356	1.086	.9205	1.1694	67° 00′
10	043	934	542	279	337	088	194	665	50
20	072	961	525	314	318	089	182	636	40
30	102	.3987	508	348	300	090	171	606	30
40	131	.4014	491	383	282	092	159	577	20
50	160	041	475	417	264	093	147	548	10
24° 00′	.4189	.4067	2.459	.4452	2.246	1.095	.9135	1.1519	66° 00′
10	218	094	443	487	229	096	124	490	50
20	247	120	427	522	211	097	112	461	40
30	276	147	411	557	194	099	100	432	30
40	305	173	396	592	177	100	088	403	20
50	334	200	381	628	161	102	075	374	10
25° 00′	.4363	.4226	2.366	.4663	2.145	1.103	.9063	1.1345	65° 00′
10	392	253	352	699	128	105	051	316	50
20	422	279	337	734	112	106	038	286	40
30	451	305	323	770	097	108	026	257	30
40	480	331	309	806	081	109	013	228	20
50	509	358	295	841	066	111	.9001	199	10
26° 00′	.4538	.4384	2.281	.4877	2.050	1.113	.8988	1.1170	64° 00′
10	567	410	268	913	035	114	975	141	50
20	596	436	254	950	020	116	962	112	40
30	625	462	241	.4986	2.006	117	949	083	30
40	654	488	228	.5022	1.991	119	936	054	20
50	683	514	215	059	977	121	923	1.1025	10
27° 00′	.4712	.4540	2.203	.5095	1.963	1.122	.8910	1.0996	63° 00′
		cos θ	sec θ	cot θ	tan θ	csc θ	sin θ	Radians	Degrees

Angle θ

Table V Values of Trigonometric Functions (continued)

Angle θ

Degrees	Radians	sin θ	csc θ	tan θ	cot θ	sec θ	cos θ		
27° 00′	.4712	.4540	2.203	.5095	1.963	1.122	.8910	1.0996	63° 00′
10	741	566	190	132	949	124	897	966	50
20	771	592	178	169	935	126	884	937	40
30	800	617	166	206	921	127	870	908	30
40	829	643	154	243	907	129	857	879	20
50	858	669	142	280	894	131	843	850	10
28° 00′	.4887	.4695	2.130	.5317	1.881	1.133	.8829	1.0821	62° 00′
10	916	720	118	354	868	134	816	792	50
20	945	746	107	392	855	136	802	763	40
30	.4974	772	096	430	842	138	788	734	30
40	.5003	797	085	467	829	140	774	705	20
50	032	823	074	505	816	142	760	676	10
29° 00′	.5061	.4848	2.063	.5543	1.804	1.143	.8746	1.0647	61° 00′
10	091	874	052	581	792	145	732	617	50
20	120	899	041	619	780	147	718	588	40
30	149	924	031	658	767	149	704	559	30
40	178	950	020	696	756	151	689	530	20
50	207	.4975	010	735	744	153	675	501	10
30° 00′	.5236	.5000	2.000	.5774	1.732	1.155	.8660	1.0472	60° 00′
10	265	025	1.990	812	720	157	646	443	50
20	294	050	980	851	709	159	631	414	40
30	323	075	970	890	698	161	616	385	30
40	352	100	961	930	686	163	601	356	20
50	381	125	951	.5969	675	165	587	327	10
31° 00′	.5411	.5150	1.942	.6009	1.664	1.167	.8572	1.0297	59° 00′
10	440	175	932	048	653	169	557	268	50
20	469	200	923	088	643	171	542	239	40
30	498	225	914	128	632	173	526	210	30
40	527	250	905	168	621	175	511	181	20
50	556	275	896	208	611	177	496	152	10
32° 00′	.5585	.5299	1.887	.6249	1.600	1.179	.8480	1.0123	58° 00′
10	614	324	878	289	590	181	465	094	50
20	643	348	870	330	580	184	450	065	40
30	672	373	861	371	570	186	434	036	30
40	701	398	853	412	560	188	418	1.0007	20
50	730	422	844	453	550	190	403	.9977	10
33° 00′	.5760	.5446	1.836	.6494	1.540	1.192	.8387	.9948	57° 00′
10	789	471	828	536	530	195	371	919	50
20	818	495	820	577	520	197	355	890	40
30	847	519	812	619	511	199	339	861	30
40	876	544	804	661	501	202	323	832	20
50	905	568	796	703	492	204	307	803	10
34° 00′	.5934	.5592	1.788	.6745	1.483	1.206	.8290	.9774	56° 00′
10	963	616	781	787	473	209	274	745	50
20	.5992	640	773	830	464	211	258	716	40
30	.6021	644	766	873	455	213	241	687	30
40	050	688	758	916	446	216	225	657	20
50	080	712	751	.6959	437	218	208	628	10
35° 00′	.6109	.5736	1.743	.7002	1.428	1.221	.8192	.9599	55° 00′
10	138	760	736	046	419	223	175	570	50
20	167	783	729	089	411	226	158	541	40
30	196	807	722	133	402	228	141	512	30
40	225	831	715	177	393	231	124	483	20
50	254	854	708	221	385	233	107	454	10
36° 00′	.6283	.5878	1.701	.7265	1.376	1.236	.8090	.9425	54° 00′
		cos θ	sec θ	cot θ	tan θ	csc θ	sin θ	Radians	Degrees

Angle θ

Table V   Values of Trigonometric Functions (continued)

Angle θ									
Degrees	Radians	sin θ	csc θ	tan θ	cot θ	sec θ	cos θ		
36° 00'	.6283	.5878	1.701	.7265	1.376	1.236	.8090	.9425	54° 00'
10	312	901	695	310	368	239	073	396	50
20	341	925	688	355	360	241	056	367	40
30	370	948	681	400	351	244	039	338	30
40	400	972	675	445	343	247	021	308	20
50	429	.5995	668	490	335	249	.8004	279	10
37° 00'	.6458	.6018	1.662	.7536	1.327	1.252	.7986	.9250	53° 00'
10	487	041	655	581	319	255	969	221	50
20	516	065	649	627	311	258	951	192	40
30	545	088	643	673	303	260	934	163	30
40	574	111	636	720	295	263	916	134	20
50	603	134	630	766	288	266	898	105	10
38° 00'	.6632	.6157	1.624	.7813	1.280	1.269	.7880	.9076	52° 00'
10	661	180	618	860	272	272	862	047	50
20	690	202	612	907	265	275	844	.9018	40
30	720	225	606	.7954	257	278	826	.8988	30
40	749	248	601	.8002	250	281	808	959	20
50	778	271	595	050	242	284	790	930	10
39° 00'	.6807	.6293	1.589	.8098	1.235	1.287	.7771	.8901	51° 00'
10	836	316	583	146	228	290	753	872	50
20	865	338	578	195	220	293	735	843	40
30	894	361	572	243	213	296	716	814	30
40	923	383	567	292	206	299	698	785	20
50	952	406	561	342	199	302	679	756	10
40° 00'	.6981	.6428	1.556	.8391	1.192	1.305	.7660	.8727	50° 00'
10	.7010	450	550	441	185	309	642	698	50
20	039	472	545	491	178	312	623	668	40
30	069	494	540	541	171	315	604	639	30
40	098	517	535	591	164	318	585	610	20
50	127	539	529	642	157	322	566	581	10
41° 00'	.7156	.6561	1.524	.8693	1.150	1.325	.7547	.8552	49° 00'
10	185	583	519	744	144	328	528	523	50
20	214	604	514	796	137	332	509	494	40
30	243	626	509	847	130	335	490	465	30
40	272	648	504	899	124	339	470	436	20
50	301	670	499	.8952	117	342	451	407	10
42° 00'	.7330	.6691	1.494	.9004	1.111	1.346	.7431	.8378	48° 00'
10	359	713	490	057	104	349	412	348	50
20	389	734	485	110	098	353	392	319	40
30	418	756	480	163	091	356	373	290	30
40	447	777	476	217	085	360	353	261	20
50	476	799	471	271	079	364	333	232	10
43° 00'	.7505	.6820	1.466	.9325	1.072	1.367	.7314	.8203	47° 00'
10	534	841	462	380	066	371	294	174	50
20	563	862	457	435	060	375	274	145	40
30	592	884	453	490	054	379	254	116	30
40	621	905	448	545	048	382	234	087	20
50	650	926	444	601	042	386	214	058	10
44° 00'	.7679	.6947	1.440	.9657	1.036	1.390	.7193	.8029	46° 00'
10	709	967	435	713	030	394	173	.7999	50
20	738	.6988	431	770	024	398	153	970	40
30	767	.7009	427	827	018	402	133	941	30
40	796	030	423	884	012	406	112	912	20
50	825	050	418	.9942	006	410	092	883	10
45° 00'	.7854	.7071	1.414	1.000	1.000	1.414	.7071	.7854	45° 00'
		cos θ	sec θ	cot θ	tan θ	csc θ	sin θ	Radians	Degrees

Angle θ

Answers to Selected Problems

Chapter 1

Exercise Set 1

1. **a.** $0.666\ldots$ **b.** 0.625 **c.** $0.181818\ldots$ **d.** -1.6 **e.** $0.370370\ldots$
3. Correct to two decimal places **5. a.** Rational **b.** Rational **c.** Irrational **d.** Rational
   e. Rational **7. a.** Rational **b.** May be either rational or irrational
   c. Rational if $x = 0$; otherwise irrational **9. a.** $\frac{2}{3}$ **b.** $\frac{28}{9}$ **11. a.** $\frac{9}{37}$ **b.** $\frac{1{,}709}{333}$
13. **a.** By the Pythagorean theorem, the diagonal of the square is $\sqrt{2}$.
    b. Construct a rectangle 2 units long by 1 unit high. The diagonal will be $\sqrt{5}$.

Exercise Set 2

1. **a.** Distributive property **b.** Definition of additive inverse **c.** Definition of multiplicative inverse
   d. Definition of additive identity **e.** Definition of multiplicative identity **f.** Associative property for addition
   g. Associative property for multiplication **h.** Commutative property for addition
   i. Commutative property for multiplication **j.** Property of additive identity **3. a.** -6 **b.** 15 **c.** 2
   d. $\frac{1}{3}$ **e.** 19 **5. a.** 0 **b.** 1 **c.** -1 **d.** No
7. **a.** No. For example, $7 - (5 - 3) = 7 - 2 = 5$, but $(7 - 5) - 3 = 2 - 3 = -1$.
   b. No. For example, $12 \div (4 \div 2) = 12 \div 2 = 6$, but $(12 \div 4) \div 2 = 3 \div 2 = \frac{3}{2}$.
9. **a.** 5 **b.** 2 **c.** 0 **d.** 5 **e.** 3
11. Taken in order, the properties used are: equality property 4, property of additive identity, associative property of addition, property of additive inverse, property of additive identity.
13. Let $a \in R$, $a \neq 0$. Suppose $a \cdot b = 1$. Then

$$a^{-1} \cdot (a \cdot b) = a^{-1} \cdot 1 \qquad \text{Equality property 4}$$
$$a^{-1} \cdot (a \cdot b) = a^{-1} \qquad \text{Property of multiplicative identity}$$
$$(a^{-1} \cdot a) \cdot b = a^{-1} \qquad \text{Associative property for multiplication}$$
$$1 \cdot b = a^{-1} \qquad \text{Property of multiplicative inverse}$$
$$b = a^{-1} \qquad \text{Property of multiplicative identity}$$

15. $a(b - c) = a[b + (-c)] = ab + a(-c) = ab + [-(ac)] = ab - ac$ **17.** $1/b = 1 \cdot b^{-1} = b^{-1}$
19. $(a^{-1} \cdot b^{-1}) \cdot (ab) = (b^{-1} \cdot a^{-1}) \cdot (ab) = b^{-1}(a^{-1} \cdot a) \cdot b = b^{-1} \cdot 1 \cdot b = b^{-1} \cdot b = 1$
    So $a^{-1}b^{-1}$ is the inverse of ab.
21. **a.** -1 **b.** 0 **c.** 1 **d.** 5 **e.** -2

Exercise Set 3

1. a. $<$ **b.** $>$ **c.** $<$ **d.** $>$ **e.** $<$ **3. a.** $7 - 3 = 4$ is positive
b. $4 - (-2) = 4 + 2 = 6$ is positive **c.** $-1 - (-4) = -1 + 4 = 3$ is positive **d.** $5 - 2 = 3$ is positive
e. $-3 - (-5) = -3 + 5 = 2$ is positive **5. a.** $2 < x < 5$ **b.** $-1 < x \le 0$ **c.** $0 \le x < 2$
d. $0 \le x \le 6$ **e.** $3 < x \le 4$ **7. a.** [number line from 0 to 7 with open circle at 1 and arrow]

b. [number line from −2 to 4]

c. [number line from −2 to 3]

d. [number line from 0 to 7]

e. [number line from −3 to 3]

9. a. $-2 < x < 2$ **b.** $y > 1$ or $y < -1$ **c.** $-3 \le t \le 3$ **d.** $w \ge 2$ or $w \le -2$ **11. a.** $\{x: \ |x| < 5\}$
b. $\{x: \ |x| > 6\}$ **c.** $\{x: \ |x| \le 4\}$ **d.** $\{x: \ |x| \ge 8\}$
13. If $a > 0$, then $a^2 = a \cdot a$ is positive since R^+ is closed under multiplication. If $a < 0$, then $(-a) > 0$, and $(-a)(-a) = a^2 > 0$.
15. If $a < b$, then $b - a$ is positive. If c is positive, then by closure of R^+ under multiplication, $(b - a) \cdot c$ is positive. But $(b - a)c = bc - ac$. So by definition, $ac < bc$. If c is negative, then $-c$ is positive. So $(b - a)(-c)$ is positive. But $(b - a)(-c) = -bc + ac = ac - bc$. So, since this is positive, $bc < ac$, or equivalently, $ac > bc$.
17. If $a < b$ and $c < d$, then by property 3, $a + c < b + c$ and $b + c < b + d$. Thus, by property 6, $a + c < b + d$.
19. If $a < b$, then $\begin{cases} a/c < b/c & \text{if } c > 0 \\ a/c > b/c & \text{if } c < 0 \end{cases}$

If $c > 0$, by Problem 18, $c^{-1} > 0$. So by property 4, $ac^{-1} < bc^{-1}$. But $ac^{-1} = a/c$ and $bc^{-1} = b/c$. If $c < 0$, then $c^{-1} < 0$, since otherwise the product of c and c^{-1} would not be positive. Thus, by property 4, $ac^{-1} > bc^{-1}$, and so $a/c > b/c$.
21. False. For example, $6 < 7$ and $3 < 5$. But $(6 - 3) > (7 - 5)$.

Exercise Set 4

1. a. 2 **b.** -3 **c.** -6 **d.** -2 **e.** 8 **3. a.** -6 **b.** -20 **c.** 30 **d.** -5 **e.** -1
5. a. 4 **b.** 7 **c.** 15 **d.** 15 **e.** 12 **7. a.** $\frac{19}{12}$ **b.** $\frac{11}{24}$ **9. a.** $\frac{13}{12}$ **b.** 1 **11. a.** $\frac{13}{5}$
b. $-\frac{11}{6}$ **13. a.** $-\frac{4}{7}$ **b.** $\frac{9}{4}$ **15. a.** $-\frac{5}{108}$ **b.** $\frac{43}{1,260}$ **17. a.** $\frac{1}{10}$ **b.** $\frac{61}{72}$ **19. a.** 16
b. 26 **c.** 28 **d.** 64 **e.** 45
21. First assume Definition 2. Then $a \div b = a \cdot b^{-1}$. Let $c = a \cdot b^{-1}$. Multiply both sides by b, giving $cb = (a \cdot b^{-1})b = a \cdot (b^{-1} \cdot b) = a \cdot 1 = a$, or $a = bc$. So c is a number which when multiplied by b gives a. Now assume the definition as stated in this problem. Then we have

$$a \div b = c$$

where $bc = a$. Since $b \ne 0$, then b^{-1} exists, and we multiply by b^{-1} to get

$$b^{-1}(bc) = b^{-1}a$$
$$(b^{-1}b)c = ab^{-1}$$
$$1 \cdot c = ab^{-1}$$
$$c = ab^{-1}$$

Therefore, $a \div b = ab^{-1}$.
23. a. $\frac{109}{30}$ **b.** $\frac{5}{6}$ **25. a.** $-\frac{9}{8}$ **b.** $-\frac{17}{30}$

Exercise Set 5

1. $\frac{2}{11}$ **3.** $-\frac{29}{8}$ **5.** $-\frac{14}{11}$ **7.** $-\frac{33}{40}$ **9.** $-\frac{1}{11}$ **11.** 2

13. $-\frac{16}{63}$ **15.** $\frac{1}{3}$ **17.** $\frac{9}{8}$ **19.** $\frac{5}{4}$

Exercise Set 6

1. a. 81 **b.** $\frac{1}{4}$ **c.** -8 **d.** 1 **e.** 4 **3. a.** $\frac{1}{16}$ **b.** $\frac{2}{3}$ **c.** -27 **d.** $\frac{243}{5}$ **e.** $\frac{2}{9}$
5. a. x^7 **b.** x^4 **c.** x^{12} **d.** x^4y^6 **e.** 1
7. a. $\dfrac{a^5}{b^5}$ **b.** $\dfrac{a^3b^6}{c^3}$ **c.** $\dfrac{2t^4}{3s^6}$ **9. a.** $\dfrac{b^6}{a^3c^6d^4}$ **b.** $\dfrac{x^{10}}{y^{16}}$
11. $\frac{7}{8}$ **13. a.** 2×10^{12} **b.** 7×10^{-6} **c.** 3.1×10^6 **d.** 1.3×10^{-1} **e.** 2.54×10^8

15. a. 6.58×10^4 **b.** $(-2.5) \times 10^3$ **17.** 2.5
19. a. 1.673×10^{-24} **b.** $602,300,000,000,000,000,000,000$
21. a. Let $n = -p, p > 0$. Then

$$(a^m)^n = (a^m)^{-p} = \frac{1}{(a^m)^p} = \frac{1}{a^{mp}} = a^{-(mp)} = a^{m(-p)} = a^{mn}$$

b. Let $m = -p, p > 0$. Then

$$(a^m)^n = (a^{-p})^n = \left(\frac{1}{a^p}\right)^n = \frac{1^n}{(a^p)^n} = \frac{1}{a^{pn}} = a^{-pn} = a^{(-p)n} = a^{mn}$$

c. Let $m = -p, n = -q, p > 0, q > 0$. Then

$$(a^m)^n = (a^{-p})^{-q} = \left(\frac{1}{a^p}\right)^{-q} = (a^p)^q = a^{pq} = a^{(-p)(-q)} = a^{mn}$$

d. If $n = 0$, then $(a^m)^n = (a^m)^0 = 1 = a^0 = a^{m \cdot 0} = a^{mn}$.
23. 1.874×10^{20}

Exercise Set 7

1. a. $5\sqrt{5}$ **b.** $2\sqrt[3]{5}$ **3. a.** $7\sqrt{5}$ **b.** $-17\sqrt{2}$ **5.** $5x^2y^3\sqrt{2y}$ **7.** $3xz^2\sqrt{2xz}$ **9.** $10ac^2\sqrt{3}/b^3$
11. $3x^3y^2\sqrt{2}$ **13.** $-3ab^2\sqrt[3]{3a^2}$ **15. a.** $\sqrt{6}/3$ **b.** $2\sqrt{10}/5$ **c.** $2\sqrt{7}/7$ **d.** $\sqrt[3]{2}/2$ **e.** $3\sqrt{6}/10$
17. $4y^5\sqrt{6}/3x^3$ **19. a.** $2^{13/12}$ **b.** $3^{1/6}$ **c.** $\frac{1}{3}$ **d.** 1 **21. a.** $6x^{5/6}$ **b.** $x^{1/12}y^{1/6}$ **c.** $4x^{5/6}$
23. a. $2x^2$ **b.** $4a^2/b$ **c.** $-5y^4/x^3$ **25. a.** $256a^8b^{12}$ **b.** $-128a^7/b^{14}$ **c.** $216a^3b^6$
27. a. $2a^{4/3}b^{2/3}$ **b.** $2^{3/2}a^{9/2}b^{15/2}$ **c.** $5^{4/3}x^{8/3}y^{16/3}$ **29.** $\sqrt[kn]{a^{km}} = a^{km/kn} = a^{m/n} = \sqrt[n]{a^m}$
31. $\sqrt[m]{\sqrt[n]{a}} = (a^{1/n})^{1/m} = a^{(1/n) \cdot (1/m)} = a^{1/nm} = a^{1/mn} = \sqrt[mn]{a}$ **33.** $\left(\frac{a}{b}\right)^{-m/n} = \left(\frac{a}{b}\right)^{(-1) \cdot (m/n)} = \left[\left(\frac{a}{b}\right)^{-1}\right]^{m/n} = \left(\frac{b}{a}\right)^{m/n}$

35. a. $\dfrac{y^3}{4x^{1/2}}$ **b.** $\dfrac{b}{64a^2c^{2/3}}$ **37.** $\dfrac{\sqrt{2}}{2}$ **39.** $\dfrac{11\sqrt[3]{4}}{6}$ **41.** $\dfrac{9a^2b^3\sqrt[3]{4c}}{8c^2}$

Review Exercise Set

1. a. 0.6875 **b.** 3.125 **c.** $0.428571428571\ldots$ **d.** $0.185185185\ldots$ **e.** $-2.1666\ldots$ **3. a.** Rational
b. Rational **c.** Irrational **d.** Rational **e.** Rational **f.** Irrational **g.** Rational **h.** Rational
i. Irrational **j.** Irrational **5. a.** 3 **b.** 5 **c.** 12 **d.** 2 **e.** 13
7. a.

b.

c.

d.

e.

9. a. $\frac{1}{18}$ **b.** $\frac{59}{72}$ **11. a.** $\frac{59}{1,680}$ **b.** $-\frac{3}{4}$ **13. a.** $\frac{39}{2}$ **b.** $\frac{5}{14}$ **15. a.** 3.94×10^5
b. -3.585×10^{-5} **c.** 5×10^{-4} **17. a.** $\dfrac{3^6z^2}{x^2y^3}$ **b.** $4a^2b^2$ **19. a.** $\dfrac{27y^9z^3}{x^6}$ **b.** $\dfrac{x^{10}}{4y^6}$
21. a. $5xy^2\sqrt{2x}$ **b.** $\dfrac{-2a^2}{b}\sqrt[3]{4}$ **23. a.** $\dfrac{\sqrt{6}}{3}$ **b.** $\dfrac{2\sqrt{5}}{5}$ **c.** $\dfrac{2\sqrt[3]{6}}{3}$ **d.** $\dfrac{\sqrt{6xy}}{4y}$
25. If $a > b$, then $a - b$ is positive. Therefore, $(a - c) - (b - c) = a - c - b + c = a - b$ is positive, so that $a - c > b - c$.
27. If $a \neq 0$, then a^{-1} exists, and from $ab = 0$, we get

$$a^{-1}(ab) = a^{-1} \cdot 0$$
$$(a^{-1}a)b = 0$$
$$1 \cdot b = 0$$
$$b = 0$$

Thus, if $a \neq 0$, then b must be 0. So either $a = 0$ or $b = 0$.

29. 4 **31.** $-\frac{33}{5}$

Chapter 2

Exercise Set 2

1. $5x^2 + 10x + 13$ **3.** $8y^3 + 2y^2 + 2y - 11$ **5.** $t^2 - 3t + 14$ **7.** $7x^2 + 4xy + 2y^2$ **9.** $2x^2 + x + 3$
11. $x^2 + 7x + 12$ **13.** $x^2 + x - 12$ **15.** $x^2 + 3xy + 2y^2$ **17.** $2 - x - x^2$ **19.** $6x^2 + 17x + 7$
21. $8t^2 - 8t - 6$ **23.** $10y^2 - 23y + 12$ **25.** $8 - 14x + 3x^2$ **27.** $x^2 - 16$ **29.** $4x^2 - 1$ **31.** $a^2 - 4b^2$
33. $9a^2 - 49b^2c^2$ **35.** $x^2 - 4x + 4$ **37.** $x^2 - 4xy + 4y^2$ **39.** $4x^2 - 12x + 9$ **41.** $a^4 - 6a^2b^2 + 9b^4$
43. $x^3 - 5x + 2$ **45.** $x^4 - 4x^3 + 9x^2 - 10x + 8$ **47.** $x^2 + 2xy + y^2 - 4$ **49.** $x^2 - y^2 + 2y - 1$
51. $16 - x^2 + 2xy - y^2$ **53.** $4x^2 + 9y^2 + 16 - 12xy + 16x - 24y$

Exercise Set 3

1. $a^7 + 7a^6b + 21a^5b^2 + 35a^4b^3 + 35a^3b^4 + 21a^2b^5 + 7ab^6 + b^7$ **3.** $a^6 - 6a^5b + 15a^4b^2 - 20a^3b^3 + 15a^2b^4 - 6ab^5 + b^6$
5. $1 + 6x + 15x^2 + 20x^3 + 15x^4 + 6x^5 + x^6$ **7.** $16x^4 + 32x^3y + 24x^2y^2 + 8xy^3 + y^4$
9. $16x^4 + 96x^3y + 216x^2y^2 + 216xy^3 + 81y^4$ **11.** $125x^3 - 75x^2y + 15xy^2 - y^6$
13. $\dfrac{32}{s^5} - \dfrac{240}{s^4} + \dfrac{720}{s^3} - \dfrac{1{,}080}{s^2} + \dfrac{810}{s} - 243$ **15.** $x^{10} - 10x^8y + 40x^6y^2 - 80x^4y^3 + 80x^2y^4 - 32y^5$ **17.** $-120a^3b^7$
19. $-48{,}384x^5y^3$ **21.** Write $(a - b)^n = [a + (-b)]^n$. The factor $(-b)$ appears alternately to even and odd powers in the
expansion, and since $(-b)^k = [(-1)b]^k = (-1)^k \cdot b^k$, the sign is that of $(-1)^k$, which is plus when k is even and minus
when k is odd.   **23.** **a.** $-25{,}344x^{10}y^7$ **b.** $61{,}236x^{3/2}$ **25.** $17{,}920a^8b^{12}$
27. $32{,}768a^{30} - 737{,}280a^{28}b^{-2} + 7{,}741{,}440a^{26}b^{-4} - 50{,}319{,}360a^{24}b^{-6}$

Exercise Set 4

1. **a.** $2ab(a - 2b)$ **b.** $3a^2b^3c(2ab + 3c)$ **3.** **a.** $8x^2y^2(3x - 5x^3yz + 6y^2z^2)$ **b.** $52a^3b(3a - 5b)$
5. **a.** $2x^{1/3}(x - 2)$ **b.** $6x^{1/3}(2 + 3x)$ **7.** **a.** $(x + 2)(y + 1)$ **b.** $(1 - a)(b - 1)$ **9.** **a.** $(x + y)(x - 1)$
    b. $(xy + 1)(x - 1)$ **11.** **a.** $(x - 1)(y - 1)$ **b.** $(a - c)(a - b)$ **13.** **a.** $(y - 4)(x - 3)$
    b. $(a - 1)(a^2 + b^2)$ **15.** **a.** $(t - 6)^2$ **b.** $(3m + 1)^2$ **17.** **a.** $(2x - 3y)^2$ **b.** $(4t - 9s)^2$
19. **a.** $(x + 1)(x + 2)$ **b.** $(x - 2)(x + 1)$ **21.** **a.** $(y + 7)(y - 2)$ **b.** $(a + 8)(a - 3)$
23. **a.** $(3x + 2)(2x - 3)$ **b.** $(4x - 3y)(2x + 3y)$ **25.** **a.** $(2x^2 + 1)(x^2 + 4)$ **b.** $(3t - 4)(8t + 9)$
27. **a.** $(2x + 3)(2x - 3)$ **b.** $(4m + 5n)(4m - 5n)$ **29.** **a.** $(a - 1)(a + 1)(a^2 + 1)(a^4 + 1)$ **b.** $y^2(xy + 1)(xy - 1)$
31. **a.** $(x - 2 + y)(x - 2 - y)$ **b.** $(x + y - 2)(x - y + 2)$ **33.** **a.** $(x + 2 + y)(x + 2 - y)$
    b. $(4a - 5 - 5b)(4a - 5 + 5b)$ **35.** **a.** $(x - 4)(x^2 + 4x + 16)$ **b.** $(3a + 2)(9a^2 - 6a + 4)$
37. **a.** $(x - 5)(x^2 + 5x + 25)$ **b.** $(a + 6)(a^2 - 6a + 36)$ **39.** $(x - 1)(x^4 + x^3 + x^2 + x + 1)$
41. $(x + y)(x^6 - x^5y + x^4y^2 - x^3y^3 + x^2y^4 - xy^5 + y^6)$ **43.** $(1 - 2x)(1 + 2x + 4x^2 + 8x^3 + 16x^4)$
45. $(x^2 - 3y)(x^8 + 3x^6y + 9x^4y^2 + 27x^2y^3 + 81y^4)$ **47.** $(2x - y)^3$ **49.** $(x - 2)^4$ **51.** $(a - b)^5$
53. $(2x + 3y^2)^3$ **55.** $x(3x + 5)(3x - 5)$ **57.** $2x(3x - 1)(x - 3)$ **59.** $(x + 2)^2(x - 2)^2$
61. $(2x + 3)(16x^4 - 24x^3 + 36x^2 - 54x + 81)$ **63.** $(x - 1 + y)(x - 1 - y)$ **65.** $4t(2t - 3)(t - 2)$
67. $(3a + 4b)(9a^2 - 12ab + 16b^2)$ **69.** $2a^2b(a + 3b - 4)$ **71.** $\dfrac{3x - 1}{(x + 1)^{1/3}(x - 1)^{2/3}}$ **73.** $\dfrac{5}{(x - 2)^{1/3}(2x + 1)^{2/3}}$
75. $2ab(4a + 9b)^2$ **77.** $(2x + 3)(20x - 21)$ **79.** $(x - 2y + 1)^2$ **81.** **a.** $(x - 2)(x^2 + 2x + 4)(x + 2)(x^2 - 2x + 4)$
    b. $(x^2 + 4)(x^2 + 4 + 2\sqrt{3}x)(x^2 + 4 - 2\sqrt{3}x)$ **83.** $(x - y + z - 2)(x - y - z + 2)$
85. $(a^2 + b^2)(a^4 - a^2b^2 + b^4)(a - b)(a^2 + ab + b^2)(a + b)(a^2 - ab + b^2)$ **87.** $(x - 3)^5$ **89.** $(a - b)(a^2 + ab + b^2 + 1)$
91. $(x - y)(x - y - 4)$

Exercise Set 5

1. **a.** $\dfrac{x - 1}{x + 2}$ **b.** $\dfrac{2x - 3}{x + 2}$ **3.** $\dfrac{2x^2 + x - 6}{x^2 - 5x + 4}$ **5.** $\dfrac{x^2 + x + 1}{x - 6}$ **7.** $\dfrac{2x - 1}{x + 3}$ **9.** $\dfrac{4x - 2}{x^2 + 2x - 8}$ **11.** $\dfrac{4}{x - 1}$
13. $\dfrac{x^2 - 7x + 4}{x^2 - 2x}$ **15.** $\dfrac{3x + 18}{x + 4}$ **17.** $\dfrac{x + 2}{x - 2}$ **19.** $\dfrac{1}{2x^2 + 3x + 1}$ **21.** $-\dfrac{x + 2}{4x^2}$ **23.** $\dfrac{1}{2 - x}$ **25.** $x - y$

27. $-\dfrac{2}{a^2 + ah}$ **29.** $\dfrac{1}{z+2}$ **31.** $\dfrac{ax + 2x + 2a}{ax + 2x + 2a + 4}$ **33.** $\dfrac{-2(2x+h)}{x^2(x+h)^2}$ **35.** $\dfrac{3-2x}{(x-1)(x^4+x)}$

37. $\dfrac{(x+4)^2(2x-25)}{(2x-3)^3}$ **39.** $\dfrac{x+a-2ax}{(1-2x)(1-2a)}$ **41.** $2x+8$

Exercise Set 6

1. $\dfrac{y-x}{xy}$ **3.** $\dfrac{b-a}{ab}$ **5.** $-\dfrac{x+4}{x^3}$ **7.** $\dfrac{x^2-2x-3}{x^2-2x+1}$ **9.** $\dfrac{x-2}{x+1}$ **11.** $\dfrac{1-x}{2\sqrt{x}}$ **13.** $\dfrac{1-x}{\sqrt{1-2x}}$

15. $\dfrac{x-1}{\sqrt{x^2+4}}$ **17.** $\dfrac{4}{(4-x^2)^{3/2}}$ **19.** $\dfrac{2y-y^3}{(1-y^2)^{3/2}}$ **21.** $\dfrac{1}{\sqrt{x+1}+1}$ **23.** $\dfrac{2}{\sqrt{2(x+h)+1}+\sqrt{2x+1}}$

25. $-(\sqrt{x+1}+\sqrt{x+2})$ **27.** $\dfrac{4(x^2-3)}{9(x^2-1)^{4/3}}$ **29.** $\dfrac{2x-x^2}{(1-2x)^{5/2}}$ **31.** $\dfrac{-1}{\sqrt{x}\,\sqrt{x+h}(\sqrt{x}+\sqrt{x+h})}$

33. $\dfrac{x^2-3}{3(x^2-1)^{4/3}}$ **35.** $\dfrac{-x(x^2+5)}{3(x^2-1)^{2/3}(x^2-4)^{3/2}}$

Review Exercise Set

1. a. $5x^2+4x-9$ **b.** $2x^2-2y^2$ **3. a.** $10x^2-7xy-12y^2$ **b.** x^2-4y^2 **c.** $9x^2-30x+25$
   d. $x^4+x^3-5x^2+7x-4$ **5. a.** $27a^3+108a^2b+144ab^2+64b^3$
   b. $x^{12}-18x^{10}+135x^8-540x^6+1{,}215x^4-1{,}458x^2+729$ **7. a.** $-336{,}798a^6b^{10}$ **b.** $2^{16}\cdot 3^5\cdot 5\cdot 17\cdot 19x^8$
9. a. $(3x+1)(x-2)$ **b.** $xy(x-6y)^2$ **11. a.** $(2x^2+3)(x+2)(x-2)$ **b.** $(x+2+y)(x+2-y)$
13. a. $x(8x+3y)(3x-4y)$ **b.** $(a-3)(a+3)^2$ **15. a.** $(2-x+y)(2+x-y)$ **b.** $(3x^2+2y)(2x^2-3y)$
17. a. $\left(x^2-\dfrac{y}{2}\right)\left(x^4+\dfrac{x^2y}{2}+\dfrac{y^2}{4}\right)$ **b.** $(x^4+1)(x^2+1)(x+1)(x-1)$ **19. a.** $3(2a-3b)^2$
   b. $(s^2+2t^2)(s^8-2s^6t^2+4s^4t^4-8s^2t^6+16t^8)$ **21. a.** $x(x+1)^{1/2}(x-1)^{1/2}(2x-1)(x+2)$
   b. $2x(1-2x)^{1/3}(3-10x)$ **23.** $\dfrac{2}{2-x}$ **25.** $\dfrac{6}{x-3}$ **27.** $-\dfrac{3x+8}{(x^2-4)^{3/2}}$ **29.** $\dfrac{1}{(1-x^2)^{3/2}}$
31. $\dfrac{-8x}{3(x^2+1)^{1/3}(x^2-1)^{5/3}}$ **33. a.** $\dfrac{\sqrt{x+3}+2}{x-1}$ **b.** $\dfrac{a-\sqrt{a^2-16}}{4}$ **35. a.** $(x-2y+2)(x-2y-2)$
   b. $(x-1)(x-2)(x^2+x+1)$ **37. a.** $(2x-3y+z+3)(2x-3y-z-3)$ **b.** $\dfrac{-5}{(2x-3)^{4/3}(x+1)^{2/3}}$
39. $\dfrac{(3x+2)^2(6x-53)}{(2x-5)^2}$ **41.** $\dfrac{x-3}{2\sqrt{x+1}\,(3x-1)^{4/3}}$

Chapter 3

Exercise Set 2

1. a. $\{4\}$ **b.** $\{2\}$ **3. a.** $\{-2\}$ **b.** $\{-1\}$ **5. a.** $\{\frac{7}{2}\}$ **b.** $\{1\}$ **7. a.** $\{\frac{15}{4}\}$ **b.** $\{-\frac{6}{11}\}$
9. a. $\{-\frac{11}{2}\}$ **b.** $\{-\frac{27}{13}\}$ **11. a.** $\{-\frac{2}{7}\}$ **b.** $\{-\frac{5}{31}\}$ **13. a.** $\{-\frac{31}{29}\}$ **b.** $\{\frac{29}{31}\}$ **15. a.** $\{-2,-3\}$
   b. $\{2,-\frac{1}{4}\}$ **17. a.** $\{\frac{2}{3},-4\}$ **b.** $\{0,\frac{3}{2}\}$ **19. a.** $\{\frac{3}{2},-\frac{3}{2}\}$ **b.** $\{0,\frac{9}{4}\}$ **21. a.** $\{-1,\frac{3}{2}\}$
   b. $\{2,-\frac{5}{3}\}$ **23. a.** $\{-1,\frac{1}{2}\}$ **b.** $\{3,-\frac{2}{3}\}$ **25. a.** $\{1,-\frac{1}{2}\}$ **b.** $\{\frac{7}{3},-\frac{5}{2}\}$ **27. a.** $\{-2,\frac{3}{2}\}$
   b. $\{1,-1\}$ **29.** $\dfrac{b+d}{a-c}$ **31.** $\dfrac{2y+7}{5y+3}$ **33.** $\{3b/2a,-5b/4a\}$ **35.** $\{m/2,-m/3\}$ **37.** $k=1$ or $k=3$

Exercise Set 3

1. $\{2,-6\}$ **3.** $\{3,-1\}$ **5.** $\{2,-5\}$ **7.** $\{8,-3\}$ **9.** $\{2\pm\sqrt{2}\}$ **11.** $\{4,-\frac{8}{3}\}$ **13.** $\left\{\dfrac{5\pm\sqrt{37}}{2}\right\}$

15. $\{3,-\frac{3}{2}\}$ **17.** $\{1\pm\sqrt{3}\}$ **19.** $\left\{\dfrac{-5\pm\sqrt{61}}{2}\right\}$ **21.** $\left\{\dfrac{1\pm\sqrt{41}}{4}\right\}$ **23.** $\left\{\dfrac{1\pm\sqrt{7}}{3}\right\}$ **25.** $\left\{\dfrac{1\pm\sqrt{61}}{10}\right\}$

27. $\left\{\dfrac{1 \pm \sqrt{13}}{4}\right\}$ **29.** $\left\{\dfrac{3 \pm \sqrt{41}}{4}\right\}$ **31.** $\left\{\dfrac{3 \pm \sqrt{33}}{4}\right\}$ **33.** $\left\{\dfrac{v_0 \pm \sqrt{v_0^2 + 2gs_0}}{g}\right\}$ **35.** $\left\{\dfrac{x(w + \sqrt{mgw})}{mg - w}\right\}$

37. $\left\{-\sqrt{3}\,k, \dfrac{3\sqrt{3}\,k}{2}\right\}$ **39.** $\left\{1, -\dfrac{3k + 2}{2}\right\}$ **41.** $\{77.88, -13.13\}$

Exercise Set 4

1. a. $9 - 2i$ **b.** $-2 - 5i$ **c.** $-1 + 9i$ **d.** $6 - 3i$ **3. a.** $18 + i$ **b.** 5 **c.** $-83 - i$

d. $15 - 5i$ **5. a.** $\dfrac{1 - i}{2}$ **b.** $\dfrac{3 + 2i}{13}$ **c.** $-i$ **d.** $\dfrac{7 - 8i}{113}$ **7. a.** i **b.** $-i$ **c.** $\dfrac{3 + 29i}{34}$

d. $\dfrac{9 + 23i}{61}$ **9. a.** $2i$ **b.** $9i\sqrt{2}$ **c.** $2i\sqrt{10}$ **d.** -6 **e.** -4 **11. a.** $-i$ **b.** i **c.** 1

d. i **e.** -1 **13. a.** $\{\pm 2i\}$ **b.** $\left\{\dfrac{1 \pm i\sqrt{7}}{2}\right\}$ **15. a.** $\left\{\dfrac{-3 \pm i\sqrt{3}}{2}\right\}$ **b.** $\left\{\dfrac{3 \pm i}{2}\right\}$

17. a. $\left\{\dfrac{2 \pm i\sqrt{14}}{3}\right\}$ **b.** $\left\{\dfrac{2 \pm i\sqrt{6}}{5}\right\}$ **19. a.** $\left\{\dfrac{3 \pm i\sqrt{5}}{2}\right\}$ **b.** $\left\{\dfrac{-3 \pm i\sqrt{11}}{5}\right\}$ **21. a.** Real and unequal

b. Imaginary **23. a.** Real and unequal **b.** Imaginary **25. a.** $-7 - 24i$ **b.** $-117 - 44i$

27. $-20 + 20i$

29. a. $\overline{z_1 z_2} = \overline{(a + bi)(c + di)} = \overline{(ac - bd) + (ad + bc)i} = (ac - bd) - (ad + bc)i$

  $\overline{z_1} \cdot \overline{z_2} = \overline{(a + bi)(c + di)} = (a - bi)(c - di) = ac - adi - bci + bdi^2 = (ac - bd) - (ad + bc)i$

b. First calculate z_1/z_2:

$$\frac{z_1}{z_2} = \frac{a + bi}{c + di} = \frac{a + bi}{c + di} \cdot \frac{c - di}{c - di} = \frac{(ac + bd) + (bc - ad)i}{c^2 + d^2}$$

$$\overline{\left(\frac{z_1}{z_2}\right)} = \frac{(ac + bd) + (ad - bc)i}{c^2 + d^2}$$

$$\frac{\overline{z_1}}{\overline{z_2}} = \frac{a - bi}{c - di} = \frac{a - bi}{c - di} \cdot \frac{c + di}{c + di} = \frac{(ac + bd) + (ad - bc)i}{c^2 + d^2}$$

Exercise Set 5

1. $\left\{\pm 1, \pm\frac{3}{2}\right\}$ **3.** $\left\{\frac{1}{3}, -\frac{1}{2}\right\}$ **5.** $\{\pm 1, \pm 2i\}$ **7.** $\left\{\pm 2, \pm\frac{3}{2}\right\}$ **9.** $\{-1, 27\}$ **11.** $\{5\}$ **13.** $\{12, -2\}$
15. $\{-8\}$ **17.** $\left\{\frac{6}{5}\right\}$ **19.** $\left\{3, \frac{5}{2}\right\}$ **21.** $\left\{4, \frac{1}{2}\right\}$ **23.** $\left\{0, \frac{17}{2}\right\}$ **25.** $\{-5\}$ **27.** No solution **29.** $\{2\}$
31. $\{-3\}$ **33.** $\{5\}$ **35.** $\left\{-3, -\frac{4}{3}\right\}$ **37.** $\left\{3, \frac{9}{5}\right\}$ **39.** $\left\{\frac{1}{3}, -2\right\}$ **41.** $\{3, -1\}$ **43.** $\{6\}$ **45.** $\{-3\}$
47. $\left\{\pm\sqrt{\sqrt{3} - 1}, \pm i\sqrt{\sqrt{3} + 1}\right\}$ **49.** $\left\{\pm\frac{1}{8}, \pm 8i/27\right\}$

Exercise Set 6

1. 90 miles from Washington **3.** 8 kilometers per hour **5.** 36.5 feet by 73 feet **7.** 13 feet by 23 feet
9. 123, 125, 127 **11.** 86 **13.** 12,234 grandstand, 10,562 bleacher **15.** 120 miles per hour
17. 50 cubic centimeters **19.** 5 seconds **21.** \$125 **23.** \$3,200 at $5\frac{1}{2}\%$, \$6,800 at 7% **25.** 645 pounds
27. $3\frac{1}{2}$ hours, $2\frac{1}{2}$ hours **29.** 4, 6 **31.** 50 **33.** 12 inches by 22 inches **35.** 350
37. 38¢ fare, 290,000 riders **39.** $133\frac{1}{3}$ gallons 96 octane, $66\frac{2}{3}$ gallons 87 octane **41.** 4 days
43. Two solutions: 50 feet or 90 feet **45. a.** $0.0024t/1,990$ **b.** 99.5 minutes

Exercise Set 7

1. $\{x: \ x < 1\}$

5. $\{x: \ x < -\frac{7}{3}\}$

9. $\{x: \ x > \frac{7}{3}\}$

3. $\{x: \ x < 2\}$

7. $\{x: \ x < 2\}$

11. $\{x: \ x > -\frac{16}{5}\}$

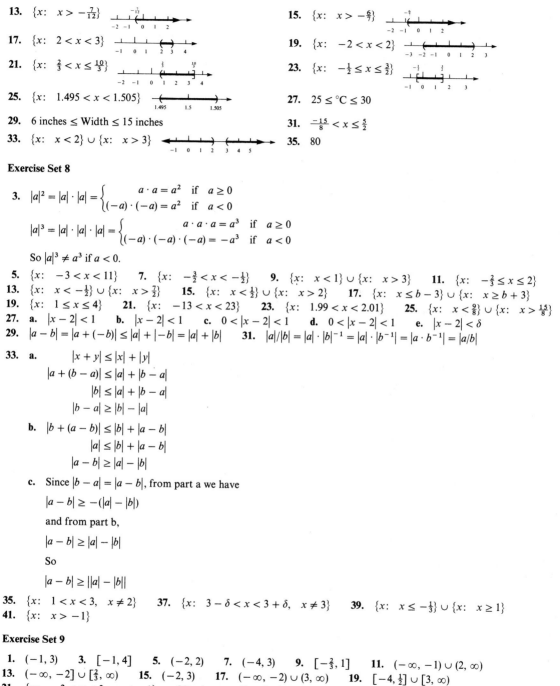

13. $\{x: \ x > -\frac{7}{12}\}$

15. $\{x: \ x > -\frac{6}{7}\}$

17. $\{x: \ 2 < x < 3\}$

19. $\{x: \ -2 < x < 2\}$

21. $\{x: \ \frac{2}{3} < x \le \frac{10}{3}\}$

23. $\{x: \ -\frac{1}{2} \le x \le \frac{3}{2}\}$

25. $\{x: \ 1.495 < x < 1.505\}$

27. $25 \le \ ^\circ C \le 30$

29. 6 inches $\le$ Width $\le$ 15 inches

31. $\frac{-15}{8} < x \le \frac{5}{2}$

33. $\{x: \ x < 2\} \cup \{x: \ x > 3\}$

35. 80

Exercise Set 8

3. $|a|^2 = |a| \cdot |a| = \begin{cases} a \cdot a = a^2 & \text{if} \quad a \ge 0 \\ (-a) \cdot (-a) = a^2 & \text{if} \quad a < 0 \end{cases}$

$|a|^3 = |a| \cdot |a| \cdot |a| = \begin{cases} a \cdot a \cdot a = a^3 & \text{if} \quad a \ge 0 \\ (-a) \cdot (-a) \cdot (-a) = -a^3 & \text{if} \quad a < 0 \end{cases}$

So $|a|^3 \ne a^3$ if $a < 0$.

5. $\{x: \ -3 < x < 11\}$ **7.** $\{x: \ -\frac{3}{2} < x < -\frac{1}{2}\}$ **9.** $\{x: \ x < 1\} \cup \{x: \ x > 3\}$ **11.** $\{x: \ -\frac{2}{3} \le x \le 2\}$
13. $\{x: \ x < -\frac{1}{2}\} \cup \{x: \ x > \frac{7}{2}\}$ **15.** $\{x: \ x < \frac{1}{2}\} \cup \{x: \ x > 2\}$ **17.** $\{x: \ x \le b - 3\} \cup \{x: \ x \ge b + 3\}$
19. $\{x: \ 1 \le x \le 4\}$ **21.** $\{x: \ -13 < x < 23\}$ **23.** $\{x: \ 1.99 < x < 2.01\}$ **25.** $\{x: \ x < \frac{9}{8}\} \cup \{x: \ x > \frac{15}{8}\}$
27. **a.** $|x - 2| < 1$ **b.** $|x - 2| < 1$ **c.** $0 < |x - 2| < 1$ **d.** $0 < |x - 2| < 1$ **e.** $|x - 2| < \delta$
29. $|a - b| = |a + (-b)| \le |a| + |-b| = |a| + |b|$ **31.** $|a|/|b| = |a| \cdot |b|^{-1} = |a| \cdot |b^{-1}| = |a \cdot b^{-1}| = |a/b|$

33. **a.**
$$|x + y| \le |x| + |y|$$
$$|a + (b - a)| \le |a| + |b - a|$$
$$|b| \le |a| + |b - a|$$
$$|b - a| \ge |b| - |a|$$

b.
$$|b + (a - b)| \le |b| + |a - b|$$
$$|a| \le |b| + |a - b|$$
$$|a - b| \ge |a| - |b|$$

c. Since $|b - a| = |a - b|$, from part a we have
$$|a - b| \ge -(|a| - |b|)$$
and from part b,
$$|a - b| \ge |a| - |b|$$
So
$$|a - b| \ge ||a| - |b||$$

35. $\{x: \ 1 < x < 3, \ x \ne 2\}$ **37.** $\{x: \ 3 - \delta < x < 3 + \delta, \ x \ne 3\}$ **39.** $\{x: \ x \le -\frac{1}{3}\} \cup \{x: \ x \ge 1\}$
41. $\{x: \ x > -1\}$

Exercise Set 9

1. $(-1, 3)$ **3.** $[-1, 4]$ **5.** $(-2, 2)$ **7.** $(-4, 3)$ **9.** $[-\frac{2}{3}, 1]$ **11.** $(-\infty, -1) \cup (2, \infty)$
13. $(-\infty, -2] \cup [\frac{2}{3}, \infty)$ **15.** $(-2, 3)$ **17.** $(-\infty, -2) \cup (3, \infty)$ **19.** $[-4, \frac{1}{2}] \cup [3, \infty)$
21. $\{x: \ -3 < x < 3, \ x \ne -1\}$ **23.** $(-\infty, -5) \cup (-1, 3)$ **25.** $(-2, -\frac{3}{2}) \cup (3, 5)$ **27.** $(-\infty, -5) \cup (2, \infty)$
29. $(-\infty, \frac{3}{2}) \cup (5, \infty)$ **31.** $(-6, 0) \cup (1, 2)$ **33.** $(-\frac{3}{2}, -1) \cup (3, 4)$ **35.** $\{k: \ k \le -\frac{8}{5}\} \cup \{k: \ k > 0\}$
37. $\{c: \ -3 < c < 1\}$ **39.** $(-\infty, -\frac{5}{3}) \cup (-1, \frac{3}{2}) \cup (2, \infty)$ **41.** $\{t: \ 0 \le t \le 3\}$ **43.** $d_{\min} = 10$ inches

Review Exercise Set

1. a. $\{\frac{14}{13}\}$ **b.** $\{\frac{29}{36}\}$ **3. a.** $\{\frac{65}{36}\}$ **b.** $\{\frac{38}{21}\}$ **5. a.** $\{\frac{3}{2}, -\frac{5}{3}\}$ **b.** $\{\frac{3}{4}, \frac{4}{3}\}$ **7. a.** $\{1, 3\}$

b. $\left\{\frac{1 \pm \sqrt{13}}{3}\right\}$ **9. a.** $\left\{\frac{5 \pm i\sqrt{7}}{4}\right\}$ **b.** $\left\{\frac{2 \pm i\sqrt{2}}{3}\right\}$ **11. a.** $\{\frac{4}{5}, -\frac{2}{3}\}$ **b.** $\left\{\frac{2 \pm \sqrt{13}}{3}\right\}$

13. a. $\{\frac{31}{14}\}$ **b.** $\{-\frac{5}{3}\}$ **15. a.** $\{4, -2\}$ **b.** $\{1, 2\}$ **17. a.** $\{-\frac{13}{2}\}$ **b.** $\{-1, -2\}$ **19. a.** $\{6\}$

b. $\{1\}$ **21. a.** $\{\pm 2i, \pm\frac{2}{3}\}$ **b.** $\{\frac{3}{2}, -\frac{4}{3}\}$ **23. a.** $23 + 2i$ **b.** $\dfrac{6 + 17i}{25}$ **c.** $-236 - 115i$

25. a. $\left\{\frac{5 \pm i\sqrt{7}}{4}\right\}$ **b.** $\left\{\frac{-1 \pm \sqrt{19}}{3}\right\}$ **27. a.** Imaginary **b.** Real and unequal **c.** Real and equal

d. Imaginary **29.** Width $= 13$ feet, Length $= 19$ feet **31.** 117 kilometers **33.** \$8,000 at 5%,
\$13,000 at 6% **35.** $4\frac{2}{3}$ gallons **37.** M's speed $= 55$ miles per hour, N's speed $= 60$ miles per hour
39. $2\frac{1}{2}$ inches **41.** 20 kilograms **43. a.** $[-1, 5)$ **b.** $(\frac{3}{8}, \frac{27}{8}]$ **45. a.** $(-3, -2] \cup (3, \infty)$
b. $(-\infty, -4) \cup (1, 2)$ **47. a.** $(-\infty, -4] \cup [2, \frac{10}{3}]$ **b.** $(-1, 3) \cup (5, \infty)$ **49. a.** $[1, \frac{5}{3}]$
b. $(-\infty, -1) \cup (5, \infty)$ **51. a.** $|x - 3| < \frac{1}{2}$ **b.** $|x - 5| < 2$ **c.** $0 < |x + 2| < 1$ **d.** $|x - a| < \delta$

e. $|y - y_0| < \varepsilon$ **53.** $15 \le °C \le 30$ **55. a.** $\{k \pm i\sqrt{k^2 - 1}\}$ **b.** $\dfrac{2ab}{a + b}$ **57.** $\{k: \ k \le \frac{2}{5}\} \cup \{k: \ k \ge 2\}$

59. $\{-2\}$ **61.** $\{\frac{1}{4}, -1\}$ **63.** $(-1, 0] \cup (1, 3) \cup [7, \infty)$ **65. a.** $(-\infty, -1)$ **b.** $(-1, 3)$
67. 3 feet by 3 feet by 5 feet **69.** $2\frac{1}{2}$ inches

Cumulative Review Exercise Set I (Chapters 1–3)

1. $\dfrac{4 - x}{2x - 3}$

3. If $a < b$ and $c > 0$, then $ac < bc$. Also, if $c < d$ and $b > 0$, $bc < bd$. By transitivity, $ac < bd$. Suppose $a = -2, b = 4$,
$c = -3, d = -2$. Then $-2 < 4$ and $-3 < -2$, but $(-2)(-3) = 6 \not< (-2)(4) = -8$.

5. a. $\frac{79}{370}$

b. Let $r_1 = \dfrac{-b + \sqrt{b^2 - 4ac}}{2a}, r_2 = \dfrac{-b - \sqrt{b^2 - 4ac}}{2a}$

$r_1 + r_2 = -b/a$, which is rational
$r_1 \cdot r_2 = c/a$, which is rational

7. $\dfrac{25(x^2 + 1)}{(3x + 4)(4x - 3)}$ **9.** $2\frac{1}{2}$ feet **11. a.** $\dfrac{2}{\sqrt{2x + 2h - 3} + \sqrt{2x - 3}}$ **b.** $\dfrac{17\sqrt{6}}{12}$ **13. a.** $(x + 2)(x - 2)(3 - xy)$

b. $2x(9x + 8y)(4x - 3y)$ **15. a.** $s = \dfrac{Fr + 1}{r - 2F}$ **b.** $t = \dfrac{-v_0 + \sqrt{v_0^2 + 2gs}}{g}$ **17. a.** $\dfrac{4a^7b}{125}$ **b.** 6×10^2

19. $-\dfrac{x^2 + 3}{3(x^2 - 1)^{5/3}}$ **21. a.** $(-3, 1)$ **b.** $(-11, -5] \cup (1, 5]$ **23.** 15 hours, 10 hours

25. 90 suits sold during the sale; regular price $= \$210$

Chapter 4

Exercise Set 2

1. **3.** **5.** **7.**

9. 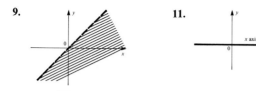 **11.** **13.** **15.**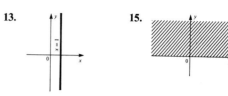

17. a. $\sqrt{13}$ **b.** 10 **19. a.** 17 **b.** $3\sqrt{2}$ **21.** $y = 0$ or -4

23. Let $A = (-7, 5)$, $B = (-3, -2)$, and $C = (4, 2)$. Then $\overline{AB} = \sqrt{65}$, $\overline{BC} = \sqrt{65}$, $\overline{AC} = \sqrt{130}$. Since $\overline{AB} = \overline{BC}$, the triangle is isosceles. Since $\overline{AB}^2 + \overline{BC}^2 = \overline{AC}^2$, it is also a right triangle.

25. Let $A = (-1, 2)$, $B = (2, 4)$, $C = (3, -4)$, and $D = (6, -2)$. Then $\overline{AB}^2 = 13$, $\overline{AC}^2 = 52$, and $\overline{BC}^2 = 65$. So $\overline{AB}^2 + \overline{AC}^2 = \overline{BC}^2$, and $\triangle ABC$ is a right triangle. Similarly, $\overline{BD}^2 = 52$ and $\overline{DC}^2 = 13$, so $\overline{BD}^2 + \overline{DC}^2 = \overline{BC}^2$, and $\triangle BDC$ is a right triangle. Since angles A and D are right angles, it follows that $ABCD$ is a rectangle.

Exercise Set 3

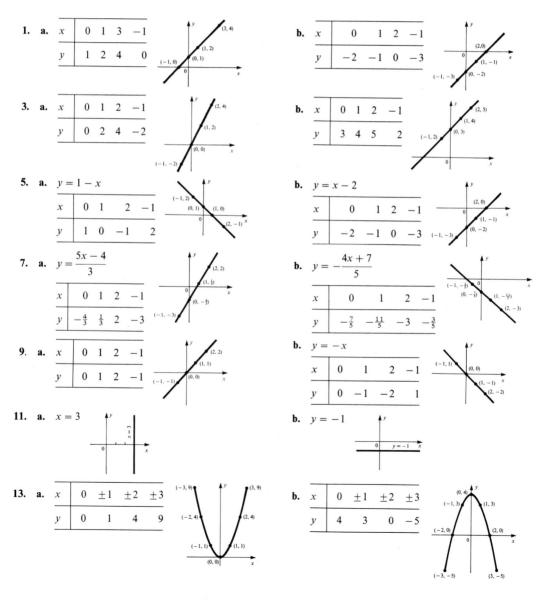

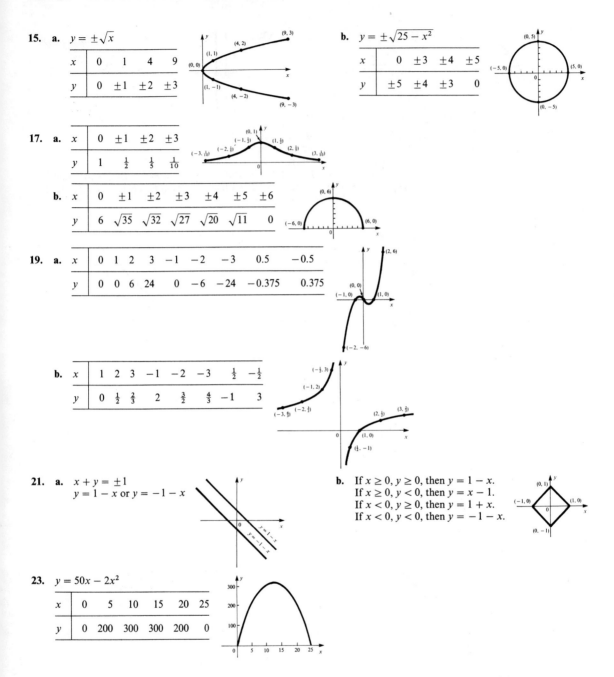

15. a. $y = \pm\sqrt{x}$

x	0	1	4	9
y	0	±1	±2	±3

b. $y = \pm\sqrt{25 - x^2}$

x	0	±3	±4	±5
y	±5	±4	±3	0

17. a.

x	0	±1	±2	±3
y	1	$\frac{1}{2}$	$\frac{1}{5}$	$\frac{1}{10}$

b.

x	0	±1	±2	±3	±4	±5	±6
y	6	$\sqrt{35}$	$\sqrt{32}$	$\sqrt{27}$	$\sqrt{20}$	$\sqrt{11}$	0

19. a.

x	0	1	2	3	-1	-2	-3	0.5	-0.5
y	0	0	6	24	0	-6	-24	-0.375	0.375

b.

x	1	2	3	-1	-2	-3	$\frac{1}{2}$	$-\frac{1}{2}$
y	0	$\frac{1}{2}$	$\frac{2}{3}$	2	$\frac{3}{2}$	$\frac{4}{3}$	-1	3

21. a. $x + y = \pm1$

$y = 1 - x$ or $y = -1 - x$

b. If $x \geq 0$, $y \geq 0$, then $y = 1 - x$.

If $x \geq 0$, $y < 0$, then $y = x - 1$.

If $x < 0$, $y \geq 0$, then $y = 1 + x$.

If $x < 0$, $y < 0$, then $y = -1 - x$.

23. $y = 50x - 2x^2$

x	0	5	10	15	20	25
y	0	200	300	300	200	0

Exercise Set 4

1. a. -1 **b.** 1   **c.** -5 **d.** 9   **3. a.** 3 **b.** -1 **c.** 0   **d.** $\frac{1}{2}$ **5. a.** 1   **b.** 2   **c.** 0

d. $\sqrt{2x - 1}$ **7. a.** $\frac{7}{2}$ **b.** $-\frac{4}{11}$ **c.** $-\frac{8}{39}$ **d.** $-\frac{13}{6}$ **9. a.** -2 **b.** 4   **c.** $2a^2 - 4$

d. $2x^2 + 4x - 2$ **e.** $2x^2 + 4hx + 2h^2 - 4$ **11. a.** 0   **b.** $\sqrt{2t - t^2}$ **c.** $\dfrac{\sqrt{t^2 - 1}}{|t|}$

d. $\sqrt{1 - t^2 - 2t(\Delta t) - (\Delta t)^2}$ **13. a.** 4   **b.** -1 **c.** -1 **d.** 0   **15. a.** 1   **b.** 1   **c.** 0

d. 0   **17.** $-\dfrac{3}{1+h}$ **19. a.** $(-\infty, -2) \cup [1, \infty)$ **b.** $\{t: |t| \ge 2\}$ **21.** $P = 2w + 24$ **23.** $C = \pi d$

25. $A = 10{,}000(1.07)^t$ **27. a.** $2\pi r^3$ **b.** $4\pi r^2$ **c.** $6\pi r^2$ **29.** $f(1/x) = \dfrac{1/x + 2}{2/x + 1} = \dfrac{1 + 2x}{2 + x} = \dfrac{1}{f(x)}$

31. a. Even   **b.** Odd   **c.** Neither   **d.** Neither   **e.** Even   **33. a.** $f(x) = 200{,}000 - 40{,}000x$

b. Domain $= \{x: \ 0 \le x \le 5\}$ **35.** $(-5, -2] \cup [4, \infty)$ **37.** $A = \dfrac{s^2\sqrt{3}}{4}$ **39.** $V = 2\pi r^2 \sqrt{a^2 - r^2}$

41. $A = \dfrac{4(10 - h)^2}{25}$ **43.** $f(x) = 0.01x + 5$ **45.** $V = \dfrac{\pi r^2}{3}(a + \sqrt{a^2 - r^2})$

47. Since we must have $x \ge 0$, it follows that $f(x) \ge 1$. So the range is contained in the given set. Now let k be any number greater than or equal to 1. Solving $1 + \sqrt{x} = k$ yields $x = (k - 1)^2$, and this is in the domain. Thus, every k in the set $\{y: \ y \ge 1\}$ is the image of an x in the domain. So $\{y: \ y \ge 1\}$ is the range.

Exercise Set 5

13. Domain $= \{x: \ x \ne 0\}$; range $= \{y: \ y \ne 0\}$; ϕ is 1–1
15. Function ϕ is not 1–1, since, for example, $\phi(2) = \phi(-2)$.

19. Domain $= \{t: \ t \ne -1\}$; f is 1–1
21. ϕ is not 1–1
23. Domain $= \{x: \ x \ge 1\}$; h is 1–1
25. b. Restricted domain $= \{x: \ x \ge 0\}$
27. $f(-1) = f(1) = f(4)$; so f is not 1–1
29. Let f be increasing. Suppose $x_1 \ne x_2$. We may suppose, in particular, that $x_1 < x_2$. Then $f(x_1) < f(x_2)$, and hence $f(x_1) \ne f(x_2)$. So f is 1–1. A similar argument applies if f is decreasing.
33. b. No. There is no x for which $f(x) = 1$.

Exercise Set 6

1. $(f \circ g)(x) = 2x$; $(g \circ f)(x) = 2x - 1$ **3.** $(f \circ g)(x) = 1 - 3x$; $(g \circ f)(x) = 7 - 3x$

5. $(f \circ g)(x) = 2x^2 + 8x + 8$; $(g \circ f)(x) = 2x^2 + 2$ **7.** $(f \circ g)(x) = \dfrac{x - 3}{x - 2}$; $(g \circ f)(x) = -\dfrac{x + 1}{3x + 2}$

9. $(f \circ g)(x) = 4x^2 - 16x + 16$; $(g \circ f)(x) = 2x^2 - 4x - 1$ **11.** $(f \circ g)(x) = \dfrac{2}{2 - x}$; $(g \circ f)(x) = \dfrac{2x - 2}{x}$

13. $(f \circ g)(x) = \sqrt{x^2 - 1}$, domain $= \{x: \ |x| \ge 1\}$; $(g \circ f)(x) = x - 1$, domain $= \{x: \ x \ge 1\}$
15. $(f \circ g)(x) = 2\sqrt{x^2 - 3x + 2}$, domain $= \{x: \ x \ge 2\} \cup \{x: \ x \le 1\}$; $(g \circ f)(x) = 2\sqrt{x^2 - 1} - 3$, domain $= \{x: \ |x| \ge 1\}$

17. a. $\dfrac{2x^2 - 9}{5}$ **b.** $\dfrac{4x^2 - 12x - 66}{25}$ **c.** x **d.** x **e.** $\dfrac{25x^2 + 30x - 3}{4}$ **19.** $f^{-1}(x) = \dfrac{x - 5}{2}$

21. $g^{-1}(x) = 2x - 8$ **23.** $G^{-1}(x) = \dfrac{7x + 4}{3}$ **25.** $g^{-1}(x) = \dfrac{1 - 2x}{x}$, $x \ne 0$ **27.** $G^{-1}(x) = x^2$, $x \ge 0$

29. $h^{-1}(t) = \dfrac{3t}{1 - t}$, $t \ne 1$ **31.** $F^{-1}(z) = \sqrt{z^2 + 4}$, $z \ge 0$ **33.** $f^{-1}(x) = \sqrt{x + 4}$ on domain $\{x: \ x \ge -4\}$

35. $h^{-1}(x) = 1 + \sqrt{x + 1}$ on domain $\{x: \ x \ge -1\}$

37. $f^{-1}(x) = \dfrac{bx - a}{1 - x}$, $x \ne 1$; domain of $f = \{x: \ x \ne -b\}$; range of $f = \{y: \ y \ne 1\}$; domain of $f^{-1} = \{x: \ x \ne 1\}$; range of $f^{-1} = \{y: \ y \ne -b\}$

39. $(h \circ f)(t) = \dfrac{62{,}784}{(6{,}400 + 4t - 0.0001t^2)^2}$

Exercise Set 7

1. a.

b.

3. a.

b.

5. a.

b.

7. a.

b.

9. a.

b.

11. a. Function; no inverse   **b.** Function; no inverse   **c.** Not a function   **d.** Function; has inverse
  e. Function; no inverse   **f.** Function; has inverse
13. a. If $f(x_1) = f(x_2)$, then $2x_1 - 5 = 2x_2 - 5$, and so $x_1 = x_2$. Thus, f is $1 - 1$, and $f^{-1}(x) = (x + 5)/2$.

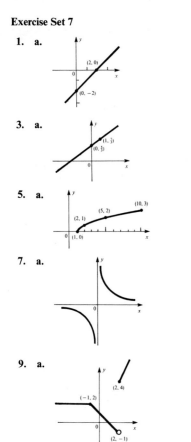

  b. If $g(x_1) = g(x_2)$, then $\sqrt{x_1 - 2} = \sqrt{x_2 - 2}$, or $x_1 - 2 = x_2 - 2$, so that $x_1 = x_2$. So g is $1-1$ and $g^{-1}(x) = x^2 + 2$, $x \geq 0$.

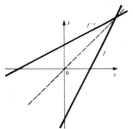

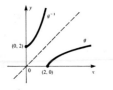

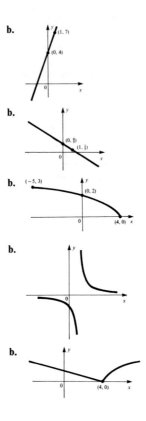

15. a.

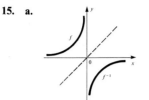

b.

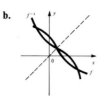

17. Increasing on $(-\infty, -10], [-6, 0], [4, 8]$; decreasing on $[-10, -6], [0, 4], [8, \infty)$

19. $y_1 = \sqrt{4 - x^2}$; $y_2 = -\sqrt{4 - x^2}$; neither function has an inverse

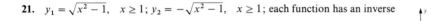

21. $y_1 = \sqrt{x^2 - 1}$, $x \geq 1$; $y_2 = -\sqrt{x^2 - 1}$, $x \geq 1$; each function has an inverse

23.

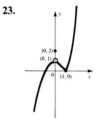

Exercise Set 8

1. $u = kv$ **3.** $z = kxy$ **5.** $s = k/\sqrt{t}$ **7.** $F = km_1m_2/r^2$ **9.** $s = kt^2$ **11.** $\frac{20}{3}$ **13.** 12 **15.** 336

17. 32 pounds **19.** 22 amperes **21.** $k = \pi/3$; $V = 1/3\pi r^2 h$; $V = 120\pi$ **23.** 6°F per minute

25. 7 cases per day **27.** Approximately 671 days **29.** 375 pounds **31.** 302.5 : 1

Review Exercise Set

1. a.

x	0	1	2	-1	-2
y	-1	1	3	-3	-5

b.

x	0	1	2	3	-1	-2
y	$\frac{4}{3}$	$\frac{2}{3}$	0	$-\frac{2}{3}$	2	$\frac{8}{3}$

3. a.

x	0	2	4	-2
y	-2	3	8	-7

b.

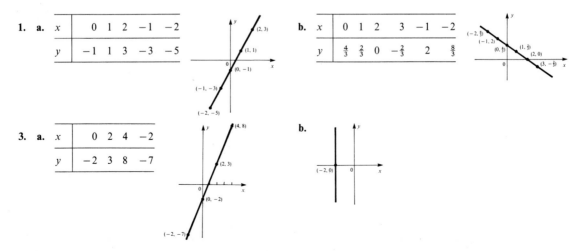

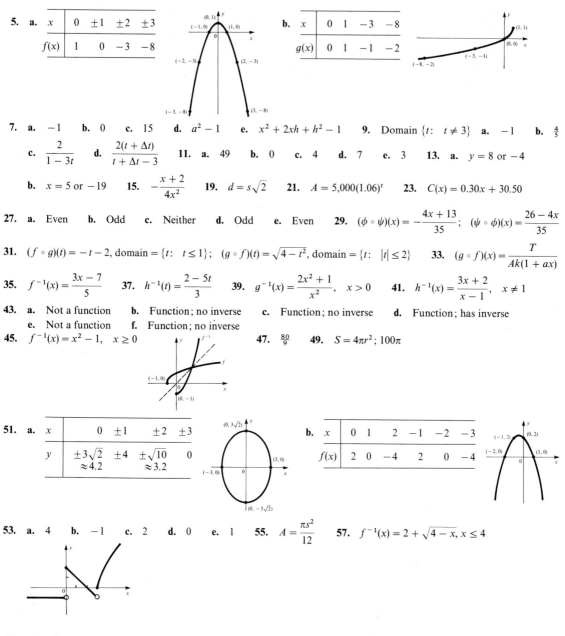

5. **a.**

x	0	± 1	± 2	± 3
$f(x)$	1	0	-3	-8

b.

x	0	1	-3	-8
$g(x)$	0	1	-1	-2

7. **a.** -1 **b.** 0   **c.** 15   **d.** $a^2 - 1$ **e.** $x^2 + 2xh + h^2 - 1$ **9.** Domain $\{t: \ t \neq 3\}$ **a.** -1 **b.** $\frac{4}{5}$

c. $\dfrac{2}{1 - 3t}$ **d.** $\dfrac{2(t + \Delta t)}{t + \Delta t - 3}$ **11.** **a.** 49   **b.** 0   **c.** 4   **d.** 7   **e.** 3   **13.** **a.** $y = 8$ or -4

b. $x = 5$ or -19 **15.** $-\dfrac{x + 2}{4x^2}$ **19.** $d = s\sqrt{2}$ **21.** $A = 5{,}000(1.06)^t$ **23.** $C(x) = 0.30x + 30.50$

27. **a.** Even   **b.** Odd   **c.** Neither   **d.** Odd   **e.** Even   **29.** $(\phi \circ \psi)(x) = -\dfrac{4x + 13}{35}$; $(\psi \circ \phi)(x) = \dfrac{26 - 4x}{35}$

31. $(f \circ g)(t) = -t - 2$, domain $= \{t: \ t \le 1\}$; $(g \circ f)(t) = \sqrt{4 - t^2}$, domain $= \{t: \ |t| \le 2\}$ **33.** $(g \circ f)(x) = \dfrac{T}{Ak(1 + ax)}$

35. $f^{-1}(x) = \dfrac{3x - 7}{5}$ **37.** $h^{-1}(t) = \dfrac{2 - 5t}{3}$ **39.** $g^{-1}(x) = \dfrac{2x^2 + 1}{x^2}$, $x > 0$ **41.** $h^{-1}(x) = \dfrac{3x + 2}{x - 1}$, $x \neq 1$

43. **a.** Not a function   **b.** Function; no inverse   **c.** Function; no inverse   **d.** Function; has inverse
   e. Not a function   **f.** Function; no inverse
45. $f^{-1}(x) = x^2 - 1$, $x \ge 0$ **47.** $\frac{80}{9}$ **49.** $S = 4\pi r^2$; 100π

51. **a.**

x	0	± 1	± 2	± 3
y	$\pm 3\sqrt{2}$ ≈ 4.2	± 4	$\pm\sqrt{10}$ ≈ 3.2	0

b.

x	0	1	2	-1	-2	-3
$f(x)$	2	0	-4	2	0	-4

53. **a.** 4   **b.** -1 **c.** 2   **d.** 0   **e.** 1   **55.** $A = \dfrac{\pi s^2}{12}$ **57.** $f^{-1}(x) = 2 + \sqrt{4 - x}$, $x \le 4$

Chapter 5

Exercise Set 2

1. **a.** 5   **b.** $\frac{3}{4}$

c. $-\frac{7}{5}$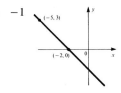

d. -1

3. $m_{AB} = -\frac{1}{4}, m_{BC} = -\frac{7}{3}, m_{AC} = 1$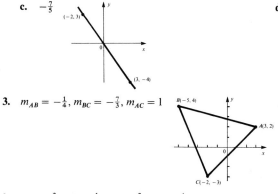

5. 5 units horizontal; 4 units vertical   **7.** $y = 7$

9. $m_{AB} = \frac{3}{4}, m_{BC} = \frac{1}{4}, m_{CD} = \frac{3}{4}, m_{AD} = \frac{1}{4}$; parallelogram, since opposite sides have same slope

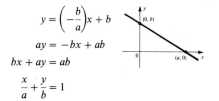

11. $y + 2x - 2 = 0$ **13.** $8x + 3y - 1 = 0$ **15.** $4x - 3y + 12 = 0$ **17.** $2x - 3y + 6 = 0$

19. $m = 3, b = -4$ **21.** $m = -\frac{1}{3}, b = \frac{7}{6}$

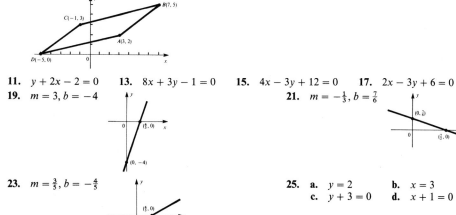

23. $m = \frac{3}{5}, b = -\frac{4}{5}$

25. **a.** $y = 2$ **b.** $x = 3$
 c. $y + 3 = 0$ **d.** $x + 1 = 0$

27. Since $(a, 0)$ and $(0, b)$ are on the line, the slope is $-b/a$. So the equation is

$$y = \left(-\frac{b}{a}\right)x + b$$

$$ay = -bx + ab$$

$$bx + ay = ab$$

$$\frac{x}{a} + \frac{y}{b} = 1$$

29. Let $A = (2, 2), B = (-1, -7), C = (3, 5)$. Method 1: $m_{AB} = 3, m_{BC} = 3$, so they are collinear. Method 2: Equation of line AB is $3x - y - 4 = 0$, and C satisfies this equation.

31. Let $A = (0, -3), B = (2, 1), C = (-1, -5)$. Method 1: $m_{AB} = 2, m_{BC} = 2$, so they are collinear. Method 2: Equation of line AB is $y = 2x - 3$, and C satisfies this equation.

33. **a.** $3x - 2y + C = 0$ has slope $\frac{3}{2}$.   **b.** $C = 17; 3x - 2y + 17 = 0$

35. Method 1 (point–slope): $y + 4 = \frac{1}{2}(x - 3)$ Method 2 (method of Problem 33): $x - 2y + C = 0$
$$x - 2y - 11 = 0$$
$$C = -11$$
$$x - 2y - 11 = 0$$

37. All have slope $\frac{2}{3}$; the lines are all parallel.

Exercise Set 3

1. $3x - 4y - 27 = 0$ **3.** $x + 3y - 5 = 0$ **5.** $5x - 2y = 0$ **7.** $3x + 4y - 22 = 0$ **9.** $3x + 4y - 8 = 0$

11. $x - y - 10 = 0$ **13.** $x - 3 = 0$ **15.** $x - 2 = 0$ **17.** $4x + 10y + 15 = 0$

19. Let $A = (2, 1)$, $B = (6, 9)$, $C = (-2, 3)$. Then $m_{AB} = 2$, $m_{AC} = -\frac{1}{2}$, so angle A is a right angle.

21. Let $A = (2, 2)$, $B = (0, -1)$, $C = (-4, 1)$, $D = (-2, 4)$. Then $m_{AB} = \frac{3}{2}$, $m_{BC} = -\frac{1}{2}$, $m_{CD} = \frac{3}{2}$, $m_{AD} = -\frac{1}{2}$, so opposite sides are parallel.

23. $3x - 4y + C = 0$; $9x - 12y + 32 = 0$ **25.** $x - 2y + 6 = 0$

27. Using similar triangles we see that $x - x_1 = x_2 - x$, so $2x = x_1 + x_2$, or $x = (x_1 + x_2)/2$. A similar construction gives y.

29. Let $A = (-5, 2)$, $B = (-7, -6)$, $C = (3, -4)$. Then the perpendicular bisector of side AB is $x + 4y + 14 = 0$, of side BC is $5x + y + 15 = 0$, of side AC is $4x - 3y + 1 = 0$.

31. $2x + 2y - 3 = 0$ **33.**   **a.** $2x - y - 5 = 0$

                                                  b. $x + 2y + 5 = 0$

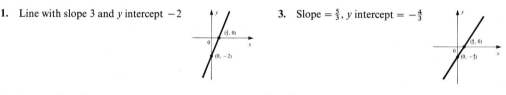

Exercise Set 4

1. Line with slope 3 and y intercept -2 **3.** Slope $= \frac{5}{3}$, y intercept $= -\frac{4}{3}$

5. $f(x) = -\frac{2}{3}x + \frac{8}{3}$ **7.** $g(x) = 2x - 1$ **9.** $h(x) = -\frac{2}{3}x + \frac{7}{3}$ **11.** No, a linear function is 1–1; $f(x) = 3$

13. $f(x) = 0.2x + 18$; $93   **15.** $F(x) = 8x$; 64 pounds   **17.** $(f \circ g)(x) = a_1 b_1 x + (a_0 + a_1 b_0)$, which is linear

19. Let $f(x) = ax + b$, $a \neq 0$. If $f(x_1) = f(x_2)$, then

$$ax_1 + b = ax_2 + b$$

$$ax_1 = ax_2$$

$$x_1 = x_2 \qquad \text{since } a \neq 0$$

So f is 1–1. If $k \in R$, then $f\left(\dfrac{k - b}{a}\right) = k$; so f is 1–1 from R onto R.

21. Let the slope of the graph of f be m. Then the slope of the graph of g is $-1/m$. So $f(x) = mx + b_1$ and $g(x) = (-1/m)x + b_2$, where b_1 and b_2 are the respective y intercepts.

$$(f \circ g)(x) = m\left(-\frac{1}{m}x + b_2\right) + b_1 = -x + (mb_2 + b_1)$$

$$(g \circ f)(x) = -\frac{1}{m}(mx + b_1) + b_2 = -x + \left(-\frac{b_1}{m} + b_2\right)$$

In each case the slope of the graph is -1.

23. $V(t) = \left(\dfrac{R - C}{N}\right)t + C$; $V(8) = \$6,200$ **25.** $C(0.6000) \approx 0.8252$

Exercise Set 5

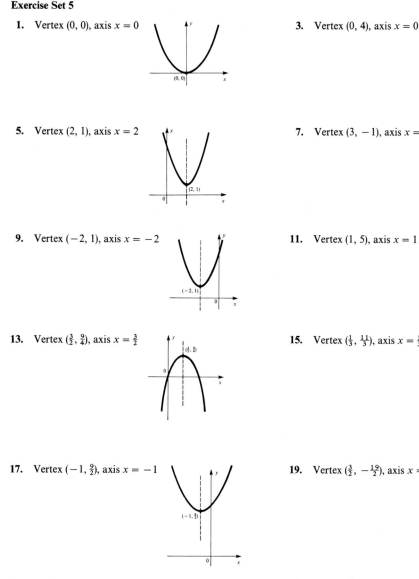

1. Vertex $(0, 0)$, axis $x = 0$

3. Vertex $(0, 4)$, axis $x = 0$

5. Vertex $(2, 1)$, axis $x = 2$

7. Vertex $(3, -1)$, axis $x = 3$

9. Vertex $(-2, 1)$, axis $x = -2$

11. Vertex $(1, 5)$, axis $x = 1$

13. Vertex $(\frac{3}{2}, \frac{9}{4})$, axis $x = \frac{3}{2}$

15. Vertex $(\frac{1}{3}, \frac{11}{3})$, axis $x = \frac{1}{3}$

17. Vertex $(-1, \frac{9}{2})$, axis $x = -1$

19. Vertex $(\frac{3}{2}, -\frac{19}{2})$, axis $x = \frac{3}{2}$

21. Maximum value 1 when $x = 1$ **23.** Minimum value $-\frac{3}{2}$ when $x = \frac{3}{2}$ **25.** Minimum value $-\frac{16}{5}$ when $x = \frac{1}{5}$

27. Maximum value $\frac{16}{3}$ when $x = \frac{4}{3}$ **29.** 60 items, minimum cost = \$400   **31.** 576 feet

33. $f(h + t) = a(h + t - h)^2 + k = at^2 + k,$
$f(h - t) = a(h - t - h)^2 + k = a(-t)^2 + k = at^2 + k$
So $f(h + t) = f(h - t)$.

35. If x is the number of 10¢ reductions, then the average attendance is $100 + 10x$ and the price of each ticket is $3.00 - 0.10x$. So the revenue $R(x) = (3.00 - 0.10x)(100 + 10x) = 300 + 20x - x^2$.

37. $P(x) = 90x - 0.05x^2 - 10{,}000$, $x = 900$ gives maximum profit of \$30,500, selling price for maximum profit = \$55

Exercise Set 6

1. $(-\infty, 0) \cup (2, \infty)$ **3.** $[-5, 1]$ **5.** $(-\frac{1}{3}, 2)$ **7.** $(-\infty, -4] \cup [\frac{3}{2}, \infty)$ **9.** $(-\infty, -1) \cup (\frac{3}{2}, \infty)$

11. $[-\frac{2}{3}, \frac{1}{2}]$ **13.** $(\frac{4}{5}, \frac{3}{2})$ **15.** $(-\infty, -\frac{5}{6}) \cup (\frac{3}{2}, \infty)$ **17.** $(-\infty, -\frac{6}{5}) \cup (1, \infty)$

19. $P(x) = 26x - x^2 - 105$, $P(x)$ is positive for $5 < x < 21$, maximum profit $= 64$ when $x = 13$

Exercise Set 7

1. Parabola

3. Circle

5. Hyperbola

7. Circle

9. Ellipse

11. Parabola

13. Parabola

15. Hyperbola

17. Upper half of circle, $x^2 + y^2 = 4$

19. Upper half of ellipse, $\dfrac{x^2}{9} + \dfrac{y^2}{4} = 1$

21. Upper half of parabola, $y^2 = x + 2$

23. $x - 2 = \frac{1}{2}(y - 3)^2$, parabola

25. $\dfrac{(x - 2)^2}{9} + \dfrac{y^2}{4} = 1$, ellipse

27. $x + 3 = -2(y - 4)^2$, parabola

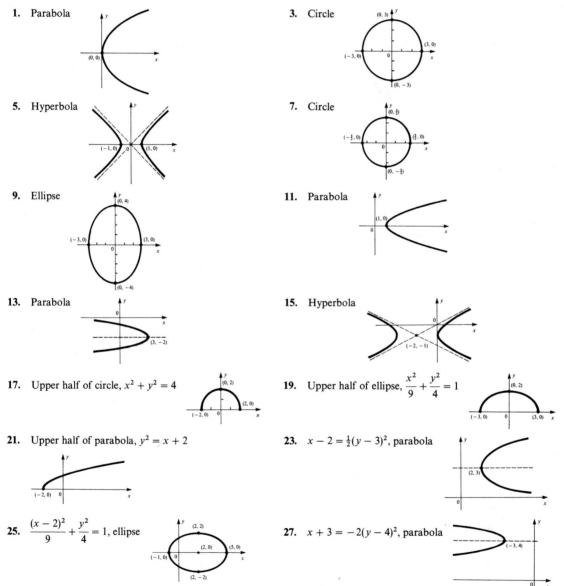

29. $\dfrac{(x+1)^2}{1} - \dfrac{(y+2)^2}{4} = 1$, hyperbola

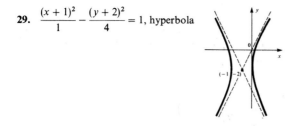

Review Exercise Set

1. a. $-\frac{7}{6}$ **b.** -4 **3. a.** $y = 3$ **b.** $x = 2$ **5. a.** $3x - 2y - 4 = 0$ **b.** $x - y + 5 = 0$
7. a. $3x - 7y + 21 = 0$ **b.** $9x + 7y - 24 = 0$ **9. a.** $m = \frac{3}{2}, a = 6, b = -9$ **b.** $m = -\frac{6}{5}, a = \frac{7}{3}, b = \frac{14}{5}$

11. a. $5x + 4y - 1 = 0$ **b.** $3x + 4y + 14 = 0$ **13. a.** $3x + 2y - 27 = 0$ **b.** $2x - 3y + 8 = 0$
15. $3x + 5y - 7 = 0$ **17. a.** $8x + 3y + C = 0, 8x + 3y - 7 = 0$ **b.** $x + 3y + C = 0, x + 3y - 14 = 0$

19. $g(x) = x - 3, m = 1$ **21.** $(f \circ g)(x) = \dfrac{10 - 9x}{5}, (g \circ f)(x) = -\dfrac{2 + 9x}{5}$

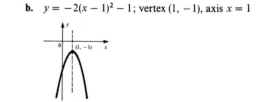

23. $T = -\dfrac{h}{500} + T_0; 36°F$

25. a. $y = (x - 3)^2 - 5$; vertex $(3, -5)$, axis $x = 3$ **b.** $y = -2(x - 1)^2 - 1$; vertex $(1, -1)$, axis $x = 1$

27. a. Minimum is $-\frac{11}{4}$ when $x = \frac{3}{2}$ **b.** Maximum is 12 when $x = -2$ **29.** 100 parts; minimum cost is \$10,000
31. a. $(-1, 2)$ **b.** $(-3, 1)$ **33. a.** $(-\infty, \frac{1}{2}] \cup [\frac{4}{3}, \infty)$ **b.** $[-2, \frac{4}{3}]$
35. a. Hyperbola **b.** Hyperbola

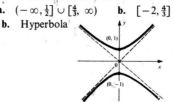

37. a. Circle   **b.** Ellipse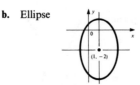

39. Let $A = (-5, 2)$, $B = (-3, -2)$, $C = (6, 1)$, $D = (4, 5)$. Then $m_{AB} = -2$, $m_{BC} = \frac{1}{3}$, $m_{CD} = -2$, $m_{AD} = \frac{1}{3}$, so opposite sides are parallel, and the figure is a parallelogram.

AB: $2x + y + 8 = 0$; BC: $x - 3y - 3 = 0$; CD: $2x + y - 13 = 0$; AD: $x - 3y + 11 = 0$

41. $3x + y - 9 = 0$ **43.** Tangent: $x + y + 2 = 0$; normal: $x - y = 0$

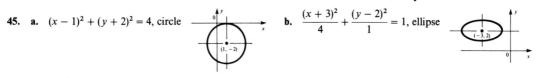

45. a. $(x - 1)^2 + (y + 2)^2 = 4$, circle   **b.** $\dfrac{(x + 3)^2}{4} + \dfrac{(y - 2)^2}{1} = 1$, ellipse

47. The center of the circle is $(2, -1)$, so the slope of the radius to the point $(-1, 3)$ is $-\frac{4}{3}$. Therefore, the slope of the tangent line is $\frac{3}{4}$. The point $(-1, 3)$ does lie on the circle. The given line has slope $\frac{3}{4}$, and the point $(-1, 3)$ lies on the line. So it is the tangent.

49. 350 people produce maximum revenue of $2,450

Chapter 6

Exercise Set 3

1. $x - 1$, $R = -5$ **3.** $x^2 + 4x + 7$, $R = 17$ **5.** $x^2 + 4x + 12$, $R = 32$ **7.** $3x^2 - 6x + 8$, $R = -11$
9. $2x^2 - 4x + 5$, $R = 0$; $x + 2$ is a factor of $2x^2 - 3x + 10$ **11.** $x^3 + x^2 + 3x + 9$, $R = 2$
13. $x^4 - x^3 - x^2 - x - 1$, $R = 2$ **15.** $3x^3 + x^2 - x + 2$, $R = -5$ **17.** $x^4 - 2x^3 + 4x^2 - 8x + 16$, $R = -113$
19. 11   **21.** 16   **23.** 2   **25.** 0   **27.** 65   **29.** 478   **31.** 202   **33.** 644   **35.** -14
37. $4x^3 - x^2 - (x/2) + \frac{3}{4}$, $R = -\frac{29}{8}$ **39.** $(x^2/2) - (9x/4) + \frac{3}{8}$, $R = -\frac{29}{8}$
41. By substitution, $P(-3) = 0$. By synthetic division, remainder on division by $x + 3$ is 0.
43. Remainder is 0 on dividing by $x + 6$.   **45.** $\frac{59}{16}$ **47.** $\frac{25}{4}$
49. Remainder on division by $x + 3$ is 0, so $x = -3$ is a root of the equation. Factors are $(x + 3)(x^2 - x + 1)$. There are no real roots of $x^2 - x + 1 = 0$ since $B^2 - 4AC = -3 < 0$.
51. a. 21.62   **b.** -39.04

Exercise Set 4

7. $x^3 - 3x^2 - 4x + 12 = 0$ **9.** $x^4 - 5x^3 + 6x^2 + 4x - 8$ **13. a.** $x^2 + 2x + 2 = 0$ **b.** $x^2 - x - 2 = 0$
   c. $x^2 - 2x + 1 = 0$ **15.** The other root is -2.   **17.** $x^4 - 10x^3 + 54x^2 - 130x + 125$

Exercise Set 5

1. a. $\pm\frac{1}{4}, \pm\frac{1}{2}, \pm\frac{3}{4}, \pm 1, \pm\frac{3}{2}, \pm 3$ **b.** $\pm\frac{1}{6}, \pm\frac{1}{3}, \pm\frac{1}{2}, \pm\frac{2}{3}, \pm\frac{5}{6}, \pm 1, \pm\frac{4}{3}, \pm\frac{5}{3}, \pm 2, \pm\frac{5}{2}, \pm\frac{10}{3}, \pm 4, \pm 5, \pm\frac{20}{3}, \pm 10, \pm 20$
3. Upper bound 3, lower bound -4; root between 2 and 3   **5.** $x = 2, -1$; -1 is a double root   **7.** $x = 4, -1 \pm i$
9. $x = -\frac{3}{4}, 1 \pm i\sqrt{3}$ **11.** $x = -3, -1, \frac{5}{2}$ **13.** $x = -1, 1 \pm i$; -1 is a double root   **15.** $x = 1, 2, 4, -2$
17. $x = -1, 3$; -1 is a triple root   **19.** $(x - 1)(x + 4)(x - 3)$ **21.** $(x - 3)(x + 1)(x^2 + 4)$

23. $(x + 1)(x - 3)(x^2 + 2x + 2)$ **25.** $(x + 1)(x + 2)(2x - 3)^2$ **27.** $x = \frac{2}{3}, -4$; -4 is a double root

29. $x = -6$; there is a sign change between $x = 1$ and $x = 2$, between $x = 0$ and $x = -1$, and between $x = -1$ and $x = -2$, so there are three irrational roots and one rational root

Exercise Set 6

1. 1.28   **3.** -1.80 **5.** 2.25   **7.** $2, -1.51$ **9.** $-3, 2.55$ **11.** $2.54, -0.69, -2.84$ **13.** $1.49, -2.36$
15. $5, -1, 1.39$

Exercise Set 7

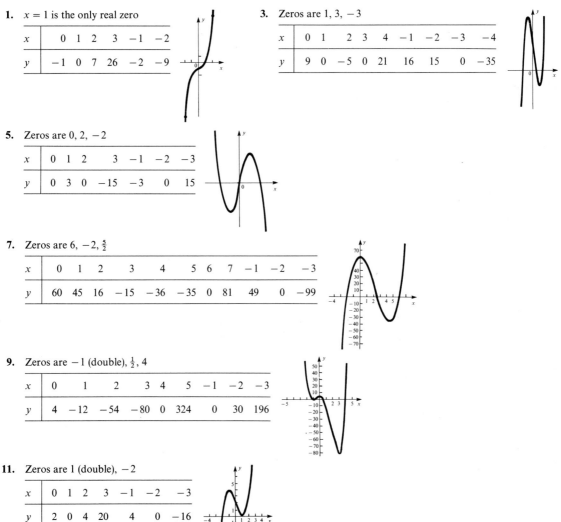

1. $x = 1$ is the only real zero

x	0	1	2	3	-1	-2
y	-1	0	7	26	-2	-9

3. Zeros are $1, 3, -3$

x	0	1	2	3	4	-1	-2	-3	-4
y	9	0	-5	0	21	16	15	0	-35

5. Zeros are $0, 2, -2$

x	0	1	2	3	-1	-2	-3
y	0	3	0	-15	-3	0	15

7. Zeros are $6, -2, \frac{5}{2}$

x	0	1	2	3	4	5	6	7	-1	-2	-3
y	60	45	16	-15	-36	-35	0	81	49	0	-99

9. Zeros are -1 (double), $\frac{1}{2}, 4$

x	0	1	2	3	4	5	-1	-2	-3
y	4	-12	-54	-80	0	324	0	30	196

11. Zeros are 1 (double), -2

x	0	1	2	3	-1	-2	-3
y	2	0	4	20	4	0	-16

13. Rational zero -3, irrational zeros between 3 and 4 and between -1 and -2; these irrational zeros are $1 + \sqrt{5}$ and $1 - \sqrt{5}$

x	0	1	2	3	4	-1	-2	-3	-4
y	-12	-20	-20	-6	28	-2	4	0	-20

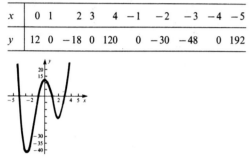

15. Zeros are 0 (double), 4, -2

x	0	1	2	3	4	5	-1	-2	-3
y	0	9	32	45	0	-175	5	0	-63

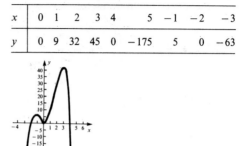

17. Zeros are 1, 3, -1, -4

x	0	1	2	3	4	-1	-2	-3	-4	-5
y	12	0	-18	0	120	0	-30	-48	0	192

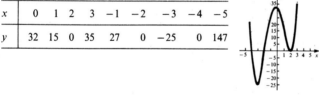

19. Real zeros are 2, $-\frac{3}{2}$

x	0	1	2	3	-1	-2
y	-12	-5	0	45	-15	40

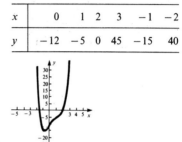

21. Zeros are 2 (double), -2, -4

x	0	1	2	3	-1	-2	-3	-4	-5
y	32	15	0	35	27	0	-25	0	147

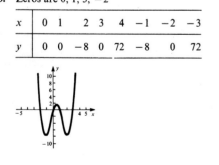

23. $(-6, \frac{4}{3}) \cup (5, \infty), x \neq -4$ **25.** $[-1, 2] \cup [4, \infty)$ **27.** $(-\infty, -2) \cup (-2, 1) \cup (3, \infty)$
29. $(-\infty, -3) \cup (-1, 2) \cup (2, \infty)$ **31.** $(-\infty, -2] \cup [1, \infty)$
33. Zeros are 0, 1, 3, -2

x	0	1	2	3	4	-1	-2	-3
y	0	0	-8	0	72	-8	0	72

35. Zeros are 2, 4, -1 (double), -3

x	0	1	2	3	4	5	-1	-2	-3	-4
y	24	48	0	-96	0	864	0	24	0	-432

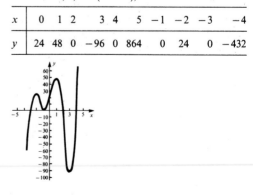

Exercise Set 8

1. **a.** $x = 4$ **b.** $x = 1, x = -1$ **3.** **a.** $x = -1, x = \frac{5}{2}$ **b.** $x = 0, x = 5$ **5.** **a.** $x = 3, x = 4, x = -3$
   b. $x = 0, x = -1$ **7.** **a.** $y = 1$ **b.** $y = \frac{1}{2}$ **9.** **a.** $y = -1$ **b.** None   **11.** **a.** None   **b.** $y = 0$

13. x intercept: $-\frac{3}{2}$; y intercept: -3; vertical
asymptote: $x = 1$; horizontal asymptote: $y = 2$

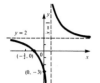

15. x intercept: 1; y intercept: $\frac{1}{9}$; vertical asymptotes:
$x = 3$, $x = -3$; horizontal asymptote: $y = 0$

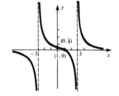

17. x intercept: 2; y intercept: $\frac{2}{3}$; vertical asymptotes:
$x = 3$, $x = -1$; horizontal asymptote: $y = 0$

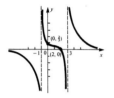

19. x intercepts: 4, -2; y intercept: none; vertical
asymptotes: $x = 0$, $x = 2$; horizontal asymptote: $y = 1$

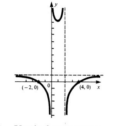

21. Vertical asymptotes: $x = 1$, $x = 4$, $x = -2$;
horizontal asymptote: $y = -2$

23. Vertical asymptotes: $x = 1$, $x = -1$, $x = 2$; horizontal
asymptote: none

25. x intercepts: 1, -3; y intercept: $-\frac{1}{2}$; vertical
asymptotes: $x = 3$, $x = -2$; horizontal
asymptote: $y = -1$

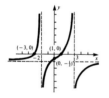

27. x intercepts: 1, -1; y intercept: $-\frac{1}{2}$; vertical asymptote:
$x = -2$; horizontal asymptote: none

Review Exercise Set

1. **a.** $2x^2 + 7x + 10, R = 25$ **b.** $x^2 - 3x + 6, R = -14$ **3.** **a.** $3x^3 - 14x^2 + 56x - 223, R = 887$
   b. $2x^4 + 4x^3 - 2x^2 - 4x - 8, R = 0$; so $x - 2$ is a factor of the polynomial   **5.** $2x^4 + x^3 + 12x^2 + 9x - 54$
7. **a.** $\pm\frac{1}{3}, \pm\frac{2}{3}, \pm 1, \pm\frac{4}{3}, \pm 2, \pm\frac{8}{3}, \pm 4, \pm 8$ **b.** $\pm\frac{1}{10}, \pm\frac{1}{5}, \pm\frac{3}{10}, \pm\frac{2}{5}, \pm\frac{1}{2}, \pm\frac{3}{5}, \pm\frac{4}{5}, \pm 1, \pm\frac{6}{5}, \pm\frac{3}{2}, \pm 2, \pm\frac{12}{5}, \pm 3,$
   $\pm 4, \pm 6, \pm 12$ **9.** $\{3, -2\}$; 3 is a double root   **11.** $\{\frac{1}{2}, -\frac{4}{3}, \pm 2i\}$ **13.** $\{-1, -2, \frac{3}{2}\}$; $\frac{3}{2}$ is a double root
15. $(2x - 5)(x^2 - 2x + 4)$
17. Zeros are 0, -1, $\frac{5}{2}$

x	0	1	2	3	-1	-2
y	0	-6	-6	12	0	-18

19. Zeros are 3 (double), $-\frac{1}{2}$, -4

x	0	1	2	3	4	-1	-2	-3	-4	-5
y	36	60	30	0	72	-48	-150	-180	0	576

21. Zeros are $\frac{1}{2}$, 3, -1, -3

x	0	1	2	3	4	-1	-2	-3	-4
y	9	-16	-45	0	245	0	-25	0	189

23. $(-\infty, -1) \cup (-1, 2)$ **25.** $[-\frac{3}{2}, \infty)$

27. x intercept: 0; y intercept: 0; vertical asymptotes: $x = 1$, $x = -1$; horizontal asymptote: $y = 0$

29. x intercepts: 2, $-\frac{1}{2}$; y intercept: none; vertical asymptotes: $x = 0$, $x = 5$; horizontal asymptote: $y = 2$

31. -2.60 **33.** $2x^4 - 13x^3 + 22x^2 + 47x - 130$ **35.** $\{\frac{4}{3}, -\frac{3}{2}, \pm\sqrt{2}i\}$

37. Rational zero -3, irrational zero between 0 and 1, two zeros are imaginary

x	0	1	2	-1	-2	-3	-4
y	-6	4	40	-14	-20	0	94

39. $(-\infty, -3) \cup (1, 2) \cup [4, \infty)$ and the isolated point $x = 0$

41. -0.366

43. x intercepts: 0, -1; y intercept: 0; vertical asymptote: $x = 1$; horizontal asymptote: none

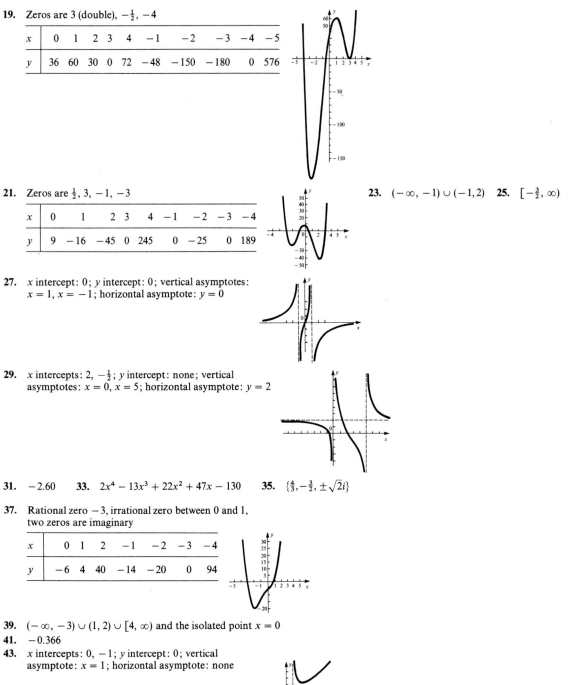

Chapter 7

Exercise Set 1

1. a. 2.6651   **b.** 31.5443   **c.** 0.0906   **d.** 7.3891   **e.** 0.3679

3.

x	0	1	2	3	-1	-2
y	1	4	16	64	0.25	0.0625

5.

x	0	1	2	-1	-2
y	1	2.7	7.39	0.37	0.14

7.

x	0	1	2	3	-1	-2	-3
y	1	$\frac{2}{3}$	$\frac{4}{9}$	$\frac{8}{27}$	$\frac{3}{2}$	$\frac{9}{4}$	$\frac{27}{8}$

9.

x	0	1	2	3	4	-1	-2
y	$\frac{1}{2}$	1	2	4	8	$\frac{1}{4}$	$\frac{1}{8}$

11.

x	0	1	2	3	-1	-2	-3
y	1.7	1	0.58	0.33	3	5.20	9

13.

x	0	1	2	3	-1	-2
y	0	0.63	0.86	0.95	-1.72	-6.39

15.

x	0	± 1	± 2	± 3
y	2	1	$\frac{1}{8}$	$\frac{1}{256}$

17.

x	0	1	2	3	4	-1	-2
y	$\frac{4}{3}$	1	$\frac{3}{4}$	$\frac{9}{16}$	$\frac{27}{64}$	$\frac{16}{9}$	$\frac{64}{27}$

19.

x	0	1	2	3	4	5	-1
y	$\frac{1}{4}$	$\frac{1}{2}$	1	2	4	8	$\frac{1}{8}$

21. a. $2,477.65   **b.** $2,488.42   **c.** $2,492.15

23. 70.71 grams   **25.** 120 years   **27.** 4 times as many, 16 times as many   **29.** $3.00   **31.** 5

33.

x	0	± 1	± 2	± 3
y	1	1.54	3.76	10.07

35.

x	0	1	2	3	-1	-2	-3
y	0	0.76	0.96	0.995	-0.76	-0.96	-0.995

37. $-6.25°$ **39.** $2(\frac{3}{2})^{7/3}$ billion ≈ 5.15 billion   **41.** 9.6 pounds per square inch

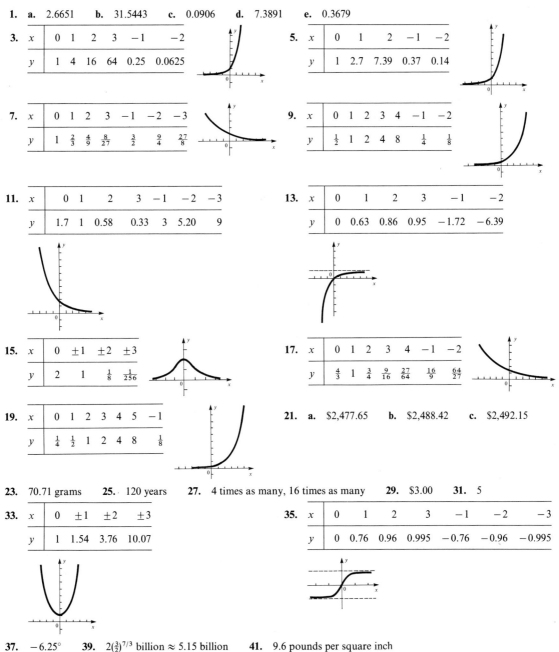

Exercise Set 2

1. a. 3 **b.** 3 **3. a.** x **b.** x **5. a.** 4 **b.** -3 **7. a.** 3 **b.** -4 **9. a.** 2 **b.** -2
11. a. 5 **b.** -5 **13. a.** 3 **b.** $\frac{1}{9}$ **15. a.** 3 **b.** 0.001 **17. a.** 0 **b.** $\frac{27}{8}$ **19. a.** -1
b. 2 **21. a.** $\frac{1}{27}$ **b.** $\frac{2}{3}$ **23. a.** 10 **b.** -3 **25. a.** $\log_4 64 = 3$ **b.** $\log_3 \frac{1}{9} = -2$
27. a. $\log_2 256 = 8$ **b.** $\log_2 0.125 = -3$ **29. a.** $2^4 = 16$ **b.** $10^2 = 100$ **31. a.** $5^{-1} = 0.2$
b. $8^{2/3} = 4$

33. $x = 2^y$

x	1	2	4	8	$\frac{1}{2}$	$\frac{1}{4}$	$\frac{1}{8}$
y	0	1	2	3	-1	-2	-3

35. $x = 2^y - 1$

x	0	1	3	7	$-\frac{1}{2}$	$-\frac{3}{4}$	$-\frac{7}{8}$
y	0	1	2	3	-1	-2	-3

37. $x = \frac{1}{2}(4^y + 3)$

x	2	$\frac{7}{2}$	$\frac{19}{2}$	$\frac{13}{8}$	$\frac{49}{32}$	$\frac{5}{2}$	$\frac{7}{4}$
y	0	1	2	-1	-2	$\frac{1}{2}$	$-\frac{1}{2}$

39. $x = \pm 2^{y/2}$

x	±1	±2	±4	±8	$\pm\frac{1}{2}$	$\pm\frac{1}{4}$	$\pm\frac{1}{8}$
y	0	2	4	6	-2	-4	-6

41. $x = -e^y$

x	-1	-2.7	-7.39	-20.09	-0.37	-0.14	-0.05
y	0	1	2	3	-1	-2	-3

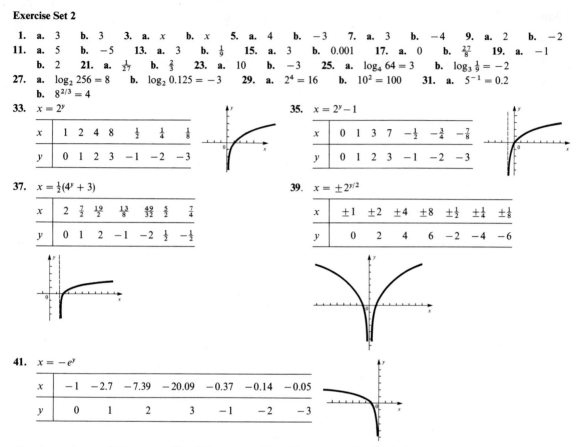

43. Approximately 6.76 years **45.** 12.6 years **47.** 6.21 years **49.** 110 decibels
51. Approximately 38.88 hours **53.** Approximately 47.25 minutes **55.** 33.22 feet

Exercise Set 3

1. a. $\log(x + 1) + \log(x - 2)$ **b.** $\log(x + 1) - \log(x - 2)$ **3. a.** $4 \log(x - 2)$ **b.** $\frac{3}{2}\log(2x + 1)$
5. a. $\log(x - 3) + \log(x + 1)$ **b.** $\log(x + 2) + \log(x - 2) - \log(x - 3)$ **7. a.** $\frac{2}{3}\log(x + 1)$
b. $-\frac{1}{2}\log(3x - 4)$ **9. a.** $\frac{1}{2}\log(x^2 + y^2)$ **b.** $\frac{1}{2}[\log(x + y) + \log(x - y)]$ **11.** $\frac{1}{3}\log(1 - 2x) - 2\log(x + 3)$
13. $\log(2x - 5) + \log(3x + 4) - [\log(x + 2) + \log(x - 1)]$ **15.** $\frac{3}{2}\log(x - 1) - \frac{1}{2}\log(x + 2)$
17. $\log 3 + 4 \log x - \frac{3}{2}\log(2x + 1)$ **19.** $2 \log x + \frac{1}{2}\log(1 - 3x) - \log(1 + 2x)$
21. $\frac{3}{2}\log(2x + 3) + 4 \log(x + 1) - [\log 3 + \log x + \frac{1}{2}\log(1 - x)]$
23. $\log 3 + \log x + 2 \log(x^2 + 4) - [\log(x + 2) + \frac{1}{2}\log(2x + 3)]$ **25.** $-\frac{1}{2}[\log(2x - 3) + \log(x + 1)]$ **27. a.** 81
b. 25 **29. a.** $\log(x/y)$ **b.** $\log x^2(x + 1)^3$ **31. a.** $\log \dfrac{2\sqrt{x - 1}}{(x + 2)^3}$ **b.** $\log \dfrac{(x + 2)^3}{\sqrt{x - x^2}}$
33. a. $\log \dfrac{\sqrt{5x + 3}}{(x + 2)^2(3x + 1)}$ **b.** $\log \dfrac{2Cx}{\sqrt{x + 1}}$ **35. a.** $x = \dfrac{\ln 3}{\ln 4}$ **b.** $x = \dfrac{\ln 5}{\ln 5 - \ln 3}$ **37. a.** $x = \dfrac{\ln 8}{1 + \ln 2}$
b. $x = \dfrac{\ln 2 - \ln 3}{\ln 48}$ **39.** $x = 4$ **41.** $x = 3$ **43.** $x = 10$ **45.** $x = \frac{9}{2}$ **47.** $x = 4$ **49.** $x = 1$

51. $y = 2x^{3/2}$ **53.** $y = \dfrac{Cx^3(2x-3)}{\sqrt{x+2}}$ **55.** $y = \frac{1}{2}[3 + C(x-1)^2]$ **57. a.** $\log_a x = \dfrac{\log_b x}{\log_b a} = \dfrac{g(x)}{g(a)}$

b. $\log_b x = \dfrac{\log_a x}{\log_a b} = \dfrac{f(x)}{f(b)}$ **c.** $\log_a b = \dfrac{1}{\log_b a}$, so $f(b) \cdot g(a) = 1$ **59. a.** 0.7925 **b.** -0.7712

61. a. 0.1606 **b.** 8.6555 **63. a.** 1.6341 **b.** 0.4033 **65.** $x = \frac{3}{2}$ **67.** $x = 1$ **69.** $x = 12$

71. $\dfrac{1}{\sqrt{x} - \sqrt{x-1}} = \dfrac{1}{\sqrt{x} - \sqrt{x-1}} \cdot \dfrac{\sqrt{x} + \sqrt{x-1}}{\sqrt{x} + \sqrt{x-1}} = \dfrac{\sqrt{x} + \sqrt{x-1}}{x - (x-1)} = \sqrt{x} + \sqrt{x-1}$, so $\log \dfrac{1}{\sqrt{x} - \sqrt{x-1}}$

$= \log(\sqrt{x} + \sqrt{x-1})$

73.   **75.** $x = \ln(2 + \sqrt{5})$ **77.** $x = 2$ or $x = 8$

Review Exercise Set

1. a.

x	0	1	2	3	-1	-2	-3
y	1	$\frac{3}{2}$	$\frac{9}{4}$	$\frac{27}{8}$	$\frac{2}{3}$	$\frac{4}{9}$	$\frac{8}{27}$

b.

x	0	1	2	3	-1	-2	-3
y	0	$\frac{1}{2}$	$\frac{3}{4}$	$\frac{7}{8}$	-1	-3	-7

c.

x	1	2	4	8	$\frac{1}{2}$	$\frac{1}{4}$	$\frac{1}{8}$
y	0	1	2	3	-1	-2	-3

d.

x	0	-1.72	-6.39	-19.09	0.63	0.86	0.95
y	0	1	2	3	-1	-2	-3

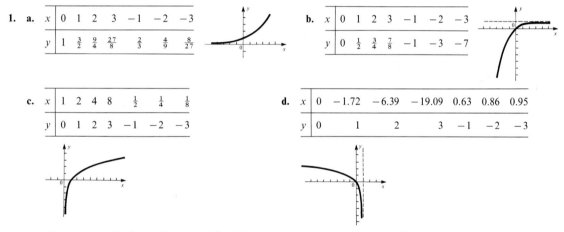

3. a. $3^2 = 9$ **b.** $10^{-3} = 0.001$ **c.** $2^8 = 256$ **d.** $y = k^v$ **e.** $s = r^z$ **5. a.** 16   **b.** $\frac{1}{4}$ **c.** 4

7. a. 3   **b.** 2   **c.** -7 **9. a.** $\log(x-1) - 3\log(x+2)$ **b.** $\frac{1}{2}[\log(2x-1) - (\log 3 + \log x)]$

11. a. $\log \dfrac{Cx\sqrt{2x-3}}{8}$ **b.** $\log \dfrac{x^4}{(x-1)\sqrt{x+2}}$ **13. a.** $\log \dfrac{2x^3}{\sqrt[3]{(x-1)(x+2)}}$ **b.** $\log \dfrac{(2x+3)^{3/2}}{2(x+4)^3}$

15. a. $\frac{3}{2}$ **b.** -3 **c.** -4 **d.** -5 **17.** $k = -\dfrac{1}{t}\ln\left(1 - \dfrac{y}{C}\right)$ **19.** $x = 3$ **21.** $x = 5$ **23.** $x = \frac{2}{3}$

25. $x = \frac{7}{2}$ **27.** Approximately 15.87 pounds   **29.** Approximately 529,750

31. a. Quarterly, \$6,734.28; continuously, \$6,749.29   **b.** Approximately 11.55 years

33. 10.0 pounds per square inch   **35.** Approximately 6.93%   **37. a.** 0.5670   **b.** 0.5900

39.   **41.** 33.22 seconds   **43.** 0.00196

Cumulative Review Exercise Set II (Chapters 4–7)

1. **a.** $(-2, \frac{1}{2}] \cup (2, \infty)$ **b.** $[-\frac{10}{3}, 1] \cup [4, \infty)$
3.   **a.** 2   **b.** -1 **c.** 1   **d.** 9   **e.** -1

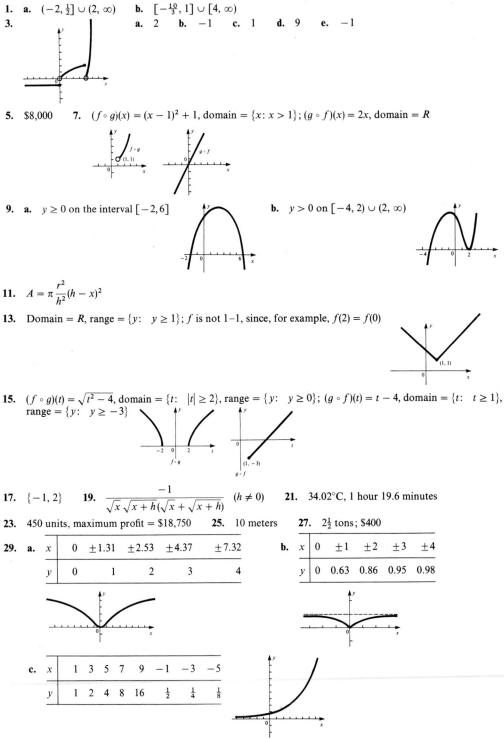

5. \$8,000   7. $(f \circ g)(x) = (x - 1)^2 + 1$, domain $= \{x: x > 1\}$; $(g \circ f)(x) = 2x$, domain $= R$

9. **a.** $y \geq 0$ on the interval $[-2, 6]$ **b.** $y > 0$ on $[-4, 2) \cup (2, \infty)$

11. $A = \pi \dfrac{r^2}{h^2}(h - x)^2$

13. Domain $= R$, range $= \{y: \ y \geq 1\}$; f is not 1–1, since, for example, $f(2) = f(0)$

15. $(f \circ g)(t) = \sqrt{t^2 - 4}$, domain $= \{t: \ |t| \geq 2\}$, range $= \{y: \ y \geq 0\}$; $(g \circ f)(t) = t - 4$, domain $= \{t: \ t \geq 1\}$,
     range $= \{y: \ y \geq -3\}$

17. $\{-1, 2\}$ 19. $\dfrac{-1}{\sqrt{x}\,\sqrt{x+h}\,(\sqrt{x} + \sqrt{x+h})}$ $(h \neq 0)$ 21. 34.02°C, 1 hour 19.6 minutes

23. 450 units, maximum profit $= \$18,750$ 25. 10 meters   27. $2\frac{1}{2}$ tons; \$400

29. **a.**

x	0	± 1.31	± 2.53	± 4.37	± 7.32
y	0	1	2	3	4

b.

x	0	± 1	± 2	± 3	± 4
y	0	0.63	0.86	0.95	0.98

c.

x	1	3	5	7	9	-1	-3	-5
y	1	2	4	8	16	$\frac{1}{2}$	$\frac{1}{4}$	$\frac{1}{8}$

d.

x	0	$\frac{1}{2}$	$\frac{3}{2}$	2	3	5	9	-1	-3	-7	$\frac{5}{4}$	$\frac{3}{4}$	$\frac{9}{8}$	$\frac{7}{8}$
y	1	0	0	1	2	3	4	2	3	4	-1	-1	-2	-2

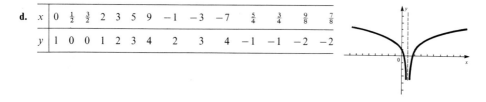

Chapter 8

Exercise Set 2

1. $a = \frac{5}{2}, b = 5\sqrt{3}/2$ **3.** $a = 10\sqrt{3}, c = 20$ **5.** $a = 2\sqrt{3}, c = 4\sqrt{3}$ **7.** $a = b = 4$ **9.** $a = 5\sqrt{3}/2, b = \frac{5}{2}$
11. $b = 3, c = 3\sqrt{2}$ **13.** $c = \sqrt{13}$ **15.** $c = 5$ **17.** $b = 24$ **19.** $c = 3$ **21. a.** $5\pi/6$ **b.** $5\pi/4$
   c. $5\pi/18$ **d.** $10\pi/3$ **e.** $2\pi/5$ **23. a.** $7\pi/4$ **b.** $-3\pi/4$ **c.** $4\pi/3$ **d.** $19\pi/6$ **e.** $4\pi/9$
25. a. $225°$ **b.** $420°$ **c.** $160°$ **d.** $-630°$ **e.** $54°$
27. a. $s = 6$ **b.** $r = \frac{4}{3}$ **c.** $\theta = \frac{12}{5}$ **d.** $s = \pi/3$ **e.** $r = 32/\pi$ **29.** $8\pi/3$ inches   **31.** 50 rpm
33. 330π miles per hour $\approx 1,036.73$ miles per hour $\approx 1,520.53$ feet per second
35. $10\pi/3$ feet per second
37. $r_B = 7.78$ inches, $r_C = 9.78$ inches
39. Denote α, β, and γ as shown. Then $\alpha + \beta = 90°$. Also, $\alpha + \beta + \gamma = 180°$. So $\gamma = 90°$. Thus, V is a square, with area c^2. Area I = Area II = Area III = Area IV = $\frac{1}{2}ab$; Area of large square = $(a + b)^2$, so

$$c^2 + 4(\tfrac{1}{2}ab) = (a + b)^2$$
$$c^2 + 2ab = a^2 + 2ab + b^2$$
$$c^2 = a^2 + b^2$$

Exercise Set 3

1. $B = 32.6°, a = 23.45, c = 27.84$ **3.** $A = 23°, a = 5.73, c = 14.67$ **5.** $B = 41°, a = 18.11, b = 15.75$
7. $B = 57.5°, b = 38.77, c = 45.97$ **9.** $A = 28.53°, B = 61.47°, b = 28.24$ **11.** $A = 67°20', a = 6.13, c = 6.64$
13. $B = 54.5°, b = 4.85, c = 5.96$ **15.** $A = 48.32°, B = 41.68°, a = 28.53$
17. a. $\sin L = s/t, \cos L = r/t, \csc L = t/s, \sec L = t/r, \tan L = s/r, \cot L = r/s$
   b. $\sin K = r/t, \cos K = s/t, \csc K = t/r, \sec K = t/s, \tan K = r/s, \cot K = s/r$
19. $29.74°; 16.12$ inches   **21.** 215.62 feet   **23.** 284.1 feet   **25.** Approximately $4\frac{1}{4}$ inches   **27.** 725.08 feet
29. 51.5 miles

Exercise Set 4

1. $\sin(5\pi/4) = -1/\sqrt{2}, \cos(5\pi/4) = -1\sqrt{2}, \csc(5\pi/4) = -\sqrt{2}, \sec(5\pi/4) = -\sqrt{2}, \tan(5\pi/4) = 1, \cot(5\pi/4) = 1$
3. $\sin(-4\pi/3) = \sqrt{3}/2, \cos(-4\pi/3) = -\frac{1}{2}, \csc(-4\pi/3) = 2/\sqrt{3}, \sec(-4\pi/3) = -2, \tan(-4\pi/3) = -\sqrt{3},$
   $\cot(-4\pi/3) = -1\sqrt{3}$
5. $\sin \pi = 0, \cos \pi = -1, \csc \pi$ undefined, $\sec \pi = -1, \tan \pi = 0, \cot \pi$ undefined
7. $\sin 270° = -1, \cos 270° = 0, \csc 270° = -1, \sec 270°$ undefined, $\tan 270°$ undefined, $\cot 270° = 0$
9. $\sin(3\pi/4) = 1/\sqrt{2}, \cos(3\pi/4) = -1/\sqrt{2}, \csc(3\pi/4) = \sqrt{2}, \sec(3\pi/4) = -\sqrt{2}, \tan(3\pi/4) = -1, \cot(3\pi/4) = -1$
11. $\sin 480° = \sqrt{3}/2, \cos 480° = -\frac{1}{2}, \csc 480° = 2/\sqrt{3}, \sec 480° = -2, \tan 480° = -\sqrt{3}, \cot 480° = -1/\sqrt{3}$
13. $\sin 3\pi = 0, \cos 3\pi = -1, \csc 3\pi$ undefined, $\sec 3\pi = -1, \tan 3\pi = 0, \cot 3\pi$ undefined   **15. a.** $\sqrt{3}/2$ **b.** $1/\sqrt{2}$
   c. $1/\sqrt{3}$ **d.** -2 **e.** -2 **17. a.** $-\sqrt{3}$ **b.** -1 **c.** 1 **d.** $-\frac{1}{2}$ **e.** 0 **19. a.** -2
   b. $\sqrt{3}$ **c.** $-\sqrt{3}/2$ **d.** $-1/\sqrt{2}$ **e.** $-2/\sqrt{3}$ **21.** -8

Exercise Set 5

1. $C = 120°$, $b = 2.66$, $c = 6.74$

3. Solution 1: $A = 82°29'$, $B = 61°01'$, $a = 141.68$; Solution 2: $A = 24°31'$, $B = 118°59'$, $a = 59.30$

5. $B = 12.92°$, $C = 139.08°$, $c = 29.30$ **7.** $A = 107°$, $a = 16.04$, $b = 12.31$ **9.** $C = 77°58'$, $a = 9.90$, $c = 15.93$

11. $A = 113.58°$, $C = 23.42°$, $a = 23.52$ **13.** $A = 33.69°$, $C = 38.31°$, $a = 65.32$

15. Solution 1: $A = 25°15'$, $B = 132°15'$, $b = 9.42$; Solution 2: $A = 154°45'$, $B = 2°45'$, $b = 0.61$ **17.** 1,029 feet

19. 66°20' west of due south; after 37 minutes

Exercise Set 6

1. $a = 13.23$, $C = 40°53'$, $B = 79°07'$ **3.** $c = 12.65$, $A = 23°05'$, $B = 59°35'$ **5.** $A = 52°50'$, $B = 32°05'$, $C = 95°05'$

7. $b = 312.79$, $A = 46.67°$, $C = 29.83°$ **9.** $A = 49.85°$, $B = 86.30°$, $C = 43.85°$ **11.** 13.56 inches

13. 53.9 pounds **15.** 477.73 miles **17.** 34.42 feet

Exercise Set 7

1. 216 **3.** 1.16 **5.** 4.45 **7.** 25.98

9. Let A be obtuse as shown in the figure. Construct the altitude h from C to side AB extended.

$\sin(180° - A) = h/b$

$h = b \sin(180° - A) = b \sin A$

Area $= \frac{1}{2}$(Base) $\cdot$ (Altitude) $= \frac{1}{2}ch = \frac{1}{2}bc \sin A$

11. 48.73 **13.** 48.92

Review Exercise Set

1. $B = 30°$, $c = 10$, $a = 5\sqrt{3}$ **3.** $A = 30°$, $c = 8$, $b = 4\sqrt{3}$ **5.** $A = 45°$, $a = 5\sqrt{2}$, $b = 5\sqrt{2}$

7. $A = 30°$, $B = 60°$, $b = 7\sqrt{3}$ **9.** $B = 57°$, $a = 7.08$, $b = 10.90$

11. $a = 111.34$, $A = 47.51°$, $B = 42.49°$ **13.** $A = 63.8°$, $a = 30.06$, $b = 14.79$

15. a. $4\pi/3$ **b.** $7\pi/4$ **c.** $\pi/12$ **d.** 3π **e.** $8\pi/9$ **17. a.** $4\pi \approx 12.57$ **b.** $(240/\pi)° \approx 76.39°$

19. a. $\frac{1}{2}$ **b.** $-\frac{1}{2}$ **c.** $1/\sqrt{3}$ **d.** 0 **e.** $-\sqrt{2}$ **f.** 2 **g.** 1 **h.** -1 **i.** -2 **j.** $-\sqrt{3}/2$

21. $b = 10.72$, $c = 20.82$, $C = 118.9°$

23. Solution 1: $A = 64.49°$. $B = 83.51°$, $a = 13.62$; Solution 2: $A = 51.51°$, $B = 96.49°$, $a = 11.82$

25. $A = 100.2°$, $B = 52.2°$, $C = 27.6°$ **27.** No solution **29.** $A = 37.36°$, $B = 43.05°$, $C = 99.59°$

31. Solution 1: $B = 104.21°$, $C = 50.59°$, $b = 34.38$; Solution 2: $B = 25.39°$, $C = 129.41°$, $b = 15.21$

33. 61.2 feet **35.** 103.12 feet, area $= 4{,}914.91$ square feet **37.** 22.86 feet **39.** 526.46 miles

41. 8.48 miles from first tower, 7.93 miles from second

Chapter 9

Exercise Set 1

1. $\sin \theta = -2\sqrt{2}/3$, $\cos \theta = -\frac{1}{3}$, $\csc \theta = -3/2\sqrt{2}$, $\sec \theta = -3$, $\tan \theta = 2\sqrt{2}$, $\cot \theta = 1/2\sqrt{2}$

3. $\sin \theta = -\frac{12}{13}$, $\cos \theta = \frac{5}{13}$, $\csc \theta = -\frac{13}{12}$, $\sec \theta = \frac{13}{5}$, $\tan \theta = -\frac{12}{5}$, $\cot \theta = -\frac{5}{12}$

5. $\sin \theta = -\sqrt{15}/4$, $\cos \theta = \frac{1}{4}$, $\csc \theta = -4/\sqrt{15}$, $\sec \theta = 4$, $\tan \theta = -\sqrt{15}$, $\cot \theta = -1/\sqrt{15}$

7. $\theta = 4\pi/3$; $\sin \theta = -\sqrt{3}/2$, $\cos \theta = -\frac{1}{2}$, $\csc \theta = -2/\sqrt{3}$, $\sec \theta = -2$, $\tan \theta = \sqrt{3}$, $\cot \theta = 1/\sqrt{3}$

9. $\sin \theta = -2\sqrt{2}/3$, $\cos \theta = \frac{1}{3}$, $\csc \theta = -3/2\sqrt{2}$, $\sec \theta = 3$, $\tan \theta = -2\sqrt{2}$, $\cot \theta = -1/2\sqrt{2}$

11. a. $(0, 1)$ **b.** $(-1, 0)$ **c.** $(0, -1)$ **d.** $(-\sqrt{3}/2, \frac{1}{2})$ **e.** $(-\frac{1}{2}, -\sqrt{3}/2)$

13. a.

b.

c.

d.

15. **a.** $\pi/6$ **b.** $\pi/4$ **c.** $\pi/3$ **d.** $\pi/2$ **e.** 0

17. **a.** $\sin 0.3 = 0.2955$, $\cos 0.3 = 0.9553$, $\tan 0.3 = 0.3093$
   b. $\sin 2.718 = 0.4110$, $\cos 2.718 = -0.9116$, $\tan 2.718 = -0.4509$
   c. $\sin 14.03 = 0.9943$, $\cos 14.03 = 0.1070$, $\tan 14.03 = 9.2955$
   d. $\sin(-3.625) = 0.4648$, $\cos(-3.625) = -0.8854$, $\tan(-3.625) = -0.5249$

19. **a.** $\pi/2$ **b.** 0   **c.** $\pi/4, 5\pi/4$ **d.** $0, \pi$ **e.** $\pi/2, 3\pi/2$ **f.** $0, \pi$ **g.** $3\pi/2$ **h.** π **i.** $3\pi/4, 7\pi/4$

21. **a.** $P(\theta)$ is either $(0, 1)$ or $(0, -1)$, so $\tan \theta$ is undefined   **b.** $P(\theta)$ is either $(1, 0)$ or $(-1, 0)$, so $\cot \theta$ is undefined
   c. $P(\theta)$ is either $(0, 1)$ or $(0, -1)$, so $\sec \theta$ is undefined   **d.** $P(\theta)$ is either $(1, 0)$ or $(-1, 0)$, so $\csc \theta$ is undefined

23. If $k \ \varepsilon \ R$, $\tan \theta = y/x = k$, with (x, y) on the unit circle if $x = 1/\sqrt{1 + k^2}$ and $y = k/\sqrt{1 + k^2}$. So if $P(\theta) = (1/\sqrt{1 + k^2}, k/\sqrt{1 + k^2})$, then $\tan \theta = k$.

Exercise Set 2

1. $\cos \theta = -\frac{4}{5}$, $\csc \theta = -\frac{5}{3}$, $\sec \theta = -\frac{5}{4}$, $\tan \theta = \frac{3}{4}$, $\cot \theta = \frac{4}{3}$

3. $\sin \theta = \frac{4}{5}$, $\cos \theta = -\frac{3}{5}$, $\csc \theta = \frac{5}{4}$, $\sec \theta = -\frac{5}{3}$, $\cot \theta = -\frac{3}{4}$

5. $\sin \theta = -1/\sqrt{5}$, $\csc \theta = -\sqrt{5}$, $\sec \theta = \sqrt{5}/2$, $\tan \theta = -\frac{1}{2}$, $\cot \theta = -2$

7. $\cos \theta = -2\sqrt{2}/3$, $\csc \theta = 3$, $\sec \theta = -3/2\sqrt{2}$, $\tan \theta = -1/2\sqrt{2}$, $\cot \theta = -2\sqrt{2}$

Exercise Set 3

1. $\sin(\alpha + \beta) = \dfrac{1 + \sqrt{3}}{2\sqrt{2}}$, $\cos(\alpha + \beta) = \dfrac{1 - \sqrt{3}}{2\sqrt{2}}$ **3.** **a.** $-\dfrac{1 + \sqrt{3}}{2\sqrt{2}}$ **b.** $-\dfrac{1 + \sqrt{3}}{2\sqrt{2}}$

5. **a.** $\dfrac{1 - \sqrt{3}}{2\sqrt{2}}$ **b.** $-\dfrac{\sqrt{3} + 1}{2\sqrt{2}}$ **7.** **a.** $\dfrac{1 + \sqrt{3}}{2\sqrt{2}}$ **b.** $\dfrac{1 + \sqrt{3}}{2\sqrt{2}}$ **9.** **a.** $-\frac{56}{65}$ **b.** $-\frac{63}{65}$

11. **a.** $\dfrac{2\sqrt{10} - 2}{9}$ **b.** $\dfrac{-\sqrt{5} + 4\sqrt{2}}{9}$ **13.** **a.** $\dfrac{31}{17\sqrt{5}}$ **b.** $\dfrac{38}{17\sqrt{5}}$ **15.** **a.** $-\frac{65}{33}$ **b.** $-\frac{65}{16}$

21. It suffices to take a particular pair of values, say $\alpha = \pi/3$ and $\beta = \pi/6$:

$$\sin(\alpha + \beta) = \sin\left(\frac{\pi}{3} + \frac{\pi}{6}\right) = \sin\frac{\pi}{2} = 1$$

$$\sin \alpha + \sin \beta = \sin\frac{\pi}{3} + \sin\frac{\pi}{6} = \frac{\sqrt{3}}{2} + \frac{1}{2} = \frac{1 + \sqrt{3}}{2} \neq 1$$

23. $\pi/3, 5\pi/3$

Exercise Set 4

1. $\sin 2\theta = -\frac{24}{25}$, $\cos 2\theta = -\frac{7}{25}$ **3.** $\sin 2\theta = \frac{120}{169}$, $\cos 2\theta = \frac{119}{169}$ **5.** $\sin 2\theta = -\frac{4}{5}$, $\cos 2\theta = \frac{3}{5}$

7. $\sin 2\theta = 2x\sqrt{1 - x^2}$, $\cos 2\theta = 1 - 2x^2$ **9.** $\sin \theta = \sqrt{3}/3$, $\cos \theta = -\sqrt{6}/3$ **11.** $\sin \theta = -\frac{3}{5}$, $\cos \theta = \frac{4}{5}$

13. **a.** $\sqrt{2 - \sqrt{2}}/2$ **b.** $\sqrt{2 + \sqrt{3}}/2$ **c.** $\sqrt{2 + \sqrt{3}}/2$ **d.** $\sqrt{2 - \sqrt{2}}/2$

15. $\sin(\alpha/2) = 3/\sqrt{13}$, $\cos(\alpha/2) = -2/\sqrt{13}$ **17.** $\sin(\alpha/2) = -1/\sqrt{3}$, $\cos(\alpha/2) = \sqrt{6}/3$ **19.** $P(\theta) = (\frac{3}{5}, \frac{4}{5})$

21. **a.** $-\sin \theta$ **b.** $-\sin \theta$ **c.** $\cos \theta$ **d.** $-\cos \theta$ **e.** $\cos \theta$

25. **a.** $\sin 2\theta = 2\sqrt{x^2 - 1}/x|x|$ **b.** $\cos 2\theta = (2 - x^2)/x^2$ **c.** $\sin \frac{1}{2}\theta = \sqrt{(x - 1)/2x}$
   d. $\cos \frac{1}{2}\theta = \sqrt{(x + 1)/2x}$

Exercise Set 5

1. **a.** $\dfrac{\sqrt{3} + 1}{\sqrt{3} - 1}$ **b.** $\dfrac{\sqrt{3} - 1}{\sqrt{3} + 1}$ **3.** $\frac{24}{7}$ **5.** $\dfrac{4\sqrt{2} + 1}{2(\sqrt{2} - 1)}$ **7.** **a.** $\frac{120}{119}$ **b.** $-\frac{240}{161}$ **c.** $-\frac{2}{3}$ **d.** 4

9. $\tan 2\theta = -\frac{24}{7}$, $\tan \frac{1}{2}\theta = 3$ **11.** **a.** $\sqrt{6}/2$ **b.** $-\sqrt{6}/2$ **13.** **a.** $1/\sqrt{2}$ **b.** $-1/\sqrt{2}$

15. a. $2 \sin 4x \cos x$ **b.** $\frac{1}{2}(\sin 8x + \sin 2x)$

19. For the second part, use $\tan(Y \pm \theta) = 1/\cot(Y \pm \theta)$ and Problem 17.

25. $\tan A = \frac{11}{3}$, so $A \approx 74.74°$; $\tan B = \frac{11}{10}$, so $B \approx 47.73°$; $\tan C = \frac{11}{7}$, so $C \approx 57.53°$

Exercise Set 8

1. $\{\pi/6, 11\pi/6\}$ **3.** $\{\pi/3, 5\pi/3\}$ **5.** $\{\pi/3, \pi/2, 3\pi/2, 5\pi/3\}$ **7.** $\{\pi/3, \pi, 5\pi/3\}$ **9.** $\{\pi/2, 3\pi/2\}$

11. $\{\pi/6, \pi/2, 5\pi/6, 3\pi/2\}$ **13.** $\{\pi/3, 2\pi/3, 4\pi/3, 5\pi/3\}$ **15.** $\{\pi/2, 3\pi/2\}$

17. $\{0, \pi/3, 2\pi/3, \pi, 4\pi/3, 5\pi/3, 2\pi/9, 4\pi/9, 8\pi/9, 10\pi/9, 14\pi/9, 16\pi/9\}$

19. $\{5\pi/24, 7\pi/24, 17\pi/24, 19\pi/24, 29\pi/24, 31\pi/24, 41\pi/24, 43\pi/24\}$ **21.** $\{2n\pi, \pi/2 + 2n\pi, n = 0 \pm 1, \pm 2, \ldots\}$

23. No solution   **25.** $\{\pi/3, 2\pi/3, 4\pi/3, 5\pi/3\}$

Exercise Set 9

1. Amplitude = 1, period = π

3. Amplitude = 2, period = 2π

5. Amplitude = 3, period = 4

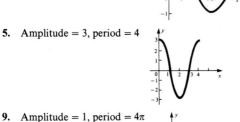

7. Amplitude = 2, period = $2\pi/3$

9. Amplitude = 1, period = 4π

11. Amplitude = 1, period = π

13. Amplitude = 2, period = $2\pi/3$

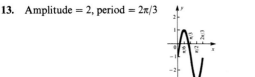

15. Period = 1

17. Amplitude = 1, period = 2π, phase shift = $\pi/3$

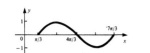

19. Amplitude = 1, period = π, phase shift = $-\pi/4$

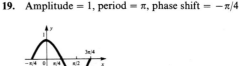

21. Amplitude = 2, period = $2\pi/3$, phase shift = $-\frac{2}{3}$

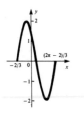

23. Amplitude = 2, period = 2, phase shift = $\frac{1}{4}$

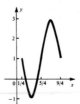

25. Amplitude $= 2$, period $= 2\pi/3$, phase shift $= \pi/6$

27. **a.** $y = \sqrt{2}\sin(x - \pi/4)$; amplitude $= \sqrt{2}$, period $= 2\pi$, phase shift $= \pi/4$

   b. $y = 5\sin(x + \theta)$, where $\theta \approx 2.5$ radians as shown; amplitude $= 5$, period $= 2\pi$, phase shift $= -\theta \approx -2.5$

29. Period $= 6\pi$

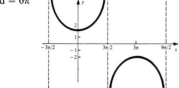

Exercise Set 10

1. **a.** $-\pi/3$ **b.** $5\pi/6$ **3.** **a.** $3\pi/4$ **b.** $\pi/2$ **5.** **a.** 0 **b.** $-\pi/6$ **7.** **a.** $\pi/2$ **b.** $7\pi/6$
9. **a.** $-2\sqrt{2}/3$ **b.** $\frac{5}{4}$ **11.** **a.** $\frac{4}{3}$ **b.** $\frac{3}{5}$ **13.** **a.** $2\sqrt{2}/3$ **b.** $-\frac{7}{24}$ **15.** $2/\sqrt{5}$ **17.** $-\frac{24}{25}$
19. $-\frac{120}{119}$ **21.** $\frac{4}{3}$ **23.** $\frac{1}{9}$ **25.** $3/\sqrt{10}$ **27.** $\sin\theta = -\frac{3}{5},\ \cos\theta = \frac{4}{5}$ **29.** $\frac{63}{65}$ **31.** $\frac{11}{10}$

Exercise Set 11

1. $\theta = 7\pi/4,\ r = \sqrt{2};\ 1 - i = \sqrt{2}[\cos(\pi/4) + i\sin(\pi/4)]$ **3.** $\theta = 11\pi/6,\ r = 2;\ \sqrt{3} - i = 2[\cos(11\pi/6) + i\sin(11\pi/6)]$

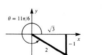

5. $\theta = \pi,\ r = 4;\ -4 = 4(\cos\pi + i\sin\pi)$ **7.** $\theta = 4\pi/3,\ r = 4;\ -2 - 2i\sqrt{3} = 4[\cos(4\pi/3) + i\sin(4\pi/3)]$

9. $\theta = 0,\ r = 5;\ 5 = 5(\cos 0 + i\sin 0)$

11. $-1 - i\sqrt{3}$ **13.** $-5i$ **15.** $3\sqrt{3} - 3i$ **17.** $1 - i$ **19.** $(-3\sqrt{3}/2) + (3i/2)$
21. $z_1 z_2 = 6[\cos(3\pi/2) + i\sin(3\pi/2)] = -6i;\ z_1/z_2 = \frac{2}{3}[\cos(-\pi) + i\sin(-\pi)] = -\frac{2}{3}$
23. $z_1 z_2 = 8[\cos(5\pi/4) + i\sin(5\pi/4)] = -4\sqrt{2} - 4i\sqrt{2};\ z_1/z_2 = 2[\cos(-\pi/4) + i\sin(-\pi/4)] = \sqrt{2} - i\sqrt{2}$
25. $z_1 z_2 = 8[\cos(13\pi/6) + i\sin(13\pi/6)] = 4\sqrt{3} + 4i;\ z_1/z_2 = \frac{1}{2}[\cos(3\pi/2) + i\sin(3\pi/2)] = -i/2$
27. $z_1 z_2 = 24(\cos 170° + i\sin 170°);\ z_1/z_2 = \frac{2}{3}(\cos 30° + i\sin 30°) = (\sqrt{3}/3) + (i/3)$
29. $729[\cos(9\pi/2) + i\sin(9\pi/2)] = 729[\cos(\pi/2) + i\sin(\pi/2)] = 729i$
31. $256[\cos(44\pi/3) + i\sin(44\pi/3)] = 256[\cos(2\pi/3) + i\sin(2\pi/3)] = -128 + 128i\sqrt{3}$
33. $256[\cos(20\pi/3) + i\sin(20\pi/3)] = 256[\cos(2\pi/3) + i\sin(2\pi/3)] = -128 + 128i\sqrt{3}$
35. $\zeta_0 = 2[\cos(\pi/6) + i\sin(\pi/6)] = \sqrt{3} + i;\ \zeta_1 = 2[\cos(5\pi/6) + i\sin(5\pi/6)] = -\sqrt{3} + i;\ \zeta_2 = 2[\cos(3\pi/2) + i\sin(3\pi/2)] = -2i$

37. $\zeta_0 = 1(\cos 0 + i \sin 0) = 1$; $\zeta_1 = 1[\cos(\pi/3) + i \sin(\pi/3)] = \frac{1}{2} + i\sqrt{3}/2$; $\zeta_2 = 1[\cos(2\pi/3) + i \sin(2\pi/3)] = -\frac{1}{2} + i\sqrt{3}/2$;
$\zeta_3 = 1(\cos \pi + i \sin \pi) = -1$; $\zeta_4 = 1[\cos(4\pi/3) + i \sin(4\pi/3)] = -\frac{1}{2} - i\sqrt{3}/2$; $\zeta_5 = 1[\cos(5\pi/3) + i \sin(5\pi/3)] = \frac{1}{2} - i\sqrt{3}/2$

39. $\zeta_0 = 4[\cos(3\pi/4) + i \sin(3\pi/4)] = -2\sqrt{2} + 2i\sqrt{2}$; $\zeta_1 = 4[\cos(7\pi/4) + i \sin(7\pi/4)] = 2\sqrt{2} - 2i\sqrt{2}$

41. $\{\sqrt{3} + i, 2i, -\sqrt{3} + i, -\sqrt{3} - i, -2i, \sqrt{3} - i\}$ **43.** $\{2, 1 + i\sqrt{3}, -1 + i\sqrt{3}, -2, -1 - i\sqrt{3}, 1 - i\sqrt{3}\}$

45. By the hint we have $\rho^n(\cos n\phi + i \sin n\phi) = r[\cos(\theta + 2k\pi) + i \sin(\theta + 2k\pi)]$. So $\rho^n = r$, $\rho = r^{1/n}$, and $n\phi = \theta + 2k\pi$.
Therefore, $\phi = (\theta + 2k\pi)/n$. By taking $k = 0, 1, 2, \ldots, n - 1$, we get all the distinct answers.

Exercise Set 12

1.

x	$-\pi/2$	-1	-0.75	-0.5	-0.25	-0.125	0.125	0.25	0.5	0.75	1	$\pi/2$
y	0.637	0.841	0.909	0.959	0.990		0.997	0.997	0.990	0.959	0.909	0.841	0.637

y approaches 1 as x approaches 0

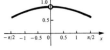

5. 2   **7.** -2 **9.** 1

Review Exercise Set

1. a. $\dfrac{1 + \sqrt{3}}{1 - \sqrt{3}}$ **b.** $\dfrac{\sqrt{2} - \sqrt{6}}{4}$ **c.** $\dfrac{\sqrt{2} + \sqrt{2}}{2}$ **d.** $\dfrac{-\sqrt{2} - \sqrt{6}}{4}$

3. a. $2\pi/3$ **b.** $-\pi/3$ **c.** $-\pi/3$ **d.** π **e.** $-\pi/4$ **5. a.** $-\frac{24}{25}$ **b.** $-\frac{7}{25}$ **c.** $2/\sqrt{5}$

7. a. $-\frac{13}{85}$ **b.** $\frac{36}{85}$ **c.** $\frac{13}{84}$ **9.** $\sin(\theta/2) = 5/\sqrt{34}$, $\cos(\theta/2) = -3/\sqrt{34}$, $\tan(\theta/2) = -\frac{5}{3}$

11. a. $-\frac{24}{7}$ **b.** $2/\sqrt{7}$ **27.** $\{0, \pi, \pi/6, 5\pi/6, 7\pi/6, 11\pi/6\}$ **29.** $\{3\pi/2, \pi/6, 5\pi/6\}$ **31.** $\{7\pi/6, 11\pi/6, \pi/2\}$

33. a. Period $= 4$

b. Amplitude $= 2$, period $= 4$

35. $z_1 \cdot z_2 = 6[\cos(13\pi/6) + i \sin(13\pi/6)] = 6[\cos(\pi/6) + i \sin(\pi/6)] = 3\sqrt{3} + 3i$; $z_1/z_2 = \frac{2}{3}[\cos(-\pi/2) + i \sin(-\pi/2)] = -2i/3$

37. $256[\cos(4\pi/3) + i \sin(4\pi/3)] = -128 - 128i\sqrt{3}$ **39.** 1   **47.** $\{\pi/3, 2\pi/3, 4\pi/3, 5\pi/3, \pi/2, 3\pi/2\}$

49. $\{\pi/2, 3\pi/2, \pi/6, 5\pi/6\}$

51. a. Amplitude $= 2$, period $= \pi$, phase shift $\pi/6$ **b.** $y = 2\left(\dfrac{\sqrt{3}}{2} \cos x - \frac{1}{2} \sin x\right) = 2 \sin\left(\dfrac{\pi}{3} - x\right) = -2 \sin\left(x - \dfrac{\pi}{3}\right)$;
amplitude $= 2$, period $= 2\pi$, phase shift $= \pi/3$

53. $\{4i, -2\sqrt{3} - 2i, 2\sqrt{3} - 2i\}$

Cumulative Review Exercise Set III (Chapters 8 and 9)

1. a. $B = 60°$, $c = 8\sqrt{3}$, $a = 4\sqrt{3}$ **b.** $A = 30°$, $c = 8$, $b = 4\sqrt{3}$ **c.** $B = 45°$, $a = b = 8\sqrt{2}$
d. $A = 30°$, $B = 60°$, $b = 10\sqrt{3}$ **e.** $A = B = 45°$, $a = 6\sqrt{2}$ **3. a.** $-\frac{24}{25}$ **b.** $\frac{33}{65}$ **c.** $-1/\sqrt{5}$
d. $-\sqrt{5}$ **5. a.** $z_1 z_2 = 8[\cos(3\pi/2) + i \sin(3\pi/2)] = -8i$ **b.** $z_1/z_2 = 2[\cos(7\pi/6) + i \sin(7\pi/6)] = -\sqrt{3} - i$
c. $z_2^4 = 16[\cos(2\pi/3) + i \sin(2\pi/3)] = -8 + 8i\sqrt{3}$
d. $\zeta_0 = 2[\cos(2\pi/3) + i \sin(2\pi/3)] = -1 + i\sqrt{3}$; $\zeta_1 = 2[\cos(5\pi/3) + i \sin(5\pi/3)] = 1 - i\sqrt{3}$ **7. a.** $\frac{7}{9}$ **b.** $-\frac{3}{5}$

c. $\frac{4}{3}$ **d.** $1/\sqrt{5}$ **9. a.** 1.97 radians   **b.** 11.82 inches   **c.** 35.46 square inches
d. 16.58 square inches   **11. a.** $r = 1.194$ kilometers   **b.** 16.66 miles per hour

15. $\{4i, -2\sqrt{3} - 2i. 2\sqrt{3} - 2i\}$

17. a. Amplitude $= 2$, period $= 4\pi$, phase shift $= 2\pi/3$ **b.** Amplitude $= \frac{1}{2}$, period $= 2$, phase shift $= -\frac{1}{4}$

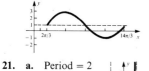

21. a. Period $= 2$ **b.** The given equation is equivalent to
$x = \frac{1}{2} \sin 2(y - \pi/3)$, with $-\pi/2 \le 2(y - \pi/3) \le \pi/2$,
or $\pi/12 \le y \le 7\pi/12$.

23. a. $\{0, 4\pi/3\}$ **b.** $\{\pi/6, 5\pi/6, 7\pi/6, 11\pi/6\}$

Chapter 10

Exercise Set 1

1. $(4, 2)$ **3.** $(-1, -3)$ **5.** $(4, 3)$ **7.** $(\frac{17}{11}, \frac{2}{11})$ **9.** $(4, -2)$ **11.** $(-5, 6)$ **13.** $(-2, 3, 4)$

15. $(2, -1, 4)$ **17.** $(-6, 2, 5)$ **19.** $(-\frac{11}{14}, -\frac{11}{24})$ **21.** $\left(\dfrac{ac + bd}{a^2 + b^2}, \dfrac{bc - ad}{a^2 + b^2} \right)$

23. If both sides of the second equation are multiplied by $-\frac{1}{2}$ and then 5 is added to both sides, the result is the same as
the first equation. So the equations are equivalent. The solution set is $\left\{ \left(c, \dfrac{3c - 5}{4} \right): \; c \in R \right\}$.

25. $8x - 6y = 3$ This is impossible, so the equations are inconsistent.
  $\underline{-8x + 6y = 4}$
     $0 = 7$

27. The solution of the system consisting of the first two equations is $(-3, 2)$, and this pair also satisfies the third equation.

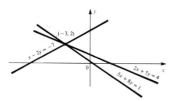

29. 95 and 32   **31.** 4 by 8

33. Airspeed $= 525$ miles per hour, wind velocity $= 75$ miles per hour

35. 3 miles per hour walking, 5 miles per hour jogging

37. 86   **39.** 6 hours and 40 minutes   **41.** $4,000 at 5%, $7,000 at 7%

43. 14 cubic centimeters of 10% solution, 16 cubic centimeters of 25% solution

45. $10\frac{2}{3}$ pounds Indian tea, $9\frac{1}{3}$ pounds Ceylon tea   **47.** Coffee is $4.25 per pound, eggs are 90¢ per dozen

49. $(\frac{1}{5}, \frac{1}{10}, -\frac{7}{10})$ **51.** $(\frac{23}{10}, -\frac{8}{5}, -\frac{9}{2}, \frac{29}{5})$

55. Since there are more unknowns than equations, the system is dependent. The solution set is
$\{(2c - 2, 5c - 5, 3c - 2, c): \; c \in R\}$.

57. $a = -1, b = 2, c = 5$ **59.** $\frac{1}{2}$ hour, 15 miles   **61.** 30 hours for A, 24 hours for B, 40 hours for C

Exercise Set 2

1. $(-4, 3, 1)$　**3.** $(2, -3, 7)$　**5.** $(-3, 4, -2)$　**7.** $\{(2 - c, 1 - c, c): \ c \in R\}$　**9.** $(2, -1, 0)$　**11.** $(5, 2, -3)$
13. $(2, -1, 4)$　**15.** $(-2, -2, 1)$　**17.** $\{(-3 + 7c, c, 4 - 3c): \ c \in R\}$　**19.** $\{(10c, 8c, c): \ c \in R\}$
21. $(\frac{10}{11}, -\frac{3}{11}, \frac{3}{11})$　**23.** Inconsistent　**25.** $(\frac{2}{3}, -\frac{22}{3}, -4)$　**27.** $\{(2, c - 5, 3c - 9, c): \ c \in R\}$
29. $\{(-5c, -2c, c): \ c \in R\}$　**31.** $(-\frac{4}{3}, \frac{2}{3}, \frac{7}{3})$　**33.** $(3, -1, 5, 6)$　**35.** $(-19, 16, -22, -9)$　**37.** $(1, -1, 3)$
39. $(2, 3, 0, -1)$　**41.** $(3, 4, -1, 2)$　**43.** Inconsistent　**45.** $\{(c, 2c, -c, 3c): \ c \in R\}$
47. $(-27, -21, 3, -6, 5)$

Exercise Set 3

1. a. 11　**b.** -78　**c.** -60　**3. a.** 0　**b.** 5　**c.** 0　**5. a.** -1　**b.** $\begin{vmatrix} 3 & 1 \\ -4 & -3 \end{vmatrix} = -5$

c. 5　**7. a.** -6　**b.** $\begin{vmatrix} -2 & 1 \\ 5 & -3 \end{vmatrix} = 1$　**c.** 1

9. a. $1 \begin{vmatrix} -4 & 5 \\ -6 & -1 \end{vmatrix} - (-3) \begin{vmatrix} 3 & -2 \\ -6 & -1 \end{vmatrix} + 7 \begin{vmatrix} 3 & -2 \\ -4 & 5 \end{vmatrix} = 34 + 3(-15) + 7(7) = 38$

b. $(-6) \begin{vmatrix} -2 & 1 \\ 5 & -3 \end{vmatrix} - (-1) \begin{vmatrix} 3 & 1 \\ -4 & -3 \end{vmatrix} + 7 \begin{vmatrix} 3 & -2 \\ -4 & 5 \end{vmatrix} = -6(1) + (-5) + 7(7) = 38$

11. 14　**13.** 118　**15.** -14　**17.** 62　**19.** $\{4, -1\}$

21. $\begin{vmatrix} a + kb & b \\ c + kd & d \end{vmatrix} = ad + kbd - bc - kbd = ad - bc = \begin{vmatrix} a & b \\ c & d \end{vmatrix}$

$\begin{vmatrix} a + kc & b + kd \\ c & d \end{vmatrix} = ad + kcd - bc - kcd = ad - bc = \begin{vmatrix} a & b \\ c & d \end{vmatrix}$

23. 44　**25.** 38
29. If x is replaced by x_1 and y is replaced by y_1, the determinant is 0, since two rows are identical (see Problem 28). Similarly, if x is replaced by x_2 and y by y_2, the determinant is 0. Thus, (x_1, y_1) and (x_2, y_2) satisfy the equation. Expanding the determinant shows that the equation is linear, and the conclusion follows.
31. $\{0, 1, -2, -\frac{3}{2}\}$

Exercise Set 4

1. $(\frac{37}{11}, -\frac{1}{11})$　**3.** $(-\frac{14}{13}, \frac{16}{13})$　**5.** $(\frac{13}{82}, -\frac{25}{41})$　**7.** $(2, \frac{1}{2})$　**9.** $(-\frac{5}{7}, \frac{13}{7})$　**11.** $(14, 8)$　**13.** $(\frac{4}{3}, \frac{22}{9}, \frac{5}{9})$

15. $(3, -2, -3)$　**17.** $(\frac{3}{2}, -\frac{1}{2}, 2)$　**19.** $(\frac{16}{11}, \frac{31}{11}, -\frac{6}{11})$　**21.** $(\frac{21}{17}, -\frac{39}{17}, -\frac{20}{17})$　**23.** $\left(\dfrac{2a - 3}{1 - 5a}, \dfrac{a + 5}{5a - 1} \right)$

25. $\left(\dfrac{2x^2 - 1}{2x^2 + 1}, \dfrac{2x}{2x^2 + 1} \right)$　**27.** $\left(\dfrac{3}{4}, \dfrac{6 - 3m}{2} \right)$　**29.** $(\frac{7}{9}, \frac{26}{9}, \frac{19}{9})$　**31.** $(\frac{23}{15}, \frac{2}{5}, -\frac{16}{15}, -\frac{2}{15})$

Exercise Set 5

1. $\begin{bmatrix} 2 & 6 & 3 \\ 10 & 4 & 16 \end{bmatrix}$　**3.** $\begin{bmatrix} -2 & 2 \\ -6 & 11 \end{bmatrix}$　**5.** $\begin{bmatrix} -2 & 14 \\ 0 & 0 \end{bmatrix}$　**7.** $\begin{bmatrix} -5 & 8 \\ 10 & -1 \\ 27 & 14 \end{bmatrix}$　**9.** $\begin{bmatrix} 21 & 14 \\ 17 & 26 \end{bmatrix}$　**11.** $\begin{bmatrix} 7 & -1 & 10 \\ 13 & 11 & 25 \end{bmatrix}$

13. $\begin{bmatrix} 1 & 9 & -7 & 7 \\ 10 & 6 & -28 & 0 \\ -7 & 15 & 10 & 16 \end{bmatrix}$　**15.** $x = 5, y = -2$　**17.** $\begin{bmatrix} 0 & 0 \\ 0 & 0 \end{bmatrix}$　Neither A nor B need be 0 in order for their product to be 0.

21. $A^2 = \begin{bmatrix} 2 & 2 \\ 2 & 2 \end{bmatrix}, A^3 = \begin{bmatrix} 4 & 4 \\ 4 & 4 \end{bmatrix}, A^4 = \begin{bmatrix} 8 & 8 \\ 8 & 8 \end{bmatrix}, A^n = \begin{bmatrix} 2^{n-1} & 2^{n-1} \\ 2^{n-1} & 2^{n-1} \end{bmatrix}$
23. $(A + B)(A - B) = A^2 - AB + BA - B^2$

25. $A = \begin{bmatrix} a_{11} & a_{12} & a_{13} & \cdots & a_{1n} \\ a_{21} & a_{22} & a_{23} & \cdots & a_{2n} \\ a_{31} & a_{32} & a_{33} & \cdots & a_{3n} \\ \vdots & \vdots & \vdots & \vdots & \vdots \\ a_{n1} & a_{n2} & a_{n3} & \cdots & a_{nn} \end{bmatrix}$, $X = \begin{bmatrix} x_1 \\ x_2 \\ x_3 \\ \vdots \\ x_n \end{bmatrix}$, $B = \begin{bmatrix} b_1 \\ b_2 \\ b_3 \\ \vdots \\ b_n \end{bmatrix}$

Exercise Set 6

1. $\{(2, 1), (-1, -2)\}$ **3.** $\{(3, 0), (-1, -2)\}$ **5.** $\{(\frac{5}{3}, -\frac{11}{9}), (-1, -3)\}$ **7.** $\{(8, 2), (\frac{17}{4}, -\frac{1}{2})\}$

9. $\{(\frac{3}{4}, \frac{3}{2}), (-\frac{3}{4}, -\frac{3}{2})\}$ **11.** $\{(1, -1), (\frac{25}{4}, \frac{5}{2})\}$ **13.** $\{(-1, -1), (\frac{7}{4}, \frac{4}{7})\}$ **15.** $\{(1, \sqrt{2}), (1, -\sqrt{2})\}$

17. $\{(4, 3), (\frac{8}{3}, \frac{1}{3})\}$ **19.** $\{(9, 6), (9, -6)\}$ **21.** $\{(0, -4), (4, 4)\}$ **23.** $\{(2, 3), (2, -3), (-2, 3), (-2, -3)\}$

25. $\{(5, 0), (-\frac{2}{3}, \frac{17}{6})\}$ **27.** $\{(\frac{3}{2}, \frac{1}{2}), (\frac{3}{2}, -\frac{1}{2}), (-\frac{3}{2}, \frac{1}{2}), (-\frac{3}{2}, -\frac{1}{2})\}$ **29.** $\{(7, 2), (4, -1)\}$ **31.** $\{(2, 1), (\frac{2}{5}, -\frac{11}{5})\}$

33. $\left\{\left(\frac{\sqrt{39}}{3}, \frac{1}{3}\right), \left(-\frac{\sqrt{39}}{3}, \frac{1}{3}\right), (\sqrt{2}, -2), (-\sqrt{2}, -2)\right\}$ **35.** 7 by 18 **37.** 12 and 8 **39.** 12 by 5

41. $2\sqrt{5}$ inches wide by $4\sqrt{5}$ inches high **43.** \$3,600 at 5%

45. $\left\{(3, -1), (-4, -8), \left(\frac{-1 + \sqrt{17}}{2}, \frac{7 + \sqrt{17}}{2}\right), \left(\frac{-1 - \sqrt{17}}{2}, \frac{7 - \sqrt{17}}{2}\right)\right\}$ **47.** $\{(1, 1), (\frac{1}{2}, 2)\}$

49. $r = 3, h = 8$; or $r = 2\sqrt{2 + \sqrt{13}} \approx 4.74, h = -4 + 2\sqrt{13} \approx 3.21$

51. $x + 2y - 8 = 0$ **53.** 8

Exercise Set 7

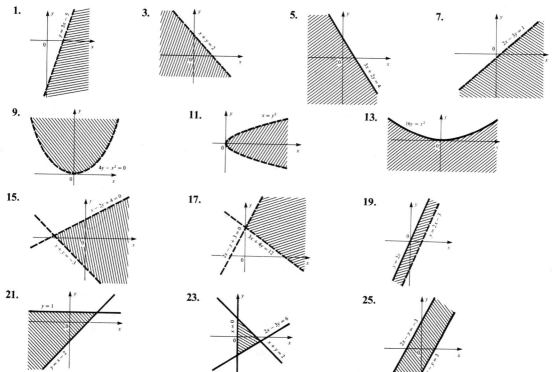

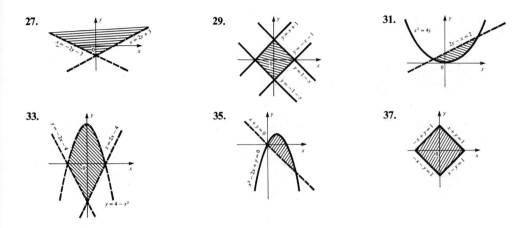

27.   **29.**   **31.**

33.   **35.**   **37.**

Exercise Set 8

1.   Max $F = 11$ at $(4, 3)$ **3.**   Max $G = 24$ at $(0, 4)$ **5.**   Min $H = 16$ at $(2, 2)$ **7.**   Min $C = 0$ at $(0, 5)$
9.   30 cases of each kind   **11.**   7 hours for A, 5 hours for B **13.**   500 gallons to each owner
15.   8 parts of type A', 10 parts of type B **17.**   20 days for each plant
19.   1 pound of feed number 1, 4 pounds of feed number 2   **21.**   350 gallons from store 2 to A, 400 gallons from store 1 to B

Review Exercise Set

1. $(3, 2)$ **3.** $(-8, 5)$ **5.** $(9, 5, 3)$ **7.** $(6, 5, -4, 3)$ **9.** $(-6, 8, -5)$ **11.** $(4, -6, -8)$
13. $\left\{ \left(\dfrac{13 - c}{7}, \dfrac{5 + 5c}{7}, c \right): \ c \in R \right\}$ **15.** $(10, 6, -4, -3)$ **17.** $\{(-6 - c, 10, 2c + 3, c): \ c \in R\}$ **19.**   Inconsistent
21. a.   1   **b.**   167   **23.** $x = 1$ or $-\frac{4}{5}$ **25. a.** $\left(-\frac{68}{3}, -43 \right)$ **b.** $\left(\frac{19}{44}, \frac{12}{11} \right)$ **27.** $(0, -1, 1)$
29. $66\frac{2}{3}$ miles, $53\frac{1}{3}$ miles   **31.**   12 cubic centimeters 40% solution, 18 cubic centimeters 15% solution   **33.**   11 by 5
35. $18\frac{3}{4}$ gallons pure water, $31\frac{1}{4}$ gallons 40% salt solution   **37.**   \$1,800 at 6% interest
39. $4\frac{1}{6}$ hours for the larger pump, $6\frac{1}{4}$ hours for the smaller one   **41.** $\left\{(-1, 2), \left(3, -\frac{2}{3}\right)\right\}$ **43.** $\{(8, -22), (-2, 8)\}$
45. $\left\{(2, 0), \left(-\frac{8}{5}, -\frac{12}{5}\right)\right\}$ **47.** $\left\{\left(\frac{3}{4}, \frac{3}{4}\right), \left(\frac{2}{3}, \frac{4}{3}\right)\right\}$ **49.**   15 and 7 or -7 and -15 **51.**   24 by 7 inches
53. a.   **b.**

55. a.   **b.**

57.   Max $F = 60$ at $(6, 0)$ **59.**   Min $G = 7$ at $\left(\frac{4}{3}, 4\right)$ **61.**   250 acres corn, 50 acres oats   **63.**   9 short trips, 1 long trip
65. $\left\{ \left(\dfrac{50 + 13c}{12}, \dfrac{2 - 3c}{4}, \dfrac{2 - 5c}{12}, c \right): \ c \in R \right\}$ **67.** $\{(-3c, -4c, c, 2c): \ c \in R\}$ **69.** $(7, 3, 9, -6, -4)$
71. $\left(-\dfrac{6}{a + 2}, -\dfrac{2a + 1}{a + 2} \right), a \neq 2$ **73.** $a = 2, b = -3, c = -5$ **75.** $\left\{(3, 4), \left(\frac{21}{34}, \frac{1}{34}\right)\right\}$
77. $2\frac{2}{3}$ gallons 40% solution, $5\frac{1}{3}$ gallons 25% solution   **79.** $(3, 7)$ and $(-1, -1)$
81.   15 miles per hour riding, 3 miles per hour walking   **83.**   82 pounds lawn fertilizer, 40 pounds garden fertilizer

Chapter 11

Exercise Set 2

1. x intercept 2; y intercept -2 **3.** x intercepts $\frac{5}{2}$, -1; y intercept -5 **5.** x intercept 1; y intercept $-\frac{1}{4}$
7. x intercept 2; y intercept $\frac{1}{2}$ **9.** x intercept $\frac{3}{8}$; no y intercept   **11.** x intercepts 4, -3; y intercepts 2, -2
13. a. y axis   **b.** Origin   **15. a.** y axis   **b.** y axis   **17. a.** Origin   **b.** Origin
19. a. x axis, y axis, origin   **b.** Origin   **21. a.** Origin   **b.** Origin   **23. a.** x axis
   b. x axis, y axis, origin   **25. a.** Origin   **b.** y axis   **27. a.** y axis   **b.** None   **29. a.** Origin
   b. y axis   **31.** x intercepts 1, 2, $-\frac{2}{3}$; y intercept $-\frac{4}{5}$ **33.** x intercepts 1, 3; y intercept -1
35. a. Origin   **b.** Origin   **37. a.** Symmetric to $x = y$ **b.** Symmetric to $x = y$ **c.** Symmetric to $x = y$
   d. Symmetric to $x = y$ **e.** Symmetric to $x = y$

Exercise Set 3

1. $x = 2, y = 0$ **3.** $x = 4, y = 0$ **5.** $x = 4, y = 2$ **7.** $x = 2, x = -2, y = 1$ **9.** $x = 2, x = -2$
11. $x = 3, x = -3, y = 2$ **13.** $x = 0, x = 1, x = -\frac{3}{2}, y = 0$ **15.** $x = 2, y = 1$ **17.** $x = 0, y = 3x - 4$
19. $x = 2, x = -2, y = x$ **21.** $x = 1, x = -1, y = 0$ **23.** $x = 1$
25. $x = 4, x = -4, y = 1$ on right, $y = -1$ on left   **27.** $y = 0$ **29.** $x = 2, x = -2, x = 3, x = -3, y = 1$
31. $x = 3, x = -3, y = 1, y = -1$ **33.** $x = 4, x = -2, y = 0$ **35.** $x = -1, y = 2$ **37.** $x = 0, y = 0$

39. $x = \dfrac{(2n + 1)\pi}{2}, n = 0, \pm 1, \pm 2, \ldots$

Exercise Set 4

1. $-2 \le x \le 2$ **3.** $-1 < x \le 2$ **5.** $-1 \le x \le 1, -1 \le y \le 1$ **7.** $-1 \le x < 1$ **9.** $x = 0, -2 < y < 2$
11. $x \le -3, 0 < x \le 3$ **13.** $x < \frac{1}{2}$ **15.** $x \ge 1, x \le -1, -2 < y < 2$ **17.** $x \le 0, 1 < x \le 2$
19. $x > \frac{1}{4}, y > 1$

Exercise Set 5

1. Intercept: x intercept -2; no y intercept
   Symmetry: not symmetric to x axis, y axis, or origin
   Asymptotes: $x = 0$; $y = 1$
   Excluded regions: none

3. Intercepts: x intercept 2; y intercept $\frac{2}{3}$
   Symmetry: not symmetric to x axis, y axis, or origin
   Asymptotes: $x = -3$; $y = -1$
   Excluded regions: none

5. Intercepts: x intercept 0; y intercept 0
   Symmetry: origin
   Asymptotes: $x = 1$; $x = -1$; $y = 0$
   Excluded regions: none

7. Intercepts: x intercepts 2, -2; y intercept $\frac{4}{9}$
   Symmetry: y axis
   Asymptotes: $x = 3$; $x = -3$; $y = 1$
   Excluded region: $\frac{4}{9} < y \le 1$

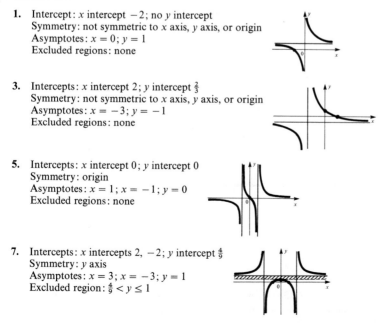

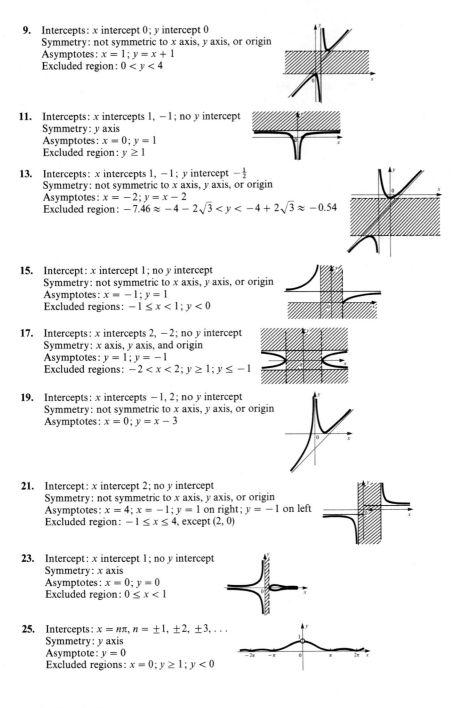

9. Intercepts: x intercept 0; y intercept 0
Symmetry: not symmetric to x axis, y axis, or origin
Asymptotes: $x = 1$; $y = x + 1$
Excluded region: $0 < y < 4$

11. Intercepts: x intercepts 1, -1; no y intercept
Symmetry: y axis
Asymptotes: $x = 0$; $y = 1$
Excluded region: $y \geq 1$

13. Intercepts: x intercepts 1, -1; y intercept $-\frac{1}{2}$
Symmetry: not symmetric to x axis, y axis, or origin
Asymptotes: $x = -2$; $y = x - 2$
Excluded region: $-7.46 \approx -4 - 2\sqrt{3} < y < -4 + 2\sqrt{3} \approx -0.54$

15. Intercept: x intercept 1; no y intercept
Symmetry: not symmetric to x axis, y axis, or origin
Asymptotes: $x = -1$; $y = 1$
Excluded regions: $-1 \leq x < 1$; $y < 0$

17. Intercepts: x intercepts 2, -2; no y intercept
Symmetry: x axis, y axis, and origin
Asymptotes: $y = 1$; $y = -1$
Excluded regions: $-2 < x < 2$; $y \geq 1$; $y \leq -1$

19. Intercepts: x intercepts -1, 2; no y intercept
Symmetry: not symmetric to x axis, y axis, or origin
Asymptotes: $x = 0$; $y = x - 3$

21. Intercept: x intercept 2; no y intercept
Symmetry: not symmetric to x axis, y axis, or origin
Asymptotes: $x = 4$; $x = -1$; $y = 1$ on right; $y = -1$ on left
Excluded region: $-1 \leq x \leq 4$, except (2, 0)

23. Intercept: x intercept 1; no y intercept
Symmetry: x axis
Asymptotes: $x = 0$; $y = 0$
Excluded region: $0 \leq x < 1$

25. Intercepts: $x = n\pi$, $n = \pm 1, \pm 2, \pm 3, \ldots$
Symmetry: y axis
Asymptote: $y = 0$
Excluded regions: $x = 0$; $y \geq 1$; $y < 0$

Review Exercise Set

1. **a.** Origin   **b.** x axis   3. **a.** y axis   **b.** x axis, y axis, origin   5. **a.** Origin   **b.** y axis
7. **a.** $x = 1$, $x = -1$, $y = 0$ **b.** $x = 2$, $x = -2$, $y = 2$ 9. **a.** $x = 1$, $x = -\frac{3}{2}$, $y = \frac{1}{2}$ **b.** $x = -2$, $y = 0$
11. **a.** $x = 1$, $y = x + 1$ **b.** $x = 2$, $x = -2$, $y = 0$ 13. **a.** $x = 1$, $x = -2$, $y = 1$ **b.** $x = 3$, $y = x + 1$
15. **a.** $x = 1$, $x = -1$, $y = 2$ on right, $y = -2$ on left   **b.** $x = -2$

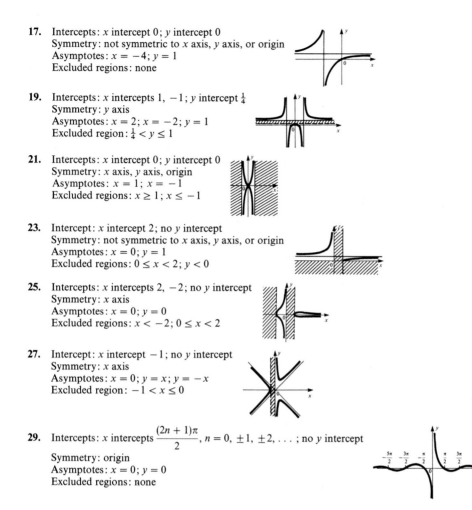

17. Intercepts: x intercept 0; y intercept 0
Symmetry: not symmetric to x axis, y axis, or origin
Asymptotes: $x = -4$; $y = 1$
Excluded regions: none

19. Intercepts: x intercepts 1, -1; y intercept $\frac{1}{4}$
Symmetry: y axis
Asymptotes: $x = 2$; $x = -2$; $y = 1$
Excluded region: $\frac{1}{4} < y \leq 1$

21. Intercepts: x intercept 0; y intercept 0
Symmetry: x axis, y axis, origin
Asymptotes: $x = 1$; $x = -1$
Excluded regions: $x \geq 1$; $x \leq -1$

23. Intercept: x intercept 2; no y intercept
Symmetry: not symmetric to x axis, y axis, or origin
Asymptotes: $x = 0$; $y = 1$
Excluded regions: $0 \leq x < 2$; $y < 0$

25. Intercepts: x intercepts 2, -2; no y intercept
Symmetry: x axis
Asymptotes: $x = 0$; $y = 0$
Excluded regions: $x < -2$; $0 \leq x < 2$

27. Intercept: x intercept -1; no y intercept
Symmetry: x axis
Asymptotes: $x = 0$; $y = x$; $y = -x$
Excluded region: $-1 < x \leq 0$

29. Intercepts: x intercepts $\dfrac{(2n + 1)\pi}{2}$, $n = 0, \pm 1, \pm 2, \ldots$; no y intercept

Symmetry: origin
Asymptotes: $x = 0$; $y = 0$
Excluded regions: none

Chapter 12

Exercise Set 2

1. $x^2 + y^2 - 2x - 4y - 4 = 0$; circle of radius 3, with center (1, 2)   **3.** $3x^2 + 3y^2 - 18x + 14y + 10 = 0$
5. $x^2 + y^2 - 8x + 12y + 32 = 0$ **7.** $y^2 = 12x$ **9.** $x^2 - 2x + 8y + 1 = 0$ **11.** $5x^2 + 9y^2 = 45$
13. $x^2 = 4y$ **15.** $4x^2 + 3y^2 - 10y + 7 = 0$ **17.** $36x^2 - 64y^2 + 36x + 128y - 199 = 0$
19. $16x^2 - 24xy + 9y^2 - 142x - 106y + 76 = 0$

Exercise Set 3

1. **a.** $x^2 + y^2 - 2x + 6y + 6 = 0$ **b.** $x^2 + y^2 + 4x + 8y + 4 = 0$
3. $(x - 2)^2 + (y + 3)^2 = 9$; circle, center (2, -3), radius 3   **5.** $(x + 1)^2 + (y - 4)^2 = 16$; circle, center $(-1, 4)$, radius 4

7. $(x + 4)^2 + (y - 2)^2 = 0$; degenerate circle—graph consists of point $(-4, 2)$ only

9. $(x + \frac{3}{4})^2 + (y - \frac{5}{4})^2 = \frac{29}{8}$; circle, center $(-\frac{3}{4}, \frac{5}{4})$, radius $\sqrt{\frac{29}{8}} \approx 1.9$

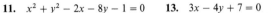

11. $x^2 + y^2 - 2x - 8y - 1 = 0$ **13.** $3x - 4y + 7 = 0$

15. $x^2 + y^2 - 12x - 2y + 1 = 0$ or $x^2 + y^2 + 12x + 10y + 25 = 0$

17. Distance between centers = sum of radii, so they are tangent; line joining centers, $12x + 5y - 16 = 0$; common tangent line, $5x - 12y + 2 = 0$ **19.** $x^2 + y^2 + 2x + 4y - 20 = 0$

21. $(0, 1)$ and $(\frac{3}{2}, -\frac{1}{2})$; common chord, $x + y - 1 = 0$; perpendicular bisector, $2x - 2y - 1 = 0$

Exercise Set 4

1. a. $y^2 = 8x$ **b.** $x^2 = -12y$ **3. a.** $(x - 1)^2 = 8(y + 2)$ **b.** $(x + 3)^2 = -8(y - 1)$

5. a. $(x - 2)^2 = 12(y + 3)$ **b.** $(y + 1)^2 = 6(x + \frac{1}{2})$ **7.** $(y - 4)^2 = -12(x - 3)$ **9.** $(y + 2)^2 = -4(x - 3)$

11. $3x^2 - 4x - 4y + 5 = 0$

13. $(x - 2)^2 = 12(y + 1)$; focus $(2, 2)$ **15.** $(x - 1)^2 = y + 1$; focus $(1, -\frac{3}{4})$ **17.** $(y + \frac{1}{2})^2 = -6(x - \frac{3}{2})$; focus $(0, -\frac{1}{2})$

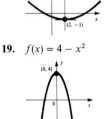

19. $f(x) = 4 - x^2$

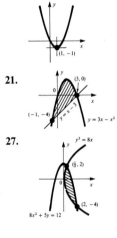

21.

23.

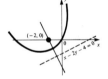

25. $(x - 1)^2 + (y - 1)^2 = 4$

27.

29. $4x^2 + 4xy + y^2 + 28x - 16y + 4 = 0$

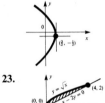

Exercise Set 5

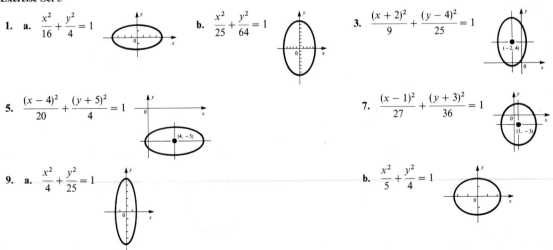

1. a. $\dfrac{x^2}{16} + \dfrac{y^2}{4} = 1$ **b.** $\dfrac{x^2}{25} + \dfrac{y^2}{64} = 1$ **3.** $\dfrac{(x + 2)^2}{9} + \dfrac{(y - 4)^2}{25} = 1$

5. $\dfrac{(x - 4)^2}{20} + \dfrac{(y + 5)^2}{4} = 1$ **7.** $\dfrac{(x - 1)^2}{27} + \dfrac{(y + 3)^2}{36} = 1$

9. a. $\dfrac{x^2}{4} + \dfrac{y^2}{25} = 1$ **b.** $\dfrac{x^2}{5} + \dfrac{y^2}{4} = 1$

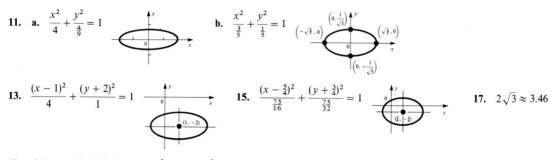

11. **a.** $\dfrac{x^2}{4} + \dfrac{y^2}{\frac{4}{9}} = 1$

b. $\dfrac{x^2}{\frac{3}{5}} + \dfrac{y^2}{\frac{1}{5}} = 1$

13. $\dfrac{(x-1)^2}{4} + \dfrac{(y+2)^2}{1} = 1$

15. $\dfrac{(x-\frac{5}{4})^2}{\frac{75}{16}} + \dfrac{(y+\frac{3}{4})^2}{\frac{75}{32}} = 1$

17. $2\sqrt{3} \approx 3.46$

19. Circumscribed circle: $(x-2)^2 + (y+3)^2 = 25$; inscribed circle: $(x-2)^2 + (y+3)^2 = \frac{25}{2}$

21. $24x^2 + 4xy + 21y^2 - 500 = 0$

Exercise Set 6

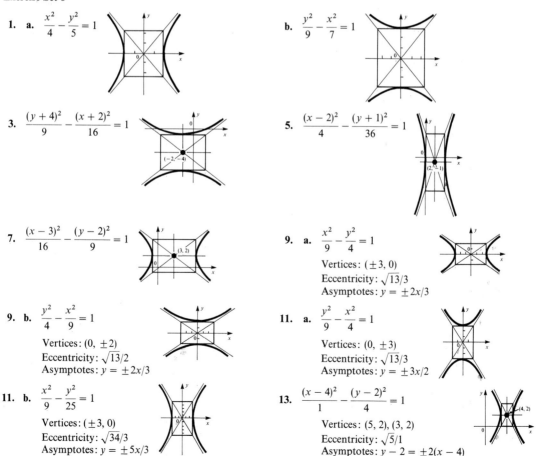

1. **a.** $\dfrac{x^2}{4} - \dfrac{y^2}{5} = 1$

b. $\dfrac{y^2}{9} - \dfrac{x^2}{7} = 1$

3. $\dfrac{(y+4)^2}{9} - \dfrac{(x+2)^2}{16} = 1$

5. $\dfrac{(x-2)^2}{4} - \dfrac{(y+1)^2}{36} = 1$

7. $\dfrac{(x-3)^2}{16} - \dfrac{(y-2)^2}{9} = 1$

9. **a.** $\dfrac{x^2}{9} - \dfrac{y^2}{4} = 1$

Vertices: $(\pm 3, 0)$
Eccentricity: $\sqrt{13}/3$
Asymptotes: $y = \pm 2x/3$

9. **b.** $\dfrac{y^2}{4} - \dfrac{x^2}{9} = 1$

Vertices: $(0, \pm 2)$
Eccentricity: $\sqrt{13}/2$
Asymptotes: $y = \pm 2x/3$

11. **a.** $\dfrac{y^2}{9} - \dfrac{x^2}{4} = 1$

Vertices: $(0, \pm 3)$
Eccentricity: $\sqrt{13}/3$
Asymptotes: $y = \pm 3x/2$

11. **b.** $\dfrac{x^2}{9} - \dfrac{y^2}{25} = 1$

Vertices: $(\pm 3, 0)$
Eccentricity: $\sqrt{34}/3$
Asymptotes: $y = \pm 5x/3$

13. $\dfrac{(x-4)^2}{1} - \dfrac{(y-2)^2}{4} = 1$

Vertices: $(5, 2), (3, 2)$
Eccentricity: $\sqrt{5}/1$
Asymptotes: $y - 2 = \pm 2(x-4)$

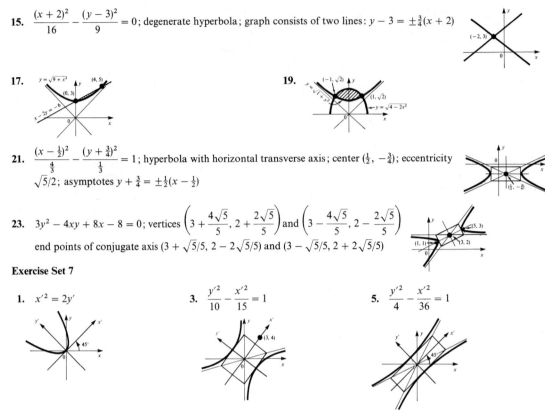

15. $\dfrac{(x+2)^2}{16} - \dfrac{(y-3)^2}{9} = 0$; degenerate hyperbola; graph consists of two lines: $y - 3 = \pm\frac{3}{4}(x+2)$

17.

19.

21. $\dfrac{(x-\frac{1}{2})^2}{\frac{4}{3}} - \dfrac{(y+\frac{3}{4})^2}{\frac{1}{3}} = 1$; hyperbola with horizontal transverse axis; center $(\frac{1}{2}, -\frac{3}{4})$; eccentricity $\sqrt{5}/2$; asymptotes $y + \frac{3}{4} = \pm\frac{1}{2}(x - \frac{1}{2})$

23. $3y^2 - 4xy + 8x - 8 = 0$; vertices $\left(3 + \dfrac{4\sqrt{5}}{5}, 2 + \dfrac{2\sqrt{5}}{5}\right)$ and $\left(3 - \dfrac{4\sqrt{5}}{5}, 2 - \dfrac{2\sqrt{5}}{5}\right)$
end points of conjugate axis $(3 + \sqrt{5}/5, 2 - 2\sqrt{5}/5)$ and $(3 - \sqrt{5}/5, 2 + 2\sqrt{5}/5)$

Exercise Set 7

1. $x'^2 = 2y'$

3. $\dfrac{y'^2}{10} - \dfrac{x'^2}{15} = 1$

5. $\dfrac{y'^2}{4} - \dfrac{x'^2}{36} = 1$

7. $y'^2 = -4(x' - 1)$

9.

11.

13. $(x+1)(y-2) = -2$

15. $(x+3)(y-4) = -8$

17. $(x'-1)^2 = -4(y'+1)$

Exercise Set 8

1. a. Hyperbola **b.** Parabola **c.** Ellipse **d.** Hyperbola **e.** Ellipse

Review Exercise Set

1. a. $3x^2 + 3y^2 + 12x - 26y + 55 = 0$ **b.** $y^2 - 8x - 2y + 1 = 0$

3. a. $x^2 + y^2 = 4$; circle

b. $\dfrac{y^2}{4} - \dfrac{x^2}{4} = 1$; hyperbola

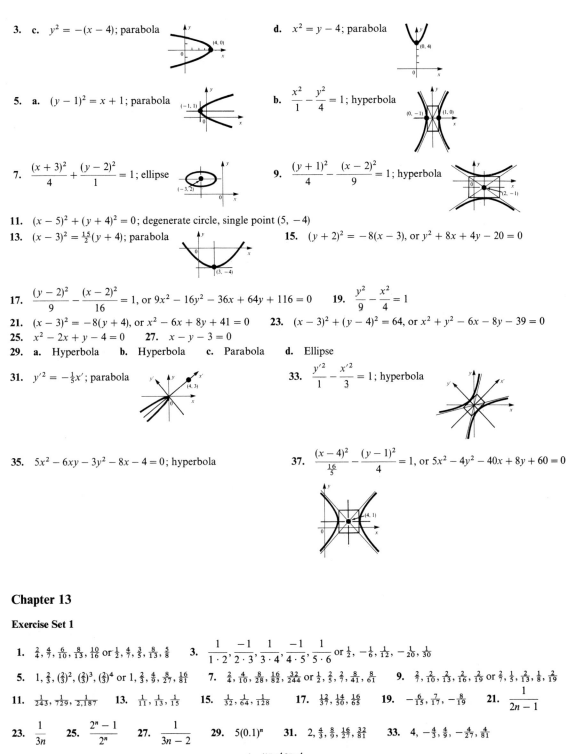

3. **c.** $y^2 = -(x - 4)$; parabola

d. $x^2 = y - 4$; parabola

5. **a.** $(y - 1)^2 = x + 1$; parabola

b. $\dfrac{x^2}{1} - \dfrac{y^2}{4} = 1$; hyperbola

7. $\dfrac{(x + 3)^2}{4} + \dfrac{(y - 2)^2}{1} = 1$; ellipse

9. $\dfrac{(y + 1)^2}{4} - \dfrac{(x - 2)^2}{9} = 1$; hyperbola

11. $(x - 5)^2 + (y + 4)^2 = 0$; degenerate circle, single point $(5, -4)$

13. $(x - 3)^2 = \frac{15}{2}(y + 4)$; parabola

15. $(y + 2)^2 = -8(x - 3)$, or $y^2 + 8x + 4y - 20 = 0$

17. $\dfrac{(y - 2)^2}{9} - \dfrac{(x - 2)^2}{16} = 1$, or $9x^2 - 16y^2 - 36x + 64y + 116 = 0$ **19.** $\dfrac{y^2}{9} - \dfrac{x^2}{4} = 1$

21. $(x - 3)^2 = -8(y + 4)$, or $x^2 - 6x + 8y + 41 = 0$ **23.** $(x - 3)^2 + (y - 4)^2 = 64$, or $x^2 + y^2 - 6x - 8y - 39 = 0$

25. $x^2 - 2x + y - 4 = 0$ **27.** $x - y - 3 = 0$

29. **a.** Hyperbola   **b.** Hyperbola   **c.** Parabola   **d.** Ellipse

31. $y'^2 = -\frac{1}{5}x'$; parabola

33. $\dfrac{y'^2}{1} - \dfrac{x'^2}{3} = 1$; hyperbola

35. $5x^2 - 6xy - 3y^2 - 8x - 4 = 0$; hyperbola

37. $\dfrac{(x - 4)^2}{\frac{16}{5}} - \dfrac{(y - 1)^2}{4} = 1$, or $5x^2 - 4y^2 - 40x + 8y + 60 = 0$

Chapter 13

Exercise Set 1

1. $\frac{2}{4}, \frac{4}{7}, \frac{6}{10}, \frac{8}{13}, \frac{10}{16}$ or $\frac{1}{2}, \frac{4}{7}, \frac{3}{5}, \frac{8}{13}, \frac{5}{8}$ **3.** $\dfrac{1}{1 \cdot 2}, \dfrac{-1}{2 \cdot 3}, \dfrac{1}{3 \cdot 4}, \dfrac{-1}{4 \cdot 5}, \dfrac{1}{5 \cdot 6}$ or $\frac{1}{2}, -\frac{1}{6}, \frac{1}{12}, -\frac{1}{20}, \frac{1}{30}$

5. $1, \frac{2}{3}, (\frac{2}{3})^2, (\frac{2}{3})^3, (\frac{2}{3})^4$ or $1, \frac{2}{3}, \frac{4}{9}, \frac{8}{27}, \frac{16}{81}$ **7.** $\frac{2}{4}, \frac{4}{10}, \frac{8}{28}, \frac{16}{82}, \frac{32}{244}$ or $\frac{1}{2}, \frac{2}{5}, \frac{2}{7}, \frac{8}{41}, \frac{8}{61}$ **9.** $\frac{2}{7}, \frac{2}{10}, \frac{2}{13}, \frac{2}{16}, \frac{2}{19}$ or $\frac{2}{7}, \frac{1}{5}, \frac{2}{13}, \frac{1}{8}, \frac{2}{19}$

11. $\frac{1}{243}, \frac{1}{729}, \frac{1}{2,187}$ **13.** $\frac{1}{11}, \frac{1}{13}, \frac{1}{15}$ **15.** $\frac{1}{32}, \frac{1}{64}, \frac{1}{128}$ **17.** $\frac{12}{37}, \frac{14}{50}, \frac{16}{65}$ **19.** $-\frac{6}{15}, \frac{7}{17}, -\frac{8}{19}$ **21.** $\dfrac{1}{2n - 1}$

23. $\dfrac{1}{3n}$ **25.** $\dfrac{2^n - 1}{2^n}$ **27.** $\dfrac{1}{3n - 2}$ **29.** $5(0.1)^n$ **31.** $2, \frac{4}{3}, \frac{8}{9}, \frac{16}{27}, \frac{32}{81}$ **33.** $4, -\frac{4}{3}, \frac{4}{9}, -\frac{4}{27}, \frac{4}{81}$

35. $2, -4, 12, -48, 240$ **37.** $\dfrac{n}{2n^2 - 3}$ **39.** $\dfrac{(-1)^{n-1}2^{n-1}}{3^n + 1}$ **41.** $1, 1, 2, 3, 5$ **43.** $1, 2, 2, 1, \frac{1}{2}$

45. For $n = 10, 100$, and $1,000$ the points are

$(\frac{7}{5}, \frac{20}{101}) = (1.4, 0.2)$

$(\frac{26}{25}, \frac{200}{10,001}) = (1.04, 0.02)$

$(\frac{251}{250}, \frac{2,000}{1,000,001}) = (1.004, 0.002)$

The limit point is $(1, 0)$.

Exercise Set 2

1. $a_{n+1} - a_n = 3(n + 1) - 3n = 3$; so $d = 3$, and the sequence is arithmetic; 15, 18, 21, 24, 27

3. $a_{n+1} - a_n = [8 - 3(n + 1)] - (8 - 3n) = -3$; so $d = -3$, and the sequence is arithmetic; $-7, -10, -13, -16, -19$

5. $a_{n+1} - a_n = \dfrac{13 - 3(n + 1)}{2} - \dfrac{13 - 3n}{2} = -\dfrac{3}{2}$; so $d = -\dfrac{3}{2}$, and the sequence is arithmetic; $-1, -\frac{5}{2}, -4, -\frac{11}{2}, -7$

7. $\dfrac{a_{n+1}}{a_n} = \dfrac{(-\frac{1}{2})^n}{(-\frac{1}{2})^{n-1}} = -\dfrac{1}{2}$; so $r = -\frac{1}{2}$, and the sequence is geometric; $-\frac{1}{8}, \frac{1}{16}, -\frac{1}{32}, \frac{1}{64}$

9. $\dfrac{a_{n+1}}{a_n} = \dfrac{3^n/9}{3^{n-1}/9} = 3$; so $r = 3$, and the sequence is geometric; 3, 9, 27, 81

11. Arithmetic, $d = 3$; $a_{16} = 44$ **13.** Geometric, $r = -\frac{1}{3}$; $a_{10} = \frac{1}{6,561}$ **15.** Arithmetic, $d = 4$; $a_{21} = 70$

17. Geometric, $r = \frac{1}{2}$ **19.** Arithmetic, $d = 4$ **21.** Neither   **23.** Neither   **25.** Neither

27. Geometric, $r = \frac{2}{3}$ **29.** Neither   **31.** $a_1 = -25$ **33.** $a_1 = -14, d = 2$ **35.** $a_1 = -1, r = -\frac{2}{3}$

37. a. $x = \frac{25}{2}$ **b.** $x = \pm 10$ **39.** 14, 17, 20, 23, 26, 29

41. $x_{n+1} - x_n = (a_{n+1} + 5) - (a_n + 5) = a_{n+1} - a_n = $ a constant; so $x_1, x_2, x_3, \ldots$ is an arithmetic sequence

43. Arithmetic sequence, $d = \ln 2$

45. Let $r = a_{n+1}/a_n$. Then $\log a_{n+1} - \log a_n = \log(a_{n+1}/a_n) = \log r$, which is a constant. So the sequence $\log a_1, \log a_2,$ $\log a_3, \ldots$ is arithmetic, with common difference $d = \log r$.

47. 1,024

49. $A = P(1 + r)^n$; the sequence is $P(1 + r), P(1 + r)^2, P(1 + r)^3, \ldots$, which is geometric, with common ratio $1 + r$

51. \$9,830.40

Exercise Set 3

1. $4 + \frac{7}{3} + \frac{10}{5} + \frac{13}{7}$ **3.** $\dfrac{1}{1 \cdot 2} - \dfrac{1}{2 \cdot 3} + \dfrac{1}{3 \cdot 4} - \dfrac{1}{4 \cdot 5} + \dfrac{1}{5 \cdot 6} - \dfrac{1}{6 \cdot 7}$ **5.** $\frac{1}{2} - \frac{1}{5} + \frac{1}{10} - \frac{1}{17} + \frac{1}{26} - \frac{1}{37} + \frac{1}{50} - \frac{1}{65}$

7. $2 + 7 + 12 + 17 + 22 + 27 + 32 + 37 + 42 + 47$ **9.** $e^{-1} + 2e^{-2} + 3e^{-3} + 4e^{-4} + 5e^{-5} + 6e^{-6}$ **11.** 1,010

13. 9,300   **15.** 0.222222222222   **17.** 34.5   **19.** 60,700   **21.** 400   **23.** $6[1 - (\frac{2}{3})^8] = \frac{12,610}{2,187}$

25. $(\frac{12}{7})[1 - (-\frac{3}{4})^6] = \frac{1,443}{1,024}$ **27.** 5,050   **29.** 322   **31.** 120

33. First job sixth year salary $= \$15,500$, accumulated earnings $= \$82,500$; second job sixth year salary $= \$14,693.28$, accumulated earnings $= \$73,359.29$; second salary exceeds first in 8 years; accumulated earnings exceed first in 13 years

35. Should take first offer; earnings are \$10,737,418.23 and \$465,000, respectively

Exercise Set 4

1. Converges to $\frac{3}{2}$ **3.** Diverges   **5.** Converges to $\frac{10}{19}$ **7.** Converges to $\frac{30}{11}$ **9.** Diverges

11. Converges to 5   **13.** Converges to 27   **15.** Diverges   **17.** Converges to $1/(\sqrt{2} - 1)$

19. Converges to $\frac{10}{11}$ **21.** $\frac{5}{33}$ **23.** $\frac{17}{11}$ **25.** $\frac{117}{37}$ **27.** $\dfrac{2}{2 - x}, |x| < 2$

29. $a_1 = 2, a_n = \dfrac{-1}{n(n - 1)}$ for $n \geq 2$, so series is $2 - \dfrac{1}{2 \cdot 1} - \dfrac{1}{3 \cdot 2} - \dfrac{1}{4 \cdot 3} - \cdots$; $S = \lim_{n \to \infty} S_n = \lim_{n \to \infty} \left[1 + \left(\dfrac{1}{n}\right)\right] = 1$

31. $\dfrac{1}{3 + x^2}$, valid for $|-x^2/3| < 1$ or $|x| < \sqrt{3}$ **33.** $\frac{2}{5}(2 + i)$

35. $1 + 2x + 4x^2 + 8x^3 + \cdots = \sum_{n=0}^{\infty} (2x)^n$, valid for $|x| < \frac{1}{2}$ **37.** $\dfrac{1}{2} + \dfrac{x}{4} + \dfrac{x^2}{8} + \cdots = \sum_{n=0}^{\infty} \dfrac{x^n}{2^{n+1}}$ valid for $|x| < 2$

39. The sum is $\lim_{n \to \infty}[4 - (1/3^n)] = 4$; $a_1 = \frac{11}{3}$ and for $n \geq 2$, $a_n = S_n - S_{n-1} = 2/3^n$. So the series is $\frac{11}{3} + \frac{2}{9} + \frac{2}{27} + \frac{2}{81}$ $+ \cdots$. Beginning with the second term, this is geometric, with $r = \frac{1}{3}$.

Exercise Set 6

1. **a.** 3   **b.** 20   **c.** 28   **d.** 120   **e.** 1   **f.** 252   **g.** 435   **h.** 4,368   **i.** 7   **j.** 28

3. **a.** $(k+1)!$ **b.** $k+1$ **c.** $n(n-1)$ **5.** $\binom{r}{r} = \dfrac{r(r-1)(r-2) \cdots (r-r+1)}{1 \cdot 2 \cdot 3 \cdots r} = \dfrac{r!}{r!} = 1$

7. $\binom{n}{n-k} = \dfrac{n!}{(n-k)![n-(n-k)]!} = \dfrac{n!}{(n-k)!k!} = \binom{n}{k}$ **9.** $(x-y)^5 = x^5 - 5x^4y + 10x^3y^2 - 10x^2y^3 + 5xy^4 - y^5$

11. $a^4 - 12a^3b + 54a^2b^2 - 108ab^3 + 81b^4$ **13.** $s^6 - 12s^5t + 60s^4t^2 - 160s^3t^3 + 240s^2t^4 - 192st^5 + 64t^6$

15. $x^{12} - 12x^{10}y + 60x^8y^2 - 160x^6y^3 + 240x^4y^4 - 192x^2y^5 + 64y^6$ **17.** $495x^8y^4$ **19.** $-2,835a^4b^3$ **21.** 1.126

23. $(1+1)^n = \binom{n}{0}1^n + \binom{n}{1}1^{n-1} \cdot 1 + \binom{n}{2}1^{n-2} \cdot 1^2 + \binom{n}{3}1^{n-3} \cdot 1^3 + \cdots + \binom{n}{n}$

or: $2^n = \binom{n}{0} + \binom{n}{1} + \binom{n}{2} + \cdots + \binom{n}{n}$

25. $\dfrac{n-r+1}{r} \cdot \binom{n}{r-1} = \dfrac{n-r+1}{r} \cdot \dfrac{n!}{(r-1)!(n-r+1)!} = \dfrac{\overline{(n-r+1)}n!}{r \cdot (r-1)!\overline{(n-r+1)}(n-r)!} = \dfrac{n!}{r!(n-r)!} = \binom{n}{r}$

27. $\sqrt{1+x} = (1+x)^{1/2} = 1 + \dfrac{x}{2} - \dfrac{x^2}{8} + \dfrac{x^3}{16} - \dfrac{5x^4}{128} + \cdots$ **29.** 0.99499

Review Exercise Set

1. **a.** $\frac{1}{2}, \frac{2}{3}, \frac{3}{4}, \frac{4}{5}$ **b.** 1, 5, 9, 13   **3.** **a.** $\frac{1}{4}, -\frac{3}{16}, \frac{9}{64}, -\frac{27}{256}$ **b.** $1/\log 2, -1/(2 \log 3), 1/(3 \log 4), -1/(4 \log 5)$

5. **a.** $\dfrac{1}{n(n+1)}$ **b.** $\dfrac{n}{n+1}$ **7.** Arithmetic, $\frac{21}{2}$ **9.** Geometric, $-\frac{1}{2,048}$ **11.** **a.** 2, 5, 11, 23, 47

b. 1, 2, 6, 24, 120   **13.** **a.** 9   **b.** $\pm 3\sqrt{5}$ **15.** **a.** $\frac{243}{64}$ inches   **b.** $\frac{3,367}{64}$ inches   **c.** 64 inches

17. 1,590   **19.** $-2,400$ **21.** $-1,200$ **23.** $\frac{33}{2}$ **25.** $\frac{3}{8}$ **27.** **a.** $\frac{6}{11}$ **b.** $\frac{4}{27}$ **c.** $\frac{24}{11}$ **d.** $\frac{1,004}{37}$

e. $\frac{101}{99}$ **29.** 182   **39.** **a.** 2,300   **b.** 84   **c.** 4   **d.** 1   **e.** 66

41. **a.** $x^7 - \dfrac{7x^6y}{2} + \dfrac{21x^5y^2}{4} - \dfrac{35x^4y^3}{8} + \dfrac{35x^3y^4}{16} - \dfrac{21x^2y^5}{32} + \dfrac{7xy^6}{64} - \dfrac{y^7}{128}$

b. $1 + 10x^2 + 45x^4 + 120x^6 + 210x^8 + 252x^{10} + 210x^{12} + 120x^{14} + 45x^{16} + 10x^{18} + x^{20}$

43. **a.** $-25,344x^5y^7$ **b.** $-85,995,520x^{11}$ **45.** $-\frac{14,581}{729}$ **47.** \$66,550

49. $23\frac{1}{4}, 22\frac{1}{2}, 21\frac{3}{4}, 21, 20\frac{1}{4}, 19\frac{1}{2}, 18\frac{3}{4}, 18, 17\frac{1}{4}, 16\frac{1}{2}, 15\frac{3}{4}$

51. $\binom{n}{k} - \binom{n-1}{k} = \dfrac{n!}{k!(n-k)!} - \dfrac{(n-1)!}{k!(n-1-k)!} = \dfrac{n! - (n-1)!(n-k)}{k!(n-k)!} = \dfrac{(n-1)!(n-n+k)}{k!(n-k)!} = \dfrac{(n-1)!k}{k!(n-k)!}$

$= \dfrac{(n-1)!}{(k-1)![(n-1)-(k-1)]!} = \binom{n-1}{k-1}$

53. **a.** $1 - x + x^2 - x^4 + \cdots = \sum\limits_{n=1}^{\infty}(-1)^{n-1}x^{n-1}$, $|x| < 1$ **b.** $\dfrac{3}{2} + \dfrac{3x^2}{4} + \dfrac{3x^4}{8} + \dfrac{3x^6}{16} + \cdots = \sum\limits_{n=1}^{\infty}\dfrac{3}{2}\left(\dfrac{x^2}{2}\right)^{n-1}$, $|x| < \sqrt{2}$

Cumulative Review Exercise Set IV (Chapters 10–13)

1. $(\frac{5}{2}, -\frac{3}{4}, \frac{1}{2})$

3. x intercept: 4
y intercepts: $\pm\frac{4}{3}$
Symmetry: x axis
Excluded regions: $(-\infty, -3] \cup [3, 4)$
Asymptotes: $x = \pm 3$, $y = 0$

5. **a.** 1985 dollar will be worth 0.53 times 1980 dollar

b. $0.1(2^0 + 2^{0.1} + 2^{0.2} + \cdots + 2^{1.9}) = \sum\limits_{n=1}^{20} 0.1(2^{0.1})^{n-1} = \dfrac{0.1(2^2 - 1)}{2^{0.1} - 1} \approx 4.18$

7. 24 pounds Colombian, 60 pounds Sumatran, 16 pounds Brazilian   **9.** 80 standard, 60 deluxe

11. **a.** 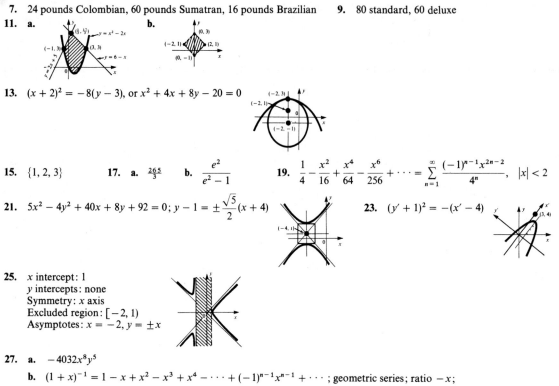   **b.**

13. $(x + 2)^2 = -8(y - 3)$, or $x^2 + 4x + 8y - 20 = 0$

15. $\{1, 2, 3\}$ **17.** **a.** $\frac{265}{3}$ **b.** $\dfrac{e^2}{e^2 - 1}$ **19.** $\dfrac{1}{4} - \dfrac{x^2}{16} + \dfrac{x^4}{64} - \dfrac{x^6}{256} + \cdots = \sum_{n=1}^{\infty} \dfrac{(-1)^{n-1} x^{2n-2}}{4^n}, \quad |x| < 2$

21. $5x^2 - 4y^2 + 40x + 8y + 92 = 0$; $y - 1 = \pm \dfrac{\sqrt{5}}{2}(x + 4)$ **23.** $(y' + 1)^2 = -(x' - 4)$

25. x intercept: 1
y intercepts: none
Symmetry: x axis
Excluded region: $[-2, 1)$
Asymptotes: $x = -2$, $y = \pm x$

27. **a.** $-4032 x^8 y^5$

b. $(1 + x)^{-1} = 1 - x + x^2 - x^3 + x^4 - \cdots + (-1)^{n-1} x^{n-1} + \cdots$; geometric series; ratio $-x$;

$\displaystyle\sum_{n=1}^{\infty} (-1)^{n-1} x^{n-1} = \dfrac{1}{1 + x} = (1 + x)^{-1}$

Index

Trigonometric Functions

For a right triangle

$$\sin A = \frac{a}{c} \qquad \csc A = \frac{c}{a}$$

$$\cos A = \frac{b}{c} \qquad \sec A = \frac{c}{b}$$

$$\tan A = \frac{a}{b} \qquad \cot A = \frac{b}{a}$$

For a general angle

$$\sin \theta = \frac{y}{r} \qquad \csc \theta = \frac{r}{y}$$

$$\cos \theta = \frac{x}{r} \qquad \sec \theta = \frac{r}{x}$$

$$\tan \theta = \frac{y}{x} \qquad \cot \theta = \frac{x}{y}$$

For a real number x

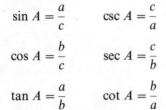

π radians $= 180°$

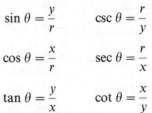

If f is any trigonometric function,
$f(\text{Real number } x) = f(\text{Angle of } x \text{ radians})$

Special Triangles

Trigonometric Form of a Complex Number

$z = a + bi = r(\cos \theta + i \sin \theta)$

De Moivre's Theorem: $z^n = r^n(\cos n\theta + i \sin n\theta)$

Law of Sines

$$\frac{a}{\sin A} = \frac{b}{\sin B} = \frac{c}{\sin C}$$

Law of Cosines

$$a^2 = b^2 + c^2 - 2bc \cos A$$
$$b^2 = a^2 + c^2 - 2ac \cos B$$
$$c^2 = a^2 + b^2 - 2ab \cos C$$

Area of a Triangle

$$A = \tfrac{1}{2}ab \sin C = \tfrac{1}{2}ac \sin B = \tfrac{1}{2}bc \sin A$$